Lecture Notes in Networks and Systems

Volume 1513

The series "Lecture Notes in Networks and Systems" publishes the latest developments in Networks and Systems—quickly, informally and with high quality. Original research reported in proceedings and post-proceedings represents the core of LNNS.

Volumes published in LNNS embrace all aspects and subfields of, as well as new challenges in, Networks and Systems.

The series contains proceedings and edited volumes in systems and networks, spanning the areas of Cyber-Physical Systems, Autonomous Systems, Sensor Networks, Control Systems, Energy Systems, Automotive Systems, Biological Systems, Vehicular Networking and Connected Vehicles, Aerospace Systems, Automation, Manufacturing, Smart Grids, Nonlinear Systems, Power Systems, Robotics, Social Systems, Economic Systems and other. Of particular value to both the contributors and the readership are the short publication timeframe and the world-wide distribution and exposure which enable both a wide and rapid dissemination of research output.

The series covers the theory, applications, and perspectives on the state of the art and future developments relevant to systems and networks, decision making, control, complex processes and related areas, as embedded in the fields of interdisciplinary and applied sciences, engineering, computer science, physics, economics, social, and life sciences, as well as the paradigms and methodologies behind them.

Indexed by SCOPUS, EI Compendex, INSPEC, WTI Frankfurt eG, zbMATH, SCImago.

All books published in the series are submitted for consideration in Web of Science.

For proposals from Asia please contact Aninda Bose (aninda.bose@springer.com).

Mohuya Chakraborty ·
Shambhu Prasad Chakrabarty · Althaf Marsoof
Editors

Proceedings of Sixth International Ethical Hacking Conference

AI and Law (eHaCON 2025)

Springer

Editors
Mohuya Chakraborty
Cybersecurity Centre of Excellence,
Institute of Engineering and Management
University of Engineering and Management
Kolkata, West Bengal, India

Shambhu Prasad Chakrabarty
University of Engineering and Management
Kolkata, West Bengal, India

Althaf Marsoof
Division of Business Law, College
of Business (Nanyang Business School)
Nanyang Technological University
Singapore, Singapore

ISSN 2367-3370 ISSN 2367-3389 (electronic)
Lecture Notes in Networks and Systems
ISBN 978-981-96-8631-5 ISBN 978-981-96-8632-2 (eBook)
https://doi.org/10.1007/978-981-96-8632-2

This Springer imprint is published by the registered company Springer Nature Singapore Pte Ltd.
The registered company address is: 152 Beach Road, #21-01/04 Gateway East, Singapore 189721, Singapore

Program Committee

Chief Patron

Prof. Banani Chakrabarti, Chancellor, University of Engineering and Management, India

Patrons

Dr. Sajal Dasgupta, Vice-Chancellor, University of Engineering and Management, India
Dr. Satyajit Chakrabarti, Pro Vice-Chancellor, University of Engineering and Management, India

General Conference Chair

Dr. Mohuya Chakraborty, Director, Cybersecurity Centre of Excellence, University of Engineering and Management, Kolkata, India

Advisory Committee

Dr. Valentina Emilia Balas, Aurel Vlaicu University of Arad, Romania
Dr. Ana Penteado, Notre Dame University, Australia
Dr. Angappa Gunasekaran, California, State University, Baskerfield, California
Dr. S. K. Ramesh, California State University, Northridge
Dr. Chandan Chakraborty, NITTR, India

Dr. Dorothea La "Chon" Abraham, Mason School of Business at William & Mary, USA
Dr. Gleuto Serafim, Ixonn Group, Australia
Mr. Marcelo Bussacarini, Cisco, Australia
Dr. Pabitra Mitra, IIT, Kharagpur, India
Dr. Mohuya Chakraborty, University of Engineering and Management, India
Dr. Rajiv Ganguly, University of Engineering and Management, India

Keynote Speakers

1. Dr. Gleuto Serafim, Director and Founder at Ixonn Group, Australia
2. Dr. Souvik Mukherjee, India
3. Dr. Ana Penteado, Adjunct Associate Professor, Notre Dame University, Australia
4. Dr. Shambhu Prasad Chakrabarty, Dean of Law, University of Engineering and Management, Kolkata, India
5. Dr. Althaf Marsoof, Deputy Head and Assistant Professor of Law, Division of Business Law, College of Business, Nanyang Technological University, Singapore
6. Mr. Marcelo Bussacarini, Global Security Brand Strategist, CISCO, Sydney, New South Wales, Australia
7. Dr. Mohuya Chakraborty, Director, Cybersecurity Centre of Excellence (CCoE), University of Engineering & Management, Kolkata, India

Editorial Board

Dr. Mohuya Chakraborty, University of Engineering and Management, Kolkata, India
Dr. Shambhu Prasad Chakrabarty, University of Engineering and Management, India
Dr. Althaf Marsoof, Nanyang Technological University, Singapore

Technical Program Committee

Dr. Mohuya Chakraborty, University of Engineering and Management, India
Dr. Chandan Chakraborty, NITTTR, India
Dr. Shambhu Prasad Chakrabarty, University of Engineering and Management, India
Dr. Ana Penteado, Notre Dame University, Australia
Dr. Sukanya Mukherjee, UEMK, India
Mr. Sudip Kumar Palit, UEMK, India

Organizing Committee

Dr. Mohuya Chakraborty, University of Engineering and Management, India
Dr. Shambhu Prasad Chakrabarty, University of Engineering and Management, India
Dr. Sukanya Mukherjee, IEM, India
Ms. Shrabana Chattopadhyay, UEMK, India
Ms. Abhishikta Basu, UEMK India
Mr. Victor Nayak, UEMK, India
Mr. Sudip Kumar Palit, UEMK, India

Preface

The progressions in the technologies of computer networks, the Internet, and artificial intelligence have changed the panorama of cybersecurity. Hackers are unceasingly in the course of discovering innovative procedures to access defenseless cyber systems and produce mayhem. Cyberlaw has a great role to play in saving the world from malicious activities on the Internet. The primary aim of this conference is to promote community consciousness of the ever-increasing cybercrimes occurring in our modern society and the lawful measures that can be adopted to avert them. Through four distinct tracks—the Evolution of AI, AI and Cybersecurity, AI and Law, and AI and Industry 6.0—keynote addresses, research paper presentations, and 3-minute scholarly thesis presentations are delivered. The keynotes covered a wide range of topics covering threat intelligence and defense, AI and blockchain integration for dApp development, AI-driven assessment of carbon storage potential in the dominant mangrove species of Indian Sundarbans, seeking the deep side of AI for humanity, AI as the attacker or AI as the defender, AI and legal education, AI in the Republic of Congo, and theories of victimology in the age of digital misinformation. The oral paper presentation tracks have covered a variety of research topics, including deep learning, machine learning, graphology, data analytics, intelligent systems, big data, prompt engineering, blockchain technology, augmented intelligence, IPR, crime prevention, and cybercrimes. To create a secure society, a platform for discussion and exchange of ideas about various topics, including data privacy, deep fakes, blockchain, cryptography, penetration testing, cyber-physical systems, digital twins, ethics, supply chain management, malware analysis, and risk management is established. Researchers from academia and industry, both domestically (India) and internationally (US, UK, Singapore, Australia, and Dubai), participated in the discussions. The proceedings present each facet of the conference in the expectation that, shortly, new research discoveries may lead to the expansion of feasible solutions for innocuous systems. This book is split into five parts. The first part presents the *Keynotes*. The second part discusses the *Evolution of AI*, the third part highlights *AI and Cybersecurity*, while the fourth and fifth parts deal with *AI and Law* and *AI and Industry 6.0*, respectively.

Keynotes

Two keynotes cover topics such as the role of AI in enhancing cybersecurity and persistent gender inequality in the video game industry.

Evolution of AI

In this part, eight papers present various application areas of AI, including health care, gig workers' contracts, automated vehicles, sustainable city planning, graphology, and personality prediction.

AI and Cybersecurity

In this part, there are nine papers including permissioned and public blockchain applications, NFT ecosystem challenges and opportunities, generative AI, AI in intrusion detection and digital forensics, the role of AI in online shopping and wearable health care, and deepfakes.

AI and Law

This part comprises nine papers that deal with ethical issues in AI, the role of AI in dispute resolution in India, anti-money laundering frameworks, autonomous weapon systems and their legality in international humanitarian law, mergers and acquisitions, regulated industrial monitoring, shared AI and IoT decisions, and convergence of AI and law.

AI and Industry 6.0

This part is all about the application of AI in industrial areas. Eight papers depict AI used for virtual power plants, blue economy in Bali, CO_2 emissions in Kolkata, mitigating acidification in coastal waters of Digha, saffron and white terror in India and America, tribal knowledge systems, relationship of forest and aboriginals in West Bengal, and insights from India and Canada about dark web's influence on web technologies and security.

The editors trust that this volume is exclusive and significant in that it delivers numerous research papers on end-to-end standpoints on the applications of AI to solve cybersecurity issues and AI-based cybercrime detection from today's perspective. Within the canopy of AI, research papers on deep learning, machine learning, neural networks, and data analytics have been included that may be quite beneficial to a large group of scientists, engineers, advocates, researchers, and cybersecurity community about the ever-evolving era of cybercrime and application of AI to mitigate them.

Kolkata, India — Mohuya Chakraborty
Kolkata, India — Shambhu Prasad Chakrabarty
Singapore — Althaf Marsoof

Message from Chief Patron

Welcome to the sixth edition of the eHaCON International Ethical Hacking Conference, held in 2025. The University of Engineering and Management's Cybersecurity Centre of Excellence in Kolkata, India, hosts eHaCON 2025 every year. Leading academic researchers and industry practitioners from India and beyond will have a forum to exchange their creative ideas, experiences, and state-of-the-art research in the fields of network security, ethical hacking, and cybersecurity. This is the primary goal of this premier event. Since the event is held this year in collaboration with the university's Law Department and the Centre of Excellence in Cyber Physical Systems, it will cover a wide range of legal issues pertaining to cybersecurity.

The National Institute of Technical Teachers Training and Research, Kolkata, India; Notre Dame University, Australia; the Indian School of Ethical Hacking, India; Cisco, Sydney, New South Wales, Australia; Nanyang Technological University, Singapore; Congo University, Congo; and Ixonn Group, Australia, have generously contributed to the technical sponsorship that has enabled eHaCON 2025. For this event's tremendous success, I would like to thank all of the sponsors, associates, supporters, and University of Engineering and Management members.

Prof. Banani Chakrabarti
Chancellor
University of Engineering and Management
Kolkata, India
February 2025

Message from Patrons

Welcome to the sixth International Ethical Hacking Conference: AI and Law, eHaCON 2025. The University of Engineering and Management's Cybersecurity Centre of Excellence hosts eHaCON 2025 every year. This year, the Department of Law has partnered. We express our profound appreciation to the eHaCON 2025 organizers for organizing such a fantastic event, where attendees from around the globe presented research papers on a range of topics, including AI and Cybersecurity, AI and Law, Evolution of AI, and Industry 6.0.

We are wishing the conference further success and motivation for the organizers' work. I hope this conference strengthens relationships and gives us all the tools we need to make a difference.

Dr. Sajal Dasgupta
Vice Chancellor
University of Engineering and Management
Kolkata, India
February 2025

Dr. Satyajit Chakrabarti
Pro Vice Chancellor
University of Engineering and Management
Kolkata, India
February 2025

Message from Conference General Chair

The Cybersecurity Centre of Excellence, in association with the Department of Law, University of Engineering and Management (UEM), Kolkata, hosted the 6th International Ethical Hacking Conference—Leveraging AI to Law during February 20–21, 2025, at the University of Engineering and Management, Kolkata, New Town campus. It was a great pleasure to greet everyone who attended. The purpose of eHaCON 2025 was to provide a forum for free discussion on the implications of emerging technology for a safe society. Technical paper presentations, live demonstrations, workshops, a gaming competition, and an online hacking coding challenge were all well-balanced throughout the conference. The objective was to begin the process of completely automating cyber defense. During a two-day instructive research paper presentation and keynote address, the most significant new findings about computer network attacks and defenses, commercial security solutions, and practical real-world security experiences are given. Many nations from all over the world contribute research papers.

The keynote speakers, Dr. Ana Penteado from Notre Dame University in Australia, Dr. Althaf Marsoof from Nanyang Technological University in Singapore, Mr. Marcelo Bussacarini from Cisco in Australia, Dr. Gleuto Serafim from Ixonn Group in Australia, Dr. Shambhu Prasad Chakrabarty from India, and Dr. Souvik Mukherjee from India, have my deepest gratitude for their presentations on a variety of innovative legal, AI, and security issues in the digital world.

Dr. Shambhu Prasad Chakrabarty organized the entire program flawlessly, for which I am incredibly thankful. To ensure the high caliber of the conference, very strict guidelines were followed to choose only the best technical papers from a vast number of submissions, with the assistance of a superb committee of national and international experts.

I am grateful to Dr. Sukanya Mukherjee and Prof. Sudip Kumar Palit for serving as reviewers of multiple papers. Additionally, I would like to thank Prof. Victor Nayak for his ongoing involvement in floor management and Prof. Shrabana Chattopadhyay for her excellent work conducting the technical programs. Professor Pallabi Sengupta and Prof. Abhishake Roy deserve special recognition for their exceptional work compiling all of the abstracts for our souvenir. Professor Abhisikta Basu did wonders in managing food and venue. Participants arrived from a variety of Indian and international schools, institutions, government agencies, and businesses. I hope the several events, such as 3MST, Dcrfr, CTF, and workshops on AI and law, were very beneficial to them.

I would like to express my gratitude to the University of Engineering and Management for co-sponsoring this conference and for their financial and infrastructural assistance. For their outstanding contribution, I am thankful to all of the academics and researchers from Romania, Australia, India, Singapore, and the USA who serve on the advisory and technical committees. I also want to express my gratitude to the local organizing committee, rapporteurs, session chairmen, and co-chairs for their tireless efforts in making this conference a huge success.

Finally, but just as importantly, thanks to all of the writers and participants, Springer for technical help, and the University of Congo for technical collaboration. In addition to enjoying our culturally vibrant city of joy, Kolkata, I hope they had a good time at the conference!

Dr. Mohuya Chakraborty
Director
Cybersecurity Centre of Excellence
University of Engineering and Management
Kolkata, India
February 2025

Message from Convenor

On behalf of the Organizing Committee of eHaCON 2025 International Ethical Hacking Conference: AI and Law, it is my pleasure to welcome the attendees to the University of Engineering and Management, Kolkata, India.

The conference consists of four technical sessions with 42 contributed papers and six keynote addresses. eHaCON 2025 Program Committee, comprising distinguished members, worked hard to organize the technical program. Following rigorous review process, out of about 295 submissions only 42 papers were accepted for presentation in the technical sessions.

Behind every successful event there lies the hard work, commitment, and dedication of many personalities. Firstly, I wish to thank the entire program committee for the excellent job it did in organizing the technical sessions. Special thanks are due to all the reviewers for their obligation in reviewing the papers within a very short time. I am indebted to the faculty members of the university for managing the conference on cybersecurity and law where participants from various schools, colleges, and industry, and government officials are benefited. A special thanks goes to the conference chair (Dr.) Mohuya Chakraborty for giving us immense support and encouragement throughout this period. Once again, I hope that all the delegates from India and abroad will find the program beneficial and will enjoy the historic city of joy—Kolkata.

I sincerely thank Late (Dr.) Satyajit Chakrabarti, Chancellor, (Dr.) Sajal Dasgupta, Vice-Chancellor, and (Dr.) Satyajit Chakrabarti, Pro Vice-Chancellor of the University of Engineering and Management, for their constant support throughout the event. Last but not the least, I thank all our delegates from various industries as well as academia for participation without whom the conference would not have been possible.

Dr. Shambhu Prasad Chakrabarty
Dean
Department of Law
University of Engineering and Management
Kolkata, India
February 2025

Contents

AI and Industry 6.0

About the Editors

Mohuya Chakraborty is a distinguished cybersecurity expert who holds various positions, including Director of the Cybersecurity Centre of Excellence, Dean of Faculty Welfare and HRDC, and Professor of the Department of Computer Science and Engineering at the University of Engineering and Management Kolkata. She holds a B.Tech. and M.Tech. from Calcutta University, a Ph.D. in Mobile Computing from Jadavpur University, and a Chartered Management Institute (CMI Level 5) certification in Leadership and Management from Dudley College, London. She has received numerous awards, including the Paresh Lal Dhar Bhowmik Award and the Rashtriya Gaurav Award. Chakraborty is a Certified Publons Academy Mentor, and she has published over 100 research papers in high-impact journals, books, book chapters, and conference proceedings, and has been involved in numerous international conferences. She has mentored several Ph.D. scholars in India and abroad.

Shambhu Prasad Chakrabarty is Dean of the Department of Law at the University of Engineering and Management in Kolkata, India. He holds an LLB and LLM in Business Law from Calcutta University and a Ph.D. from the University of North Bengal. He has taught and researched at Calcutta University and West Bengal National University of Juridical Sciences for over 20 years. Chakrabarty has edited and published nine books and has written over 50 research papers in various fields, including Climate Change, Traditional Knowledge, Intellectual Property Rights, and Technology Law. He has supervised three Doctoral Scholars and has presented his work at over 70 conferences. He has been invited to numerous national and international events, including the University of Jakarta and University of Padjadjaran, Indonesia. Chakrabarty is a member of the Law and Society Association, the Commonwealth Lawyers Association, and IPIRA, where he engages with research and policy initiatives.

Althaf Marsoof is Associate Professor at Nanyang Business School, Nanyang Technological University, Singapore, and Deputy Head of the Business Law Division. He holds a B.Sc. in Information Technology from Curtin University, an LLM from the University of Cambridge, and an M.Phil. from the University of Queensland. Althaf's doctoral thesis focused on the liability of Internet intermediaries for trade-mark infringements, which was funded by the Dickson Poon Ph.D. Scholarship. His book, *Internet Intermediaries and Trade Mark Rights* was published by Routledge in 2019. His book was cited in the European Court of Justice's Opinion. Althaf has won several awards at NTU, including the John Cheung Endowment Social Media Award in 2018, the NBS Research Award for the Division of Business Law in 2020, and the REP Teacher of the Year Award in 2021. He is qualified to practice law in Sri Lanka.

Keynotes

The Role of Artificial Intelligence in Enhancing Cybersecurity: Trends, Challenges, and Ethical Considerations

Gleuto M. Serafim

Abstract The increasing integration of Artificial Intelligence (AI) in the Cybersecurity domain has changed the way that countries defend themselves against ever-evolving and dynamic threats. This paper provides a broad survey of the use of AI in Cybersecurity, focusing on the new and emerging trends, the main issues, and the need for ethical steering (Harshith J, Gill MS, Jothimani M (2023) Evaluating the Vulnerabilities in ML systems in terms of adversarial attacks. arXiv preprint arXiv:2308.12918. https://arxiv.org/pdf/2308.12918). Based on a multi-disciplinary analysis, it reviews the development of AI-enhanced threat detection, real-time threat intelligence, automated incident response, and predictive analytics, with a critical assessment of adversarial machine learning and the use of AI in cyber weapons (Rafique SH, Abdallah A, Musa NS, Murugan T (2024) Machine learning and deep learning techniques for internet of things network anomaly detection—current research trends. Sensors 24(6):1968. https://doi.org/10.3390/s24061968). The research identifies the weaknesses of conventional security measures and explains how AI, especially deep learning and behavior-based models, improve detection and swift response (Schmit (2023) J Ind Inf Integr 36). In addition, the study looks at systemic issues such as data scarcity, explainability, systemic bias, and regulatory gaps. It suggests that a human-AI teaming approach is a viable way to achieve cyber resilience (Sarker IH, Janicke H, Mohammad N, Watters P, Nepal S (2023) AI potentiality and awareness: a position paper from the perspective of human-AI teaming in cybersecurity. Int Conf Intell Comput Optimiz 140–149. https://arxiv.org/pdf/2310.12162). The discussion is informed by current empirical evidence and policy, legal, and ethical frameworks for AI-driven cybersecurity practices, which emphasize transparency, fairness, and accountability (Ibid.; Bernardez Molina S, Nespoli P, Gómez Mármol F (2023) Tackling cyberattacks through ai-based reactive systems: a holistic review and future vision. arXiv e-prints, arXiv: 2312.06229. https://doi.org/10.48550/arXiv.2312.06229). Finally, this paper argues for the need for cooperation across sectors, investment in the security of AI, and the

G. M. Serafim (✉)
Ixonn Group, Sydney, NSW, Australia
e-mail: gleuto.serafim@ixonn.com

M. Chakraborty et al. (eds.), *Proceedings of Sixth International Ethical Hacking Conference*, Lecture Notes in Networks and Systems 1513,
https://doi.org/10.1007/978-981-96-8632-2_1

creation of adaptive and ethically aligned AI structures to protect the digital environment in a connected world with increasing cyber risks (Andrada et al. (2023) AI & Soc 38:1321–1331; Musser M, Lohn A, Dempsey JX, Spring J, Kumar RSS, Leong B, Liaghati C, Martinez C, Grant CD, Rohrer D (2023) Adversarial machine learning and cybersecurity: risks, challenges, and legal implications. arXiv preprint arXiv:2305.14553. https://arxiv.org/abs/2305.14553; Oseni A, Moustafa N, Janicke H, Liu P, Tari Z, Vasilakos A (2021) Security and privacy for artificial intelligence: opportunities and challenges. arXiv preprint arXiv:2102.04661. https://arxiv.org/pdf/2102.04661).

Keywords AI-powered cyber-attacks · Cyber threat intelligence · Ethical AI governance human-AI teaming · Predictive security analysis · SOCs

1 Introduction

The watershed invention of computers and later the rise of the Internet have revolutionized various aspects of information technology (IT), business, industry, communication, and, eventually, almost every aspect of human life. This digital revolution impacts the global economy and has opened boundless opportunities in investments and innovation [1]. However, malicious actors also used this background to exploit IT systems in harmful ways for personal gain or to cast harm on governments, critical infrastructures, organizations, and public. As a result, the number and sophistication of cyber-attacks have skyrocketed and posed significant challenges to the security and resilience of today's digital systems and interconnected networks [2].

The traditional cyber defense approach that only relies on human expertise and conventional security mechanisms could not cope with such a diverse and sophisticated threat landscape [3]. To maintain the cyber-physical integrity, confidentiality, and the performance of these digital systems and networks, there is an acute need to develop more competent, robust, and adaptive cyber defense solutions which can repulse existing cyber threat vectors and are able to adapt new ones. In this context, innovative security models that are underpinned by advanced artificial intelligence (AI) and machine learning (ML) are particularly getting increasing attention from academia, industry, government and policymakers, law enforcement, and military [4]. The underlying argument for adopting AI solutions into cybersecurity (CS, for brevity, used interchangeably in this text) benches on the assumption that it can enhance the security of the various digital systems and networks [5]. Furthermore, AI-based solutions can support human cyber experts to predict, detect, analyze, mitigate, and respond to cyber-attacks in an efficient manner, and assist them in decision-making while dealing with an incident-driven scenario.

1.1 Background and Rationale

In the last decade, the pace of technology penetration into human society has been rising, and AI technologies have been very active in this acceleration. Even with the protean potentiality and the promising awareness of AI, awareness of the feasibility and beneficial functionalities of Human-Aligned Type AI (HAType AI) in the domain of cybersecurity is thought to be limited compared to several other application areas, e.g., the cognitive sciences, environment, medical sciences, the philosophy of mind, the social media services, and Internet shopping [1, 6]. This position paper is proposed to overview AI potentiality in terms of the transformation of thoughtful behavior, to raise awareness of anticipated AI functionalities concerning sophistication in cybersecurity, and to appeal to regulatory bodies of government to foster financial support within the realization of expected comprehensive and long intense computer-based HAType AI applications [4].

Despite the exploitation of AI in craftsmanship cybersecurity solutions so as to assist security and possible appropriate behaviors for security enhancement, training and teaching nonetheless require significant performance data extracted from schemes or resources [7]. Thus, production domain security modeling becomes a national need. Nevertheless, the conditions stated and the norms concerning the industry static games generally utilized to design frameworks for security infrastructure are somewhat superficial and do not meet legitimacy matters that may be addressed by cybersecurity experts [8, 9].

1.2 Scope and Objectives

The focus of this research is on artificial intelligence (AI)-based cybersecurity solutions, in general, and specifically upon the trends and some specific challenges including systemic bias, the lack of explainable AI, and the availability of labeled data [6]. A departure from current mainstream AI is taken looking particularly for methods based on associationist psychology. The main goals, within a broader design research context, include developing a deep learning back-end for an experimental cybersecurity analysis tool, to generate synthetic datasets by reverse engineering training data, to run an automated framework that intelligently selects and fine-tunes AI algorithms, and to adapt AI methods to portico-adequate models of reasoning [5].

AI-based cybersecurity solutions are an ever-increasing presence in the literature and industry. They have the potential to either take any human cyber defense to its logical conclusion by fully automating most aspects of attack detection, defense, and response, or else to represent a significant force multiplier that goes beyond signature-based intrusion detection [4, 6] Nonetheless, perhaps the grandparent of all challenges, the one regarding how to effectively model adversaries, remains largely unaddressed [10]. Instead of addressing this gargantuan task, this paper aims, that is, to foster thought and debate around key points in order to advance both the field of

AI-based cybersecurity and the general security posture of today's digital systems and interconnected networks.

2 The Evolution of Cybersecurity

Modern society's dependency on the Internet has outpaced cybersecurity. The weaponization of cyber attacks through ransomware, worm-like wiper, botnets, and other means spells a complicated and chaotic challenge for the Internet security, which directly threatens the reliability, robustness, and availability of both the national and enterprise network infrastructure, as well as end-users' data and devices [1, 2]. Cyber threats are constantly evolving and adapting. Apart from regular patch updates, a new, more flexible, and robust system should be developed against growing sophisticated and stealthy threats [9]. Studying and developing intelligent systems that are able to learn, think, and act like humans to deal with the sophisticated cyber-offensive, meaning to defend in an automatic cognitive fashion, are considered as an urgent need and global exigency of the hour [3, 5].

2.1 Traditional Approaches Versus AI-Enhanced Security

Traditional cybersecurity approaches, predominantly based on rule-based systems, static firewalls, and human-centered monitoring, are increasingly proving inadequate in the face of today's rapidly evolving and stealthy threat landscape. Signature-based intrusion detection systems (IDS), once the gold standard in cyber defense, often fail to detect novel or polymorphic threats that do not match known patterns [9, 11]. As adversaries adopt more sophisticated tactics—such as zero-day exploits, AI-enabled malware, and multi-vector attacks—there is a critical need to develop intelligent systems that mimic human cognitive functions to autonomously detect, respond to, and adapt against threats in real time.

Artificial intelligence (AI) and machine learning (ML) offer transformative capabilities in enhancing the speed, accuracy, and depth of threat detection. These intelligent systems are designed to learn from historical and real-time data, identify anomalies, and continuously improve their performance without explicit reprogramming [5]. Unlike traditional models, AI-enabled systems can generalize from incomplete or ambiguous data, making them especially useful in dynamic threat environments where patterns evolve quickly and unpredictably [7].

Understanding the concept of intelligence, human or artificial, is fundamental in framing this technological transition. Intelligence, broadly defined as the ability to learn, reason, adapt, and solve complex problems, underpins both human cognition and the design of AI systems [12, 13]. In cybersecurity, this translates to the ability of systems to proactively identify potential threats, reason through the severity and scope of an attack, and act swiftly—often autonomously—to neutralize risk. The implementation of deep learning algorithms and neural networks exemplifies this capacity for abstract reasoning, mirroring aspects of human intelligence in digital defense contexts [14]. Nevertheless, despite these advantages, the deployment of AI-based cybersecurity tools must contend with several limitations, including data bias, lack of transparency, and difficulties in explaining decisions, especially in critical infrastructure and defense settings [8]. Therefore, a hybrid model that combines human expertise with AI's computational power—often referred to as Human-AI Teaming—is increasingly recommended for optimal cyber resilience.

3 AI for Threat Detection and Prevention

A malware is present in the network. Once the malware is identified, the Network Detection System (NDS) sends an alarm to the Incident Response System (IRS). The IRS would select the most effective countermeasure from the options received by the Heuristic Inference Component, or Controller (HIC), and each node would implement it with the help of the Node Blockade Manager (NBM). An Intrusion Prevention System (IPS) is constantly analyzing the network and proactively generating security measures. The network is split into three categories: the protected infrastructure, the NDS infrastructure, and a set of homogeneous, secure and reliable nodes of the network, named nodes. The NDS is responsible for detecting and analyzing malicious traffic. The HIC takes the results obtained from the detection carried out by the NDS and decides the strategies to counteract the threats. The NBM is responsible for the rapid and robust blockade strategy from the selected nodes [4].

Cybersecurity has been at the center of public interest earning notable concerns from individuals and organizations across the globe. Most of the current AI applications in the field of cybersecurity are focused on improving detection, prevention, and response systems. AI can greatly benefit cybersecurity by enabling more sophisticated detection, prevention, and response systems. One of the main advantages of AI in cybersecurity is its ability to analyze large volumes of data and identify patterns and anomalies that may be indicative of a cyber-attack [11]. Popular machine learning algorithms, such as support vector machines, neural networks, and decision trees, can be trained on datasets to recognize patterns of known attacks and detect previously unseen threats [14]. Traditional approaches to security, such as signatures or heuristics for malware detection, rely on static, rule-based systems that are clearly not suited to the vast quantity and complexity of modern cyber threats [11]. On the other hand, AI has an excellent capacity to learn and adapt to new threats, making it particularly suitable for the constantly evolving threat landscape. The autonomous nature of AI-based threat response systems gives them a significant advantage over traditional systems. Being constantly on, AI systems can monitor network activity around the clock, and quickly pinpoint any vulnerabilities that arise in the system. Furthermore, in the case of an actual cyber-attack, AI can automatically draft and carry out an event response strategy, which in many cases can reduce resolution times from hours to seconds [7, 9].

3.1 Threat Hunting Techniques

The data and information sharing of companies are rapidly prevailing. Due to the data-driven necessity, various data analysis solutions are established. Thus, these systems contain critical data for cybersecurity. Since these systems contain critical data and the popularity of data analysis, these kinds of networks are becoming increasingly common for Tesla Transportation Systems. Therefore, it gains more attention from attackers. Thus, the cybersecurity of these networks is crucial. To achieve this goal, the first and possibly the most important step is to accurately detect these kinds of cyber threats. Therefore, the design of the Cyber Threat Hunting Tool is crucial for the security of Tesla Transportation Systems. As an additional task of this thesis, the Cyber Threat hunting is further discussed, and possible methodology is examined. Three different methodologies, Data-Driven Hunting, Intel-Driven Hunting, and TTP-Driven Hunting (Tactics, Techniques, and Procedures), were developed. Small-scaled experiments are also performed with the test datasets of these three approaches [15].

Cyber threat hunting is defined as a proactive cyber defense activity. Instead of waiting for the cyber threat to compromise the system, the cybersecurity personnel will proactively hunt for the potential cyber threat. It is important for a threat analysis system to adapt ML scheme for increasing speed, size, and performance of the data in term of flow data [5, 16]. Generally, the most commonly known categories are data-driven hunting, Intel-driven hunting, and TTP-driven hunting. While the data-driven hunting way starts with data in order to hunt for a threat, Intel-driven hunting starts with threat information. Similarly, TTP-driven hunting applies threat information as well, but this information particularly consists of tactics, techniques, and procedures of attack [17].

There are also various definitions of cyber threat hunting. Another group describes cyber threat hunting as a baseline for next-generation security operation centers. Additionally, some studies provide a new way to apply cyber threat hunting to the industrial control system domains employing state-based system-strong reasoning. Recently, several showcase new methods, including the evidence-based classification of cyber threat hunting, hybrid cyber threat hunting, sports selection algorithms based on feature-based tracking networks, deep learning models, and the application of various machine learning techniques based on an organization's data structure to improve a systematic understanding [14, 18]. Companies and security operation centers are similar. These attempts have increased interest in cyber threat hunting. On the other hand, researchers seek the necessity to investigate these new research domains with this approach. Therefore, the purpose of this research is to study existing studies of cyber threat hunting, describe the basics for understanding of this model, and analyze these studies to find modern trends.

3.2 *Real-Time Threat Intelligence*

Some of the most sophisticated early warning and protection systems are those used in cybersecurity. Cyberspace is a complex environment. The main problems that arise are the diverse, periodically emerging, constantly evolving, and potentially invisible, sophisticated nature of threats. Recent reports of Advanced Persistent Threat (APT), Operation Aurora, and Stuxnet have shown the significant capabilities of attacks and the extensive damage that can occur [19, 20]. In such an environment, the speed and ability to process information of human operators, who are responsible for analyzing the threat, are limited [21]. However, the use of technologies and systems where this limitation can be overcome is increasing.

In this context, more advanced, real-time detection mechanisms are used instead of traditional anti-virus software, and are complemented by more sophisticated systems, providing early warning and threat intelligence [22]. They are capable of analyzing events and trajectories that go beyond individual attacks. While over 290 thousand new malicious files are detected daily [23], most are still using known threats. New attacks are more likely. There is a need for systems that can help the analyst in the early stages of new and evolving attacks.

As an AI-based solution, a new cybersecurity platform that can significantly help analysts to use their time effectively and reduce the time between attack detection and attack prevention has been developed. An alternative-likelihood representation of security data (ARD) has been proposed that can support more sophisticated queries than traditional correlation rule templates when analyzing security data and network traffic. In a case study with this data representation and a minimum Bayesian network learning algorithm, a general model of the computer network of an extremely large organization in terms of security has been learned [24]. A larger version of this organization comprising also energy plants and refineries comprises thousands of private and public IP addresses, several firewalls, IDSs, and honeypots. For this "Application to Real Data" task, edited, filtered public network data of 12 days of this larger organization has been provided. Along with the data a table is provided encoding a simple network configuration. This network consists of 4 sub-networks, protected by 3 firewalls, and one in-line IDS [4].

4 Predictive Security with Machine Learning

In a digital, connected world, where connectivity between machines is a key enabler, there are many opportunities for bad actors to find the weakest link in a network or a software system to gain access to sensitive information, carry out espionage, or conduct sabotage. Varying trends in intrusion techniques, together with broader issues such as legacy systems, geographically distributed installations, remote operations, and tension between organizational layers and stakeholders, create a rich context for efforts to increase cyber risk [25]. This discusses and describes the work carried out to develop a dynamic and self-adapting system that can accept data from multiple security information and event management (SIEM) systems and other security systems and translate relevant events and recognized patterns first into probabilistic models and then yield predictions [26].

With the information society and the development of new technologies, the importance and complexity of securing data and cyber assets are increasing. In a technically connected world, a company, a city, or a country's Smart Infrastructure increases vulnerability to attacks. Contemporary digital industries are characterized by complex cyber-physical systems consisting of interconnected devices. Artificial Intelligence is essential for the development of autonomous network security applications and systems and is one of the core building blocks for next-generation network protection [27]. Machine Learning-based methods are transforming the field of intrusion detection with the ability to assess risks, notify events of interest, and adapt mechanisms against unexpected threats in real time [26].

Regarding the continuous increase in cybersecurity incidents in cyber-physical systems and interconnected devices, the need for Intrusion Detection Systems with high detection rates and specificity is increasing, as they are crucial to ensuring Smart Infrastructure operation and safety [28]. Undertaking aims at examining how intrusion trends influence the development of smart infrastructures, and the digital industries that secure it, with a focus on AI-enabled solutions including multiple ML algorithms [1]. A particular focus is on the application of Deep Learning and Stacking Ensemble algorithms producing and testing ML models on both, real production data [25]. Taking into account the specificity of smart industries and infrastructures, an anomaly and Signature-based ML models are developed [26].

4.1 Anomaly Detection

Protecting networks and devices from cyber threats is an increasing financial issue for both individuals and organizations. Currently, the most popular innovation in data networks and computerized devices is the Internet of Things (IoT), which greatly adds to the prevalence of cyber-attacks. Numerous studies have been conducted over the years to strengthen the security of computer infrastructure and systems, and many practical approaches have been developed, yet it is not a complete solution. The rampant nature of the array of assaults from cyber-attackers and the ability of defending information systems opposing innovative tools and techniques are not preserved by many proposed approaches. The usage of smart and creative machine learning so-called Artificial Intelligence (AI) systems for cyber defense is an important research study direction to address these complexities.

Machine learning is now automatically utilized to develop intelligent systems for detecting and intervening on cyber-attacks. As the prevalence of Internet of Things (IoT) gadgets increases every day, there is also a potential growth in the variety of cyber-attacks on them. In IoT systems, challenges for anomaly identification rise exponentially compared to traditional systems due to the vast amount of data accumulated. Therefore, in the field of cyber-attack detection, the use of AI methods has become necessary, mainly machine learning, where machine learning performs excellently in detecting anomalies. In that perspective, cyber-attack and anomaly identification research efforts in IoT structures that use AI (especially machine learning) algorithms have been explored in this particular context [7].

4.2 Behavioral Analysis

Cybercriminals are constantly on the lookout for new techniques and approaches for gaining unauthorized access to corporate networks and sensitive business data. Using traditional cybersecurity systems is no longer effective in detecting and mitigating modern-day threats [29]. This scenario requires a more proactive approach to cybersecurity, and this is where behavioral analysis comes into play. Instead of considering individual threats, changes in behavior provide a clearer picture for detecting threat patterns. Monitoring every action taken on the network is a substantial challenge for cybersecurity experts in large companies. This is why modern security information and event management systems use algorithms to identify and sort useful or worrying behaviors [30]. In this way, the responses of malicious entities will stand out more easily against a background of legitimate activity, which could be inattentive clicks on links to malware or attempts to gain unauthorized access to network resources [31].

5 AI-Powered Security Operation Centers (SOCs)

With increasing global connectivity, dependency on information communication technology (ICT) in homes, enterprises, across nations, and on a global scale, a rapid rise in cyber-attacks has been witnessed most notably with sophisticated attacks against high profile organizations, critical infrastructures, and nation-states [32]. The number of attack vectors and adversaries is ever-evolving, growing, and changing rapidly indubitably. In light of these maladies, the role of AI-driven cybersecurity and risk management is paramount and has attracted significant interest in a bid to edge cybersecurity landscape [33].

AI currently ranks among the top technology trends and advancements. Leveraging on advances in computing power, data storage space, and velocity, machine learning models, namely deep learning/artificial neural networks have become state of the art in a number of domains including speech, image recognition, and machine translatability. AI has transcended in IT security, and novel security processes are emerging. AI is now a key enabler in next-level security automation in Security

Operation Centers (SOCS) which are ever more becoming intelligent and automated. These are powered to better and proactively detect, investigate, prioritize, and mitigate threats at a given pace and accuracy besides decision-making [3].

5.1 Benefits and Limitations

In order to assist human experts in decision-making and fill the gap in security skills and knowledge, many methods based on AI and data analysis have been studied and employed to develop various security solutions [34]. AI technologies have shown great potential to significantly enhance security in the broad terms of cybersecurity or physical security. Although a variety of studies have been conducted on security systems based on AI technologies, there are still several issues that need to be dealt with to widely spread the benefits of AI technologies to security [35].

Although different AI methodologies are used, anomaly-based approaches dominate the current development in security, which leads to an overcrowded and intricate family of anomaly detectors. Considering that not all possible artifacts from a domain will be benign in a hostile context, a comprehensively continuing side effect might be the unanticipated creation of security skills in persistent threats. Aiming to model the security of software systems, AI technologies have been exploited. Especially in recent years, AI has shown great potential for improving the security of software systems in different respects. However, in terms of the effectiveness of the security model, AI technologies are still in a formative stage and confront several challenges to achieve superior security modeling capabilities [36].

6 Automating Incident Response with AI

The complexity and velocity of today's cyber threats have made traditional intrusion detection and prevention mechanisms inefficient. New reactive cyber defense systems using AI are therefore attracting widespread attention. This effort has included taking advantage of real-time analysis, machine learning, and adaptive response with the help of AI capabilities and intelligent agents [37]. AI technology is being used to support timely detection and analysis of potential threats [38]. Moreover, AI can now be used to automate various aspects of network defense, such as designing secure systems, verifying systems for compliance with security policies, and mining large volumes of log and sensor data from security devices to better detect intrusions.

Incident response systems traditionally consist of human operators that are guided by fixed procedures to detect, delay, and recover from cyber-attacks [5]. Such procedures are usually ineffective in the face of highly sophisticated cyber-attackers using stealthy, evasive, and tailored attack strategies. AI can help to detect, anticipate, and react to such threats, by automating many actions traditionally carried out by human operators [4]. These systems are fed by intelligent analyses of attacker behaviors, collected by agent systems running in both attacker and defender domains, whose artificial intelligence is used to predict future evolutions of the conflict and autonomously adapt defensive and counter-defensive strategies.

6.1 Challenges and Opportunities

Artificial intelligence (AI) has demonstrated transformative societal impact. Its intersection with cybersecurity is a logical corollary in the context of the current technology-infused digital world. AI mechanisms are being designed for making autonomous cybersecurity decisions, including automation of proactive defense while enhancing the situational awareness of ongoing cyber threats for human cyber experts [39]. However, such a paradigm shift in leveraging AI for security seems to be concurrently existing with several yet unaddressed challenges. While a plethora of research aims to enhance the cyber defense and situational awareness capabilities through AI-based systems, it is also crucial to addressing AI-generated adversarial actions and thwarting adversarial use of AI by cyber-attackers, which follow a "cat and mouse" pattern. The realization of AI-generated adversarial actions (against security) is founded in competitive optimization challenges and requires in-depth investigations at the theoretical level, particularly in implementing AI systems compatible with security standards.

In this work, a multi-disciplinary conjoining of AI and cybersecurity domains has been adopted to delineate or conceptualize AI "events" and their implications for security in a structured manner. It provides surveys, taxonomies, and development investigations, and explores likely future areas. This is vital for driving research agendas and implementing resilient AI-based security infrastructures [5]. A diverse range of "events" largely focusing on the use and implementation of AI in the context of, or affecting, cybersecurity has been considered. This permits an extended or holistic research consideration as well as insight into broader issues, which may be less obviously connected with security and are not typically addressed in specialized AI journals or security conferences.

7 The Dark Side of AI: Weaponizing Cyber Threats

On December 23, 2015, the regional power grid in Ukraine was manipulated in an unprecedented cyber-attack. The attack was coordinated in nature and resulted in widespread power outages that affected 225,000 customers. BlackEnergy and KillDisk malware were employed to incapacitate 32 substations in the northwestern Ivano-Frankivsk region of Ukraine. Communications and hard disk systems were wiped, effectively preventing the substations from receiving instruction from the central grid. Technicians were forced to manually reset infected systems, resulting in a large-scale blackout. It was found that the malware was implanted at least six months in advance, spread to other networks throughout Europe, and displayed a high level of expertise. In short the operation was highly coordinated with extensive reconnaissance, skilled implementation, and complex actions. CyberSecurity Ventures predicted that by 2021 the cost of cybercrime would have risen to $6 trillion globally, and from 2016 to 2025 the global cybersecurity market would grow from $106 billion to $167 billion [40]. As corporations and states evolve their capabilities to conduct cyber operations, Artificial Intelligence (AI) technology will play an influential role in this domain [41]. As AI capabilities continue to advance they will be widely applied, and this provides a strategic analysis of AI and its likely interaction with cyber power. It is discussed that AI will likely influence cyber operations in at least three distinct ways: (1) AI-powered autonomous systems for civilian use will introduce new security vulnerabilities; (2) Military autonomous weapon systems, likewise powered by artificial intelligence, are being developed, which will establish new types of digital vulnerabilities in conflicts; (3) Offensive AI-driven cyber weapons will generate new challenges to the security of digital networks, as well as novel opportunities for conducting cyber operations against adversarial networks [8].

7.1 Adversarial Machine Learning

Artificial Intelligence (AI) technologies have rapidly developed over recent years and are presently being used in numerous domains. As AI technologies automate previously manual tasks, they are assisting individuals in various areas of their everyday routine. Security vulnerabilities lie in the widespread applications and unknown/untold consequences of development [42]. The old story of Dr. Frankenstein and his artificial creature could serve as a benchmark for the use of AI in the modern cybersecurity world. Numerous AI-based systems are trained as security mechanisms, such as botnets, malware, and intrusion detection systems [10]. However, these security mechanisms might be overwhelmed with adversarially constructed inputs that do not contain noxious content and are not recognized as such by conventional security mechanisms. It is no wonder that cybercriminals are applying AI tools to both safeguard their criminal operations against opponents and attack other AI systems forged as defensive tools. As the boundaries between the two groups of systems (cybersecurity mechanisms) become thinner, the differences blur, making it difficult to distinguish between them. At this point, it is important to switch from applying offensives to developing defenses to secure AI mechanisms and guard them against adversaries [42]. There is a growing interest in examining and developing robust AI mechanisms that can be used safely. It is also vital to understand the potential attack surfaces of AI mechanisms comprehensively. If system developers are better aware of the dangers that AI transport, initiatives could be taken to deploy them more securely, mitigating the security dangers [9].

8 AI-Powered Cyber-Attacks

As Artificial Intelligence (AI) advances, it will have a growing influence on cyber-physical systems (CPS). AI may assist in either security or attacks, and the growth in CPS driven by AI is projected to produce AI integrated into a variety of new markets, including mechanical systems, infrastructure systems, and military applications. The use of AI in cyber defense become broader to address the spread of AI in cyber operations, and the protection of AI-controlled devices from tampering [6]. The task of AI-secured systems is to use AI to enhance the protection of AI and the devices it controls from disruption. An overview of emerging developments is provided, including new threats from use of AI in cyber operations, the work of establishing resilient CPS against AI-enabled threats, the rise of AI tooling for cyber-physical systems, efforts to ensure the trustworthiness of such systems, and the need for the evolvement of defense strategies to cope with AI [43].

With the proliferation of cyber-physical systems controlled by AI, a growing trend of using AI to protect those systems from cyber-attacks is noticed. There have been works focusing on trustworthiness issues of AI supervised control, which emphasizes attacks on the AI controller that enable an adversary to control the system. The AI-enabled protection of CPS from cyber intrusion is of growing interest for research and policy, but most existing approaches rely on new AI advances [1]. Therefore, the protection of advanced, AI-operated agents that compete in an adversarial environment gains substantial relevance. A formal framework is introduced that is versatile in representing a wide range of security settings of the learned AI model. This framework is then employed to identify best practices in the protection of AI agents and to develop an improved defense strategy [44].

8.1 Types of AI-Powered Attacks

Cybersecurity has significantly evolved with an array of Artificial Intelligence (AI)-powered attacks including polymorphic malware, zero-day vulnerabilities, spear phishing, and denial-of-service attacks [45]. Malware that allows an unauthorized person to control an affected device is known as a Computer Virus. The captured device then becomes part of a botnet, a collection of compromised devices. This botnet is used in coordinated attacks, such as Distributed Denial-of-Service attacks, which are often used to intimidate governments and corporations [46]. Ransomware encrypts the local system and requires a ransom to decrypt the data, typically in digital currency. Polymorphic malware changes its code structure and internal functions with each variant release, thereby complicating signature-based detection. Spear phishing is a targeted type of phishing in which messages are personalized to the victim. Spear phishing has become easier with the advent of social media and is often used to gain access to critical systems due to its high success rates [47]. A company was hacked by the attackers starting with a successful spear-phishing campaign against its employees. Viewing employee details based on accounts, a pitch to the employees was sent offering a free service, but prompting them to login with their work credentials instead.

Business applications for Internet traffic flow predict the attack effect range, informing the user or network admins. Similarly, AI-enabled attacks could be developed to evade routing, bypassing standard defenses [48]. AI cybersecurity could move away from traditional network structures, such as isolating sensitive networks, to distillate or invisible nets. AI hackers could also take advantage of less secure networks, such as surveillance cameras or power grids. In addition, cyber-physical systems, such as pacemakers that wirelessly transmit telemetry, could be attacked [49]. Researchers can try to improve the accuracy of threat and anomaly detection. Increasing the resilience of AI in cybersecurity involves combining models and human operators [50]. Another idea would look at the trust in AI cybersecurity. Major attacks could occur if the AI cybersecurity manipulates data fed into a network. This trust problem is most notable for AI cybersecurity as it becomes more widely adopted and may be the simplest to resolve. The framework is about the maturity of AI cybersecurity. The AI cybersecurity sector is quickly varying. Hardening systems against adversarial attacks, for example, ensures the security model is not using inputs to produce radiant results [51]. This could be carried out in AI cybersecurity experimenting with harmful data interpolation capable of producing unexpected malware signatures, a setback to attackers using heavy AI in their software. These would be handled with the right planning and research and limit AI cybersecurity enhancements.

9 Ethical Considerations in AI-Driven Cybersecurity

Artificial intelligence (AI) methods have escalated in popularity in the domain of cybersecurity over the years, aiming to automate quickened defensive countermeasures. Machine learning-based AI formulas are prominent in this area. However, as with any double-edged sword, such algorithms are similarly being adopted to automate and expedite assaults [52]. Here, crucial realms of AI applied to cybersecurity are elaborated via existing, modern, and forthcoming appliances. Predicaments, hypotheses, trials, resolutions, and attempted assessments in employing AI software on both the defense and attack fronts are also delineated. Cyberspace is the fabric for the contemporaneous information era, where millions of entities, including individuals, organizations, and machines, transmit, store, and utilize data. All are interconnected to improve fiscal and social development, and the innermost workings rely extensively on infallible operation. Artificial intelligence applications in cybersecurity in recent, current, and future viewpoints are explained. Modern applications handle present advancements of AI, predominantly in machine learning [53]. AI for automated information safety danger identification systems utilizing constraint learning methods are the seeds of growth in cyber defense technology.

The forthcoming realm is constituted of the most theoretically pertinent yet to be deciphered field, covering anti-science through to future AI combatant methods, as well as a number of present and unproven trials and hypotheses. Research reservations of the effect and procedure of AI-operated assailant algorithms and defenses appreciate the vital truth that, to control an entity, it is indispensable to comprehend it: so the emphasis is chiefly on basic AI and AI-applied assault appraisal along with the counterbalance analyses [54]. Yet, since such AI-operated assailants and defenses are both multifaceted infrastructures and the body of related work is constrained, attendant appraises are included on intimately related but somewhat more traditional topics, such as the psychology of AI utility in cybersecurity. Using AI fortifies outcomes in a revised methodical concept for AI-driven attacker and defensive algorithms in today's confrontational engagements over tactics and weapon choice. Accordingly, evaluation design is fashioned and initial exemplar study results reviewed [5, 55].

9.1 Bias and Fairness Issues

Artificial Intelligence (AI) has been used increasingly in a broad spectrum of applications over the past two decades. With multiple recent advancements in AI technologies and an increased proliferation of data through Internet-connected devices, AI research and development as well as its applications are expected to grow significantly in the next decade [56]. Cybersecurity is becoming an increasingly important and challenging task due to the rapidly growing number of attacks by cybercriminals. In recent years, the Internet of Things (IoT) has played an important role in monitoring and supporting various aspects of human health. Conversational systems are important applications in smart assistants and robots used in health care. Smart assistants, intelligent tutoring systems, and e-learning platforms have been used increasingly in educational applications to help learners. In health care, computer vision and deep learning are being used increasingly for medical image analysis. Blockchain technology has many interesting applications in cybersecurity, energy systems, medical services, IoT, and other systems. However, blockchain technology also faces many challenges, and possible research directions in the future are discussed. Social robotics is a rapidly evolving domain in multiple application areas, including special education in schools and special education services. Furthermore, the pandemic has drawn attention to the challenges faced by family carers, particularly in how they access support services. Machine learning algorithms have been widely applied in various applications. The development of deep learning in AI technology has created new opportunities in different fields. In reinforcement learning, an AI agent learns how to behave in an environment to achieve certain goals. Transmission power control mechanisms play an important role in improving the energy efficiency of wireless networks. Military and surveillance applications require the interception of signals from desired objects. Internet of Vehicles has extensive application coverage in intelligent transportation systems. Social Influencing Simulation

System is proposed for decision makers to address public concern and managing the outbreak of diseases [6].

9.2 Transparency and Accountability

When it comes to AI and ML in the context of cybersecurity, transparency and accountability are under-researched [57]. Demands for transparency, accountability, and auditability are becoming louder, and regulations and technical solutions are posed in response. However, many seem to misunderstand what transparency, accountability, and auditability actually are or entail in the case of cybersecurity involving AI- and ML-enhanced decision-making. Therefore, this chapter disambiguates these three concepts as relevant to the context of cybersecurity and stresses what challenges can be expected.

Accountability and transparency are often lumped together. Nevertheless, there is an important distinction to be made. Transparency efforts often stand at the beginning of an accountability chain. The idea is to provide clear visibility to the public and to controllers, auditors, and supervisory authorities, why a decision was made. Often, porting the actual decision-making process is not an option; sometimes it may even appear impossible since in some cases the data is sensitive or classified. This brings us to accountability which, however, is after all different from transparency. Accountability denotes responsibility for a decision or an action taken. It refers to the capacity to explain and justify decisions, actions, and the processes leading to them. It can be seen as the overarching ambition to manage societal demands for trust and fairness in automated and autonomously taken decisions [58].

10 The Future of AI in Cybersecurity

Artificial Intelligence (AI) is one of today's most popular buzzwords. Despite the fact that it has its roots in the 1960s, modern AI is starting to be explored from approximately 2010. AI is becoming a powerful tool to assist and augment the capacity of cybersecurity analysts in an ever-growing cyber-threat landscape [59]. While a tremendous amount of effort is being put to develop AI-based cybersecurity solutions, there are already several challenges to be overcome before AI cybersecurity fulfils its potentiality [5]. One of them is the high expectation of security experts without knowing the limitations of AI. This essay thus identifies potential trends in AI security modeling and tailors cybersecurity domains from a more realistic perspective with what extent AI technology can be maximized [60].

Cybersecurity is prominently looming on the risk management radar of businesses. Compromising sensitive data through cyber-attacks can have devastating consequences on consumer trust for a brand or organization, opening it up to liability and destroying customer confidence in doing business. Humans designing cybersecurity systems have for the longest time approached security concerns of various installation paradigms using iterative rules to extend the level of protection and deterrence to attacks. This big-data immersed, hyper-connected world is making today's security landscape more complicated [6]. This security posture is incredibly dynamic, with the complexity of the selected installation. These fast technological advances can bring about the increasing number of different types of potential threats, with existing organized cybercriminals leveraging more sophisticated tools and methodologies. This evolving cyber-threat environment necessitates the design and development of sophisticated AI-driven cybersecurity mechanisms that can understand, predict, and respond automatically [59].

10.1 Emerging Trends and Technologies

In today's hyper-connected world, with the emergence of the Internet of Things (IoT), the true potentials of a smart society are beginning to come to life. With forecasts of about 30 billion connected devices reaching 50 billion in a smart society by 2025, technological innovations are advancing the world of cyber-physical systems [61]. The true power of the cyber-physical system is the exchange of real-time information between the physical elements, which enhances machine-to-machine communication. In turn, this leads to the advancement of critical infrastructural areas, which are the key capabilities for a smart society benefit. However, with the advent of this enormous progress, numerous basic questions arise. The two fundamental issues that arise are the security and safety of the cyber-infrastructure entities. Secure and readily available cyber-infrastructure is absolutely imperative for the safe use of crucial infrastructures. The easily distributed denial-of-service attack might inhibit the work of the embattled servers or even hit the whole network. The network traffic might be distributed among the most critical links using the load-balancing attack. Thus, the network might be cascaded into the uncontrolled behavior of the nodes using the cascading attack, using the power grid as an illustration. This incapacitates the cyber part of the infrastructure from controlling the physical part and so seriously affects the infrastructures from functioning properly. Due to the relevance of the challenges, there arises an urgency for handling such a security issue [60].

11 Strategic Recommendations

The need for security in computer systems has grown exponentially with the amount of electronic data processed and transmitted by these systems. The use of artificial intelligence (AI) in the intrusion detection field, especially the machine learning technique such as deep learning, is speculated to have a significant positive impact. AI would improve the functionality of existing intrusion detection systems by enabling users to analyze audit logs of large network traffic and other data collected from computer systems. Furthermore, this technology would help users experiment with different intrusion detection methodologies. While there are many potential benefits of deep learning in cybersecurity, there are a range of potential risks; relevant challenges also exist. Many of the vulnerabilities, threats, and harms have legal implications and require policy consideration [8]. Therefore, this section presents 17 recommendations to increase the security of AI in cybersecurity and promote broader awareness and preparation for adversarial AI threats. Some of the following recommendations emphasize opportunities for industry and government policymakers to promote AI security by investing in technical research. However, the majority of the recommendations focus on changes to processes, institutional culture, and awareness among AI developers and users. These recommendations are organized around broadly anticipated deep learning cybersecurity use cases and practical measures to address them.

11.1 Policy Implications

The use of artificial intelligence (AI) in cybersecurity is continuously growing. This growth is driven by the vast capacity of AI and its varieties, making it adaptive to embracing the ever-growing landscape of cyber threats. In cybersecurity, AI's main applications are focused on identifying, reacting to, and mitigating cybersecurity attacks. AI has proven to be very efficient in pattern recognition, which is at the heart of monitoring tons of data. AI-based system monitoring can stabilize data and inform an analyst about potential security issues in data. AI is able to identify irregularities or discrepancies automatically and in advance to trigger with human intervention or in an automated manner [5]. Furthermore, AI has the capacity to reconstruct and associate security-related events to gain visibility and help reconstruct a wider picture. This paper presents arguments and counterarguments on the effectiveness and relevance of employing AI in enhancing cybersecurity. Trends are scrutinized, challenges are confronted, and the ethical side of the equation is analyzed. As far as it is concerned, it is expected that this paper will seed a necessary discussion to profoundly think and investigate earlier embracing the utilization of AI in cybersecurity which is inevitably to develop further, however, in an efficient and ethical manner. To this end, six research question specifications were set to operate on [62].

11.2 Investment Strategies

Besides the positive impact that Artificial Intelligence (AI) brings to cybersecurity, investment strategies remain a challenging issue. Thus, the potentiality and challenges of AI-based cybersecurity were investigated. This requires that a position paper is presented to highlight the current research progress, underpinning fundamental issues, and unveil perspectives on AI-aware cyber defense collaboration in the coming era of Controllable-AI. The main cited issues and potential directions concern a mismatch of assumptions about security modeling, the need for new models of security, and the prospects for new means of automated cyber defense. It is hoped that addressing the identified issues and pursuing the potential directions might lay a foundation for the development of next-generation cybersecurity solutions [8]. The widespread use of the Internet has made daily life highly dependent on networks and cyberspace services, which have made cyberspace the most welcome hiding environment for unwanted and illegal activities. With the development of cyberspace, both normal and illegal activities have expanded through the Internet. On the one hand, traditional attacks, such as computer viruses and malware, have become more common, and the damage caused has become more serious. On the other hand, cyberterrorism and cyber war activities have gradually surfaced, which pose the greatest security threat to all walks of life. Cybersecurity needs to have robust collaboration and a fusion of people's cognitive capabilities of analyzing, understanding, and resolving security occurrences, and machines automated pervasive cognition capabilities, including sharing contextual and actionable awareness of security occurrences and inducing effective awareness of trustworthiness on command execution [4]. From this perspective, awareness plays a pivotal role in the cyber defense chain mechanism of secure mission-oriented models manifesting operational activities, understanding/representing system states, running security operations, etc.

Bibliography

1. Schmitt M (2023) Securing the digital world: Protecting smart infrastructures and digital industries with artificial intelligence (AI)-enabled malware and intrusion detection. J Ind Inf Integr 36:100520. https://doi.org/10.1016/j.jii.2023.100520
2. Mayer M (2018) Artificial intelligence and cyber power from a strategic perspective. https://fhs.brage.unit.no/fhs-xmlui/bitstream/handle/11250/2497514/IFS%20Insights_4_2018_Mayer.pdf?sequence=1
3. Onwubiko C, Ouazzane K (2022) Challenges towards building an effective cyber security operations centre. arXiv preprint arXiv:2202.03691. https://arxiv.org/pdf/2202.03691
4. Bernardez Molina S, Nespoli P, Gómez Mármol F (2023) Tackling cyberattacks through AI-based reactive systems: a holistic review and future vision. *arXiv e-prints*, arXiv: 2312.06229. https://doi.org/10.48550/arXiv.2312.06229
5. Sarker IH, Janicke H, Mohammad N, Watters P, Nepal S (2023) AI potentiality and awareness: a position paper from the perspective of human-AI teaming in cybersecurity. Int Conf Intell Comput Optimiz 140–149. https://arxiv.org/pdf/2310.12162
6. Ibid
7. Rafique SH, Abdallah A, Musa NS, Murugan T (2024) Machine learning and deep learning techniques for internet of things network anomaly detection—current research trends. Sensors 24(6):1968. https://doi.org/10.3390/s24061968
8. Musser M, Lohn A, Dempsey JX, Spring J, Kumar RSS, Leong B, Liaghati C, Martinez C, Grant CD, Rohrer D (2023) Adversarial machine learning and cybersecurity: risks, challenges, and legal implications. *arXiv preprint*arXiv:2305.14553. https://arxiv.org/abs/2305.14553

9. Oseni A, Moustafa N, Janicke H, Liu P, Tari Z, Vasilakos A (2021) Security and privacy for artificial intelligence: Opportunities and challenges. arXiv preprint arXiv:2102.04661. https://arxiv.org/pdf/2102.04661
10. Harshith J, Gill MS, Jothimani M (2023) Evaluating the vulnerabilities in ML systems in terms of adversarial attacks. *arXiv preprint*arXiv:2308.12918. https://arxiv.org/pdf/2308.12918
11. Sommer R, Paxson V (2010) Outside the closed world: on using machine learning for network intrusion detection 2010 IEEE symposium on security and privacy. https://citeseerx.ist.psu.edu/document?repid=rep1&type=pdf&doi=b46bd10ab7a2de4edd6ca6bd712155ff27432592
12. Legg S, Hutter M (2007) A collection of definitions of intelligence. Front Artificial Intell Appli 157:17. https://arxiv.org/pdf/0706.3639
13. Sternberg RJ (2005) The theory of successful intelligence. Revista Interamericana de Psicología/Interamerican Journal of Psychology 39(2):189–202. https://www.redalyc.org/pdf/284/28439202.pdf
14. Alazab M, Soman K, Srinivasan S, Venkatraman S, Pham VQ (2023) Deep learning for cyber security applications: a comprehensive survey. Authorea Preprints. https://www.techrxiv.org/doi/full/https://doi.org/10.36227/techrxiv.16748161
15. Shaukat K, Luo S, Chen S, Liu D (2020) Cyber threat detection using machine learning techniques: a performance evaluation perspective. In: 2020 international conference on cyber warfare and security (ICCWS), pp 1–6. https://doi.org/10.1109/ICCWS48432.2020.9292388
16. Yuan S, Wu X (2021) Deep learning for insider threat detection: review, challenges and opportunities. Comput Security 104:102221. https://doi.org/10.1016/j.cose.2021.102221
17. Kavitha D, Thejas S (2024) AI enabled threat detection: Leveraging artificial intelligence for advanced security and cyber threat mitigation. IEEE Access 12:173127. https://ieeexplore.ieee.org/document/10747338
18. Mahboubi A, Luong K, Aboutorab H, Bui HT, Jarrad G, Bahutair M, Camtepe S, Pogrebna G, Ahmed E, Barry B (2024) Evolving techniques in cyber threat hunting: a systematic review. J Network Comput Appl 104004. https://doi.org/10.1016/j.jnca.2024.104004
19. Baker K (2025) What is an advanced persistent threat (APT)? CrowdStrike. Crowdstrike.com. https://www.crowdstrike.com/en-us/cybersecurity-101/threat-intelligence/advanced-persistent-threat-apt/
20. Kushner D (2013) The real story of stuxnet. IEEE Spectr 50(3):48–53. https://doi.org/10.1109/mspec.2013.6471059
21. Sfetcu N (2024) Advanced persistent threats in cybersecurity – cyber warfare. https://doi.org/10.58679/mm28378
22. Jain V, Mitra A (2024) Real-time threat detection in cybersecurity: leveraging machine learning algorithms for enhanced anomaly detection. In (pp 315–344). https://doi.org/10.4018/979-8-3693-7540-2.ch014
23. Şeker E (2019) Use of artificial intelligence techniques/applications in cyber defense. arXiv preprint arXiv:1905.12556. https://arxiv.org/pdf/1905.12556
24. Wang J, Neil M (2021) A Bayesian-network-based cybersecurity adversarial risk analysis framework with numerical examples. arXiv (Cornell University). https://doi.org/10.48550/arxiv.2106.00471
25. Moon PS, Deshmukh AB, Sarode HJ, Sharma S, Birare KM, Anandpwar WN (2024) Implementation of machine learning techniques for predictive security analytics. international conference on frontiers of intelligent computing: theory and applications
26. Uccello F, Pawlicki M, D'Antonio S, Kozik R, Choraś M (2024) Towards hybrid NIDS: combining rule-based SIEM with AI-based intrusion detectors. In: International conference on advances in computing research, pp 244-255. https://doi.org/10.1007/978-3-031-56950-0_21
27. Camacho NG (2024) The role of AI in cybersecurity: addressing threats in the digital age. J Artif Intell General Sci (JAIGS) ISSN: 3006-4023, 3(1):143–154. https://doi.org/10.60087/jaigs.v3i1.75
28. Ahmed U, Nazir M, Sarwar A, Ali T, Aggoune E-HM, Shahzad T, Khan MA (2025) Signature-based intrusion detection using machine learning and deep learning approaches empowered

with fuzzy clustering. Scientific Reports 15(1):1726. https://www.nature.com/articles/s41598-025-85866-7
29. Ahmed M, Mahmood AN, Hu J (2016) A survey of network anomaly detection techniques. J Netw Comput Appl 60:19–31. https://doi.org/10.1016/j.jnca.2015.11.016
30. Conti M, Dehghantanha A, Franke K, Watson S (2018) Internet of Things security and forensics: challenges and opportunities 78:544–546. https://doi.org/10.1016/j.future.2017.07.060
31. Delwar TS, Aras U, Mukhopadhyay S, Kumar A, Kshirsagar U, Lee Y, Singh M, Ryu J-Y (2024) The intersection of machine learning and wireless sensor network security for cyber-attack detection: a detailed analysis. Sensors 24(19):6377. https://doi.org/10.3390/s24196377
32. Weng Y, Wu J (2024) Leveraging artificial intelligence to enhance data security and combat cyber attacks. J Artif Intell General Sci (JAIGS) 5(1):392–399, ISSN: 3006-4023. https://doi.org/10.60087/jaigs.v5i1.211
33. Kaur R, Gabrijelčič D, Klobučar T (2023) Artificial intelligence for cybersecurity: literature review and future research directions. Informat Fusion 97:101804. https://doi.org/10.1016/j.inffus.2023.101804
34. Chen L, Ye Y, Bourlai T (2017) Adversarial machine learning in malware detection: arms race between evasion attack and defense. In: 2017 European intelligence and security informatics conference (EISIC), pp 99–106. https://ieeexplore.ieee.org/abstract/document/8240775
35. Sarker IH (2021) Deep cybersecurity: a comprehensive overview from neural network and deep learning perspective. SN Comput Sci 2(3):154. https://doi.org/10.1007/s42979-021-00535-6
36. Khayat M, Barka E, Serhani MA, Sallabi F, Shuaib K, Khater HM (2025) Empowering security operation center with artificial intelligence and machine learning–a systematic literature review. IEEE access. https://ieeexplore.ieee.org/abstract/document/10850912
37. da Silva Siqueira JC (2022) Autonomous Incident Response Universidade de Lisboa (Portugal)]. https://www.proquest.com/openview/b26b70e8858bc35b37c80d423a061128/1?cbl=2026366&diss=y&pq-origsite=gscholar
38. Abdul S. AI for cyber security: automated incident response systems. https://www.researchgate.net/publication/383825151
39. Dash B, Ansari MF, Sharma P, Ali A (2022) Threats and opportunities with AI-based cyber security intrusion detection: a review. Int J Softw Eng Appl (IJSEA) 13(5). https://ssrn.com/abstract=4323258
40. Malik V (2024) Artificial intelligence and cybersecurity: navigating the complexities of a data-driven future. *Isc2.org*. https://www.isc2.org/Insights/2024/07/AI-and-Cybersecurity-Navigating-a-Data-Driven-Future
41. Gusher T (2025) AI and ML are cybersecurity problems—and solutions. *Ey.com*. https://www.ey.com/en_us/insights/cybersecurity/ai-and-ml-are-cybersecurity-problems-and-solutions
42. Magazine C (2018) Cybercrime to cost the World $10.5 trillion annually By 2025. Cybercrime Magazine. https://cybersecurityventures.com/cybercrime-damages-6-trillion-by-2021/
43. Zhou J, Cao Z, Dong X, Vasilakos AV (2017) Security and privacy for cloud-based IoT: challenges. IEEE Commun Magazine 55(1):26–33. https://ieeexplore.ieee.org/abstract/document/7823334
44. Mothukuri V, Parizi RM, Pouriyeh S, Huang Y, Dehghantanha A, Srivastava G (2021) A survey on security and privacy of federated learning. Future Generat Comput Syst 115:619–640. https://www.sciencedirect.com/science/article/abs/pii/S0167739X20329848
45. Sahani N, Zhu R, Cho J-H, Liu C-C (2023) Machine learning-based intrusion detection for smart grid computing: a survey. ACM Trans Cyber-Phys Syst 7(2):1–31. https://doi.org/10.1145/3578366
46. Buczak AL, Guven E (2015) A survey of data mining and machine learning methods for cyber security intrusion detection. IEEE Commun Surveys Tutorials 18(2):1153–1176. https://ieeexplore.ieee.org/abstract/document/7307098
47. Chiew KL, Yong KSC, Tan CL (2018) A survey of phishing attacks: their types, vectors and technical approaches. Expert Syst Appl 106:1–20. https://www.sciencedirect.com/science/article/abs/pii/S0957417418302070

48. Khan MI, Arif A, Khan ARA (2024) The most recent advances and uses of AI in cybersecurity. BULLET: J Multidisiplin Ilmu 3(4):566–578. https://www.neliti.com/publications/592396/the-most-recent-advances-and-uses-of-ai-in-cybersecurity
49. Humayed A, Lin J, Li F, Luo B (2017) Cyber-physical systems security—a survey. IEEE Internet Things J 4(6):1802–1831. https://ieeexplore.ieee.org/abstract/document/7924372
50. Abeshu A, Chilamkurti N (2018) Deep learning: the frontier for distributed attack detection in fog-to-things computing. IEEE Commun Magazine 56(2):169–175. https://ieeexplore.ieee.org/abstract/document/8291134
51. Goodfellow IJ, Shlens J, Szegedy C (2014) Explaining and harnessing adversarial examples. arXiv preprint arXiv:1412.6572. https://arxiv.org/abs/1412.6572
52. Kaushik K, Khan A, Kumari A, Sharma I, Dubey R (2024) Ethical considerations in AI-based cybersecurity. In: Next-generation cybersecurity: AI, ML, and Blockchain, pp 437–470. Springer. https://doi.org/10.1007/978-981-97-1249-6_19
53. Perez A, Anwar S (2024) Risks and ethical concerns in cyber security with advancements of artificial intelligence-a systematic review. In: 2024 IEEE frontiers in education conference (FIE), 1–5. https://ieeexplore.ieee.org/abstract/document/10893551
54. Kulothungan V (2024) Securing the AI Frontier: urgent ethical and regulatory imperatives for AI-driven cybersecurity. In: 2024 IEEE international conference on big data (BigData), pp 5602–5609. https://ieeexplore.ieee.org/abstract/document/10826010
55. Balantrapu SS (2022) Ethical considerations in AI-powered cybersecurity. Int Machine Learn J Comput Eng 5(5)
56. Li B, Qi P, Liu B, Di S, Liu J, Pei J, Yi J, Zhou B (2023) Trustworthy AI: from principles to practices. ACM Comput Surveys 55(9):1–46. https://doi.org/10.1145/3555803
57. Memarian B, Doleck T (2023) Fairness, accountability, transparency, and ethics (FATE) in artificial intelligence (AI) and higher education: a systematic review. Comput Educ Artif Intell 5:100152. https://www.sciencedirect.com/science/article/pii/S2666920X23000310
58. Baker S, Xiang W (2023) Explainable AI is responsible AI: How explainability creates trustworthy and socially responsible artificial intelligence. arXiv preprint arXiv:2312.01555. https://arxiv.org/abs/2312.01555
59. Gupta M, Akiri C, Aryal K, Parker E, Praharaj L (2023) From chatgpt to threatgpt: Impact of generative AI in cybersecurity and privacy. IEEE Access 11:80218–80245. https://ieeexplore.ieee.org/abstract/document/10198233
60. Admass WS, Munaye YY, Diro AA (2024) Cyber security: State of the art, challenges and future directions. Cyber Security Appl 2:100031. https://www.sciencedirect.com/science/article/pii/S2772918423000188
61. Chui KT, Gupta BB, Liu J, Arya V, Nedjah N, Almomani A, Chaurasia P (2023) A survey of internet of things and cyber-physical systems: Standards, algorithms, applications, security, challenges, and future directions. Information 14(7):388. https://www.mdpi.com/2078-2489/14/7/388
62. What are the risks and benefits of artificial intelligence (AI) in Cybersecurity? (2015) Palo Alto Networks. https://www.paloaltonetworks.com/cyberpedia/ai-risks-and-benefits-in-cybersecurity
63. Andrada G, Clowes RW, Smart PR (2023) Varieties of transparency: exploring agency within AI systems. AI & Soc 38(4):1321–1331. https://doi.org/10.1007/s00146-021-01326-6

Hidden Figures: The Missing Connection in Women, Careers and the Video Game Industry

Ana Penteado and Shambhu Prasad Chakrabarty

Abstract This paper is about the persistent gender inequality in the video game industry. This gender bias has been exacerbated by large language model applications (LLMs) that mimic human emotions. Reality intertwines with digital imagery, such as virtual reality (VR) or monitor, which affects memory and human perception of both spaces. There is a struggle in human memory that is akin to deception. These deceptive patterns integrate the environmental experiences that make the video game industry profitable as required retained attention to the game is the main asset. The alignment of a profitable industry and emerging technology to dominate a digital market may alienate other genders when job opportunities are scarce and discriminatory, cementing a lack of social and economic mobility for future generations. That articulates a narrative of exclusion in the video game experience largely informed by cisgender identity, namely by men, which will eventually disempower women, who are also video game players and aspiring female video game designers, being marginalised in the profitable market of entertainment. Thus, artificial intelligence, a technology dominated also by one gender identity, may allow further gender segregation for decades ahead of us. We conclude that corporate and social incentives in this industry are still shy to integrate genders, leading to the creation of video games. In this paper's research, we found a necessity for more multicultural and pluralistic business strategies and focused female training programmes so that a joint effort to foster a culture of inclusion and belonging throughout the entire video game ecosystem will welcome women, particularly at senior levels, with the participation of video game developers and designers in the decision-making process, would be positive. Thus, the representation of women in this industry would permit a rapid career progression in the video game corporate space, which is much needed and justified for a better enjoyment of an inclusive and lucrative game experience.

A. Penteado (✉)
University of Notre Dame, Notre Dame, Australia
e-mail: penteado.ana2008@gmail.com

S. P. Chakrabarty
University of Engineering and Management, Kolkata, India

M. Chakraborty et al. (eds.), *Proceedings of Sixth International Ethical Hacking Conference*, Lecture Notes in Networks and Systems 1513,
https://doi.org/10.1007/978-981-96-8632-2_2

Keywords Artificial intelligence · Gender equality · Women · Video game industry · Video game (studios) · Video game market · Video game (art) · Segregation · Career progression · Digital economy · Shah Rukh Khan · Movie FAN · Economic growth · Quality education · Gaming · Game environment · Game engine · Large language models (LLMs) · Barks · Climate change · Sustainable Development Goals (SDGs) · United Nations SDG Goal 1 · United Nations SDG Goal 4 · United Nations SDG Goal 5 · United Nations SDG Goal 8 · United Nations SDG Goal 10 · United Nations Goal 13 · United Nations 17 · Electricity grids · Tertiary sector · Quaternary sector · Asian Development Bank · The World Bank · Slash Plan · Graphic processing unit (GPU) · Entertainment industry · Australia · China · Japan · South Korea · India · United States · Europe · United Kingdom · France · High-end gaming consoles · Game consoles · Sound design · Immerse effects · Self-learning · Hallucinations · ARISTOCRAT · NINTENDO · ELECTRONIC ARTS · EPIC GAMES · SONY INTERACTIVE ENTERTAINMENT · PLAYSTATION · PLAYSTATION 4 · MICROSOFT · TAKE TWO INTERACTIVE SOFTWARE · ROCKSTAR GAMES · ZYNGA · FORTNITE · BATTLE ROYALE · INFINITE BLADE · ACTIVISION BLIZZARD · CALL OF DUTY · WORLD OF WARCRAFT · TENCENT · NETEASE · UBISOFT ENTERTAINMENT · ARISTOCRAT · NAZARA · GENERATIVE ADVERSARIAL NETWORKS (GANs) · MIDJOURNEY · ASSASSINS' CREED · RED DEAD REDEMPTION II · GRAND THEFT · GAME FOR PEACE · HONOUR OF KINGS · KIDDOPIA · WORLD CRICKET CHAMPIONSHIP · Metaverse · Mental disorder · World Health Organization · CHUKA · Break the Silence · WeRise · Edutainment · Social platforms · Player Unknown's Battlegrounds (PUBG) · Computer graphics · Female developers · Compliance officer · Training data risk assessments · CEO female · Parallel AI hybrid · Lawsuit harassment · Energy consumption · Semiconductors · California Department of Fair Employment and Housing v Activision Blizzard (2021) · Benjamin Bell et al. (2012) · Correctional Chamber of Paris Court of Appeals · Detroit-Become Human · Quantic Dream · MICROSOFT WINDOWS · Violence against Women and Girls (VAWG) · THE LEGEND OF ZELDA: TEARS OF THE KINGDOM · MARIO BROS: WONDER · THE SIMS · STAR WARS BATTLEFRONT · BATTLE ROYALE FORTNITE · Children's Privacy Law (US) · BLOOD STRIKE

1 Introduction

The video game industry is predominantly a *grey area*, as in Sofia Hulten's installation, where she dressed up in a grey camouflage, and disappeared into a dustbin. Hunten art sends a message that the office is not a space for females. On the contrary, a working place is the art form of concealed gender, which we explore in this paper in the entertainment sector, the video game industry. We argue in this paper that this sector is heavily composed of a cisgender men environment, monopolistic in its ethnicity and social background.

We suggest that further research must be undertaken in this demographic to understand the motivation for an apparent diversity discouragement policy in this industry.

We observed rare work mobility to senior positions for women, namely in the video game development and in design to senior positions. Generally, these females opt for entrepreneurship.

We argue that an industry averse to diversity carries biases towards the other missing genders, in this case females, which in the video game industry may be a misunderstanding of its consumer base, divided almost equally between genders. We have insufficient data to identify other genders [1].

Consequently, it may result in discriminatory and preconceived stereotype acceptance of females as only consumers of video game products.

Having an isolated and labelled gender as consumers only illustrates a consistent disregard for females' gaming ability to improve the field of computer design and diversified thematic narratives in the video game environment.

We also argue that the absence of diversity in the video game industry should be raised as a probable link between gendered violence in games and gender-related attitudes and behaviours being perpetuated in our society.

The first part, entitled "video game industry", is divided into three sub-sections in alphabetic letters. Section *a* focuses on the video game sector in the entertainment industry, describing the national and international trade of products in the digital economy, the economics and market value of the video game industry, and the future profits within this industry with artificial intelligence improvements in the gaming experience forecasting a digital revenue of massive proportions.

Section *b* delves into the moral ambiguity of video games as a product and tool for marketing-centred attention to games, which could lead to mental illness, stress, and relaxation devices.

Section *c* deals with the climate change and environmental concerns the video game industry is consistently bringing about with more data consumption and speed for players to enjoy the immersive experience, which is in confrontation with the Sustainable Development Goals, namely SDGs, because of the huge use of electricity grids.

The second part briefly studies female representation in the video game industry, fair pay, and a safe workplace. It focuses on the video game workplace as a safe and healthy space for females to create and develop video games for the industry. Whether there is a concerted effort to recruit from other fields of knowledge, such as computer science, the game designer and game developer professionals in the career marketplace deserve further academic research in a field survey.

The third part is a selected compiled survey compiled for this paper by the authors, to consolidate the points exposed in this paper. It enumerates the most well-known and profitable video games globally surveyed for a gender diversity study inside these multinational companies. On the table, some selected products are sold internationally and commercialised for a digital community. The conclusion follows in the last part.

2 The Video Game Industry

2.1 Connection Bhi Na Kamaal Ki Cheez Hai, Connection Is All You Need

In 2016, the celebrated Indian actor, film producer, and entertainer Shah Rukh Khan's character Gaurav in the movie Fan [2] was obsessed with a Bollywood star called Aryan Khanna. The movie forecasts the digital industry's aim for any user: an eternal connection with a hero. In analogy, the video game is the hero using a tool called technology to retrieve the attention of its users. Without a steady source of electricity, especially for high-end gaming consoles, video game systems cannot be enjoyed to the full extent. Components such as graphic processing units (GPUs) [3], monitors, and audio equipment fail to provide gamers with the best video game experience without a reliable and power electricity provider.

The video game industry is financially successful and highly lucrative worldwide. Between 2019 and 2024, the video game industry has consistently raised its revenue, except for 2022 [5], due to significant economic and geopolitical turmoil in post-pandemic Europe. The video game industry is part of the digital economic sector that surpasses any other form of entertainment in profits, with a huge number of game players rising from 2021 to 2025 in China, Japan, South Korea, India in Asia, and the United States, the United Kingdom, and France leading in North America and Europe, respectively [6].

When compared to the consumption of music, television, and films, these sectors of the entertainment industry cannot match the lucrative market of video games, which has globally more than three billion players [7], and a global market value of US $184 billion dollars for 2023 [7]. Other forms in the entertainment industry have profits in the global box office of music and films around US 33.9 billion and US 28.6 billion which combined does not add more than US 62.5 billion.

The gaming industry has embraced innovative technologies, such as large language models in artificial intelligence (AI), to enhance the game environment with immersion algorithms that search for bugs and adjust the game for users worldwide.

2.2 *The Moral Ambiguity of Video Games: An Art, an Entertainment Space, and a Mental Disorder*

Everything in life has moral ambiguities, and this philosophical statement means that using some products for entertainment may also be a source of disease for humans. Video games are a source of facilitation for a dynamic creative space for designers and users [8]. Conversely, they are also the source of a disease called "gaming disorder" [9] or video game addiction, defined as "a pattern of gaming behaviour that causes significant impairment in a person's life" [9]. The World Health Organization was first included in the 11th edition of the International Classification of Diseases and Related Health Problems (ICD-11) in June 2018 [10]. Nevertheless, on the positive side, gaming is a stress relief tool activity for a society that gradually is more focused on individuality than in the outside space or may function as an escape opportunity from the problems of the real world [11].

Concerning the art of video games, the amalgamation of artificial intelligence and the creation of in-game environments has innovated exponentially the gaming experience. The use of generative artificial intelligence (GAI) allowed us to generate images from text descriptions, refining them to create artistic content. In short, text, images, and multi-prompts create new art for in-game environments as Midjourney accomplishes remarkable results, albeit there is potentially room for biases [12].

Further, by creating dialogue options with generic voicelines, a user can hear from their experience between open worlds, called *barks*, which recall the chatter of a bunch of humans as a background defined in the gaming language as non-player characters, NPCs. Ubisoft [13] developed this technology to fully immerse players into gaming environments as a tool to alleviate scriptwriters from writing *barks*, which is controversial [14].

2.3 *Sustainable Development Goals (SDGs) and the Power-Hungry Video Game Industry*

For the future of humanity, the United Nations members adopted the 2030 Agenda for Sustainable Development [15]. It was approved by all nations in 2015 and introduced the 17 Goals embedded in the Sustainable Development Goals [15]. These goals aim for a global partnership to facilitate sustainable economic growth [16] as well as to end poverty [17] and reduce inequality within countries [18]; bearing in mind gender equality [19], quality education [20], and effective action to combat climate change [21], all sectors of industry (primary, secondary, and tertiary) must also incorporate the 17th Goal, "Partnerships for the Goals" [22] to work. The video game industry is considered to be inserted into the entertainment and technology industries, therefore, as a provider of services in the tertiary sector. There are some that consider technology as an industry into the quaternary sector, or knowledge-based economy that also includes retail, information technology, research and development, and highly specialised professional services; nonetheless, the World Bank does not define these activities in a separate sector or categorisation. This classification adding as quaternary sector appears to be not espoused by also the Asian Development Bank. The World Bank classifies economies by income groups: low, lower middle, upper middle, and high income. This paper then concludes, whether technologies and information knowledge based are in another novel sector is of less importance than the income, which the technology sector is one of the more productive and able to promote rapidly social and economic mobility [23].

Concerning sustainability policies, the video game industry has been unsuccessful in energy conservation, managing power use, or gender equality. Let us examine conservation policies in energy consumption, for example. Playing on a game console such as PlayStation® or Xbox®, due to the complexity and graphic details in games, demands a greater power consumption on electricity grids. It also depends on the console model used for these gaming platforms. Generally, an electricity usage of a game console for a "PlayStation 4 consumes 85 watts of power. If a game player uses between 4 and 10 h a week, the consumption will be 0.85 kWh (kilowatt per hour) per week, which is equivalent to 3.64 kWh per month or 44.32 kWh per year." [24]. A sustainable policy should involve eco-friendly hardware components such as the CPU and/or GPU or monitors with a low energy consumption rating. As technology improves, its increased power demands affect this delicate balance between human energy consumption and the stability of our environment. The conventional hardware industry, including semiconductors and the DNA of computers, generates a substantial carbon footprint that can be changed to be more sustainable with integrated design materials and recycling e-waste [25].

2.4 A Brief Study of Female Representation in the Video Game Industry, Fair Pay, and Safe Workplace

As the video game industry grew to be a global industry with more responsibilities, including sustainability policies, issues with gender equality appeared to be sidelined. One may overlook that it is mainly a male-dominated operations field. The industry is mainly led by male CEOs worldwide; most of the Board of Directors and executive teams are male.

It is not far-fetched to assume that the video game industry is a man's world, specifically a cis-male space managed by a dominant group of middle-aged men. With the recent exception of a female Chief Executive Officer (CEO) added to a troubled American video game company, females are in more traditional positions; see below in the table.

Global video game companies follow more inclusive corporation laws and regulations, and there is an apparent balance of genders at the highest levels of the decision-making process. Still, there are some important exceptions to gender balance, especially when the video game companies (studios) are not publicly listed in the stock market. Accountability is, therefore, fractioned among many levels in the corporate ladder and a person is not tasked with accountability, being responsibility undetermined.

It is observed that most female professionals in the video game industry are recruited from a small number of educational backgrounds. These do not generally include computer science, game design, game development, game art, game programming, or game animation.

Female game developers are absent in the senior and higher positions of video game directorship, where decisions are made about products. Generally, multinational companies advertise posts for female independent executives appointed to the board. More often, their expertise in finance, legal, accounting, public relations, and other disciplines, unlike computer science and related areas, may add to the narrative of games and the decision-making process for CEOs.

These women may have bachelor's degrees in other fields of knowledge than computer science, such as Literature, Philosophy, Mythology, Archaeology, and Medicine.

In non-listed public companies, video game companies hire women to fill traditional business and administration, accounting, finance, legal, public relations, hospitality, general counsel, compliance officer, human resources, sports, economics, psychology, and offshore drilling roles. Further academic research and field surveys are needed to determine whether there is a concerted effort to recruit computer scientists, game designers, and game developer professionals for senior positions inside these companies.

At the highest corporate level of the video game industry, where decisions about future products and marketing of these companies are made, women are outnumbered by men. One must observe that a working space in which few women are present is the video game industry. It might indicate a difficult space for female careers, such as game developer, design, and computer science. There seems to be a disconnection between the creation of video games, gender equality, and the CEO decision-making framework. It is suggested that these professionals may not be supported in seeking vertical mobility in the video game industry. Whether core skills such as game concept and development are secondary to the game business of marketing and making money, it remains to be seen that no female game creator is at the top of these transnational companies (TNCs). That is a bad strategy.

Women may support the video game industry in expanding better gaming experiences. Good games have, interesting characters, attract users to pursue goals, allow them to make decisions about their actions and have rewards for good performance. However, what makes a good game is the immersion and constant presence in front of monitors or virtual reality. Female creativity works better by creating female characters that have more depth and less sexuality. Male video game creators may give more weight to sexualising female characters and less strong female characters, which is a perpetuation of unrealistic standards of beauty and female cisgender objectification and stereotypes.

With the advent of games incorporating machine learning, specifically generative adversarial networks, a question about control may be addressed, as the core of artificial intelligence is to become autonomous. This suggests that a game developer/ compliance officer may be a necessary role in which women may be trained for this task, illustrating the risks beyond software creation. More authentic training data sets also mean generating new images from an existing database without human intervention or unsupervised learning tasks.

Generative adversarial networks (GANs) create a perception of reality humans have never experienced, which can be exciting for some video game players. However, GANs are unstable applications and extremely competitive for results. GANs also need a large amount of training data and may create inaccurate face recognition hallucinations [26]. In the game industry, GAN use is recent, and we may experience other kinds of hallucinations with unknown results for users.

At the end of the day, there are no female video game creators, software designers, or game developers at the helm of these profitable companies. It is suggested that at the senior management level, where risk assessment of new products is reported, both genders may be a positive differential for gamification. We found a couple of women at that level.

In the video game industry, the power of diversity is confined to female consumers, who make up almost half of video game users worldwide. Recent research has observed an interesting diversity of initiatives in companies to boost women's participation in the board of directors, executive teams, and management teams of the more influential video game companies (Table 1). These initiatives are short-lived and restricted to some areas described above for female independent directors, board members and executive teams, and compliance officers.

The table is divided into the biggest video game companies, revenue, CEOs by gender, and products advertised to the public. We included a column dedicated to lawsuits related to diversity in the video game industry, as it deals with sexual harassment and fair compensation at the workplace, which is the topic of interest for this research.

This table is based on a simple online search and an analytical framework composed of news media sources, legal intelligence news sources, and social and professional platforms such as Google, LinkedIn®, Wall Street Journal®, Statista®, specialised game news sources, local newspapers, and the companies' official websites to classify the most profitable video game companies in the world. Product examples are best-selling video games in the selected companies, which may or may not be selling violent content against women or violence in general.

Most of these companies are located in the American State of California, a hub for other tech companies; consequently, we may defer some attention to the American jurisdiction, particularly on corporate and compliance regulatory framework, followed by other jurisdictions: Japan, China, and France. We added two other video game companies: one based in India, as this research unveils Indian female game developers, and another in Australia, an online gambling company with mobile free-to-play games, to illustrate the gender gap alluded to in this research.

Our analysis depicts an unregulated and male-only oriented industry, led by one gender, even in the recent artificial intelligence technologies used in online games. For instance, the "Parallel AI Hybrid Gaming System" under patent application number

Table 1 Video-Game Industry Landscape on Key Figures, Including Lawsuits

Company	Products example	Revenue	CEO & board of directors (female) & Executive Teams/ Risk Management (female)	Compliance officer (female)	Lawsuits on harassment and workplace safety, privacy rights, COPPA violation
NINTENDO	THE LEGEND OF ZELDA: TEARS OF THE KINGDOM® SUPER MARIO BROS: WONDER®	1.67 T	3 (three)	Regulation is available at the official website	–
ELECTRONIC ARTS	THE SIMS® STAR WARS™ BATTLEFRONT™	7.51 B	4 (four)	–	–

(continued)

Table 1 (continued)

Company	Products example	Revenue	CEO & board of directors (female) & Executive Teams/ Risk Management (female)	Compliance officer (female)	Lawsuits on harassment and workplace safety, privacy rights, COPPA violation
EPIC GAMES	BATTLE ROYALE® FORTNITE®	≥5.8 B	Information not available for the Board of Directors at the senior level 2 (two)	–	Violation of Children's Privacy Laws (US), dark patterns, sleep mode transgression, lax consumer protection in online purchases, collection of children's personal data without consent [34] "Fortnite" Creator Agrees to Pay a Record Penalty for Violating Children's Privacy Laws, available at https://www.jdsupra.com/legalnews/fortnite-creator-agrees-to-pay-a-record-6733560/ It could be argued that this is a company that is defective in protection of privacy rights for children and teens. So, to have this company in digital games with an AI parallel hybrid game which allows a player to be competing with an autonomous and capable of self-learning system without any supervision from adults may be raising the opportunities for more harm to children and teen players. To get more acquainted with AI and Metaverse applications including in games and present caution with avatars for public use [35]

(continued)

Table 1 (continued)

Company	Products example	Revenue	CEO & board of directors (female) & Executive Teams/ Risk Management (female)	Compliance officer (female)	Lawsuits on harassment and workplace safety, privacy rights, COPPA violation
SONY INTERACTIVE ENTERTAINMENT	PLAYSTATION®	13.02 T	4 (four)	–	–
TAKE-TWO INTERACTIVE	ROCKSTAR GAMES® ZYNGA®	5.35 B	4 (four)	–	–
MICROSOFT	–	245.12 B	6 (six) females	–	–

(continued)

Table 1 (continued)

Company	Products example	Revenue	CEO & board of directors (female) & Executive Teams/ Risk Management (female)	Compliance officer (female)	Lawsuits on harassment and workplace safety, privacy rights, COPPA violation
ACTIVISION BLIZZARD/ MICROSOFT (since 2023)	CALL OF DUTY: BLACK OPS 6® WORLD OF WARCRAFT®	–	Female CEO (appointed 2024)	Resigned. (Frances Townsend was the Compliance Officer at Blizzard when the harassment court case was initiated. This resignation is allegedly connected to the harassment court case. See available at https://www.business-standard.com/article/international/activision-blizzard-s-chief-compliance-officer-resigns-on-assault-charges-122100300212_1.html)	*California Department of Fair Employment and Housing v Activision Blizzard* (2021) *Benjamin Bell*et al. *v Blizzard Entertainment Inc.*et al. (2012) (class action) *Angelilli v Activision Blizzard, Inc.* (2025)

(continued)

Table 1 (continued)

Company	Products example	Revenue	CEO & board of directors (female) & Executive Teams/ Risk Management (female)	Compliance officer (female)	Lawsuits on harassment and workplace safety, privacy rights, COPPA violation
TENCENT HOLDINGS	HONOUR OF KINGS® PUBG MOBILE or GAME FOR PEACE (David Curry, PUBG Mobile Revenue and Usage Statistics (2025), in Business of Apps, published 22 January 2025. PUBG Mobile profits were 1.1 billion revenues in 2024, whereas in China, Game for Peace profits were 3 billion globally in July 2020, available at https://www.businessofapps.com/data/pubg-mobile-statistics/)	672.96 B	2 (two)	–	–

(continued)

Table 1 (continued)

Company	Products example	Revenue	CEO & board of directors (female) & Executive Teams/Risk Management (female)	Compliance officer (female)	Lawsuits on harassment and workplace safety, privacy rights, COPPA violation
UBISOFT ENTERTAINMENT	ASSASSIN'S CREED® FAR CRY® (The creation of this game as well as Far Cry 4 is credited to Jade Raymond. She started at Sony, then worked for Electronic Arts and is the producer of the Sims Online. She worked with Ubisoft Montreal where she led the development of first Assassin's Creed game. Currently, she leads her own independent studio.)	2.30 B	5 (five)	–	–
NETEASE	BLOOD STRIKE®—FPS for all IDENTITY V®	11.6 B [36]	3 (three)	–	–
NAZARA	KIDDOPIA® WORLD CRICKET CHAMPIONSHIP 3™	11.38 B	Not available	1 (one) male	
ARISTOCRAT LEISURE	Online gambling, free-to-play games	6.30 B	9 (nine)	1 (one) female	–

(content researched and selected by Ana Penteado). I state that this is my own work on this co-authored paper

US 10,885,739 B2, filed on January 5, 2021, is an invention that permits players to train their skills against an Artificial Intelligence Non-Player Character (AI NPC), which is a male-dominated scenario from one gender perspective. The Grand Theft Auto's company that is using AI NPCs, Rockstar Games, has been named as one of the video game companies with a consistent gender gap pay that has been improved slowly towards a better environment to attract women developers, albeit it remains to be seen since 2021 [27]. Examples of games using artificial intelligence/non-player character are Inworld Origins, owned by three former Google male employees, and the game Red Dead Redemption II. This last game was created by two brothers founders and developers at Rockstar Games, all men-only on senior roles [27] to illustrate this socially and economically disengaged sector and a poor environment for women to thrive their creativity. A full report from the World Economic Forum in 2023 has found that the sector of Information and Technology and Entertainment Providers has been thriving with more inclusion of female workers at senior levels; however, this cannot be interpreted as applicable to the video game industry.

From our analysis, women are fully underrepresented in the video game industry, which is a men-only space. Further studies must be conducted in this sub-sector of the entertainment industry to educate about gender equality, focusing on the gaming industry. Numbers can deceive research if we look at the entertainment industry as a homogenous sector, which is a false statement [28]. Gaming and domestic violence exposure may be part of the narrative that children may witness in their own homes, if normalised as entertainment by online games such as Detroit: Become Human developed by Quantic Dream represent a consolidation of a break on social fabric, the real and the digital intertwined to erase ethics, and values from communities. This perception of two worlds in one may be disorientating for teens and children, albeit adults may be affected by emotions in control of player's realistic ecosystems and actions towards other players. In mimicking in the digital world, a plausible reality Quantic Dream technologies become a facilitator of violence, available on PlayStation and Microsoft Windows platforms. It is a good example of domestic violence used as entertainment or alleged art in game narrative. It is worth noting that Quantic Dream S A has been involved in proceedings at the 17th Correctional Chamber of Paris Court of Appeals; the French court rulings were later overturned for unfair dismissal of some former employees, which alleged a toxic workplace in the company. Coincidentally, back in January 2018, this company was working on Detroit: Become Human for release at PlayStation console in partnership with Sony Entertainment.

Coincidentally, it appears that there is a recognisable trend of domestic violence games [29] popularly commercialised as online entertainment. These violent narratives perpetuate a lack of awareness of violence against women and girls (VAWG) [30]. Conversely, games such as Chuka, Break the Silence [31], and WeRise [32] may have contributed positively to educating both genders about rights and obligations in their local communities, empowering and advocating for gender equality in the local communities. The American multinational technology corporation Microsoft has provided cloud services to Sony Entertainment for the streaming services of Detroit: Become Human for Windows. Due to these commercial arrangements, the United

Kingdom Competition and Markets Authority (CMA) initiated an anti-trust lawsuit to protect the domestic cloud gaming market allocated to Ubisoft Entertainment [33].

Moreover, such *edutainment* games may raise awareness and empowerment for the next generation about our delicate social fabric that may often be exploited in the commercial gaming industry, a sub-sector of the entertainment industry. On the other hand, the billionaire game industry should lift their game to be more gender-inclusive and a safe place for women developers to contribute to the social and economic growth of nations in the algorithm age.

3 The Power of Weak Diversity: Companies Surveyed for This Paper

a. Nintendo (Japan)
b. Electronic Arts (USA)
c. Epic Games (USA)
d. Sony Interactive Entertainment (Japan)
e. Take-Two Interactive (USA)
f. Microsoft (USA)
g. Activision Blizzard/Microsoft (USA)
h. Tencent (China)
i. NetEase (China)
j. Nazara (India)
k. Aristocrat Leisure (Australia)

4 Conclusion

This paper surveys the purposeful and sustained exclusion of women, specifically female developers, in the video game industry. In short, women have limited influence on the final products of these large and profitable video game companies. This gender bias has been exacerbated by artificial intelligence used by many studios that create these products for further licensing to global digital platforms, mimicking human emotions with an argument of artistic content. Both genders hardly conceive that women in senior roles are so rare in this industry that it may be a suggestion of an unsafe and unhealthy working environment. It also suggests a lack of interest in pursuing a career in video game companies. This furthers gender segregation neither assists economic growth nor the creation of jobs for reducing economic disparity and inequality. Potentially, it can lead to a long-biased policy in that industry, which may extend for centuries ahead of us. In a more individualistic sense, exclusion policies benefit a current stereotype of female players being in a secondary place compared to their male counterparts. The representation of both genders is an urgent necessity to be addressed for equal pay and gender equality in policymaking in the local and global economy.

Closing the gender gap will ensure higher earning positions for females and mutual respect for creative work. If the economic gap persists in the video game industry, as it is now, that will reduce women's earnings over their lifetime and eventually disempower them. Thus, gender disparity does not become a sustainable policy and may relate to social inequalities, resulting in mental illnesses such as depression and anxiety. Whether aspiring female video game designers and developers may thrive in their independent initiatives, these companies should not marginalise them in the more lucrative mainstream market.

References

1. Chakrabarty SP (2019) LGBTQ rights in India: some philosophical and legal reflections. Commonwealth Lawyer 29(1):17–22
2. FAN (2025) Official website. https://www.youtube.com/watch?v=nkS_Ar0Yad0. Accessed 20 May 2025
3. Bridges RA, Imam N, Mintz TM (2016) Understanding GPU power: a survey of profiling, modeling, and simulation methods. ACM Comput Survey (CSUR) 49(3):1–27
4. Clement J (2024) Video game market revenue worldwide from 2019 to 2029. Statistica. https://www.statista.com/forecasts/1344668/revenue-video-game-worldwide. Accessed 21 May 2025
5. Wolf MJ (ed) (2015) Video games around the world. MIT Press, Cambridge, MA, pp 1–16
6. International Trade Administration, Media and Entertainment: Video Games Sector. https://www.trade.gov/media-entertainment-video-games-sector. Accessed 20 May 2025
7. World Economic Forum with The Conversation, Playing more video games could increase creative thinking at work, published originally in 30th September in 2019. https://www.weforum.org/stories/2019/09/playing-video-games-increase-creativity/. Accessed 20 May 2025
8. Darvesh N, Radhakrishnan A, Lachance CC, Nincic V, Sharpe JP, Ghassemi M, Straus SE, Tricco AC (2020) Exploring the prevalence of gaming disorder and Internet gaming disorder: a rapid scoping review. Systematic Rev 9:1–10. https://link.springer.com/content/pdf/https://doi.org/10.1186/s13643-020-01329-2.pdf. Accessed 20 May 2025
9. World Health Organisation. https://www.who.int/standards/classifications/frequently-asked-questions/gaming-disorder. Accessed 20 May 2025
10. United Nations, Video Games and Mental Health: Surprising Ally. https://unric.org/en/video-games-and-mental-health-a-surprising-ally/. See also, Pallavicini F, Pepe A, Mantovani F (2021) Commercial off-the-shelf video games for reducing stress and anxiety: systematic review. Published at JMIR Mental Health, 16 August 2021, 8(8). https://pmc.ncbi.nlm.nih.gov/articles/PMC8406113/. Accessed 20 May 2025. See also Lebamovski P, Gospodinova E (2025) Investigating stress during a virtual reality game through fractal and multifractal analysis of heart rate variability, applied system innovation, 8(1):16. https://www.mdpi.com/2571-5577/8/1/16. Accessed 20 May 2025
11. Jaruga-Rozdolska A (2022) Artificial intelligence as part of future practices in the architect's work: MidJourney generative tool as part of a process of creating an architectural form. Architectus 3(71): 95–104
12. Robertson J (2023) Ubisoft has created an AI tool that will generate NPC dialogue. published by THE GAMER, on 21 March 2023. https://www.thegamer.com/ubisoft-ai-tool-npc-dialogue/. Accessed 20 May 2025
13. Williams LJ (2023) Ubisoft faces major backlash after announcing AI writing tool, in GAMES HUB, News, Games Development, published on 23th March

2023. https://www.gameshub.com/news/news/ubisoft-backlash-ai-dialogue-writing-tool-ghostwriter-2610382/. Accessed 20 May 2025
14. United Nations, Department of Economic and Social Affairs, the 17 Goals. https://sdgs.un.org/goals. Accessed 20 May 2025
15. United Nations, Department of Economic and Social Affairs, Goal 8. https://sdgs.un.org/goals/goal8. Accessed 20 May 2025
16. United Nations, Department of Economic and Social Affairs, Goal 1. https://sdgs.un.org/goals/goal1. Accessed 20 May 2025
17. United Nations, Department of economic and Social Affairs, Goal 10. Available at https://sdgs.un.org/goals/goal10. Accessed 20 May 2025
18. United Nations, Department of Economic and Social Affairs, Goal 5. https://sdgs.un.org/goals/goal5. Accessed 20 May 2025
19. United Nations, Department of Economic and Social Affairs, Goal 4. https://sdgs.un.org/goals/goal4. Accessed 20 May 2025
20. United Nations, Department of Economic and Social Affairs, Goal 13. https://sdgs.un.org/goals/goal13. Accessed 20 May 2025
21. United Nations, Department of Economic and Social Affairs, Goal 17. https://sdgs.un.org/goals/goal17. Accessed 20 May 2025
22. Métreau E, Young KE, SE Eapen (2025) World Bank Group IBRD-IDA, World Development Indicators, World Bank country classifications by income level for 2024–2025. https://blogs.worldbank.org/en/opendata/world-bank-country-classifications-by-income-level-for-2024-2025. Accessed 20 May 2025
23. SlashPlan, PlayStation 4 Energy Calculator (cost and kWh Usage). Available at https://www.slashplan.com/playstation-4-energy-calculator-cost-and-kwh-usage/#:~:text=per%20kilowatt%2Dhour.-,Energy%20Consumption,or%2044.32%20kWh%20per%20year. Accessed 20 May 2025. For comparison, a domestic fridge power consumption over a full day is around 1 to 2 kWh, however, that will depend on the type, size, location, year, season, condition, usage and temperature set point. See Ryan McCarthy, How Much Power a Fridge Uses – in Watts, Cost & kWh, Reduction Revolution, published on 13 November 2019, available at https://reductionrevolution.com.au/blogs/how-to/fridge-power-consumption, accessed on 20 May 2025
24. Electronics can be more sustainable (2023) Nat. Mater. 22:1273. https://doi.org/10.1038/s41563-023-01731-w. Accessed 20 May 2025
25. Google Patents, United States Patent Office, Patent number US10, 885, 739 B2, Parallel AI Hybrid Gaming System. https://patentimages.storage.googleapis.com/e2/21/30/cecb42cf7e2c38/US10885739.pdf. Accessed 20 May 2025
26. McPhillips A (2022) Rockstar games improves its gender pay gap issues. Published on 6 April 2022. https://gamerant.com/rockstar-games-gender-pay-gap-improvements/. Accessed 20 May 2025
27. World Economic Forum (2023) Global gender gap report 2023, Full Report, published on 20 June 2023. https://www.weforum.org/publications/global-gender-gap-report-2023/in-full/gender-gaps-in-the-workforce/#evolving-gender-gaps-in-the-global-labour-market. Accessed 20 May 2025
28. Phillips T. Quantic Dream Bosses Successfully Sue French Newspaper Le Monde for Libel But Lose Case Against Second French Outlet. https://www.eurogamer.net/heavy-rain-developer-quantic-dream-successfully-managed-to-sue-french-newspaper-le-monde. See also, Glagowski P (2019) Quantic dream has no plans to appeal fine it received for circulating vulgar images, THEGAMER, published on 4 December 2019. https://www.thegamer.com/quantic-dream-fined/. Accessed 20 May 2025
29. See also, Kerr C (2021) French court overturns unfair dismissal ruling for ex-Quantic Dream developer, Eurogamer, published on 12 April, 2021. https://www.gamedeveloper.com/business/french-court-overturns-unfair-dismissal-ruling-for-ex-quantic-dream-dev. See also Issy van der Velde, Quantic Dream Claims It's A "Supportive, Safe, and Respectful" Studio, THEGAMER, published on 27 April, 2022. https://www.thegamer.com/quantic-dream-studio-statement/. Accessed 20 May 2025

30. United Nations, regional information centre for western Europe, in video games and violence against women and girld: a double-edged sword in the digital age, published in 8 December 2023. https://unric.org/en/video-games-and-violence-against-women-and-girls/. Accessed 20 May 2025
31. United Nations Office on Drugs and Crime, Primary, E4J Tools & Lessons Plans, Chuka, Break the Silence. https://www.unodc.org/e4j/zh/primary/e4j-tools-and-materials/chuka.html. Accessed 20 Feb 2025
32. United Nations Regional Information Centre for Western Europe, WeRise app. https://unric.org/en/werise-mobile-app/. Accessed 20 February 2025
33. Alexis Keenan and Daniel Howley, in Microsoft closes $69 billion takeover of Activision Blizzard, updated on 14 October, 2023, available at https://finance.yahoo.com/news/microsoft-closes-69-billion-takeover-of-activision-blizzard-130549285.html. We selected a couple of lawsuits against this company. For a complete list of all lawsuits and other legal issues concerning this company, see also, https://www.gamespot.com/articles/activision-blizzard-lawsuits-and-investigations-timeline-of-events/1100-6494785/. Accessed 20 May 2025
34. See Clement J (2025) Epic Games annual gross revenue 2018–2026, published by Statista, 13 November 2024. \https://www.statista.com/statistics/1234106/epic-games-annual-revenue/. Accessed 20 May 2025
35. Dwivedi YK, Hughes L et al. (2022) Metaverse beyond the hype: Multidisciplinary perspectives on emerging challenges, opportunities, and agenda for research, practice and policy. Int J Inform Manage 66:102542. published by Elsevier. https://doi.org/10.1016/j.ijinfomgt.2022.102542. Accessed 15 January 2025
36. Yahoo Finance, Bloomberg News, NetEase's Billionaire CEO Slashes Jobs and Games in Profit Push, published 22 February, 2025. https://finance.yahoo.com/news/netease-billionaire-ceo-slashes-jobs-160402024.html. Accessed 15 Jan 2025

Evolution of AI

Generative AI to Bridge the Gap and Improve Medical Reporting Interpretation

Trishita Ghosh, Sudeep Ghosh, Suparna Biswas, Animesh kairi, Soumen Ghosh, and Shekhar Maiti

Abstract The aim and scope of this "research paper" is to design an approach for medical report generation for patients using generative artificial intelligence (GAI). The main model used in this approach is based on holistic artificial intelligence in medicine (HAIM). The integration of generative artificial intelligence (GAI) in healthcare communication promises precise diagnoses, streamlined reporting, and efficient integration with electronic health records (EHRs), thereby revolutionizing patient care to create an advanced artificial intelligence system that can analyze medical data and produce comprehensive, easy reports for patients. Additionally, the system aims to prompt immediate precautionary measures, when necessary, facilitated by the GAI. This initiative seeks to improve communication between "healthcare providers" and "patients" by delivering clear and accessible information regarding their health status, diagnosis, and treatment plans. Ultimately, the project aimed to empower patients with clearer insights into their health, thereby fostering a more collaborative decision-making process. By leveraging AI technology, this endeavor represents a major advancement in healthcare communication,

T. Ghosh · S. Ghosh · S. Maiti (✉)
Department of Information Technology, Guru Nanak Institute of Technology, Kolkata, India
e-mail: shekharmaiti2000@gmail.com

T. Ghosh
e-mail: trishita.ghosh@gnit.ac.in

S. Ghosh
e-mail: sudeep.ghosh@gnit.ac.in

S. Biswas
Department of Electronics and Communication Engineering, Guru Nanak Institute of Technology, Kolkata, India
e-mail: suparna.biswas@gnit.ac.in

A. kairi
Department of Information Technology, IEM Saltlake Campus, Kolkata, India
e-mail: animesh.kairi@iem.edu.in

S. Ghosh
Department of Information Technology, Narula Institute of Technology, Kolkata, India
e-mail: soumen.ghosh@nit.ac.in

M. Chakraborty et al. (eds.), *Proceedings of Sixth International Ethical Hacking Conference*, Lecture Notes in Networks and Systems 1513,
https://doi.org/10.1007/978-981-96-8632-2_3

addressing the challenge of complex "medical reports" and enabling individuals to make informed health decisions. It also offers numerous advantages to patients, including improved accuracy, enhanced safety measures, and tailored insights for more effective healthcare management. Furthermore, proactive health maintenance is encouraged by delivering timely and actionable recommendations.

Keywords Generative artificial intelligence (GAI) · Holistic artificial intelligence for medicine (HAIM) · Electronic health record (EHRs) · Pathology education informational resource (PEIR) · Artificial intelligence (AI)

1 Introduction

Generative artificial intelligence is a computer technology that can produce text such as imagery, audio, and synthetic data. The recent popularity of generative AI stems from how easy it is for its user interfaces to creating the best text, graphics, and videos in no time. Generative AI is anticipated to be a key component of AI-based medical healthcare systems, which are designed to improve healthcare management and patient safety. The healthcare sector is experiencing a revolutionary shift with the advent of "artificial intelligence (AI)," particularly in the innovative domain of AI-driven medical report generation. Technological integration is reshaping the healthcare landscape, marking a significant transformation in the processing and presentation of medical information. This innovative project represents a paradigm shift in patient care by harnessing the ability of AI to deliver accurate and comprehensive medical reports. Electronic health records (EHR) are mainly used to assist in reducing medical errors, improving patient safety, and supporting better outcomes. Even though EHRs have and transport data, there are also some meaningful changes in patient information, and that information is supplied to the provider at the point of care. The holistic healthcare approach (HAIM) is manifested in its dedication to fostering a patient-centric environment. The generated reports not only provide accurate diagnoses, but also serve as educational tools, offering insights into health conditions and promoting a proactive approach to healthcare management. Image analysis plays a crucial role in generating automated medical reports. This step enables the extraction of meaning from visual data, which can then be translated into textual descriptions of image content. "Convolutional neural networks (ConvNets)" have emerged as an effective "machine learning (ML)" technique for "image analysis," particularly in areas such as "hyperspectral image classification," "spatial channel extraction" [1, 2], and textual image analysis [3]. The structure of ConvNets consists of interconnected layers of neurons [4], allowing efficient and precise image processing, which significantly enhances the quality and precision of medical image interpretation and reporting (Fig. 1).

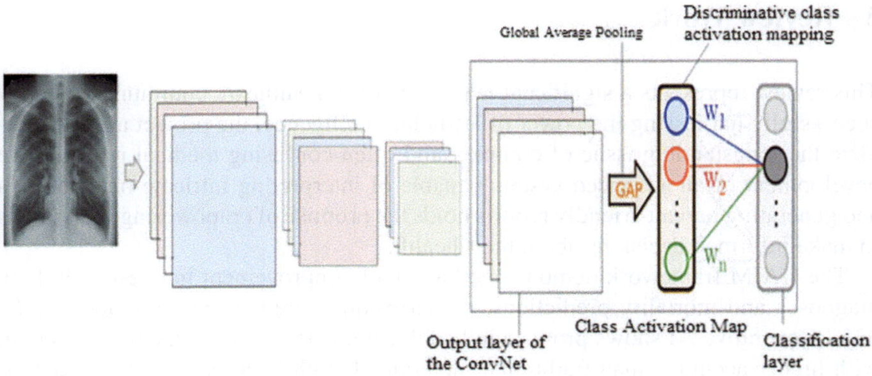

Fig. 1 The conceptual framework of class activation mapping

2 Challenges

The architecture of Generative AI is intricately designed by incorporating multiple layers of technology to ensure seamless execution. Within this complex framework, several key challenges arise.

i. *Lack of Quality Medical Data*: High-quality datasets are essential for validating AI models, both clinically and technically. However, fragmentation of medical data across various EHRs and platforms complicates the collection of comprehensive patient information and images required for testing AI algorithms.
ii. *Data Accuracy and Security Issues:* Electronic health records need to be updated immediately after each patient visit to ensure accurate data for treatment protocols. There is also a significant risk to user data privacy, necessitating systems that allow users to control the exchange and privacy of their information.
iii. *Implementation and Maintenance Requirements:* The high cost and complexity of implementing and maintaining EHR systems can slow down operations. EHRs require regular training, updates, and robust support to handle new technological demands and policy changes.

We have worked together with the AI medical report-generation approach. Extensive research has been dedicated to identifying optimal solutions to address this challenge. However, due to limited mainstream adoption of HAIM, ongoing research are still being conducted. Our focus was to confront this challenge, as it represents a crucial aspect of GAI and HAIM.

3 Review Work

This review represents a significant leap forward in healthcare communication. By successfully harnessing the power of artificial intelligence, the project aims to alleviate the longstanding issue of complex and often confusing medical reports. The development of an advanced system capable of interpreting intricate medical data and generating patient-friendly reports holds the promise of empowering individuals to make informed decisions about their health.

The HAIM framework demonstrated a notable improvement in chest pathology diagnoses and mortality predictions, outperforming single-source methods by 6–33%. Generative AI shows promise in health care, particularly in medical imaging, with higher accuracy than traditional methods, though it faces challenges such as trust, safety, and privacy concerns. GANs have enhanced medical education and research but also raise issues like the creation of fake images. The paper highlights the significant role of generative AI in radiology and the importance of transparency for trust-building. SciBERT struggles with queries related to "race" and health disparities, underscoring the need for more nuanced AI models. There is also a need for further research to optimize these technologies for clinical use and address ethical concerns.

4 Dataset

To evaluate the proposed model, we used two publicly available datasets of medical disease information and images.

The "Indiana University Chest X-Ray" Collection, also known as the "IU X-Ray" Collection, comprises 7470 pairs of chest radiographs and corresponding diagnostic reports. The dataset's target captions, which need to be generated, are derived from the content found in the impression and findings sections. Medical Text Indexer (MTI) annotation tags are utilized to predict target tags. Additionally, we have developed our own dataset by gathering medical information from various sources and incorporating X-ray images.

4.1 PEIR Gross

The "Pathology Education Informational Resource (PEIR)" "digital library3" functions as an open-access collection of medical images for educational use. We created the "PEIR Gross" dataset by extracting images and their corresponding descriptions from the Gross sub-collection, resulting in approximately "7,440 image-caption pairs" across "21" different sub-categories. In contrast to the "IU X-Ray" dataset, "PEIR Gross" captions are composed of individual sentences. We employed this dataset to evaluate our model's ability to generate "single-sentence" reports. The preprocessing methods applied to PEIR Gross were identical to those used for IU X-Ray, yielding approximately "4,450" unique words. Each image is linked to an average of "12.0 words." Furthermore, we selected "5 words" with the highest "tf-idf" scores as "tags" for each caption.

4.2 Implementation Details

For our implementation, we employed the complete "HAIM-model." We approached the multilabel classification (MLC) task's training loss as a "multilabel retrieval" problem, using a "SoftMax-cross entropy" loss (a multilabel ranking loss), since the number of tags for "semantic attention" was set at "10." In the process of paragraph generation, all hidden states and word embeddings were configured to "512-dimensions." We implemented separate embedding matrices for words and tags, recognizing that individual tags could comprise multiple words. The embeddings of the top "10" most probable tags functioned as the "semantic feature vectors" {am}M = 10 m = 1. To extract visual features, we utilized the final "convolutional layer of the VGG-19" network, which produced a "14 × 14 × 512" feature-map (Fig. 2).

For the "CNN (VGG-19)" and "Hierarchical LSTM," learning rates were established at "1e−5" and "5e−4" respectively. The various loss weights ("λtag," "λsent," "λword," and "λreg") were all set to "1.0." A stop control limit of "0.5" was implemented. To prevent overfitting, the training process was terminated prematurely.

The proposed scheme outlines the structure and flow of the chat application, detailing the roles and interactions of each component. The development and maintenance of the application can be effectively managed by following this scheme.

5 Proposed Work

A comprehensive medical diagnostic report incorporates detailed textual descriptions and relevant tags. Our proposed approach employs a hierarchical multitask model with co-attention to generate extensive paragraphs and predict key terms. We utilized a "convolutional neural network (CNN)" to extract "visual" features from image segments divided into regions. Subsequently, a "multilabel classification (MLC)" system was employed to forecast the applicable tags. Each tag was encoded as a "word-embedding vector" within the tag lexicon. The predicted tags for an image are incorporated into the corresponding "word-embedding vectors," which we consider to be the "semantic features" of this text and image. A thorough analysis of the image's attributes is conducted, and the resulting feature vector is input into a "co-attention" model. This model employs an "attention mechanism" to produce a "context vector." At this point, the encoding process is finalized.

```python
python

from langchain.chains import RetrievalQA
from langchain.llms import OpenAI
from langchain.vectorstores import FAISS

retriever = vectorstore.as_retriever(search_kwargs={"k": 2})

qa = RetrievalQA.from_chain_type(
    llm=OpenAI(),
    chain_type="stuff",
    retriever=retriever,
    chain_type_kwargs={"input_key": "query"}
)

while True:
    query = input("Input Prompt: ")
    result = qa({"query": query})
    print("Response: ", result["result"])
```

Fig. 2 Implementation code of the system

The decoding process generates text descriptions by starting with the "context" vector. Typically, a medical image description is composed of multiple sentences, each focusing on a distinct topic. Our model utilizes this structure to create reports in a "hierarchical" fashion. Initially, we produce a series of "topic vectors" at a higher level, representing individual sentences. Subsequently, we derive a sentence from each "topic vector." In essence, we feed the "context vector" into the "sentence LSTM," which then unfolds over several steps. Thus, at each step we will get a topic vector. A topic vector indicates what a generated sentence should mean or what it should be related to. The word LSTM gets input of a topic vector and provides a sequence of words which can be used to form a sentence. The sentence LSTM controls when the unrolling process comes to an end.

The Multimodal Local Chat Application was designed to provide a comprehensive chat experience with support for text, audio, image, and PDF interactions. The application leverages advanced language models and various utilities to process and respond to user input in a conversational manner.

Our model's initial objective is to identify tags for a given image, which we approach as a "multilabel classification" problem. For an "input-image"—"I," we first extract "Features"—"{v_n}," "N n = 1 ∈ R D" from a middle layer of a "convolutional neural network (CNN)." These features are then input into a "multilabel classification (MLC)-Network" to create a probability distribution across all "L-Tags":

$$P(\ell i = 1 \mid \{vn\}_{n=1}^{N}) \propto e^{(MLC_i(\{vn\}_{n=1}^{N}))} \tag{1}$$

Here, "l" ∈ R L represents a "tag vector," with "li = 1 or 0" indicating the "presence or absence" of the "i-th tag," respectively. "MLC_i" denotes the "i_{th}" output of the MLC network.

For ease of implementation, we extract "Visual Features" from the final "convolutional layer of the VGG-19" model [4] and utilize its last two fully connected layers for "MLC."

Ultimately, the embeddings of the "M" most probable "tags {am}" ["M m = 1 ∈ R E"] are employed as "semantic features" for generating topics.

Here is a proposed scheme for the chat application project, based on the provided file structure and code snippets. The scheme is structured into sections detailing the project overview, directory structure, application flow, and key functionalities. This document was designed to fit within two to four A4 pages.

The key functionalities of the application include the main application, which sets up the Streamlit interface with title, sidebar, and main chat container, and initializes session states for managing different chat sessions and database connections, handling text input, audio recording, and file uploads, triggering appropriate handlers. Audio handling involves converting audio bytes to text using a pre-trained Whisper model and utilizing librosa for converting audio bytes to array format. Database operations manage connections to a SQLite database, storing and retrieving chat messages, and implementing functions to save and delete chat messages, as well as retrieve chat history.

PDF handling uses "pypdfium2" to extract text from uploaded PDF files and splits the extracted text into chunks for easier processing and retrieval. Image handling converts image bytes to base64 format for processing and uses a loaded Llava model to generate responses based on images and user queries. LLM chains define and create various chains for handling normal chat and PDF-based interactions and load and interact with a vector database for document retrieval. Utilities include loading configuration settings from "config.yaml," saving and loading chat history to/from JSON files, and generating unique timestamps for session management. The configuration defines paths to models and embeddings, specifies parameters for different models and handlers, and sets database paths and other application settings. Prompt templates define the templates used in the chat chains to structure LLM queries and responses. Testing provides a script to verify the vector database search functionality.

5.1 Sentence LSTM

The "sentence LSTM," a "single-layer" architecture, utilizes the merged "context vector" "ctx" ∈ R C as its input. It generates a "topic vector"—**"t" ∈ R, K** for the word "LSTM" through a "topic generator," and employs a stop control mechanism to decide whether to proceed with or terminate the captioning process.

To achieve such an enhancement of contextual information inside the "topic vector t (s)," we require a "Deep-Output" layer. This layer incorporates both the "context vector ctx(s)" and sent by the hidden state h (s) from the step:

$$t^{(s)} = tanh(W_{t,hsent} \cdot h_{sent}^{(S)} + W_{t,ctx} \cdot ctx^{(s)}) \quad (2)$$

"$t^{(s)}$" is the resulting vector or value at step s, tanh(·) is the hyperbolic tangent activation function, and $W_{t,hsent}$ is a "weight matrix" or "vector" applied to $h_{sent}^{(S)}$, which is the hidden state vector at step s. $W_{t,ctx}$ is a weight matrix or vector applied to $ctx^{(s)}$ which represents the context vector or value at step s.

The probability of termination can be derived from the relation:

$$p(STOP \mid h_{sent}^{(s-1)}, h_{sent}^{(s)}) \propto exp\{W_{stop} \cdot tanh(W_{stop,s-1} \cdot h_{sent}^{(s-1)} + W_{stop,s} \cdot h_{sent}^{(s)})\} \quad (3)$$

"$p(STOP \mid h_{sent}^{(s-1)}, h_{sent}^{(s)})$" is the conditional probability of stopping, W_{stop} is a weight scalar or vector. $W_{stop,s-1}$, $W_{stop,s}$ are weight matrices or vectors for the $(s-1)^{th}$ and s^{th} time steps, respectively. "$h_{sent}^{(s-1)} h_{sent}^{(s)}$)" are hidden state vectors at the $(s-1)^{th}$ and s^{th} steps.

5.2 Word LSTM

A "word LSTM" produces each "sentence's" words. Following the approach of (Krause et al., 2017), the word LSTM's initial inputs are the "topic vector t," produced by the "sentence LSTM," and the "special START" token. Subsequent inputs consist of the word sequence. The word LSTM's hidden state hword ∈ R H is utilized directly to predict word distribution:

$$p(word \mid h_{word}) \propto exp(W_{out} h_{word}) \quad (4)$$

where W_{out} is the parameter matrix. The final report is created by concatenating all the word sequences generated by each word LSTM.

5.3 Flowchart

See Fig. 3.

6 Result

The implementation or result of this research shows the complete overview of the work that we have done throughout in this research work. The results for paragraph generation and "single-sentence generation" (shown in the bottom section of "Table 1") are presented using a standard evaluation tool for image captioning. This tool assesses performance based on several metrics: BLEU, METEOR, ROUGE, and CIDER.

Examining paragraph generation, the upper section of Table 2 clearly demonstrates that models utilizing a hierarchical LSTM decoder significantly outperform those with a single LSTM decoder. The primary distinction between Ours-"no-Attention" and CNN-RNN lies in their decoder structure: the former employs a "hierarchical LSTM," while the latter uses a "single-layer LSTM." Comparing these "two models" directly highlights "the hierarchical LSTM's" efficacy. This outcome is expected, given the well-established difficulty single-layer LSTMs face in effectively modeling lengthy sequences. Furthermore, the use of either "semantic attention (Ours-Semantic-only)" or "visual attention" (Ours-Visual-only) alone in

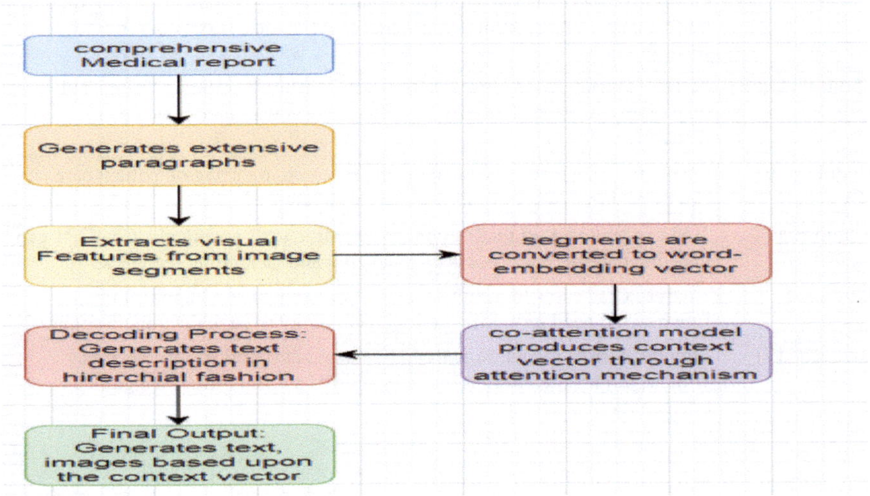

Fig. 3 Flowchart of the system

Table 1 Review work details with results

Works	Result
Soenksen [5]	HAIM framework consistently outperforms single-source approaches by 6–33%. Demonstrated improved performance in 10 chest pathology diagnoses and mortality predictions. Shapley values show importance of multimodal inputs for healthcare tasks
Zhang et al. [6]	Generative AI will have a significant role in medicine and health care. Challenges include trust, veracity, safety, privacy, and regulatory frameworks
Arora et al. [7]	"Generative Adversarial Networks (GANs)" contribute to the advancement of medical training and investigation, with an emphasis on improving patient treatment. However, issues arise regarding the attribution of scientific publications and the production of artificial imagery and video content
Musalamadugu [8]	In the field of "medical imaging," generative AI demonstrates superior accuracy compared to conventional techniques. This advanced AI technology creates new images for various purposes, including registration and segmentation. However, generative AI necessitates extensive training datasets and can be computationally demanding. Additional studies are necessary to enhance these methods for practical clinical application and to address associated ethical concerns
Munuera et al. [9]	Generative AI will have a significant role in medicine and health care. Challenges include trust, veracity, safety, privacy, and regulatory frameworks
Jadon et al. [10]	Paper discusses potential of "Generative AI" in radiology for health care emphasizing improved accuracy in diagnoses using "Gen AI" models
Bautista et al. [11]	SciBERT struggles to segregate between queries mentioning "race" and disparities
Khawaja [12]	Generative AI impacts data integration with benefits and challenges. Generative AI improves data augmentation, aiding machine learning model performance. Transparency in generative AI models is crucial for trust-building
Shoja et al. [13]	SciBERT struggles to segregate between queries mentioning "race" and disparities. "Cosine similarity" used to analyze text queries on health disparities
Huddle et al. [14]	GANs enhance medical education and research, focusing on patient care. Lower accuracy without full abstract availability
Stiglic et al. [15]	GAI can generate original, technical scientific texts accurately. ChatGPT can expedite peer-review process by analyzing and generating feedback
Inuwa [16]	Analyzed use cases in health segments for generative AI applications. Emphasized need for understanding technology to capitalize on its potential
Ambush [17]	Significant time reduction with AI generated abstracts. Lower accuracy without full abstract availability. US and EU students showed significant differences in accuracy

generating "Topic vectors" appears to have minimal impact on caption generation. This may be attributed to visual attention's limitation in capturing only subregional visual information, failing to accurately represent the entire image's semantics. Conversely, semantic attention struggles to pinpoint small abnormal image areas. Ultimately, our comprehensive model (Ours-Co-Attention) yields superior results across all evaluation metrics, validating the effectiveness of the proposed "co-attention" mechanism.

Regarding the results for generating "single sentences" (displayed in the bottom section of Table 1), the modified versions of our model ("Ours-Semantic-only" and "Ours-Visual-only") demonstrate performance comparable to leading methods. Our complete model ("Ours Co-Attention") surpasses all baseline approaches, highlighting the efficacy of the introduced co-attention mechanism.

Paragraph Generation for Text Input:

An illustration of paragraph generation for text input is shown in the picture. We found that output for the given input text is shown. But sometimes the output for given text input is not the exact output that is in the dataset. While the system gets Internet connectivity, it also has the capabilities to fetch the data from the Internet to choose the correct predicted output for the entered input (Figs. 4 and 5).

Generation for Image and Audio Input:

This figure shows the output of three paragraphs from three different models: "Ours Co-Attention," "Ours-no-Attention," and "Soft Attention." The task of each sentence is to report on a different topic. Normally, the first sentence describes the general description of the image, and the succeeding sentences describe different parts of it, such as "lung" or "heart." Only a few anomalies are detected with both "Soft Attention" and "Ours-no-Attention" models and are mistaken. On the other hand, the "Ours Co-Attention" model catches many true anomalies as shown in the "Top Three images." Thus, the contrast would represent how "co-attention" significantly beats alone at anomaly detection (Fig. 6).

The highlighted sentences describe identified abnormalities. A lateral X-ray image is shown in the second picture. The "top two" images display positive outcomes, the third demonstrates a "Partial Failure," and the "bottom image" represents a complete failure. These visuals are from the test dataset.

In the "Third image," the "Ours Co-Attention" model successfully identifies the abnormal area ("right lower lobe") and its condition ("eventration"), but fails to accurately describe the abnormality. Furthermore, the model detects "interstitial opacities" and "atherosclerotic calcification," which human experts do not consider true abnormalities. This mischaracterization may be due to the darker appearance of this X-ray compared to the previous images, suggesting the model's heightened sensitivity to such variations.

The bottom image illustrates a failure case of "Ours Co-Attention." Despite incorrectly assessing the major abnormalities, the model does identify some unusual regions: "lateral lucency" and "left lower lobe." In order to gain a better insight into the model's abnormality detection capability, Table 2 reports percentages of sentences

Table 2 Main results for paragraph generation on the IU X-Ray dataset (upper part), and single-sentence generation on the PEIR Gross dataset (lower part)

Dataset	Methods	BLUE-1	BLUE-2	BLUE-3	BLUE-4	METEOR	ROUGE	CIDER
IU X-RAY	CNN-RNN	0.316	0.211	0.140	0.095	0.159	0.267	0.111
	LRCN	0.369	0.229	0.149	0.099	0.155	0.278	0.190
	Soft ATT ATT-RK	0.399	0.251	0.168	0.118	0.167	0.323	0.302
	Ours-no-Attention	0.505	0.383	0.290	0.224	0.200	0.420	0.259
	Ours-Semantic-only	0.504	0.371	0.291	0.230	0.207	0.418	0.286
	Ours-Visual-only	0.507	0.373	0.297	0.236	0.211	0.426	0.300
	Ours-Co-Attention	0.517	0.386	0.306	0.247	0.217	0.227	0.327
	CNN-RNN	0.247‘	0.178	0.134	0.092	0.129	0.247	0.205
	LRCN	0.261	0.184	0.136	0.088	0.135	0.254	0.203
	Soft ATT	0.283	0.212	0.163	0.113	0.147	0.271	0.276
	ATT-RK	0.274	0.201	0.154	0.104	0.141	0.264	0.279
PEIR gross	Ours-No attention	0.248	0.180	0.133	0.093	0.131	0.242	0.206
	Ours-semantic-only	0.263	0.191	0.1	0.098	0.138	0.261	0.274
	Ours-visual-only	0.284	0.209	0.156	0.105	0.149	0.274	0.280
	Our-co-attention	0.300	0.218	0.165	0.113	0.149	0.279	0.329

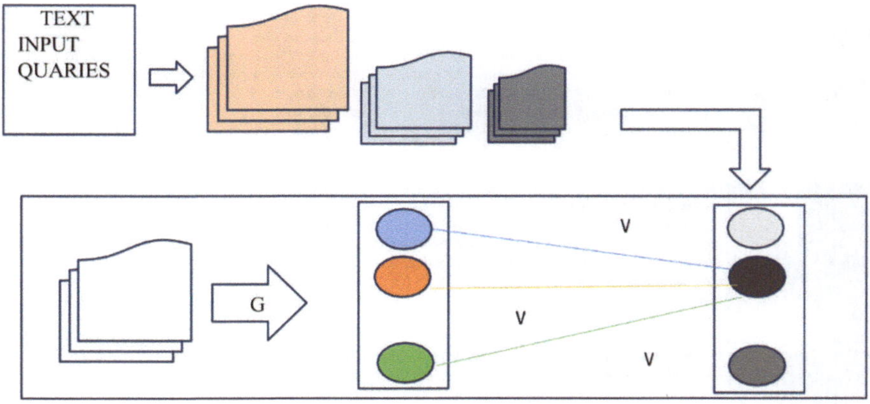

Fig. 4 Illustration of paragraph generation for text input with class activation

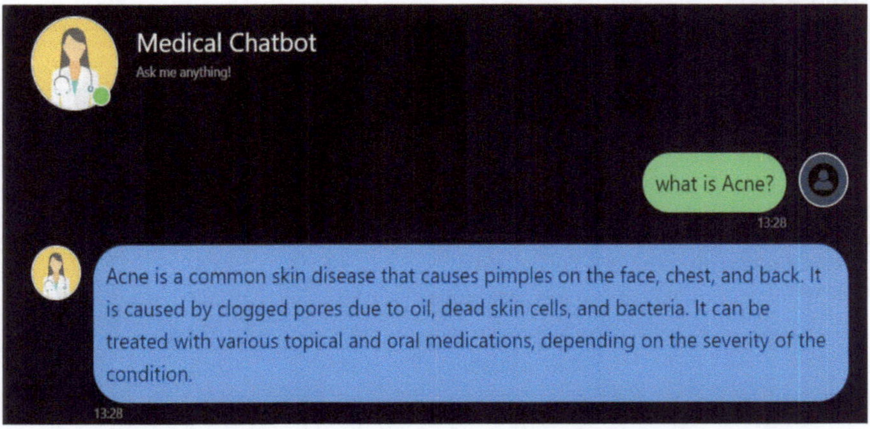

Fig. 5 Paragraph generation for text input and then prediction of the output

describing normalities and abnormalities. Sentences containing "no," "normal," "clear," or "stable" are labeled as "describing normalities." Clearly, "Ours Co-Attention" best captures the ground truth distribution of "normality and abnormality" (Table 3).

Co-Attention Learning:

The visualizations of co-attention are displayed in Fig. 7. This figure demonstrates several key points. Firstly, it shows that the sentence LSTM can produce various topics at different stages, as evidenced by the model's focus on distinct image areas and tags for each sentence. Secondly, the visual attention mechanism helps guide the model to focus on relevant image sections. This is exhibited in the "Third sentence" of the "first example" regarding "cardio," where "visual attention" dwells on areas

Image	Ground Truth	Ours-CxAttention	Ours-on-Attention	Soft Attention
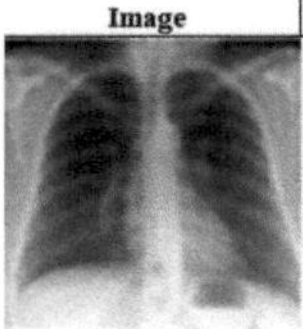	No active disease. The heart and lungs are normal in size and contour. The mediastinum is unremarkable. Heart and mediastinum normal.	No acute cardiopulmonary abnormality. The heart and mediastinum are normal. No pleural effusion or pneumothorax.	The lungs are clear bilaterally. The heart and mediastinum are normal. No pleural effusion or pneumothorax.	No acute cardiopulmonary abnormality. The lungs are clear bilaterally. The heart and mediastinum are normal. No pleural effusion or pneumothorax. Visualized osseous structures are unremarkable.
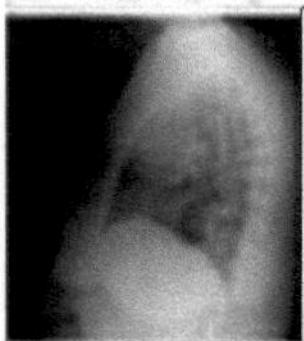	No evidence of active disease. The lungs are clear. No effusion or pneumothorax. The heart is normal in size. Mediastinum and pulmonary vessels are within normal limits. No hilar or mediastinal adenopathy.	No acute cardiopulmonary findings. The heart and mediastinum are normal. The lungs are clear. No pleural effusion or pneumothorax.	The lungs are clear bilaterally. The heart and mediastinum are normal. No pleural effusion or pneumothorax. There is no acute bony abnormality.	No acute cardiopulmonary abnormality. The lungs are clear bilaterally. The heart and mediastinum are normal. No pleural effusion or pneumothorax. Visualized osseous structures are unremarkable.
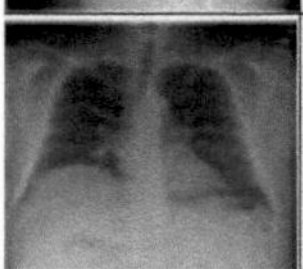	No acute cardiopulmonary abnormality. The heart size is normal. The mediastinum is unremarkable. No pneumothorax or pleural effusion.	No acute cardiopulmonary abnormality. The heart and mediastinum are normal. The lungs are clear. No pleural effusion or pneumothorax.	The lungs are clear bilaterally. The heart and mediastinum are normal. No pleural effusion or pneumothorax. There is no acute air space opacity.	No acute cardiopulmonary abnormality. The lungs are clear bilaterally. The heart and mediastinum are normal. No pleural effusion or pneumothorax. No acute air space opacity is present.
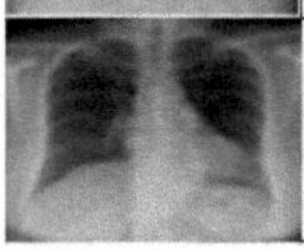	No acute cardiopulmonary abnormality. The heart and mediastinum are within normal limits. The lungs are clear. No pleural effusion or pneumothorax.	No acute disease. The heart and lungs are normal. No pleural effusion or pneumothorax. No acute bony abnormality.	No acute cardiopulmonary abnormality. The heart and mediastinum are normal. The lungs are clear. No pleural effusion or pneumothorax.	No acute cardiopulmonary abnormality. The lungs are clear bilaterally. The heart and mediastinum are normal. No pleural effusion or pneumothorax.

Fig. 6 Paragraph generation for image input—visual representation models

Table 3 Segments of text that detail the normal and abnormal features observed in the image

Method	Normality	Abnormality	Total
Soft attention	0.510	0.490	1.0
Ours noattention	0.753	0.247	1.0
Ours co-attention	0.471	0.529	1.0
Ground truth	0.385	0.615	1.0

around the heart. Semantic attention also works similarly; in the last sentence of the first example, the model appropriately pays attention to "degenerative change," which is the focus of the sentence. Finally, the "First sentence" in the "Last example" is a mischaracterization because of the inappropriate "semantic attention" of tags. Such errors may be reduced by the improvement of the "tag prediction" component. Now at last in the result part of our research, we have got from our system and the actual result we got from various sites, so we are going to compare the result accuracy and correctness of our result to the actual result and plot a graph of the results (Figs. 8 and 9).

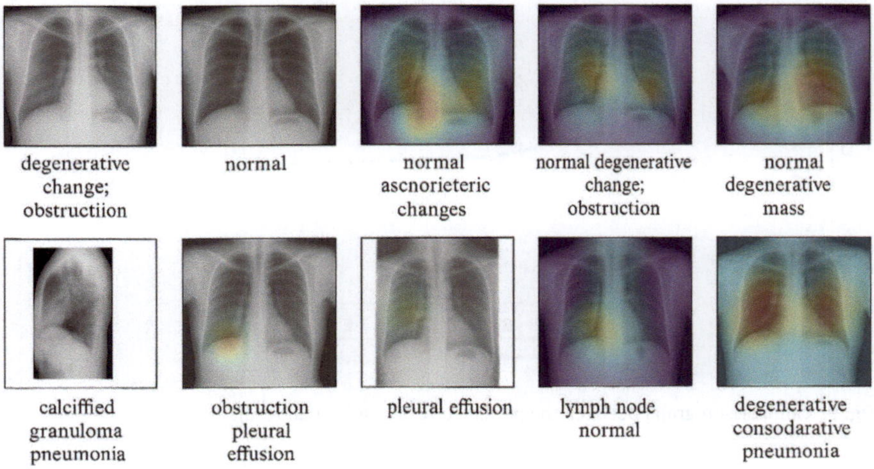

Fig.7 Illustration of co-attention for samples

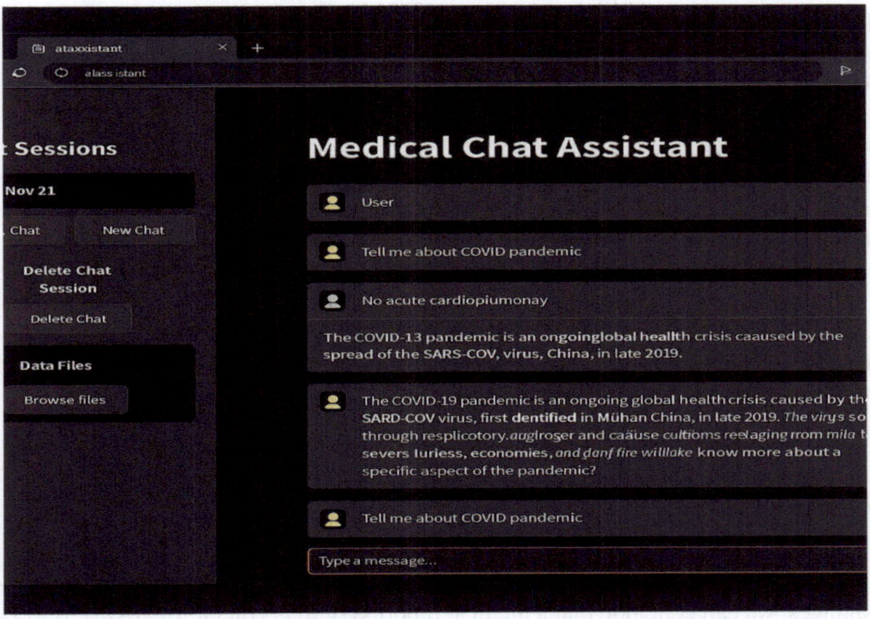

Fig. 8 Final design of the medical chat assistant system

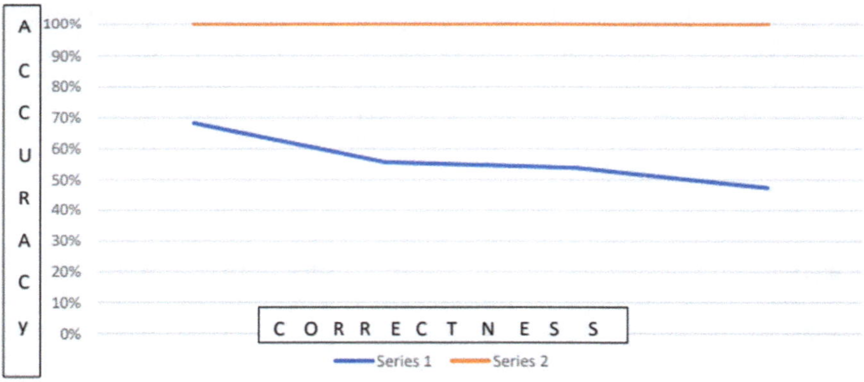

Fig. 9 Comparison graph between the predicted result and actual result

6.1 Comparison of Result

The "Comparison of Result" graph displays the accuracy levels of two distinct series across various correctness thresholds. A comprehensive analysis reveals the following.

For Series 1 (Depicted by the Blue Line):

The accuracy begins at roughly 70% and steadily declines to about 50% as correctness increases. This trend indicates that Series 1's precision diminishes when faced with more rigorous correctness standards, potentially signifying the model's difficulty in sustaining high accuracy under demanding conditions.

For Series 2 (Represented by the Orange Line):

The accuracy remains steady at 100% throughout all correctness levels. This consistency suggests that Series 2 exhibits flawless accuracy regardless of the correctness criteria, demonstrating a dependable and resilient performance that is unaffected by fluctuations in correctness standards.

7 Conclusion

By analyzing various medical systems and understanding their advantages and disadvantages, we can conclude that a medical report system is crucial for HAIM to work efficiently. It is necessary to continuously develop and find solutions to challenges for HAIM to become mainstream, as it ensures proper data analysis, problem prediction, solution provision, time-saving, and significant benefits for human health. Our approach to AI medical report generation for patients achieves high efficiency, but despite effective strategies reported in the literature, challenges and open research problems remain. Ongoing research in HAIM and GAI aims to optimize this technology for health care, addressing crucial aspects from orthology to neurology, thereby predicting and solving health issues. In contrast to conventional systems, our approach incorporates cutting-edge AI methodologies beyond patient data analysis. It offers practical recommendations and educational resources, thereby enhancing the capabilities of both the patients and medical professionals.

References

1. Cao J, Li X (2021) A 3D 2D convolutional neural network model for hyperspectral image classification. arXiv. arXiv:2111.10293
2. Shamsolmoali P, Zareapoor M, Yang J (2019) Convolutional neural network in network (cnnin): hyperspectral image classification and dimensionality reduction. IET Image Process 13:246–253
3. Ahmed SB, Naz S, Razzak MI, Yousaf R (2017) Deep learning based isolated Arabic scene character recognition. In Proceedings of the 1st international workshop on arabic script analysis and recognition (ASAR), Nancy, France, 3–5 April 2017, pp 46–51
4. Simonyan K, Vedaldi A, Zisserman A (2014) Deep inside convolutional networks: visualizing image classification models and saliency maps. In Proceedings of the 2nd international conference on learning representations, ICLR, Banff, AB, Canada, 14–16 April
5. Lusi R (2022) Soenksen partnership with Seoul National University Medical School. Integrated multimodal artificial intelligence framework for healthcare application, p 10. https://doi.org/10.1038/s41746-022-00689-4
6. Zhang P, Kamel Boulos MN (2023) Generative AI in medicine and healthcare: promises, opportunities and challenges, pp 15. https://doi.org/10.3390/fi15090286
7. Arora A, Arora A (2022) Generative adversarial networks and synthetic patient data: current challenges and future perspectives, p 5. https://doi.org/10.7861/fhj.(2022-0013)
8. Musalamadugu TS, Kannan H (2023) Generative AI for medical imaging analysis and applications, p 8. https://doi.org/10.2217/fmai-2023-0004
9. Munuera J, Esteve AV (2023) How imaging generative AI will transform medical radiological practice, p 12. https://doi.org/10.7717/hm2.2023-0012
10. Jose AJS, Kumar S (2023) Leveraging generative AI models for synthetic data generation in healthcare: balancing research and privacy, p 20. https://doi.org/10.7421/medi-2023-0006
11. Bautista YJP, Lima V, Theran C, Alo R (2023) Health disparities through generative AI models: a comprehensive study using a domain specific large language model, p 10. https://doi.org/10.48550/arXiv.2310.18355
12. Khawaja R (2023) Generative AI in healthcare: the promise, the perils and the top 10 use cases, p 4
13. Bautista YJP, Lima V, Theran C, Alo R (2023) Health disparities through generative AI models: a comprehensive study using a domain specific large language model, p 10. https://doi.org/10.48550/arXiv.2310.18355
14. Generative Artificial Intelligence in Medical Education, Research and Practice (2023) p 12. https://doi.org/10.7759/cureus.40883
15. Huddle M, Kellar J, Srikumar K, Deepak K, Martines D (2023) Generative AI will transform health care sooner, p 12
16. Stiglic G, Kopitar L, He Z, Kocbek P, Gosal L, Chakraborty P, Meyer P, Bian J (2023) Improving primary healthcare workflow using extreme summarization of scientific literature based on generative AI, p 06
17. Ambush A (2023) Impact of generative AI on data integration, p 6. https://doi.org/10.14445/22312803/IJCTT-V71I6P109

Electroencephalography-Based Autism Detection: A Comprehensive Review

Suparna Karmakar, Suparna Biswas, Nanda Dulal Jana, Aveek Chattopadhyaya, and Bishal Sarkar

Abstract Autism is a neuro developmental condition that affects how a person perceives and interacts with the world with a challenging social communication and interaction, as well as restricted and repetitive behaviors. It is called Autism "spectrum" disorder (ASD) because it affects individuals differently and to varying degrees. Some people with autism may have mild symptoms, while others may have more severe challenges. Individuals with autism may have difficulty in understanding social cues, maintaining eye contact, and engaging in reciprocal conversations and also may have difficulty with verbal and nonverbal communication. To diagnose ASD, researchers have worked with Electroencephalography (EEG), Magnetic Resonance Imaging (MRI), and genetic and sociodemographic data. In this work we have reviewed the existing literature based on EEG for the diagnosis of ASD. EEG is a non-invasive technique used to record electrical activity in the brain. Small metal electrodes are placed on the scalp, and they detect electrical impulses generated by the firing of neurons in the brain. These impulses are then amplified and recorded, producing a visual representation of brain activity known as an EEG. It is commonly used in clinical settings to diagnose neurological disorders such as epilepsy, sleep disorders, and brain injuries. It's also used in research to study brain function. Here

S. Karmakar (✉)
National Institute of Technology, Durgapur, India
e-mail: suparna.karnakar@gnit.ac.in; sk.24cs1511@nitdgp.ac.in

Department of Information Technology, GuruNanak Institute of Technology, Kolkata, India

S. Biswas · B. Sarkar
Department of Electronics & Communication Engineering, GuruNanak Institute of Technology, Kolkata, India
e-mail: suparna.biswas@gnit.ac.in

N. D. Jana
Department of Computer Science & Engineering. NIT, Durgapur, India
e-mail: nandadulal@cse.nitdgp.ac.in

A. Chattopadhyaya
Department of Electrical Engineering, GuruNanak Institute of Technology, Kolkata, India
e-mail: aveek.chattopadhyaya@gnit.ac.in

M. Chakraborty et al. (eds.), *Proceedings of Sixth International Ethical Hacking Conference*, Lecture Notes in Networks and Systems 1513,
https://doi.org/10.1007/978-981-96-8632-2_4

we have reviewed mainly EEG-based machine learning approaches to detect Autism. Future directions of research related to Autism are also mentioned.

Keywords Autism spectrum disorder (ASD) · Electroencephalography (EEG) · Magnetic resonance imaging (MRI)

1 Introduction

Communication, language, and behavior skills can all be improved with an early detection of ASD. Early diagnosis is achieved by brain signals for proper diagnosis of ASD which is a neuro developmental disorder. EEG is the most important and effective of the various brain signals. These signals from the human scalp are utilized to assess electrical brain activity and are employed to study critical neuropsychiatric brain issues. CNN, a popular deep learning (DL) technique for image classification and analysis, is implemented in this work to examine EEG-based ASD analysis. "Pearson's Correlation Coefficient" is employed to first transform the EEG channel recording into 2D form prior to classification using the popular residual neural network based on CNN model [1].

A useful tool such as ASD Genus prototype, which is used nowadays to assist in the clinical diagnosis of this disease, has been proposed in this research [2]. In this proposed model, the authors have proposed three main modules: pre-processing, feature extraction, and classification. EEG raw data is filtered by pre-processing modules to remove eye blink noise, AC powerline, and baseline drift. The extracted statistics features like mean, standard deviation, and entropy data have been computed for feature extraction and feature selection. Finally, four distinct machine learning (ML) methods were used in the classification process and "leave-one-out cross-validation" was performed to assess the models. Figure 1 shows a child having Autism.

EEG is one technique for recording the brain electrical action. Numerous frequency signals in EEG can be used to investigate the brain and its behavior. One of the brain abnormalities that can be identified by EEG is autism spectrum disorder (ASD). To ascertain whether or not people with ASD could respond to instructions to observe a particular thing, the EEG signal data was implemented in Ref. [3]. The BCIAUTP-300 dataset is linearly not separable having a 1:8 ratio and imbalanced class. When working with uneven data, undersampling was employed. Wavelet and principal component analysis were the feature extraction methods that were contrasted. The results of the investigation indicate that for classifying the Joint Attention of ASD utilizing EEG P-300, data optimal channel has been used. The highest accuracy is obtained using "GRLVQ," whereas the best G-mean is obtained from support vector machine (SVM). Based on "accuracy," "g-mean," "training time," and "testing time," the results depicted that "GRLVQ" outperforms the rest of existing methods, with "SVM" coming in second.

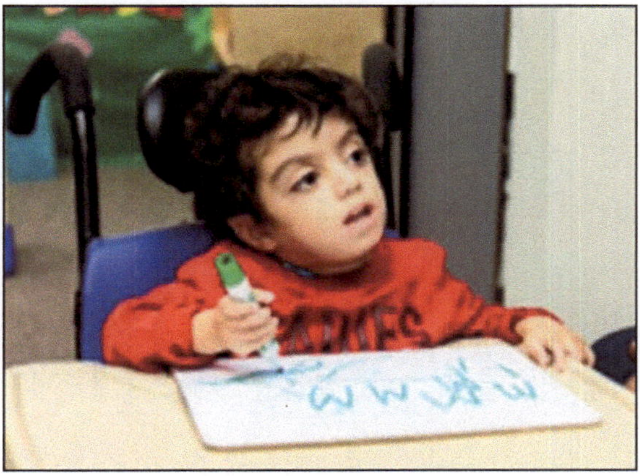

Fig. 1 A child having ASD [www.eparent.com]

Social interaction and communication abilities have been severely affected by ASD. Proper diagnosis of ASD at a very early stage is the most challenging issue facing by researchers. In [4], a research was carried out to use a deep learning algorithm to identify patterns in electroencephalograms (EEGs) that differ between children with autism and those without. Pattern recognition was employed in the brain signal database, and the multi-layer perceptron network would classify the features that were extracted. DL algorithm, which is now a popular and effective technique in the field of pattern classification and recognition, is the potential way to carry out the classification and analysis. Accuracy would be the classification's performance metric. The more effective the ASD diagnosis, the higher the percentage was observed [4].

The whole paper is organized as follows: Sect. 2 explains the different existing research work. Next Sect. 3 provides the information about the available databases of EEG to detect ASD followed by Sect. 4 giving the information of future research directions on ASD, and finally Sect. 5 provides the conclusion.

2 Related Research Work

In this section, a detailed comparative study of different works has been demonstrated in a tabular form in Table 1.

In summary, for this section we may conclude that detection of ASD based on EEG signals is done using (1) Wavelet transform and principal component analysis for the extraction of features, (2) ML algorithms such as SVM and KNN, and (3) DL algorithms. In most of the cases, it has observed that deep learning-based models outperformed machine learning-based algorithms.

AI technique has been applied on EEG data for autism detection. Prediction of risk-prone to autism genes and various medical data analysis reports are being

Table 1 Comparative study of different works

Details of work	Data used	Technique used	Result	Works
In order to diagnose ASD, authors have looked into several EEG feature extraction and various classifiers	Different EEG datasets are used: King Abdulaziz University dataset, Bonn and MIT dataset. Also used their own EEG dataset of 46 subjects	Integration of DWT, Shannon entropy Classified with KNN classifier	Classification accuracy of 94.6% (for 3-class) utilizing multi-channel signal	[5]
Generalized-Relevance-Learning-Vector-Quantization (GRLVQ) Method has been applied by authors on EEG data to assess the Autism Spectrum Disorder	BCIAUT P300	Utilized Wavelet transform and principal component analysis for the extraction of features They have normalized the data using Z-score method	87% (Highest) accuracy is obtained from GRLVQ. Best performance is given by GRLVQ	[3]
Here authors reported EEG-based technique to monitor neural development of children. Here a binary classification ASD or non-ASD had been carried out using nonlinear features and a leave-one-out cross-validation technique	EEG signals	SVM Classifier	The values of Specificity and Sensitivity were more than 95%	[6]
EEG-based technique implemented to identify abnormal activity to diagnose ASD. This method used Power Spectral Density (PSD) and machine learning	EEG signals	PCA was used to reduce the feature dimensions and classified by XGBoost and extra trees classifiers	ExtraTrees classifier performs better than XGBoost classifier with Recall value 83.3%, Specificity value 54.3%, Precision value 61%, and Accuracy of 67.7%	[7]

(continued)

Table 1 (continued)

Details of work	Data used	Technique used	Result	Works
Authors reported EEG-based ML algorithms integrating Phase-Locking Value and graph theoretic metrics for the detection of ASD	Database consists of twenty-four children's EEG recordings	SVM	Achieved 95.8% maximum accuracy	[8]
Authors reported a EEG-based time saving ASD detection technique for children	EEG signals	Used SVM and ANN classifiers	ANN method provides the maximum accuracy of 90.5%	[9]
Authors reported a spectrogram, ML and DL-based ASD detection technique using EEG signals	Dataset of King Abdulaziz University Hospital	9-different ML algorithms were used	DL-based model achieved the maximum accuracy of 99.15%	[10]
Authors used physiological measures, facial thermographic data, and EEG data to classify Autism	The EEG data has been collected using a 32-channel EEG recording machine	Authors have used Naive Bayes classifier, random forest, logistic regression, and multi-layer perceptron (MLP) models	Accuracy for various classifiers are as follows: a. Naive Bayes 82%, b. random forest 88%, c. logistic regression 94%, and d. MLP 94%	[2]
ASD was identified from EEG data. At first converted EEG signal into 2D and then classified	Applied on their own Dataset	CNN and residual neural network model	CNN provided the accuracy of 100%	[1]
DL method was used on EEG signal to detect ASD of children	EEG signals with five types of frequency band alpha, beta, gamma, theta and delta has been used	A six-layer CNN model	Achieved accuracy of 80%	[4]

(continued)

Table 1 (continued)

Details of work	Data used	Technique used	Result	Works
An EEG and DL-based technique was used to detect ASD	Database consists of ten autistic and ten normal children	Different DL models have been used: SqueezeNet, AlexNet, ShuffleNet, VGGNet, InceptionNet, DenseNet, and ResNet deep architectures	ResNet50 achieved the maximum accuracy of 81.91%	[11]
Frequency-time power measures extracted from EEG to detect ASD by authors	ADOS-2	Various ML-based algorithms are used such as Bagging, Random Forest, logistic regression, JRip, AdaBoostM1, and CNN model	Nonlinear models achieved accuracy above than 92%	[12]
Authors presented EEG-based technique to detect the epilepsy and ASD	a. Bonn dataset, b. the MIT dataset, c. the King Abdulaziz University dataset, and their own created EEG dataset	The combination of DWT, Shannon entropy, and KNN technique	Achieved the accuracy 94.6%	[13]
Identification of potential differences among children with mild and severe ASD based on EEG signal dataset was analyzed using empirical data	EEG data recorded (Participant: 36 children at resting state)	Artificial neural network (ANN)	Achieved sensitivity 100% and specificity 94.7%, and 97.2% overall accuracy	[14]

explored as potential solution for early detection of ASD symptoms. Integrating facial thermography with EEG signals is being used for ASD detection. Still application of AI in ASD detection is in its early stage. AI-based approach needs to be considered for the following:

i. AI-Based diagnosis system must be transparent and understandable to doctors and clinicians; hence application of Explainable AI (XAI) need to incorporate. Many popular XAI techniques such as LIME, SHAP, Rule-based system, and Saliency maps may be used to make ML/DL-based algorithm understandable. Among this, application of LIME has potential benefits.
ii. Application of AutoML: AutoML is the technique that provides ML algorithms for automatic selection and procedure to optimize models for a specific dataset that can improve clinical diagnosis of ASD.
iii. Multimodal approaches: Researchers may use multimodal-deep learning approaches which is a very promising technique for the detection of ASD.

3 Dataset

This section provides a EEG Dataset for ASD in a tabular form in Table 2.

4 Future Directions of Research

Although many researches are going on, still there are different challenges in the diagnosis of ASD. One of the significant challenges in the detection of Autism is the lack of efficient technique to identify or detect patients having autism based on their sociodemographic and medical characteristics. This hierarchization technique is very pivotal for early diagnosis. Recently AI is playing a big role in the detection of Autism in the early stage. But still there are many issues which need to be addressed. These are mentioned below.

i. Different methods are applied on different databases so selection of the best method is a difficult job for the researchers.
ii. Secondly there is a lack of standardization of method during the analysis and collection of data.

Multimodal approaches are now a very promising technique for the detection of ASD. Researchers may use multimodal approaches.

Table 2 EEG Dataset for ASD

Name of dataset	Details	Reference/link
EEG-BNCI Horizon 2020 dataset	Collected EEG data from both normally developing people and people with ASD	[15]
ABIDE dataset	ABIDE dataset consists of EEG data of 539 people with ASD and 573 normal controls (ages 7–64, median age 14.7 years) make up the 1112 dataset. The TUH EEG Corpus provides access to EEG data of people with autism	[16]
National Database for Autism Research (NDAR)	This dataset consists of EEG recordings along with other information from individuals suffering from ASD	[17]
EEG-TUH dataset	EEG data from individuals with autism were collected. It can be accessed from TUH EEG Corpus	[18]
Temple University Autism BrainNet	This dataset consists of EEG recordings of individuals suffering from ASD	[19]
Pediatric EEG Data Consortium (PEPS)	Consist of EEG recording	[20]

5 Conclusion

This review of EEG-based ASD detection method provides a brief idea of different approaches for the future researchers. Comparative study in Table 1 provides a clear view of the performances of various techniques. We have also shared the link of different EEG databases which are available publicly. Different gaps of the various techniques are also mentioned which will provide a path for the direction of research.

In this paper, we have reviewed only EEG-based ASD detection methods. In our future work, we want to study different ASD detection techniques based on Genetic datasets and MRI datasets.

References

1. Peya ZJ, Akhand MAH, Srabonee JF, Siddique N (2020) EEG based autism detection using CNN through correlation based transformation of channels' data, In: 2020 IEEE region 10 symposium (TENSYMP), 5–7 June 2020, Dhaka, Bangladesh, pp 1278–1281
2. Haputhanthri D, Brihadiswaran G, Gunathilaka S, Meedeniya D, Jayarathna S, Jaime M, Harshaw C (2020) Integration of facial thermography in EEG-based classification of ASD.

In: International journal of automation and computing, 17(6), pp 837–854, https://doi.org/10.1007/s11633-020-1231-6
3. MatulImah E, Dewil ES, Asto Buditjahjanto IGP (2021) A comparative analysis of machine learning methods for joint attention classification in autism spectrum disorder using electroencephalography brain computer interface. Int J Intell Eng Syst 14(3):412–424. https://doi.org/10.22266/ijies2021.0630.34
4. Ali NA, Syafeeza AR, Jaafar AS, Mohd Fitri Alif MK (2020) Autism spectrum disorder classification on electroencephalogram signal using deep learning algorithm. In: IAES international journal of artificial intelligence (IJ-AI), 9(1):91–99. ISSN 2252-8938. https://doi.org/10.11591/ijai.v9.i1
5. Ibrahim S, Djemal R, Alsuwailem A (2018) Electroencephalography (EEG) signal processing for epilepsy and autism spectrum disorder diagnosis. Biocybern Biomed Eng 38(1):16–26. https://doi.org/10.1016/j.bbe.2017.08.006
6. Bosl WJ, Tager-Flusberg H, Nelson CA (2018) EEG analytics for early detection of autism spectrum disorder: a data-driven approach. Sci. Rep. 8(1):6828. https://doi.org/10.1038/s41598-018-24318-x
7. Bhaskarachary C, Najafabadi AJ, Godde B (2020) Machine learning supervised classification methodology for autism spectrum disorder based on resting-state electroencephalography (EEG) signals. In: 2020 IEEE signal processing in medicine and biology symposium, SPMB 2020a—Proceedings, pp 1–4. https://doi.org/10.1109/SPMB50085.2020.9353626
8. Alotaibi N, Maharatna K (2021) Classification of autism spectrum disorder from eeg-basefunctional brain connectivity analysis. Neural Comput 33(7):1914–1941. https://doi.org/10.1162/neco_a_01394
9. Harun NF, Hamzah N, Zaini N, Sani MM, Norhazman H, Yassin IM (2022) EEG classification analysis for diagnosing autism spectrum disorder based on emotions. J Telecommun Electron Comput Eng 10(1–2):87–93
10. Tawhid MNA, Siuly S, Wang H, Whittaker F, Wang K, Zhang Y (2021) A spectrogram image based intelligent technique for automatic detection of autism spectrum disorder from EEG. PLoS ONE 16(6). https://doi.org/10.1371/journal.pone.0253094
11. Radhakrishnan M, Ramamurthy K, Choudhury KK, Won D andManoharan TA (2021) Performance analysis of deep learning models for detection of autism spectrum disorder from EEG signals. Trait Du Signal 38(3):853–863. https://doi.org/10.18280/ts.380332
12. Jayawardana Y, Jaime M, Jayarathna S (2019) Analysis of temporal relationships between ASD and brain activity through EEG and machine learning. In: 2019 IEEE 20th international conference on information reuse and integration for data science (IRI), pp 151–158. https://doi.org/10.1109/IRI.2019.00035
13. Ibrahim S, Djemal R, Alsuwailem A (2018) Electroencephalography (EEG) signal processing for epilepsy and autism spectrum disorder diagnosis *Biocybern.* Biomed Eng 38(01):16–26. https://doi.org/10.1016/j.bbe.2017.08.006
14. Hadoush H, Alafeef M, Abdulhay E (2019) Automated identification for autism severity level: EEG analysis using empirical mode decomposition and second order difference plot. Behav Brain Res 362:240–248. https://doi.org/10.1016/j.bbr.2019.01.018
15. https://www.bnci-horizon-2020.eu
16. https://fcon_1000.projects.nitrc.org/indi/abide
17. https://ndar.nih.gov/data.html
18. https://www.isip.piconepress. com/projects/tuh_eeg
19. https://www.autismbrainnet.org/data-access/
20. https://www.biorxiv.org/content/https://doi.org/10.1101/579550v1

The Role of AI in Revolutionizing Fairness in Platform Gig Worker Contracts

Oyndrila Ganguly

Abstract The Fourth Industrial Revolution is characterized by rapid advancements in automation and artificial intelligence (AI), which are reshaping labor markets and significantly impacting platform gig workers. While the technologies such as platform economy offer great solutions, their effectiveness depends on existing policies. The government plays a crucial role in establishing labor conditions and mediating relationships between employers and workers. However, inadequate regulatory involvement often leads to the perpetuation of informal and precarious work, leaving gig workers deprived of job security, benefits, and fair wages. In this context, a contractual analysis of platform gig economy with the technologies reveals significant challenges faced by these workers under current contractual arrangements. The lack of adequate protections within these contracts allows exploitative practices to persist, rendering workers vulnerable with limited opportunities for skill development and career advancements. This highlights the critical need to examine the implications of contractual agreements on workers' rights in an AI-driven labor landscape. Addressing the informal labor issues and improving skill development are essential for managing the effects of AI on employment; by scrutinizing the contracts governing the platform gig work, stakeholders can identify the weaknesses and advocate for stronger labor protection. This integrated approach ensures that the benefits of new technologies do not exacerbate existing inequalities, ultimately creating a more equitable environment for gig workers. The findings confirm the importance of contractual analysis as a tool to empower workers and safeguard their rights in the evolving platform gig economy.

Keywords Platform gig workers · Artificial Intelligence · Labor market · Contract · Inequalities · Workers' rights

O. Ganguly (✉)
St Xavier's College (Autonomous), Kolkata, India
e-mail: oyndrilaganguly@gmail.com

M. Chakraborty et al. (eds.), *Proceedings of Sixth International Ethical Hacking Conference*, Lecture Notes in Networks and Systems 1513,
https://doi.org/10.1007/978-981-96-8632-2_5

1 Introduction

Artificial intelligence refers to the development of machines that demonstrates intelligence. Narrow AI in particular describes the AI systems we have today, capable of performing specific tasks at or above human proficiency and potentially surpassing the value of a human worker. Currently, narrow AI is advancing rapidly, raising concerns about its integration into nearly every aspect of labor and the potential displacement of human jobs. These concerns are especially relevant within the platform gig economy, where the AI-driven automation and algorithms increasingly shape the job availability, worker management, and compensation. As media outlets extensively cover future of work, they often emphasize the risks of technological unemployment and income loss. However, this coverage tends to exaggerate the scope of these issues. The purpose of the study is to analyze the contract of platform gig workers and involvement of AI in terms of their status and job security. "To analyse the impact of Al on the labour market, a review of prior technical patterns is necessary, as this enables the mapping of past regularities to the current situation. Utilizing new technology to boost corporate efficiency while simultaneously contributing to the evolution of human capital has accelerated significantly since England's First Industrial Revolution began in the 1760s. Since that time, rapid expansion, changes in social and economic conditions and the introduction of new employment opportunities have all been conceivable. The manner in which these changes materialize is essential for determining the influence of Al. Windmill innovations and the discovery of the steam engine during the First Industrial Revolution, followed by advances in electricity and the utilization of petroleum, resulted in a tremendous rise in production and accessible resources for all sectors. Employment in urban sectors increased dramatically as low-productivity agricultural labourers shifted to industrial sectors to engage in more productive positions linked with the manufacture of completed goods." [1].

The Fourth Industrial Revolution is seen as a technological shift that will increase automation in industrial and service sectors, potentially reshaping labor markets and displacing skilled workers. Rapid innovation is creating new industries and work methods. Concerns about the future of work are divided—some fear digitalization will reduce worker autonomy and increase managerial control, while others argue the opposite. The platform gig economy, including platforms like Zomato, Swiggy, Uber, and Ola, has also grown rapidly, though it remains largely unregulated and offers little job security. Despite the fear of widespread job loss due to automation, research suggests its impact on employment may be overstated. Many jobs involve complex tasks that are difficult to automate, and technological adoption often occurs slowly, influenced by various economic and social factors. Collective worker actions and differing business needs shape how new technologies are implemented, with diverse outcomes across workplaces.

Artificial Intelligence is transforming the employment landscape, particularly in the platform-based gig economy. While AI-driven automation enhances efficiency and creates new job opportunities, it also raises concerns about job security, algorithmic control, and fair compensation. Platforms like Uber, Zomato, Ola and Swiggy have capitalized on AI to streamline operations, yet gig workers often find themselves in unregulated employment contracts that provide flexibility without stability.

The labor code has recently introduced definitions for gig and platform work within its social security regulations. However, the definitions provided in the unorganized worker social security act were insufficient in addressing these emerging categories of workers. As a result, immediate regulatory adjustments were deemed necessary. Moreover the inclusion of the definition of the self employed workers in the Unorganised Worker's Social Security Act, 2008 defined as those lacking an employer employee relationship, which resulted in misclassification and the denial of social security benefits for certain individuals because the platform work like Zomato started in the year 2008, Ola in 2010, Uber began its service in 2013 followed by Swiggy in 2014.

The rise of Narrow AI which excels at specific tasks has significantly influenced the labor market, especially in on-demand platform work. AI-driven algorithms now manage workforce allocation, pricing strategies, and performance evaluation, fundamentally altering how jobs are structured. AI does not simply eliminate jobs but reshapes employment structures, shifting work from traditional full-time roles to independent gig employment structures.

The definition of the unorganised workers that is provided in Factories Act or Employee's Provident fund Act as of the employee to be any person employed for wages in any kind of work, whether manual or otherwise, directly or indirectly receiving wages from the employer provides that all workmen are employees but not all employees are considered workman under these definitions. The term worker is broader encompassing a wider range of employees, including those not specifically covered under the definition of workman. The Workman Compensation Act does not clearly define the sector to which it applies. Additionally, it specifies that a workman under the Act is someone employed for the purpose of employer's business as per Schedule II which means that it has excluded unorganized and self-employed workers. Therefore, it cannot be said that this Act was drafted to provide significant protections to workers in unorganized sector or in Employees State Insurance Act, 1948; we can see that the Act is applicable only to factories and left out the workers of the unorganized sector. This Act has incorporated beneficial schemes, but the Government has failed to spread the awareness of this Act among the workers of the unorganized sector who are basically ignorant and illiterate and live their life under the shadow of caprices of the dominating class. The Maternity Benefit Act,1961, fails to define clearly the meaning of other performances in a way that includes workers in the unorganized sector. The Payment of Gratuity Act, 1972, does not recognize the continuous labor of those persons who had been working for 5 or more years in such establishments or firms where the number of employees is less than 10 or where the work is done on cyber platform or by help of artificial intelligence.

Despite the lack of explicit definitions in these laws, an employer-employee relationship is essential for workers to access benefits outlined in legislations, so Indian jurisprudence has made several attempts to determine the relation between master and servant through various judgements.

"A careful and close analysis of the terms of engagement of an App based worker with the Aggregator will reveal that the App-based workers in fact share a regular Employer-Employee relationship with the Aggregators. This will become evident from the following examples of agreement executed between the Aggregators and the drivers/driver partners." [2]. First of all, Ola collects user information such as IP addresses, locations, and demographic details to understand Internet behavior and personalize services. Supported by Softbank's technological solutions, Ola enhances navigation and pick-up/drop-off services through maps and geo-tagging. Leased vehicles are equipped with GPS devices for real-time location monitoring and data collection on operational hours and speed. Drivers must pass thorough verification and background checks to lease vehicles from Ola, ensuring safety and reliability which is cited in the contract that the cab aggregators, i.e., Uber and Ola operate through different categorization of car ownership agreements which are as follows:

a. "Partner vehicles: These are those which are owned by the drivers themselves and by registering with Ola or Uber through their online platform and after going through a process of physical verification of relevant documents and the vehicle, drivers can utilize their ride sharing service for earnings for a certain commission claimed by the Aggregators from their fare.
b. Fleet operators/owners registered with Ola: Certain fleet operators/owners of vehicles also have a number of vehicles registered with either of the companies. Ola also provides assistance in securing drivers for fleet vehicles if the fleet operators/owners are unable to onboard them.

Under this option, a driver has to lease the vehicle from Ola under a lease Agreement, the key terms of which (that demonstrate control by Ola) are as follows" [1].

In clause 3—use of information, subclause number 3.3.3 provides that "To obtain information regarding 'Your IP address, location and other demographic information such as age range and gender in order to ascertain user's internet behavior' and 3.3.5. it is mentioned that "For allowing Us to provide You with personalized services by remembering the choices made by You with respect to language You prefer, geographical location where You are etc. which means that every vehicle that is rented or leased from Ola "has introduced a range of innovative technological solutions aimed at enhancing the experience for both its drivers and customers specially in navigation and pick-up/drop/off services. The technology is supported by the Softbank company that holds largest market share in the ride hailing sector which has introduced leveraging maps, geo-tagging and extensive customer data." [1]. As a result, this approach improves the accuracy with which drivers can locate customers and enables users to give more precise location details to drivers. As detailed in the clauses, Ola gathers information such as IP addresses, locations, and other demographic details like the age and gender to better understand user behavior online. Moreover, subclause 3.3.5 outlines how Ola personalizes services by recalling user preferences like the language and geographical locations. Consequently, Ola mandates the installation of a GPS device in every leased vehicle to continuously share its location with Ola, always ensuring it remains active. This enables real-time monitoring of the vehicle's location, operational hours, and speed, and collects user feedback and ratings. To lease a vehicle from Ola, drivers must successfully undergo police verification, background checks, and any additional screenings deemed necessary by Ola.

Ola's terms of service specify that any vehicle leased from them must be used exclusively for Ola's business purposes. This means the driver cannot use the vehicle for any personal or other business activities. There is a limit on the number of kilometers the driver can drive the vehicle, as outlined in the Lease Agreement. If the driver goes beyond this set number of kilometers, they will have to pay extra fees to Ola. For vehicle maintenance, the driver cannot perform any repairs or maintenance tasks without first getting written permission from Ola. They must follow Ola's specific instructions for maintaining the vehicle. If the driver does not comply with these maintenance rules, it could lead to the lease being terminated. Additionally, if the vehicle sustains damage more than twice, Ola will terminate the lease agreement. In cases where the lease is terminated due to the driver's fault—whether because of excessive mileage, improper maintenance, or repeated damage—the driver will lose their right to reclaim the security deposit they initially paid. This ensures that drivers adhere strictly to Ola's rules and maintain the vehicle properly to avoid financial penalties and lease termination, which are elaborated in the contract as follows: "There are conditions imposed by the App and its terms of service (type of vehicle/ equipment used), the sequencing of tasks, control over access to customers etc. that the vehicle so leased, can only be used exclusively for the business of Ola and the driver is not allowed to use the vehicle beyond the contracted kilometers specified in the Lease Agreement. If he/she uses the vehicle beyond the contracted kilometers, then additional amounts are to be paid to Ola. Also The driver is not allowed to undertake maintenance activity on the vehicle without the prior written consent from Ola, the maintenance of the vehicle is to be carried out by the driver as per instructions of Ola, and any failure in doing so would result in a termination of the lease agreement. If there is any damage occurred to the vehicle more than 2 (two) times, would result in a termination of the lease agreement. In the event of a termination of the lease owing to some default of the driver, the driver is not entitled to even reclaim the security deposit." [2].

Under the Delivery Partner Terms and Conditions executed between Zomato and delivery partners, the terms outline that when a delivery partner first starts using Zomato's platform, they may be required to either pay an onboarding fee or a refundable security deposit, the amount and method of which will be determined by Zomato. Additionally, Zomato can charge extra fees for any additional training and support services provided to the delivery partner. To help delivery partners perform their delivery services, Zomato may provide certain assets such as the Zomato app for their smartphone and delivery bags. If the delivery partner ends their agreement with Zomato, the security deposit will be refunded, provided that the assets are returned in a usable condition. If the assets are not in usable condition, Zomato will deduct an amount from the security deposit as specified in Annexure A. The condition of the returned assets will be assessed by Zomato, and their decision on whether an asset is usable or not will be final and binding on the delivery partner which is elaborated in the contract as follow Section 3 clause iii states that "At the time of availing the Platform Services for the first time from Zomato, the Delivery Partner shall, at the discretion of Zomato, be required to pay either (i) an onboarding fee ("Onboarding fee") to Zomato of such amount and in such manner as prescribed by Zomato from time to time, or (ii) an interest free refundable security deposit ("Security deposit") to Zomato, of such amount and in the manner prescribed by Zomato from time to time". In addition, Zomato shall charge additional fees from the Delivery Partner for any additional training and support services provided by Zomato ("Training and Support Fees"). To enable the Delivery Partner to render Delivery Services, Zomato may provide the Delivery Partner with certain assets including, but not limited to, the Zomato DP Platform on the Delivery Partner's smart phone; bags etc. Upon termination of these Delivery Partner T&C and the Delivery Service Agreement, the Security Deposit, if any, shall be refunded to the Delivery Partner by the Zomato after the assets are returned to the Zomato in usable condition. In the event the asset returned by the Delivery Partner is not in usable condition, Zomato shall deduct such amount as provided in Annexure A hereto. The Parties understand that the decision of Zomato whether an asset is usable or not shall be final and binding on the Delivery Partner." [3]. Further Clause vi states that "The amount of Training and Support Fees, Onboarding Fee, Security Deposit and the Platform Charges shall be determined by Zomato and may be revised from time to time at the sole discretion of Zomato.

Therefore, as per the above-mentioned clauses it can be inferred that "Under this Agreement, driver partners have to compulsorily undergo a training course and support service to its driver partners. Correspondingly Zomato provides assets to its delivery partners (phone; bags etc.) which the driver partner has to necessarily use for providing services." [1].

Further interpretation of the terms clarifies that Zomato has the authority to determine the exact amount Delivery Partners can charge customers as Delivery Charges. This authority is absolute and cannot be revoked by the Delivery Partners. Zomato can adjust these charges based on various factors such as how many people are ordering deliveries at that moment, how many delivery partners are available to fulfill orders, and even external factors like traffic or weather conditions. These adjustments may happen regularly, and Zomato will inform Delivery Partners about any changes through appropriate communication channels as cited in Section 3 clause xiii which provides that "Notwithstanding anything to the contrary under Clause 3(ix) above, the Delivery Partner hereby absolutely, unconditionally and irrevocably authorizes Zomato to determine the amounts chargeable by the Delivery Partner as Delivery Charges, through such means as Zomato may determine, including real-time analysis of the demand for Delivery Services, the availability of delivery partners on the Zomato DP Platform to provide Delivery Services to the Users, traffic and weather conditions, seasonal peaks and such other factors as Zomato may deem fit, and you agree that the Delivery Charges may be revised from time to time on the basis of one or more of these factors. Zomato shall, from time to time and through such means as Zomato may deem fit, notify You the Delivery Charges that You may charge the Users."

Thus, it follows from the foregoing that the driver partners have to compulsorily use the Zomato DP platform that regulates the manner in which the driver partner is to provide services to customers and the Driver partner is required to abide by the timelines issued by Zomato for delivering orders to customers, and even the delivery charges levied by the driver partner are decided by Zomato.

The clauses further explain how Zomato compensates Delivery Partners for their availability to fulfill food delivery orders on its platform. In addition to the fees collected from customers for deliveries, Zomato may choose to provide an extra payment called an "Availability Fee" to encourage Delivery Partners to stay ready and connected to the platform for orders. However, Zomato also reserves the right to deduct any amounts owed by the Delivery Partner from the total funds collected, including both the Delivery Charges and the Availability Fee. This deduction ensures that any outstanding debts or fees owed by the Delivery Partner to Zomato, such as Platform Charges, are settled before Zomato pays out the remaining balance to the Delivery Partner. Moreover, Zomato incentivizes Delivery Partners with an Availability Fee, aiming to ensure they remain ready and connected to the platform for orders always, and the deduction is the control mechanism that allows Zomato to manage financial transactions efficiently and settle any outstanding debts owed by partners before distributing their earnings. Furthermore, Zomato's dominance extends to setting performance standards and behavioral guidelines through its Delivery Partner Terms and Conditions. These guidelines encompass expectations for service quality, adherence to delivery schedules, and compliance with platform rules. Partners failing to meet these standards may face penalties or even termination of their partnership with Zomato, underscoring Zomato's influence over how partners conduct their operations within its ecosystem. In essence, Zomato's control over Delivery Partners is evident in its ability to regulate pricing, manage financial transactions, and enforce operational standards through a structured contractual framework and operational guidelines which is elucidated in Clauses xii and xvii as it provides that "In order to incentivize the Delivery Partner for the availability committed by him/her on the Zomato DP Platform to undertake Delivery Services as and when he is connected to a User, Zomato may, at its discretion, pay the Delivery Partner a fee ('Availability Fee'), in addition to the Delivery Charges collected by the Delivery Partner from the Users. And, also Zomato may, at any time: (a) set off and deduct any amounts due from, payable by or proposed to be paid by Zomato to the Delivery Partner, including the Delivery Charges collected by Zomato from the Users on behalf of the Delivery Partner and the Availability Fee; and (b) apply such amounts towards any amounts due from, or payable by the Delivery Partner under or in connection with the Agreement and the Delivery Partner T&C, including without limitation the Platform Charges. Zomato shall have the right and the obligation to pay only such amounts due and payable by Zomato to the Delivery Partner, if any amount so remains due and payable after such set off, deduction and application as set out under this Clause 3(xvii). Nothing in this Clause 3(xvii) shall prejudice any right or remedy available to Zomato, whether under contract, law or equity to recover any amounts due from, or payable by the Delivery Partner under or in connection with the Agreement and the Delivery Partner T&C, whether such amounts arise under contract, tort, statute or equity." [4] which concludes that Zomato provides discretionary additional "availability fees" to its driver partners and the partners are not allowed to charge users an amount above the delivery charges fixed by Zomato. Zomato is also authorized to collect from the users, the delivery charges which are then remitted to the driver partners on a weekly basis and even entitled to deduct charges from the amount collected in lieu of services provided by the driver partners to users.

Section 4 Clause i states that "partners authorize Zomato to make Tax Deduction at Source (TDS) as per the Income Tax Act, 1961, where applicable, and other applicable taxes from the amount paid or remitted to under the Agreement or these Delivery Partner T&C, including the Delivery Charges collected by Zomato on driver's behalf from the Restaurant Partners and the Users," that authorizes Zomato company entitled to make tax deduction at source from the amounts payable to the delivery partners.

The definition of the "delivery services" [5] in the contract requires delivery partners to make themselves available as and when a request is placed upon them for delivery, through the Zomato platform. Under section 5 of the obligations of the delivery partners it states that according to "Clauses i and iii," [6] Delivery partners are required to ensure that their vehicles used for delivery are in well-maintained condition; and there is no delay in delivery and are required to conduct themselves in accordance with the policies and instructions of Zomato.

Additionally, Zomato's control over its Delivery Partners is evident in the following clause of the contract because it allows Zomato to deduct any money owed by Delivery Partners directly from the payments it collects on their behalf. This includes the fees customers pay for deliveries and any additional fees like the Availability Fee explained before that Zomato might pay to encourage partners to stay available for orders all the time. This deduction is used to cover any debts or fees that Delivery Partners owe to Zomato, such as charges for using the platform. Only after deducting these amounts does Zomato pay the remaining balance to the Delivery Partners. Therefore, this clause highlights Zomato's dominating authority in managing financial transactions within its partnership. It ensures that Zomato can enforce financial agreements and recover any outstanding debts owed by Delivery Partners under the terms of their agreement. Essentially, it gives Zomato control over the financial aspects of its relationship with Delivery Partners, allowing it to ensure all financial obligations are met before distributing earnings which are evident from the following.

Clause xviii of the contract states that "Zomato may, at any time: (a) set off and deduct any amounts due from, payable by or proposed to be paid by Zomato to the Delivery Partner, including the Delivery Charges collected by Zomato from the Users on behalf of the Delivery Partner and the Availability Fee; and (b) apply such amounts towards any amounts due from, or payable by the Delivery Partner under or in connection with the Agreement and the Delivery Partner T&C, including without limitation the Platform Charges. Zomato shall have the right and the obligation to pay only such amounts due and payable by Zomato to the Delivery Partner, if any amount so remains due and payable after such set off, deduction and application as set out under this Clause 3(xvii). Nothing in this Clause 3(xvii) shall prejudice any right or remedy available to Zomato, whether under contract, law or equity to recover any amounts due from, or payable by the Delivery Partner under or in connection with the Agreement and the Delivery Partner T&C, whether such amounts arise under contract, tort, statute or equity. Added with previous clauses Section 3 clause vii mentions that 'Delivery Partner shall create a log-in ID on the Zomato DP Platform to be able to render Delivery Services. [7] The Delivery Partner acknowledges that the sign-in details, including the username and password, are confidential and, accordingly, shall not share them with any third party, without written approval from Zomato.' And section 5 clause xiii states clearly that the Delivery Partner shall undertake the Delivery Services by himself and shall not delegate the same to any individual or third party." "Section 10.1 mentions that Zomato is at liberty to terminate the contract with a driver partner at its discretion if its finds that the delivery partner's conduct is not in consonance with the terms spelt out in the Agreement and finally 'section 19' states that Delivery partners are liable to pay penalties to Zomato in case they operate outside the terms and conditions of the Agreement." [8].

At this juncture, it is submitted that "Zomato Ltd. has undertaken its Initial Public Offering ('IPO') of Rs. 93,75,00,00,000/- and has filed a Red Herring Prospectus with the Registrar of Companies ('ROC'). Pursuant to the IPO, Zomato Ltd. is raising upto Rs. 9,00,00,000/- as a fresh issue component on the IPO, which is to be utilized towards organic and inorganic growth of business. The organic growth involves increasing delivery infrastructure, which essentially means that Zomato Ltd. will be creating additional employment for lakhs of people, thereby increasing the unorganized workforce by a large magnitude. It is therefore, all the more important that conditions of unorganized workers are regulated by bringing them within the regulations that govern and safeguard the mainstream workforce of the country" [7] and this initiative of Zomato's IPO and expansion plans indirectly relates to controlling delivery partners through its operational and contractual frameworks for the reason that are as follows.

Scale of Operations: With funds raised from the IPO, Zomato plans to expand its delivery infrastructure significantly. This expansion means more delivery partners will be required to fulfill orders, thereby increasing Zomato's control over the delivery network.

Contractual Terms: As Zomato expands, it continues to set and enforce contractual terms with its delivery partners. These terms dictate how much partners can charge customers (Delivery Charges), how funds are handled (deductions for fees and charges), and what performance standards they must meet [8].

Regulation and Compliance: The call to regulate unorganized workers highlighted in the paragraph reflects broader concerns about fair treatment and regulation. Similarly, Zomato's control over delivery partners involves setting operational standards and ensuring compliance with its terms and conditions, thereby exerting control over how partners operate within its platform [9].

In the Pick-up and Delivery Executive Agreement of Bundl Technologies Pvt. Ltd. ("Swiggy"), the consideration payable to delivery executives is as per the Payout Scheme of Swiggy, which is subject to revision by Swiggy to which the delivery executive is not entitled to raise any objection, and the amount payable to a delivery executive is calculated by Swiggy depending on the number of hours logged by the delivery executive and the number of orders picked up and successfully delivered. If the delivery executive breaches any guidelines issued by Swiggy, the same would lead to a termination of Agreement as it is mentioned in the agreement under Clause 2 a and b [10, 11] that the delivery partner will receive the delivery fees from the customers which will be collected by Swiggy, only on delivery partner's behalf and will be paid to them every week after subtracting applicable taxes and amounts which are allowed to subtract according to the agreement [12].

The delivery executives are compulsorily required to carry bags and wear uniforms prescribed by Swiggy, and delivery executives are not allowed to uninstall the Swiggy application without notice to Swiggy as in the agreement under Annexure 1 Serial number 11 it is provided clearly that "wear the Swiggy T-shirt or jacket and carry the field bags bearing our logo provided by us, while performing the services." Delivery executives have to undergo training upon execution of Agreement in addition to onboarding costs as in the annexure serial number 6 it is mentioned that "attend all training session as and when required." The delivery executives are under an obligation to disclose to Swiggy all information with regard to the services and activities performed by the delivery executive under the Agreement and are penalized if they reject an order that is assigned to them through the mobile application, and if the delivery executive is not able to deliver the order within 120 min after pick up, then he is not entitled to receive any payment for that order which is cited in annexure 2 of the agreement serial number 3. The delivery executives are expected to adhere to a code of conduct prescribed by Swiggy that consists of dress, appearance, and hygiene codes.

2 Key Findings

We can see that AI governs every aspect of their work, from task allocation to earnings and penalties. Despite being labeled as independent contractors, gig workers function under strict platform-imposed guidelines; for instance Uber and Ola mandate real-time GPS tracking, limit how drivers use their leased vehicles, and enforce pricing structures without driver input. Despite these challenges, AI has introduced several advantages in the gig economy making work more accessible and efficient. One of the biggest benefits of work is flexibility as workers can choose when and how much they work, which is particularly beneficial for students, part-time workers, and those needing supplemental income [13].

The new technologies like this platform economy can overcome challenges but their growth depends on how they align with the existing policies. The government plays a vital role in shaping labor conditions and mediating between employers and workers.

3 Conclusion

When it reduces regulatory involvement or disengages entirely, it perpetuates informal and insecure work while failing to provide adequate skill development infrastructure and employee status confirmation. These gaps have a major influence on how people adopt new technologies and manage power dynamics in the workplace. Tackling informal employment and improving skills are essential to addressing the challenges posed by AI and automation, which have the potential to address India's demographic issues and strengthen the workforce. These platform gig workers are at a disadvantage under their current contracts. Without adequate regulation and protections, workers are deprived of job security, benefits, and fair wages. The lack of oversight allows an exploitative system to thrive, leaving gig workers in precarious positions with limited opportunities for skill development or career growth. This situation emphasizes the need for stronger labor policies to ensure that technologies driving the gig economy benefit workers fairly, rather than deepening inequality [14].

References

1. Mukherjee S, Sharma K (2022) Artificial intelligence and loss of jobs. https://3fdef50c-add3-4615-a675-a91741bcb5c0.usrfiles.com/ugd/3fdef5_12462a43336f4449a9ee40c24722c5ac.pdf
2. Indian Fed'n of App-Based Transp. Workers v. Union of India, Writ Petition (Civil) No. 001068 of 2021 (India)
3. Indian Journal of Law and Legal Research 4:1–9
4. Driver cum owner, Uber. https://www.uber.com/in/en/drive/requirements/?city=kolkata. Last visited 10 Mar 2024
5. Ola Fleet. Available at: https://partners.olacabs.com/attach. Last accessed 10 May 2024
6. NN, Ola Unveils New Tech Solutions to Help Drivers, Customers. Times of India (Sept. 29, 2016, 8:18 AM IST). https://timesofindia.indiatimes.com/business/startups/companies/ola-unveils-new-tech-solutions-to-help-drivers-customers/articleshow/54575790.cms. Last visited 10 Mar 2024
7. Indian Fed'n of App-Based Transp. Workers v. Union of India, Writ Petition (Civil) No. 1068 of 2021 (India)
8. Delivery Partner Terms & Conditions, Runnr. https://www.runnr.in/delivery-partner-tandc.html. Last visited 10 Mar 2024
9. Delivery Partner Terms & Conditions, Runnr. https://www.runnr.in/delivery-partner-tandc.html. Last visited 10 Mar 2024
10. Delivery Partner Terms & Conditions, Runnr. https://www.runnr.in/delivery-partner-tandc.html. Last visited 10 Mar 2024
11. Delivery Partner Terms & Conditions, Runnr. https://www.runnr.in/delivery-partner-tandc.html. Last visited 10 Mar 2024
12. Delivery Partner Terms & Conditions, Runnr. https://www.runnr.in/delivery-partner-tandc.html. Last visited 10 Mar 2024
13. Mercader Uguina JR (2024) Artificial intelligence and labor relations. 16 Cuadernos de Derecho Transnacional 1114
14. Ajunwa I (2024) Artificial intelligence, afrofuturism, and economic justice, 112 Georgetown L.J. 1267

Artificial Intelligence and the Rights of Persons with Disabilities in India: Opportunities, Challenges and the Future

Pallabi Sengupta

Abstract Technological development has attempted to help improve the lives of persons with disabilities. This chapter will firstly attempt to analyze how Artificial Intelligence (AI)-enabled tools and gadgets are helping persons with different kinds of disabilities. Nowadays, people with different types of disabilities have different platforms offering them AI-based tools and applications tailored to suit their specific needs. In this context, various AI tools and gadgets suited to help persons with physical or neurological or psychiatric disabilities will be deliberated upon. Next, the chapter will explore how AI helps in realizing the various rights of persons with disabilities as enumerated under different laws on disability like the UN Convention on the Rights of Persons with Disabilities, the Rights of Persons with Disabilities Act, 2016, etc. AI has helped in realizing key fundamental rights of persons with disabilities—from the right to live with dignity to the right of non-discrimination to the freedom of movement. However, use of AI tools has sometimes posed a challenge for these people too—like the over-reliance on AI-based diagnostic tests to determine whether one is eligible for housing, or for some benefit under a disability scheme, or the use of discriminatory AI-based tests for job recruitment. In such a context, this chapter will delve into the negative impact of AI on persons with disabilities. Finally, the chapter will look into how, with more and more advancement in AI, the same can be used for the betterment of persons with disabilities.

Keywords Artificial intelligence · AI · Persons with disabilities · Impairments · Right to live with dignity · Accessibility · Discrimination · Physical disabilities · Mental disabilities · Neurological disabilities · Reasonable accommodation · Personal mobility · AI and privacy · Data bias · Ethical concerns of AI

P. Sengupta (✉)
National Law University Odisha, Cuttack, India
e-mail: pallabi.sengupta@uem.edu.in

Department of Law, IEM's International Institute of Juridical Sciences (IIIJS), University of Engineering and Management, Kolkata, India

M. Chakraborty et al. (eds.), *Proceedings of Sixth International Ethical Hacking Conference*, Lecture Notes in Networks and Systems 1513,
https://doi.org/10.1007/978-981-96-8632-2_6

1 Introduction

Amid the cacophony of recent developments on Artificial Intelligence (AI), there was news that researchers of Monash University, Australia, have developed the world's first bionic eye. The "*Gennaris Bionic Vision System*" has been developed to provide AI-powered vision to people with vision impairments bringing hopes to thousands of people with disabilities that the time has come for them to no longer be kept behind solely because of their disabilities. If the clinical trials are successful, this will bring a revolutionary change in how AI influences the lives of persons with disabilities.

Persons with disabilities are often relegated as part of marginalized communities. They are generally isolated from the mainstream. Since the United Nations Convention on the Rights of Persons with Disabilities (UNCRPD) was finalized and opened for signature and ratification, 192 countries have become State Parties to the Convention making it that human rights treaty which has one of the highest ratifications. It is not that rights of persons with disabilities were not there before the UNCRPD. However, those rights were equally applicable for and protected all people under general human rights treaties. The UNCRPD, while not providing any specific new right, strengthened the existing rights of persons with disabilities and provided a streamlined forum in case such rights were violated. In 2016, recalling the contributions of the UNCRPD, India adopted the Rights of Persons with Disabilities Act (hereinafter called the "2016 Act") to realize its international obligations and incorporate the same into its domestic sphere. Over the last few years, the UN Committee on the Rights of Persons with Disabilities (the UNCRPD Committee) and national courts have attempted to give full effect to all the rights of persons with disabilities enshrined under our international and national laws and to prohibit the encroachment of these rights.

AI, in simple understanding, is based on development of modern technology that allows machines to perform a variety of functions like data analysis, problem solving, decision-making, providing creative outputs, etc. When such AI is used to help realize the rights of people with disabilities, the potential is huge. AI can be used to bring positive changes in the lives of persons with disabilities with the development of applications and products to cater to the specific needs of people with disabilities. However, ethical concerns surrounding AI and its data bias can also result in isolating and further alienating people with disabilities from the mainstream. Thus, AI should be used cautiously. This chapter will attempt to see how AI can be used for providing better quality of lives to persons with disabilities while trying to address the concerns of AI's negative effects on persons with disabilities.

2 Role of AI in Helping Persons with Disabilities

A disability might be the result of a physical impairment. It can also be because of a neurological disorder. Some people have intellectual disabilities. Other people live with psychological disabilities. This part of the chapter will delve into various impairments/disabilities of people and how different AI tools and assistive devices are helping in the daily lives of persons with disabilities.

2.1 *Physical Disabilities and AI*

Many disabilities have a physical basis. Most common physical impairments are because of a defect since birth or because of bodily malfunction or are the result of an injury. Common physical impairments are vision impairment (blindness and low vision), hearing impairment (deafness or hard of hearing), speech and language disability, or some kind of locomotor disability. The Right to Persons with Disabilities Act, 2016 discusses various examples of locomotor disabilities such as disabilities that are a result of cerebral palsy or muscular dystrophy or loss of sensation as a result of leprosy [1]. A person who loses a limb or the use of a limb due to any accident also becomes a victim of a physical disability.

2.1.1 AI Tools/Applications Assisting People with Physical Disabilities

Technology has advanced a lot to help people with physical disabilities. There are various AI tools that are used by such people in their daily lives. Some of these tools are discussed hereinafter:

a. **Aipoly Vision:** Developed in 2015, this application was the first of its kind. It helps recognize objects and colors for persons with vision impairments enabling them to understand their surroundings better [2].
b. **Google Lookout:** A Google tool, it uses the device camera and other sensors to describe objects nearby.
c. **Seeing AI by Microsoft:** Developed by Microsoft, this is an application that is used by people having vision impairment to identify other people and objects. This application uses the camera of the laptop to identify people and objects and describes the same to its users.
d. **Be My Eyes:** This application connects people with vision impairment with volunteers (without vision impairment) so that the latter can provide sighted assistance mainly through live video [3].
e. **RogerVoice:** Enabled with a call caption feature, people with hearing impairments use this application which helps them by converting the speech of the person they are talking to into text during the call. The users can then reply using speech or text [4].

f. **Google Live Transcribe, Ava, etc.:** Applications like Google Live Transcribe and Ava have voice-to-speech transcription features which help users with total or limited hearing impairments by making their daily conversations easy.
g. **SoundHound's ASR:** SoundHound's Automatic Speech Recognition (ASR) helps a person with stuttering issue or any speech impairment to better communicate with others by transcribing words more accurately [5]. It has helped many persons with speech impairments compete better in the business world.
h. **Amazon Echo and Voiceitt:** Amazon's Tap to Alexa feature can be used by people with speech impairments to make calls without speech [6]. The speech recognition application, Voiceitt also helps people with atypical speech communicate by translating non-standard speech to clear speech [7].
i. **Smart Highways:** Smart highways or AI-enabled intelligent highways are a blend of physical infrastructure like solar panels or sensors with AI and big data [8]. Persons with mobility impairments can use these AI-enabled highways to navigate better as they receive real-time traffic updates.
j. **Autonomous Wheelchairs:** Designed in an innovative way, these wheelchairs can be operated by people without arms or legs or those who have paralysis in their limbs (including quadriplegia). These are aimed to be powered wheelchairs with environmental sensors. On-board real-time AI helps a person with physical disability move these wheelchairs in any direction he/she wishes [9].
k. **Wheelmap:** An online map to locate wheelchair accessible places, Wheelmap helps persons with physical disabilities to better navigate in public places especially in crowded places [10].

The above-mentioned are some of the AI applications and gadgets that are helpful for persons with physical disabilities in navigating through their day-to-day lives.

2.2 Intellectual/Neurological Disabilities and AI

A limitation in intellectual/neurological functioning like learning, reasoning, or problem solving is called an intellectual/neurological disability [1]. A common example of one having an intellectual disability is a person being on the autism spectrum. Typically diagnosed in a child within the first three years of his/her life, autism may affect a person's ability to understand the world around him/her, the concept of relationships and how to relate to others, and what is accepted as "normal" behavior [1]. Learning disabilities like dyslexia, dyspraxia, aphasia [11], dysgraphia [12], etc. are also included under intellectual disabilities where a person has difficulties in thinking, learning, reading, writing, spelling, memorizing, communicating [12], etc. Sometimes, a person's physical coordination is also affected in such cases [13].

Fig. 1 Steps for training a machine learning model to diagnose neurological disorders [16]

2.2.1 Use of AI in Diagnosing Neurological Disability

AI is based on machine learning where the model is trained on data sets, and the model is able to complete tasks automatically without human interference. The task is to predict results. Deep learning is a specialized form of machine learning. Both models (conventional machine learning and deep learning) are used to diagnose various neurological disabilities like autism [14], dyslexia, attention deficiency hyperactivity disorder (ADHD) [15], etc. In order for AI to help diagnose any neurological disability, the models are fed with "*images obtained from magnetic resonance imaging (MRI), computerized tomography (CT), electroencephalogram (EEG) signals or positron emission tomography (PET) scans*" [16].

Figure 1 demonstrates the series of steps that are necessary for training a machine learning model to provide a diagnosis. First, there is the input of images or signals; next, preprocessing of the input occurs in order to clear the data; after that, extraction of features takes place, followed by the ranking of the features in order to find the most prominent features (selection). The last step is classification (diagnosis) into "normal" or "abnormal". While deep learning models carry out all the steps automatically, in conventional machine learning only the first two steps are carried out automatically, while feature extraction, selection, and classification can only be done manually.

2.2.2 AI Tools/Gadgets Used to Assist People (Children) with Neurological Disabilities

Children with neurological disabilities often need specific supporting tools to learn or even think efficiently. If the same is not provided the child will get isolated. If the same support is provided within the mainstream, it helps the child reach his/her fullest possible potential. In this context, some of the AI tools and gadgets that are specifically there in today's market for helping children with different neurological disabilities like autism, dyslexia, or ADHD are discussed hereinafter:

a. **Kaspar Robot:** A humanoid with human-like features and body, Kaspar robot is an assistive technology that helps enhance the social interaction skills of children with autism [17].
b. **LIFEisGAME Application:** This digital application is used to teach students on the autism spectrum facial emotions [18].

c. **Kiwi Robot:** An assistive technology, Kiwi robot is intended to provide personalized care and help improve the social and calculation (mathematical) skills of children having autism [19].
d. **Emotify:** Another digital application, Emotify, as the name suggests, helps children having autism understand and express feelings and learn behavioral norms [20].
e. **Squizzy Software:** Using Scrum technology, this software is valuable for cognitive development of children having autism [21].
f. **NAO Robot:** Social robots like NAO help build eye contact, better pronunciation skills, and concentration among children having autism [22].
g. **Dyslexi.co:** Claiming to be the first of its kind, this accessibility tool helps children having dyslexia learn grammar and spelling. It helps these children to also learn effective and inclusive communication skills [23].
h. **Dyslexia.ai:** Using Phonics, this application helps people with dyslexia learn reading and spelling at their own paces [24].
i. **DytectiveU:** Another digital application, DytectiveU helps children with dyslexia learn reading and writing skills [25].
j. **Dragon Naturallyspeaking:** Speech recognition software like Dragon Naturallyspeaking are helpful for children with ADHD as they encourage children to speak more efficiently resulting in better reading and spelling skills also [26].
k. **WatchMinder:** This gadget (watch) has been helpful as a memory aid for children having ADHD. Equipped with auditory and vibrating alarm features, this watch is extremely useful for children setting them reminders from time to time to focus on their work [27, 28].
l. **Kar Robot:** This assistive technology helps children with ADHD improve their cognitive abilities [29].

The above-mentioned are some of the AI applications and gadgets that are helpful for children with neurological disabilities in learning skills that they otherwise find difficult to learn.

2.3 Psychiatric Disabilities and AI

Mental illness or psychological disorders can impact mood, perception, or even memory. They can severely impair one's judgment, behavior, or capacity to recognize reality or the ability to meet the ordinary demands of life [1]. Common psychiatric disabilities include schizophrenia, bipolar disorder, identity disorder, panic disorder, depression, severe anxiety, phobia, obsessive compulsive disorder (OCD), and post-traumatic stress disorder (PTSD). Technological development and AI in the recent years have attempted to help people with chronic mental disabilities and debilitating conditions. Some of those developments are discussed in the following sub-part of the chapter. Before that, the scope of AI in the diagnosis of mental illnesses is deliberated upon.

2.3.1 Use of AI in Diagnosing Mental Disability

Unlike the positive and oftentimes revolutionary applications of AI in the physical healthcare system especially physical healthcare diagnostics, mental health care has been quite slow in adopting AI [30, 31]. Mental healthcare professionals often rely on a more hands on approach preferring to build relationships with their patients while directly interacting with the individuals (which AI obviously cannot do). Treatment also is generally more individual-centric as no two people may have similar mental issues or the reasons and circumstances for such issues. Hence, even mental health clinical data is very qualitative and individual-centric.

This does not mean that there has been no role played by AI in the mental healthcare domain. AI definitely has a huge role in the diagnosis of mental illness. Over the years, machine learning and deep learning have both been utilized to predict depression [32], probability of suicide [33], chances of developing bipolar disorder [34], etc. Thus, AI has had some success in the pre-diagnosis screening of potential risks for severe mental illnesses. However, the biggest challenge in this area is the limitation pertaining to clinical validation. Finally, the overall issue remains that diagnosis of mental illness cannot be an automated process because of the large subjectivity involved in such cases.

2.3.2 Use of AI in Supporting and Assisting People with Mental Disability

In the recent past, during the COVID-19 pandemic era, people all over the world had to practice social distancing which resulted in a feeling of isolation. This led to various mental health problems cropping up among the general masses. In such a situation, it was technology that helped bridge the physical gap among people. Various AI-enabled platforms and tools have proved their efficacy in helping people to maintain their sanity and lead better lives. Some of the recently developed AI-enabled tools, gadgets, and applications that provide mental healthcare solutions and are there to help people with mental disabilities are discussed hereinafter:

a. **Woebot:** Developed in 2017, Woebot was the first therapy chatbot that offers "*structured, manualized Cognitive Behavioral Therapy for people experiencing depression, anxiety* etc." [35]. Other platforms like Ellie, Wysa, or AI-Therapist also offer similar services.
b. **MoodKit and eMoods:** Based also on *Cognitive Behavioral Therapy*, MoodKit offers a variety of services helping people to manage their moods and improve their overall mental well-being, including addressing depression and anxiety issues [36]. On the other hand, eMoods is used to help an individual track his/her mood and provide treatment simultaneously in case of bipolar disorders [37].
c. **Sanvello and Calm:** While the Sanvello application uses *Cognitive Behavioral Therapy* to relieve stress and decrease anxiety and panic attacks [38], the Calm

application achieves the same by providing meditation techniques and relaxation therapies [39].

d. **Talkspace and BetterHelp:** Talkspace and BetterHelp are online therapy platforms that match persons with mental disorders with licensed therapists [40, 41] so that the patients can seek professional help at the earliest.

The above-mentioned are some of the AI applications and tools that are helpful for people with psychiatric disabilities in getting online therapeutic solutions and managing their symptoms better.

3 When AI Helps Realize the Rights of Persons with Disabilities

The next part of the chapter includes a detailed analysis of a few of the rights enumerated under the international and national legal frameworks for persons with disabilities. Here, there will be a deliberation of how the above-mentioned AI-enabled tools/gadgets/platforms/programs help in better guaranteeing the rights of persons with disabilities.

3.1 Right to Accessibility

Both the UNCRPD and the 2016 Act discuss the right to accessibility of persons with disabilities. Article 9 of the UNCRPD mentions that States shall take appropriate measures to ensure that persons with disabilities have access "*to physical environment, to transportation, and to information and communication*" on an equal basis with others (persons without disabilities). A similar right is also provided under Section 40 of the 2016 Act. Section 41 of the Act further talks about providing "*accessible roads*" and "*personal mobility assistance*" for persons with disabilities (especially those with physical disabilities). In this context, technological development has been the biggest support in ensuring better realization of this right to accessibility for persons with disabilities. People with physical disabilities especially in cases of paralysis or the loss of limbs can use Automated Wheelchairs. Navigating in busy public spaces has become so much easier for persons with disabilities now that they have access to Smart Highways and Wheelmaps. Many of these applications are free to download and use, making them affordable. However, sometimes, accessibility can become a concern for poor people (having a physical disability) as they may not be able to afford the latest AI-enabled gadgets (like automated wheelchairs) in the market.

Having access to information on an equal basis with persons without disabilities has also become easier for persons with disabilities as they can now use various AI-enabled tools and applications as discussed earlier (under Part 2 of this chapter).

A person with vision impairment can use Aipoly Vision, or the Google Lookout feature or the Seeing AI software and identify their surroundings with the aid of AI. People with hearing impairments can use RogerVoice or Google Live Transcribe to communicate better with the rest of the world in real time from any part of the globe. Access to immediate medical support and therapy has become easier for persons with psychiatric disabilities as they now have recourse to platforms like Woebot, Talkspace, BetterHelp, etc.

In the famous matter of *Szilvia Nyusti and Péter Takács v Hungary* [decided by the UN Committee on the Rights of Persons with Disabilities (hereinafter referred to as the UNCRPD Committee)] [42], two Hungarian nationals had approached the national courts and finally the Committee to seek remedy for the violation of their right to accessibility. Nyutsi and Takács were persons with vision impairment who had become customers of the OTP bank (a private bank) in Hungary. However, they were unable to operate the ATMs of the said bank by themselves despite paying for the same services equally as other customers (persons without disabilities) as the OTP bank's ATMs in their locality were not technologically advanced and did not support voice assistance or have braille feature. After getting no remedy in the domestic courts, when the matter was taken before the UNCRPD Committee, the Committee opined that the State of Hungary was liable for failing to prevent the discrimination that took place when these two individuals with vision impairment were unable to access the banking services at par with other customers because of the barriers that were posed as the ATMs of OTP bank did not have modern technologically advanced features. This was a clear violation of their right to accessibility. Today, with more AI-enabled gadgets and features, accessibility is one right that is being realized more and more for persons with disabilities every day.

3.2 Right to Live with Dignity and Right to Live Independently

Article 21 of the Constitution of India upholds the fundamental right of every individual to live his/life with utmost dignity. Article 10 of the UNCRPD ensures the same for all persons with disabilities on an equal basis with others. Article 19 of the said Convention also guarantees the right to live independently and within a community. All the advancements in technology have helped to ensure the proper realization of this right. With automated wheelchairs, persons having mobility issues can navigate the wheelchairs themselves without having to rely on others to move their wheelchairs. This also ensures Freedom of Movement, another fundamental right guaranteed under Article 19 of the Constitution. The AI-enabled tools discussed hereinbefore that support a person with vision impairment or hearing loss or speech impediment have further helped realize this right of independent living for persons with such physical disabilities. Article 19 (a) of the UNCRPD talks about providing the opportunity to persons with disabilities to "*choose their own place of residence*" without being forced to live in any "*particular living arrangement.*" Further, Article 19 (b) talks about providing personal assistance necessary to ensure inclusion within community life. AI-empowered tools which act as an artificial eye or ear for these persons with disabilities have helped them from feeling excluded and secluded from community living. With the help of platforms like Woebot, Talkspace, BetterHelp, etc., no longer are persons with psychiatric disabilities forced to submit to substitute decision-making (some well-wisher deciding on their behalf). Persons with mental disabilities are now in charge of their own therapy with easy access to supported decision-making (provided by AI-enabled mental healthcare applications and websites) that they can refer to or seek help from whenever needed.

In the matter of *Jeeja Ghosh v Union of India* decided in 2016, the apex court ruled in favor of Ms. Ghosh when she claimed that her fundamental right to live with dignity and her right to equality had been violated by the treatment meted out to her by the airline, SpiceJet, and the airport authorities [43]. Ms. Ghosh who has cerebral palsy was flying to Goa to attend an International Conference where she had been invited as a distinguished resource person. However, before the flight could take off from Kolkata airport, Ms. Ghosh was forcibly deboarded from the plane on account of her disability as the pilot had insisted on such removal. The main contention of SpiceJet had been that Ms. Ghosh being aware of her physical ailment should have had a companion and asked for assistance. She was seen as a flight risk. Ms. Ghosh who won the case said that she had felt like a criminal when she was forcefully removed from the plane. She said that for months she suffered from the trauma of the treatment that was meted out to her by the flight and airport authorities.

The Supreme Court, while upholding the violation of Ms. Ghosh's fundamental right to live with dignity, her right to equal treatment (with other passengers), and her right to independence, mentioned in this case that the flight and airport authorities had acted in a "callous manner" without any "sensitivity" and "positive attitude". When society acts in such a manner it results in isolation. Persons with disabilities were never treated as self-reliant people who could be useful members who could contribute to the development of the country despite their limitations [43]. With AI-enabled applications bringing persons with disabilities at par with those without disabilities, it is now a more equal world. Persons with disabilities can now participate equally in mainstream activities. The Supreme Court had observed in the Jeeja Ghosh matter that "*to most disabled persons, the society they live in is a closed door which has been locked and the key to which has been thrown away by the others*" [43]. With more technological advancements, this is no longer the case or rather it can be said that a key (through the positive intervention of AI-enabled tools) has been located or even specially made for these persons with disabilities.

3.3 Right to Equality, Universal Design, and Reasonable Accommodation

Article 14 of the Constitution of India guarantees the fundamental right to equality and non-discrimination. Disability often becomes a ground for discrimination, even unintentional discrimination as was seen in the matter of Jeeja Ghosh also. Article 5 of the UNCRPD prohibits discrimination on the ground of disability. Section 3 of the 2016 Act further states that no person shall be deprived of their liberty on the discriminatory ground of disability. Both the international instrument and the national legislation also talk about providing "universal design" and "reasonable accommodation." Universal design means "*the design of products, environments, programmes and services to be usable by all people to the greatest extent possible*" [44, 45], and "reasonable accommodation" means "*necessary and appropriate modification and adjustments, without imposing a disproportionate or undue burden in a particular case, to ensure to persons with disabilities the enjoyment or exercise of rights equally with others*" [44, 46]. These rights of persons with disabilities envisaged under the UNCRPD and the 2016 Act, are fulfilled by AI-enhanced gadgets and features like Automated Wheelchairs and Wheelmaps as they provide universal design and reasonable accommodation for these persons whenever necessary.

3.4 Right to Education (Including Inclusive Education)

Article 7 of the UNCRPD incorporates the concept of "intersectionality" and talks about ensuring various rights of children with disabilities. State Parties have the duty to ensure fulfilment of all human rights of children having disabilities on an equal basis with other children. Aim of education is to provide a child the platform to develop to his/her fullest potential [47]. Quality education helps a person develop academically and intellectually as well as in a holistic manner. Both Article 24 (2) of the UNCRPD as well as Section 16 of the 2016 Act incorporates the concept of inclusive education. It is no longer considered best for a child with disability to study in an exclusive educational institution specially catering to other children with "similar needs". Earlier such schools like "school for the blind", "school for the hearing impaired", etc. were preferred. However, it is now encouraged that children with different needs (including different disabilities) get the opportunity to study in mainstream educational institutions. This puts a stop to isolation, allowing children with disabilities the companionship of every other child. The laws (both international and national) provide for "reasonable accommodation" whenever necessary to cater to the requirement of individual need of a child with any specific disability [48]. Such individualized support is guaranteed through the use of AI-assisted tools like Kaspar Robots, Emotify, Kiwi Robots, Squizzy, Dyslexi.co. Dyslexia.ai, etc., which help children with disabilities to read, write, learn, think, and do any other activities that other children of their age do, however at their own pace, but without losing the companionship of their peers.

3.5 *Right to Personal Mobility and Assistive Technology*

Apart from the right to accessibility, perhaps the one specific right of persons with disabilities that has been most ensured through the advancement of technology is the right to personal mobility. Article 20 of the UNCRPD makes it a duty of State Parties to take effective measures to look into the mass production of mobility aids and assistive technologies. With various companies and brands bringing in different AI-enabled products to assist persons with disabilities, these people now have several options to choose from in terms of the product/software they want to use or feel the most comfortable using. In fact, persons with disabilities are seen as a robust community of customers by these brands, which specifically devote their time, effort, research, and development potentials to produce better quality products for this customer base (the persons with disabilities). Many of the applications that were discussed earlier under Part 2 of this chapter are free to use (at least have free versions) that support the basic needs of persons with vision/hearing/speech impairment or intellectual or psychiatric disabilities. This makes them affordable. However, not all are free of cost. Automated wheelchairs or WatchMinders are gadgets that people with disabilities have to purchase, and, hence, not all people (especially poor people) are able to get benefit of their latest modifications.

Thus, as proved from the above-mentioned points, many of the rights envisaged for persons with disabilities as enumerated under the UNCRPD and the 2016 Act have been more fully realized with the positive use of various AI-based innovations.

3.6 *International Law on Use of AI*

The United Nations Educational, Scientific and Cultural Organization (UNESCO) at the 41st session of its General Conference at Paris gave the Recommendation on the Ethics of AI, 2021. According to the said Recommendation of 2021, no human being shall be harmed by the use of AI systems [49]. Protection of persons with disability and guarantee of inclusiveness are under the special focus of the Recommendation [50]. Recommendation 28 emphasizes on ensuring fairness and non-discrimination while relying on AI tools and applications. The Recommendation further mentions that AI technologies should be available and accessible to all [51]. If implemented systematically by different nations, including India, this law can help better realize the rights of persons with disabilities when using AI tools and applications.

4 Challenges Faced by People with Disabilities Because of AI

The previous part of the chapter discussed the positive aspects of AI innovations in guaranteeing different rights of persons with disabilities. However, many a time AI has hindered the fulfilment of many rights of persons with disabilities. In the era of technology, the challenges that overuse (without exercising proper caution) of AI poses on persons with disabilities are discussed hereinafter.

4.1 Violation of the Right to Privacy and Confidentiality

According to the Supreme Court of India, the right to privacy is "intrinsic to life and personal liberty" and is "inherently protected" under Article 21 of the Constitution of India. However, with persons with disabilities, ensuring their right to privacy generally becomes a huge concern. Even during their day-to-day lives well-meaning family members often encroach upon this right of persons with disabilities, albeit unintentionally, while taking care of them. Persons with severe disabilities who require more support have almost no privacy guaranteed to them. When persons with disabilities use AI-enabled applications that help them function at par with others, these applications often ask their users to give access to different features in their phones/devices and the users have no choice but to accept to do so in order to download/use these applications. Further, a lot of times, these applications ask for a full disclosure of the disability status of the users in order to serve their customers better, thereby again encroaching upon the privacy of the individuals with any disability. AI often requires access to massive amount of private data in order to function effectively specifically those information relating to health and behavioral pattern. Because of the sensitive nature of their data, it is the persons with disabilities who are at the greatest risk of data privacy breaches. Further, a person with disability because of the very nature of his/her disability (say, complete vision impairment) might be more prone to mistakenly allowing unauthorized access to some sensitive information. In such cases there is an even greater risk of their data being stolen or misused [52]. In this context, the Recommendation on the Ethics of AI, 2021, should be stringently followed, especially Recommendation 32, which mentions that data for AI systems should be "collected, used, and shared" in such a way that they ensure no violation of privacy of any individual [53].

4.2 Discriminatory Algorithms and Impact of Data Bias

As discussed earlier, AI relies on machine learning, and machine learning models are trained on data sets. This data is provided by humans. If incorrect data or insufficient data is provided, the end result will be incorrect as the AI then relies on an algorithm, i.e., a basic set of parameters and guidelines to predict the outcome. Many a time, vulnerable communities like persons with disabilities are under-represented while the machines are trained, meaning that no or lesser number of data are provided about these communities during training. Thus, when predicting outcomes often the machine/AI does not see such people as the suitable candidates. This generally has a negative impact in cases of applications by persons with disabilities for housing, social security, or even health insurance, as their chances of rejection are higher than that of persons without disabilities. This data bias can ultimately result in further exclusion and re-marginalization of persons with disabilities. This is why the ultimate decision-making should be by humans, and human responsibility and accountability need to be ensured [54].

4.3 Violation of Equality of Opportunity

As discussed under the above point of data bias, because of this reason the machine learning model may ignore a person with disability during the recruitment process if the decision of the machine is not supervised and screened manually. Further, a lot of times, AI-enabled interviews like chatbots conducting the interviews might be inaccessible for a person with a particular disability. Again, an online test for job recruitment using many AI features might be inaccessible for a person with specific disabilities as that test has not been created or checked keeping particular disabilities in mind [55]. This results in a loss of opportunity for persons with disabilities in job applications, etc., despite being equally qualified as those people without any disability.

4.4 Violation of the Access to Healthcare Services

As discussed under the data bias point, when automated AI decisions are relied upon to decide who can access healthcare services or who are eligible for health insurance, persons with disabilities become the victims as they are often denied the benefits as they are not seen by the machines as ideal candidates. Often this leads to further perpetration of discriminatory practices.

4.5 Lack of Transparency

Recommendation 37 of the Recommendation on the Ethics of AI, 2021, categorically mentions that transparency needs to be ensured at every stage of use of AI tools. However, an AI's inner workings and training data are often protected by Intellectual

Property Laws, thereby making it very difficult to figure out whether machine bias is taking place. This phenomenon is popularly called the "black box" of the AI.

4.6 Technological Divide

As earlier looked at, many of the AI applications found nowadays for the benefit of persons with disabilities allow free access (at least have free versions) to its customers. However, some of the gadgets like WatchMinders or Automated Wheelchairs can be costly. Affordability then becomes a key issue. Often lack of equal access to the latest technologies (over the globe as more developed countries have better access) or insufficient digital literacy can widen the existing gap. Sometimes, such AI-enabled gadgets are not disability friendly. Many a time, AI software not designed specifically keeping the needs of persons with disabilities in mind can lead to further inequalities.

4.7 Re-Marginalization

According to the United Nations Special Rapporteur, AI algorithms are known to discriminate against individuals with "*facial differences or asymmetry, speech impairment, different gestures, communication styles,* etc." Many a time, a person who has a limb missing have a different body shape and posture. People with vision impairments do not understand hand signals. People with hearing impairments cannot respond to verbal warnings. People with intellectual disabilities act and communicate differently. In all such cases, the apprehension of the Special Rapporteur is valid as the AI algorithms do not know how to behave and, hence, discriminate against the persons with disabilities in these scenarios, leading to instances of further marginalization perpetrated by the very AI systems that were supposed to aid them [56].

4.8 Ethical Concerns

Ethical concerns in the interplay of technology and disability can be several. Firstly, ensuring fair technology access for persons with disabilities can be a challenge. The other concern is that of the risk of over-reliance on AI. This can lead to reduction in human interaction. The need of the hour is to continuously monitor that developments in technology and AI correspond with the core values of human rights (and disability rights) and that AI does not actively or unintentionally harm the ones it was supposed to benefit. Applying the core values and principles mentioned under the Recommendation on the Ethics of AI, 2021, is a must in this regard.

5 Conclusion with Recommendations

Imagine a world where there were no innovations—no AI supported tools, gadgets, or platforms were there to ease the lives of persons with disabilities. Then, such people would forever be dependent on others (people without disabilities). When AI was not there, persons with physical disabilities would have to rely on another individual to discover the world. Today, they can be live independent lives with the help of efficient tools like Google Lookout or Automated Wheelchairs. Children having autism were never seen as ones who could learn everything by themselves. Their parents and educators deemed them "hopeless". That is no longer the case as these children with the help of AI can learn most things that other children do. With applications like eMoods, Moodkit, and Calm, people with mental issues can get the help they need at a much lesser expense, that too from the sanctity of their homes, a lot of times. However, one must be conscious of the pitfalls of over-reliance on AI and try to incorporate the use of AI in such a way that it gives the best effect to persons with disabilities.

General and Specific Recommendations:

Keeping the above-mentioned deliberations in mind, the following suggestions may be looked at:

- At the international level, discussions on AI should be included in the UN Disability Inclusion Strategy.
- The States (including India) need to ensure that AI is not allowed to negatively impact persons with disabilities, encroach upon their privacy, or result in violation of their opportunities. The States also need to encourage reasonable changes to be made to AI applications and programs to guarantee that people with disability are treated at par with others.
- There should be ethical impact assessment conducted by the Governments to figure out the risks for and concerns of persons with disabilities when it comes to use of AI tools and applications. There should be constant monitoring and risk prevention strategies need to be developed to ensure that persons with disabilities are not further marginalized because of over-reliance on AI tools and applications in day-to-day life.
- The private sector should ensure transparency whenever they use AI and should actively guarantee that the data set that is used to train their models is diverse. Companies building and using AI tools and gadgets should ensure due diligence and mechanisms to decrease oversight.
- There should be data governance strategies developed for continuous evaluation of the quality of training data for AI systems. There should be data security, and privacy of persons with disabilities should be respected, protected, and promoted.
- There should be more policy debates on the engagement of AI in different human rights issues.
- More and more AI developers who have lived experience of disability should be encouraged to build applications for persons with disabilities.

All of these would in effect help make the lives of persons with disabilities smoother. AI then will be a true ally for persons with disabilities, rather than being another hurdle for them to overcome.

References

1. Right of Persons with Disabilities Act, 2016, Section 2 (zc) and the Schedule
2. Aiploy vision application. https://www.aipoly.com
3. Be My Eyes application. https://www.bemyeyes.com
4. RogerVoice application. https://rogervoice.com/us/
5. SoundHound's ASR feature. https://www.soundhound.com/voice-ai-products/asr/#:~:text=Unparalleled%20Understanding%20and%20Accuracy,complex%20speech%20with%20greater%20precision
6. Ways in which amazon simplifies shopping and entertainment for persons with disabilities. https://www.aboutamazon.in/news/innovation/amazon-accessibility-features-for-persons-with-disabilities#:~:text=Tap%20to%20Alexa%20empowers%20individuals,of%20relying%20on%20voice%20commands
7. Voiceitt inclusive voice AI with impact. https://www.voiceitt.com/
8. The role of artificial intelligence in transforming smart highways. https://highways.today/2024/09/29/ai-smart-highways/
9. Gabriel Pires et al. (1997). Autonomous wheelchair for disabled people Proc. IEEE Int. symposium on industrial electronics (ISIE97), Guimarães, 797–801. https://www.researchgate.net/publication/3731481_Autonomous_wheelchair_for_disabled_people
10. About Wheelmap. https://news.wheelmap.org/en/about-wheelmap/
11. What is Aphasia? https://www.hopkinsmedicine.org/health/conditions-and-diseases/aphasia#:~:text=Aphasia%20is%20a%20language%20disorder,speaking%2C%20reading%2C%20or%20writing
12. Dysgraphia: what it is, symptoms, diagnosis and treatment. https://my.clevelandclinic.org/health/diseases/23294-dysgraphia
13. Developmental co-ordination disorder (dyspraxia) in children. https://www.nhs.uk/conditions/developmental-coordination-disorder-dyspraxia/#:~:text=Developmental%20co%2Dordination%20disorder%20(DCD,and%20appear%20to%20move%20clumsily
14. Pham TH et al. (2020). Autism spectrum disorder diagnostic system using HOS bispectrum with EEG signals. Int J Environ Res Public Health 17:971. https://www.mdpi.com/1660-4601/17/3/971
15. Tor HT et al. (2021). Automated detection of conduct disorder and attention deficit hyperactivity disorder using decomposition and nonlinear techniques with EEG signals. Comput. Method. Program. Biomed 200:105941. https://www.sciencedirect.com/science/article/abs/pii/S0169260721000158?via%3Dihub
16. Barua PD et al. (2022) Artificial intelligence enabled personalised assistive tools to enhance education of children with neurodevelopmental disorders—a review. Int J Environ Res Public Health 19(3):1192. https://www.mdpi.com/1660-4601/19/3/1192
17. Papakostas G et al. (2021) Social robots in special education: a systematic review. Electronics 10(12):1398. https://www.mdpi.com/2079-9292/10/12/1398
18. Abirached B et al. (2011) Improving communication skills of children with ASDs through interaction with virtual characters. In: Proceedings of the 2011 IEEE 1st international conference on serious games and applications for health, Braga, Portugal, Nov 16–18, pp 1–4. https://ieeexplore.ieee.org/document/6165464
19. Arshad NI et al. (2020) Robots as assistive technology tools to enhance cognitive abilities and foster valuable learning experiences among young children with autism spectrum disorder. IEEE Access 8:116279–116291. https://ieeexplore.ieee.org/document/9114974

20. Rouhi A et al. (2019) Emotify: emotional game for children with autism spectrum disorder based on machine learning. In: Proceedings of the 24th international conference on intelligent user interfaces: companion, Marina del Ray, CA, USA. Mar 16–20, pp 31–32. https://doi.org/10.1145/3308557.3308688
21. Purnama Y et al. (2021) Educational software as assistive technologies for children with autism spectrum disorder. Proc Comput Sci 179:6–16. https://www.sciencedirect.com/science/article/pii/S1877050920324352?via%3Dihub
22. Rakhymbayeva N et al. (2020) A long-term study of robot-assisted therapy for children with severe autism and ADHD. In: Proceedings of the companion of the 2020 ACM/IEEE international conference on human–robot interaction; Cambridge, UK, pp 401–402. https://doi.org/10.1145/3371382.3378356
23. Dyslexico Accessibility Built for Dyslexia. https://dyslexi.co/
24. Dyslexia.ai Dyslexic Literacy Learning. https://dyslexia.ai/
25. Rello LK et al. (2016) Dytective: towards detecting dyslexia across languages using an online game. In: Proceedings of the 13th international web for all conference, Montreal, QC, Canada, Apr 11–13. https://doi.org/10.1145/2899475.2899491
26. Huang X et al. (2014). A historical perspective of speech recognition. Commun ACM 57:94–103. https://doi.org/10.1145/2500887
27. Somma F et al. (2019). Artificial intelligence-powered cognitive training applications for children with attention deficit hyperactivity disorder: a brief review. Child Worldw 2:4. https://www.researchgate.net/profile/Federica-Somma-2/publication/338360930_Artificial_Intelligence-powered_cognitive_training_applications_for_children_with_attention_deficit_hyperactivity_disorder_a_brief_review/links/5e0f33f8a6fdcc2837536582/Artificial-Intelligence-powered-cognitive-training-applications-for-children-with-attention-deficit-hyperactivity-disorder-a-brief-review.pdf
28. LoPresti E (2015) Rehabilitation and prosthetic services; device review: Watchminder by Watchminder. https://www.prosthetics.va.gov/AssistiveTechnology/reviews/Device_Review_WatchMinder.asp
29. Fridin M (2014) Storytelling by a kindergarten social assistive robot: a tool for constructive learning in preschool education. Comput Educ 70:53–64. https://www.sciencedirect.com/science/article/abs/pii/S036013151300225X?via%3Dihub
30. Jiang F et al. (2017) Artificial intelligence in healthcare: past, present and future. Stroke Vasc Neurol 2(4):230–243. https://svn.bmj.com/content/2/4/230
31. Miller DD et al. (2018) Artificial intelligence in medical practice: the question to the answer? Am J Med 131(2):129–133. https://linkinghub.elsevier.com/retrieve/pii/S0002934317311178
32. Arun V et al. (2018) A boosted machine learning approach for detection of depression. In: Proc IEEE Symp Ser Comput Intell SSCI 41–7. https://ieeexplore.ieee.org/document/8628945/
33. Choi SB et al. (2018). Ten-year prediction of suicide death using cox regression and machine learning in a nationwide retrospective cohort study in South Korea. J Affect Disord 23:8–14. https://www.sciencedirect.com/science/article/abs/pii/S0165032717313150?via%3Dihub
34. Erguzel TT et al. (2016) Artificial intelligence approach to classify unipolar and bipolar depressive disorders. Neural Comput Appl 27(6):1607–1616. https://doi.org/10.1007/s00521-015-1959-z
35. Fitzpatrick KK et al. (2017). Delivering cognitive behavior therapy to young adults with symptoms of depression and anxiety using a fully automated conversational agent (Woebot): a randomized controlled trial JMIR mental health 4(2):e19. https://mental.jmir.org/2017/2/e19/
36. MoodKit Review. https://onemindpsyberguide.org/guide/apps/moodkit/
37. MoodKit vs. eMoods. https://emoodtracker.com/other-mood-trackers/moodkit-vs-emoods#:~:text=MoodKit%20is%20a%20very%20impressive,your%20mood%20and%20treatment%20simultaneously
38. Bautista J et al. (2024). Multi-media field test: cognitive-behavioral therapy at our fingertips: sanvello provides on-demand support for mental health. Cognit Behav Pract. https://www.sciencedirect.com/science/article/pii/S107772292400004X

39. Panic attacks: how meditation and mindfulness can help. https://www.calm.com/blog/panic-attack-meditation
40. Talkspace. https://www.talkspace.com/
41. BetterHelp. https://www.betterhelp.com/
42. Committee on the Rights of Persons with Disabilities (2013) Nyusti and Takács v Hungary, Communication No. 1/2010, Views Adopted by the Committee at Its Ninth Session (15–19 Apr 2013) June 21. https://juris.ohchr.org/casedetails/1986/en-US
43. *Jeeja Ghosh v Union of India* (2016) 7 SCC 761
44. Convention on the Rights of Persons with Disabilities, Article 2
45. Right of Persons with Disabilities Act, 2016, Section 2 (ze)
46. Right of Persons with Disabilities Act, 2016, Section 2 (y)
47. Convention on the Rights of Persons with Disabilities, Article 24
48. Convention on the Rights of Persons with Disabilities, Article 24 (2) (c)
49. Recommendation on the ethics of artificial intelligence, 2021, Recommendation 14
50. Recommendation on the ethics of artificial intelligence, 2021, Recommendation 13
51. Recommendation on the ethics of artificial intelligence, 2021, Recommendation 28
52. AI and privacy: the privacy concerns surrounding AI, its potential impact on personal data, ET online, April 25, 2003. https://economictimes.indiatimes.com/news/how-to/ai-and-privacy-the-privacy-concerns-surrounding-ai-its-potential-impact-on-personal-data/articleshow/99738234.cms?from=mdr
53. Recommendation on the ethics of artificial intelligence, 2021, Recommendation 32
54. Recommendation on the ethics of artificial intelligence, 2021, Recommendation 36
55. Special Rapporteur on the Rights of Persons with Disabilities (2021) Report of the special rapporteur on the rights of persons with disabilities on artificial intelligence and the rights of persons with disabilities. https://www.ohchr.org/en/calls-for-input/2021/report-special-rapporteur-rights-persons-disabilities-artificial-intelligence
56. Welker Y (2023) Generative AI holds great potential for those with disabilities—but it needs policy to shape it, AI governance summit, World Economic Forum. https://www.weforum.org/stories/2023/11/generative-ai-holds-potential-disabilities/

Mobilizing Automated Vehicles: Harmonizing the Intersection of Technological Innovations and Legal Regulations

Shrabana Chattopadhyay, Ripon Bhattacharjee, Aritra Jana, and Pubali De

Abstract There will be significant shifts in transportation with the introduction of autonomous vehicles (AVs), which will increase efficiency, safety, and environmental friendliness. But these technologies can only be used to their full potential if technological breakthroughs are seamlessly integrated with strong legal regulations. The article delves into the ways in which technological advances and legal frameworks meet, highlighting how important regulatory measures are for ensuring the secure use of AVs. In addition, the article delves into the current legal framework around AVs, drawing attention to the difficulties caused by inconsistent regulations and the necessity for flexible rules that can stay up with the fast-paced advancements in technology. The purpose of this article is to examine current policies and case studies to shed light on how to effectively integrate technology advancements with legal requirements to create conditions that are favorable to the broad use of autonomous vehicles. The results highlight the need for manufacturers, lawmakers, and the general public to work together for the sake of society's safety and well-being during the shift to autonomous transportation.

Keywords Automated vehicle · Technological innovation · Legal regulation · Infrastructure · Policy measures · Sustainability

---

S. Chattopadhyay
Department of Law (IIIJS), University of Engineering and Management, Kolkata, India

R. Bhattacharjee (✉)
National Law University, Tripura, India
e-mail: ripon.bhattacharjee69@gmail.com

P. De
Institute of Engineering and Management (IEM), Kolkata, University of Engineering and Management, Kolkata, India

A. Jana
Christ University, Bangalore, India
e-mail: aritra99jana@gmail.com

M. Chakraborty et al. (eds.), *Proceedings of Sixth International Ethical Hacking Conference*, Lecture Notes in Networks and Systems 1513,
https://doi.org/10.1007/978-981-96-8632-2_7

1 Introduction

1.1 Background on Automated Vehicles (AVs) and Their Significance in the Transportation Sector

The automotive sector is vital to the economic growth, especially in specific EU member states. This significant industry is often the target of lobbying initiatives, typically aimed at maintaining current power structures within the automobile sector. The initial wheel and axle vehicles emerged in 3500 BC, after which individuals began to domesticate animals to assist in transporting these wheeled vehicles. The concept of autonomous automobiles has existed for an extended period, with one of the earliest suggestions being from Norman Bel Geddes, an American theatrical and industrial designer and futurist. In 1940, he wrote a book titled "Magic Motorways," in which he conceptualized an automated highway system where vehicles would be governed by radio signals integrated into the roadways. This initial vision established the foundational concepts for what we today refer to as autonomous or self-driving vehicles. Tesla, the most valued automotive firm globally, announced its entry into the Indian market and registered an office in Bengaluru, Karnataka, in January 2021. The current issue is not the suitability of self-driving cars on Indian roads, but rather the preparedness of Indian laws to address the challenges posed by self-driving vehicles.

The primary aim of automated vehicles or automated driving systems (ADSs) targets to improve the driving safety. European Commission's data underscores the necessity of this objective, which shows that about 90% of road accidents in Europe each year are caused by human error. Furthermore, these incidents lead to the unfortunate demise of over 40,000 individuals and inflict injuries on more than 1.5 million others [1]. In addition to potentially enhancing driving safety, advanced driver assistance systems (ADSs) may provide numerous societal advantages, such as reducing traffic congestion, streamlining parking processes, and promoting ecologically sustainable driving practices. Furthermore, they possess the capacity to enhance the mobility of old and disabled individuals, along with other marginalized communities, thereby fostering inclusion in mobility [2].

A heightened level of safety attributable to the utilization of ADSs may be achieved only when a sufficient number of these vehicles are integrated into traffic [3]. The incorporation of autonomous vehicles (AVs) into traffic systems that also comprise traditional cars complicates liability concerns. Consequently, notwithstanding several advantages, ADSs present various legal and ethical dilemmas that require regulation at both the EU and national levels. Currently, new restrictions appear to lag behind the rapid advancement of technology. Numerous authors [4, 5] emphasize the imperative for new regulations, asserting that current legal frameworks are only partially effective.

Numerous issues require resolution, including liability in the event of incidents, privacy protection, and security threats related to system intrusions. We require policies that prioritize personal safety while without obstructing the future advancement of automated vehicles. This necessitates a uniform certification system and a defined array of safety assessments for the approval of automated cars [6]. There exist ethical, in addition to legal, inquiries: decisions that may influence human life or death. Moreover, there exist technical impediments, including artificial intelligence (hereinafter AI) accurately interpreting environmental occurrences (behavior of cyclists, pedestrians, other vehicles, animals) [7]; correctly interpreting signage (e.g., damaged or defaced signs); the dubious adequacy of road infrastructure; the questionable currency of high-definition maps; and accurately detecting the environment [8]. In the scenario of a conventional, human-operated vehicle, the pedestrian and driver establish eye contact, allowing the pedestrian to ascertain the driver's intention to stop [9, 10]. The driver may also signal a person to walk across the street [11].

ADSs are characterized as "autistic and narcissistic" [12]. A pedestrian will struggle to engage with a self-driving vehicle on an equivalent "intuitive" level. In 2018, an Uber autonomous car fatally struck a person who was traversing a roadway outside of a designated crosswalk (National Transportation Safety Board, 2018). Several potential solutions exist for these issues. For instance, Drive.ai implemented LED displays outside the car to communicate with pedestrians, indicating messages such as "proceeding now, do not cross" and "awaiting your arrival." [13]

Another inquiry is to the ability of these vehicles to respond to unforeseen road scenarios, such as the presence of ambulances or fire engines [14]. Human–machine interaction presents various issues, including misinterpretation of automation, excessive dependence on automation, and inadequate feedback regarding system states delivered by the automation [15]. Ultimately, automation can yield both beneficial and detrimental impacts on human performance. Parasuraman et al. [16] examine the following facets of human performance: mental workload, situational awareness, complacency, and skill degradation.

Legal personhood is a notion that designates a legal entity with rights and corresponding responsibilities and liabilities [17]. Foreseeability, or carelessness, is an essential element of tortious acts; nonetheless, it insufficiently addresses culpability in the context of a self-driving automobile collision.

The notion of intent in artificial intelligence, especially concerning autonomous vehicles, is intricate due to the insufficient comprehension of human brain function. The purpose is a pivotal factor in numerous liability lawsuits, and the challenge of ascertaining purpose in artificial intelligence results in an insufficient liability framework.

ADSs exert environmental and social impacts as well. The environmental ramifications of ADSs are significantly shaped by the conduct of the driver and system implementation. Conversely, the societal impact pertains to the possible job substitution of drivers professional in this field (such as truck drivers, taxi drivers, and bus drivers). Although near future hold new possibilities in sectors such as technical support for ADSs, it is questionable if they will offset the jobs that have been lost.

According to Parasuraman et al. [16], an excessive emphasis on total automation may result in an inflated perception of self-sufficiency, which can falter in non-routine situations [18]. Increased machine autonomy does not inherently negate the necessity for human competencies. Conversely, it alters the dynamics of human–machine interdependent operations, re-evaluating human competencies [18]. Notwithstanding these apprehensions, ADSs are becoming more prevalent in our culture and require a robust regulatory framework.

This article presents an overview of difficulties with autonomous vehicles, including ethical dilemmas and legal questions necessitating new regulations, such as responsibility, cybersecurity, and protection of personal data. Ultimately, this offers a relative examination of the legislations from states that have partially passed this content, including the United States, United Kingdom, Norway, Sweden, Germany, Italy, and Slovenia, along with recent advancements at the EU level.

1.2 Overview of the Transformative Potential of AVs for Safety, Efficiency, and Sustainability

SAE rules indicate that the term "autonomous" is unclear, as it fails to specify whether the system depends on external organizations for communication and/or cooperation in order to do essential tasks like data collection. While autonomy in law refers to the capacity for self-governance, ADS is generally not regarded as "selfgoverning." As per SAE International [19], the appropriate terminology is "driving automation system," which denotes a system that executes part or all of the dynamic driving task consistently. SAE proposes a more precise description of an automated driving system (ADS) as the integrated hardware and software that can consistently execute the complete dynamic driving task.

The ultimate objective of manufacturers is to create a car that operates autonomously, without human intervention. The range from no automation to full automation was designed under SAE on a six-level scale of driving automation (refer to Fig. 1).

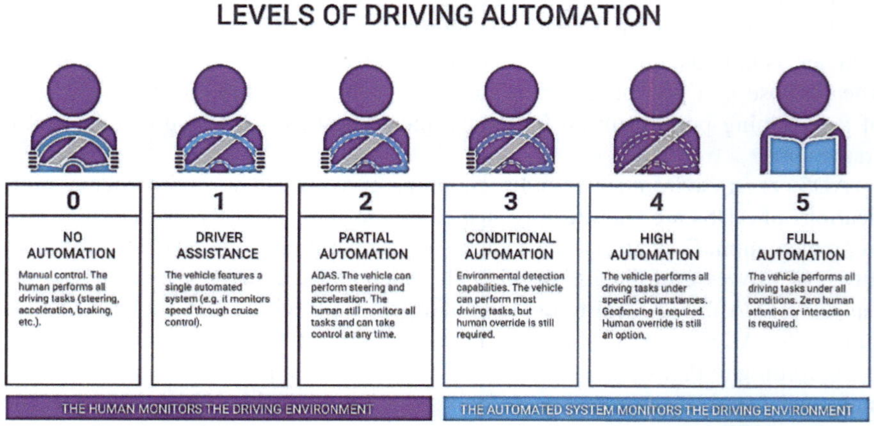

Fig. 1 6 Levels of autonomy driving (*Source* Ref. [19])

Level 0: Absence of Driving Automation: The operator is responsible for driving. Nevertheless, certain functions, including autonomous blind spot detection, emergency brake, lane change alarm aids the driver.

Level 1: Guidance to driver: This category includes features like adaptive cruise control and lane centering that provides the driver assistance.

Level 2: Substantial Driving Automation: The system is equipped with elements that provides driver assistance such as steering, brake, and support to acceleration (e.g., the car can simultaneously execute adaptation to control over cruise and lane centering).

Level 3: Limited Driving Automating tasks: The vehicle possesses capabilities to operate under specific constrained situations and will not function independently unless these conditions are met. The driver must be vigilant and assume control if necessary.

Level 4: Extreme Driving Automating tasks: The system is capable of executing all driving activities autonomously under designated conditions; automated driving functionalities will not necessitate driver intervention.

Level 5: Complete Driving Automating tasks: Vehicles execute absolute driving functions under any situations without requiring human concentration or involvement. Consequently, integrating levels into vehicles equipped with manual controls, allowing the user to drive at any time; however, only levels 4 and 5 can be applied in vehicles lacking manual controls.

The most substantial advancement occurs from level 2 to level 3, wherein the driver is permitted to divert attention from the road, and under specific situations, the system maker may bear liability for any resultant incidents.

Mercedes' Drive Pilot became the inaugural fully certified Level-3 automated driving system globally in 2022. The vehicle requires a human operator at the steering wheel to assume control if necessary for safety purposes. Furthermore, the utilization of the automated feature is typically advised solely under favorable weather or advantageous driving conditions. Level-4 vehicles are currently operational on designated routes and under particular conditions, such as shuttle buses at airports and university campuses [20]. At level four, the system is capable of executing all driving tasks autonomously under designated conditions. Level 5 (complete automation) vehicles remain unavailable on the market.

We are presently in a transitional phase from manually operated to automated driving, necessitating that drivers maintain vigilance for an extended period until full automation is achieved, which may jeopardize safety [21].

Two strategies for achieving full autonomy seem to be under consideration: progressive advancement through the SAE levels and the direct-to-level-four method (see Fig. 1). Mercedes, Tesla and BMW have a methodology which is known for their incremental technology. Conversely, start-ups such as Google's Waymo [22] adopt a direct-to-level-four strategy, concentrating on automated trials and aspiring to deliver mobility as a service [23]. Notwithstanding variances, all these cars depend on technical attributes and data processing included within the vehicle.

Alternatively, one might leverage not only the capabilities of the ADS but also the roadways and infrastructure. An instance of this type of practical test-track was conducted by Professor Sahin Albayrak, executive director of the Distributed Artificial Intelligence Lab in Berlin [20]. At present, a prevalent method is enhancing Automated Driving Systems (ADS) by incorporating oversight from a human external. The function of a human outside a self-driving car can be defined as a "remote assistant," rather than a "remote driver."

An ADS can be provided with strategic or tactical direction by a **remote assistant**. The assistant does not participate in operational driving tasks but aids in decision-making, including route planning, managing uncertainties (e.g., new patterns), and emergency response.

The direct control for executing the operations of the systems is of **remote driver**.

Nonetheless, the presence of a **remote driver** to assist drivers or passengers during technical difficulties can enhance trust in the utilization of these technologies [24]. From a technical standpoint, Autonomous Driving Systems (ADSs) require numerous technical components to operate effectively, including sensors, radars, global positioning systems, on-board cameras for navigation and environmental detection, and servo mechanisms to regulate brakes, steering, acceleration, gears (if applicable), among others (a concise overview is illustrated in Figs. 2 and 3). According to the collected data, they consistently enhance performance. Examining stages three and four, wherein the system exerts complete control over the vehicle (at least temporarily), the program makes determinations regarding life and death [25].

Dynamic Driving Task (DDT): SAE defines: "All real-time operating and tactical functions necessary for vehicle operation in on-road traffic, excluding strategic tasks such as travel planning and selecting destinations and waypoints" [19].

Minimal risk condition (MRC): According to SAE: For situations where continuing a trip is not safe, a user or ADS can bring the vehicle to a stable, stopped condition and use it as a fallback to lessen the danger of a crash [19]. To put it simply, an MRC

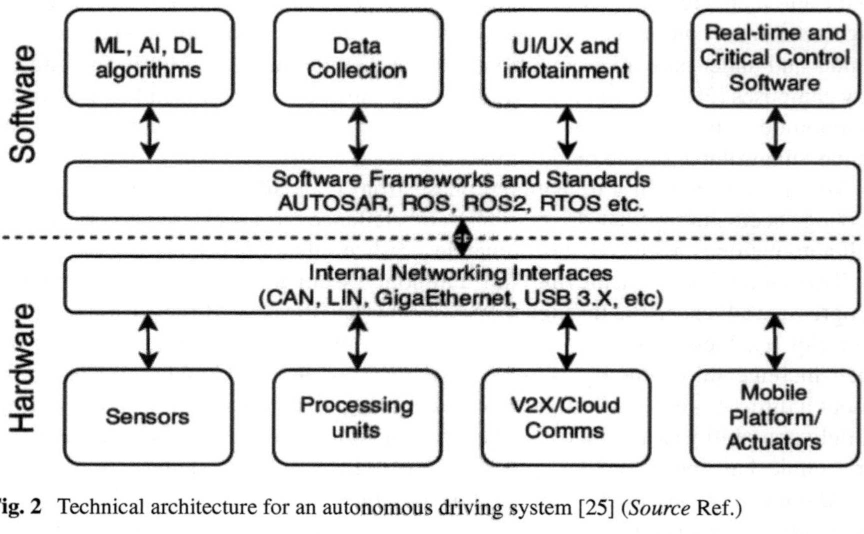

Fig. 2 Technical architecture for an autonomous driving system [25] (*Source* Ref.)

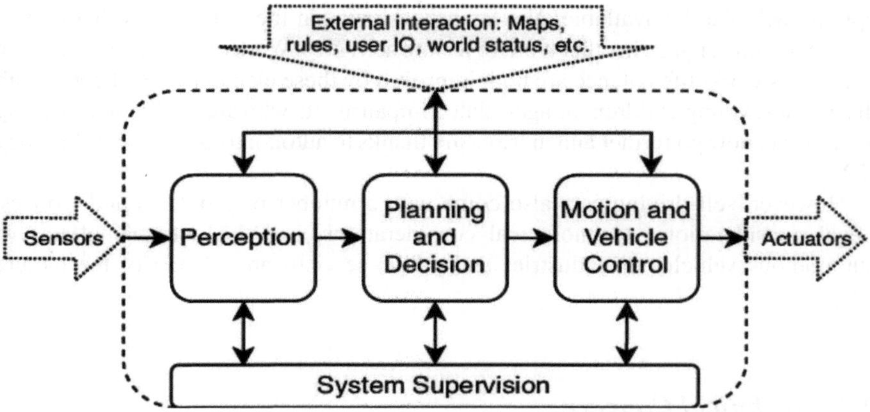

Fig. 3 Functional architecture for an autonomous driving system [25] (*Source* Ref.)

is a safe mode that the vehicle can enter when continuing the trip is no longer viable or safe. When it's safe to do so, it may entail pulling over to the side of the road to park. It is envisaged that the human driver will acquire an MRC as needed at Levels 1 and 2. Level 3 requires the DDT fallback-ready user to either perform the DDT if the vehicle is functioning or acquire an MRC when needed.

Dynamic Driving Task fallback: In cases where the ADS cannot proceed with the DDT or where a human driver fails to adequately react to a takeover request, the steps necessary to restore the automobile to a safe condition are referred to as MRC. [20, 25]

2 Issues Concerned for Automated Vehicles

The concept of automated vehicles and autonomous vehicle can be easily distinguished. The driver is rendered assistance from automated vehicle is a motor vehicle (car, bus, truck), which has elements of the driving functions as are conveyed to a computer system. The autonomous vehicle is on the other hand a fully automated vehicle, which has technologies that can execute all driving tasks without human intervention [26]. Further, we can distinguish connected vehicle, which is a motor vehicle equipped with devices to communicate with other vehicles or the infrastructure via the internet [26].

Autonomous vehicle technology's meteoric rise has several advantages. Before anything else, having software control the vehicle would make traffic a lot safer. In most cases, computers outperform humans when it comes to driving, which means fewer accidents. For example, when using software systems, your concentration won't wane because you're drunk, tired, or distracted by social media. Software has a far higher reaction time and accuracy than humans [27]. More effective transportation options will also be available. Not having to focus on the road means that human passengers can engage in other work or leisure activities. Work output and satisfaction with one's work–life balance are both improved by these elements [28]. People with disabilities, young children, or age-related impairments who are unable or unwilling to drive can now go further and more easily thanks to autonomous vehicle technology [29].

However, self-driving cars also come with a number of problems and worries. Legal considerations, technological considerations, and the potential effects of autonomous vehicles on industries and public services are all part of the picture [30].

2.1 Technical Concern

i. Autonomous technology is far from perfect, as shown by recent mishaps involving Google and Tesla automobiles [31]. In one tragic case, a white tractor-trailer was obscured by the bright sky, and the autopilot sensors of a Tesla vehicle failed to detect it, causing a deadly collision [32]. There will be a long transition phase when autonomous vehicles will likely have to coexist on the road alongside "regular" non-autonomous vehicles and other road users. Accidents will occur as a result of this interplay between software and humans. A Google driverless car collided with a bus in February 2016 because to its ignorance of the fact that longer vehicles are less likely to yield to shorter ones [33]. The commercialization of self-driving cars is also less appealing due to the high expense of the technology utilized in them [34].
ii. Many parts of our society, including the job market, transportation, and public works, will be affected by autonomous vehicles [35]. Software has the potential to displace many of the current workers in the transportation industry and allied fields [36].

The benefits of autonomous vehicles may also influence how people utilize other modes of transportation, such as aircraft, trains, and subways. It has previously been stated that people who utilize autonomous cars can do work, play, or even sleep while in the vehicle. Because of this, fewer people may want to use alternative forms of transportation, even when they provide the same benefits, like aircraft, trains, or buses. Furthermore, rumble strips, traffic lights, and guardrails can become unnecessary when cars are completely driverless. Rather, public funds will need to be allocated to upgrade infrastructure and develop safety measures specifically for autonomous vehicles [37].

2.2 *Legal Concern*

A complete overhaul of the judicial system may be necessary [31]. For instance, the Belgian Highway Code still mandates the presence of a "driver" in all vehicles, which is not compatible with the advent of autonomous automobile technology. [38] The driver's ability to handle the car and execute all required driving maneuvers is of the utmost importance. With the advent of driverless cars, this stipulation seems untenable.

Aside from concerns about privacy, another issue with self-driving cars [39] is the question of who should pay for repairs when they wreck [40]. In the future of driverless vehicles, the reliability of fault-based liability will be called into question. When an autonomous car breaks the law (for example, by running a red light), it will not be easy to identify who the "driver" is and if that person may be held responsible. The current system of fault-based liability will so give way to strict liability in traffic-related cases. Consequently, victims will need to pursue claims against other parties, like the software or vehicle manufacturer, to get their money back.

On the other hand, studies reveal that current rigorous liability laws may also be difficult to apply to the situation of autonomous vehicles. For example, it is not known whether software is a product subject to the Product Liability Directive [41] or if the damage-causing fault was present when the vehicle was released into circulation [42]. The emergence of autonomous vehicles raises concerns about potential international disputes, elevating the profile of private international law concerns.

2.3 *Ethical Concern*

A wide range of ethical considerations are brought up when automated systems make judgments that impact human life. On a more practical level, and on a more basic, philosophical level, there are others. As an example, consider the well-known "trolley problem" [43], in which an autonomous system must decide between the lives of individuals at risk of death or injury. The algorithm must choose between protecting the most lives or providing the most benefit to the most people to adhere to utilitarian principles. Nevertheless, when faced with complex circumstances like potentially life-or-death accident decisions, this method proves to be overly simplistic.

Kantian deontology is an alternative that rejects the utilitarian objectification of people. No one should have to give their life up for another. In this view, there is no room to consider which option will provide the greatest return on investment; rather, a single rule must be defined and applied in all cases.

The primary inquiry pertains to the determination of the ethical approach. Should a uniform ethical framework be integrated into all Autonomous Decision Systems (ADSs)? Contissa et al. [45] propose an ethical knob that allows the driver to determine the ethical framework that the automated driving system (ADS) should adhere to. The impartial strategy gives equal weight to passengers and third parties, the egoist approach concentrates on saving the passengers, and the altruist approach prioritizes saving others.

The selection of ethical guidelines for the ADS may not be determined by users; however, the concept of adjustable ethical settings, even if not directly modifiable by the end user, could prove beneficial for various applications, particularly in contexts with differing regulatory frameworks.

In the West, which demands high value on equality before law, the pertinent question arises: How should the algorithm in the Automated Decision System (ADS) be devised to avoid discrimination? His proposal is arbitrary. For instance, when faced with a decision to strike an younger or older individual, a random selection would yield a 50% probability of targeting either one. This approach would uphold the principle of equality before the law.

Despite the ethical discourse capturing public interest, ADS sensors, in their present configuration, encounter difficulties in accurately identifying the personal characteristics of individuals near the vehicle or occupants of other cars [45, 46].

Ethical issues must be incorporated into the conceptualization of the ADS [20, 45]. Consequently, the state ought to implement effective policies and regulations on the issue, grounded in a universally agreed agreement. Currently, laws and regulations are predominantly managed at the national level in certain states, resulting in varying restrictions. This is expected to change in the future.

3 Private International Laws and Regulations

3.1 The Vienna Convention and the Geneva Convention Provide the International Framework Upon Which Traffic Regulations are Founded

With the goal of enhancing and securing international road traffic, 95 states ratified the "1949 Geneva Convention on Road Traffic [47]."

With the goal of standardizing traffic laws and making international road travel easier, seventy-four nations have accepted the "1968 Vienna Convention on Road Traffic [48]."

It is obvious that European authorities intend to amend the Vienna Convention to incorporate AVs into road traffic, primarily to address the requirement that a "driver" maintain constant control of the vehicle. "Driver" is defined as "any person who drives a motor vehicle or other vehicle (including a bicycle), or who guides livestock, individually or in herds, or draft, pack, or saddle animals in a road" found in "Article 1, paragraph (v)". "Any motorist may, at any moment, command his vehicle or lead his animals," states "Article 8, paragraph 5."

With this goal in mind, in April 2016, numerous automakers and EU transport ministers signed the Amsterdam Declaration. The goal of this declaration is to encourage research and development in connected and automated driving as well as to foster collaboration between industries and governments in order to establish a regulatory framework.

The "Working Group on Road Traffic (WP.1)" and "the World Forum for Harmonization of Vehicle Regulations (WP.29) of the United Nations Economic Commission for Europe (UNECE)" are also engaged in discussions regarding the analysis and development of regulations for the AV [49].

By advocating for the international agreements and conventions that regulate the requirements and utilization of these unmanned technologies, the UNECE helps to facilitate automated driving technologies. UNECE also oversees the implementation of new technologies while maintaining the safety and other advancements that have been achieved in recent decades, such as interoperability and cross-border capabilities.

UNECE is affiliated with all interested parties who are interested in this work, including the automotive, computer, telecommunications, and insurance industries, governments, the "Regional Economic Integration Organization (REIO), consumer organizations, and international organizations such as the OECD/ITF/CPB, the International Telecommunication Union (ITU), and the G7 transport ministers."

Numerous international agreements establish a legal structure that ratifying countries are required to adhere to in the context of road traffic. The "Geneva Convention on Road Traffic, [47]" is dedicated to the advancement and safety of international road traffic, whereas the "Vienna Convention on Road Traffic, [48]" prioritizes road safety, among other factors.

3.2 Amendment to "Vienna Convention on Road Traffic (United Nations, 1968)"

Article 8 of the Convention 1968 was revised to permit the transfer of driving responsibilities to the system of the vehicle, provided that the technologies employed comply with the "United Nations Vehicle Regulations" or can be overridden or deactivated by the driver. The Vienna Convention provides of a vehicle to be operated by a driver, despite this amendment. Nevertheless, there is not constraint under "Vienna Convention" for the experimentation or use of automated vehicles, as demonstrated by the practices of numerous countries.

The motorist being physically present in the vehicle; rather, it mandates that they maintain control over it has not been mandated under Vienna Convention. Nevertheless, this control is not explicitly defined. This suggests that the driver can be deemed to be in control of the vehicle, given that they select the destination and route, regardless of whether they aren't present in person [50].

Consequently, the Vienna Convention underwent an additional amendment in 2020, which was drafted by the "Global Forum for Road Traffic Safety of the United Nations Economic Commission for Europe (Amendment proposal to the 1968 Convention on Road Traffic)." This amendment was implemented in July 2022.

It describes automated driving systems and establishes that the presumed competence of a driver is satisfied when the vehicle is operating an automated driving system that adheres to national technical regulations, other regulations, and pertinent international legal instruments [51].

In addition, three UN Regulations concerning autonomous vehicles were approved: Cyber-security standards for vehicle certification are mandated by "UN Regulation No. 155 on Cyber Security and Cyber Security Management Systems [52]." A monitoring and mitigating cyber hazard is the subject of this paper, which details the approval process, necessary documentation, and conformance checks for ADSs. A distinct approval mark is shown on automobiles to indicate compliance.

Regarding "Software Updates and Software Update Management Systems" (SUMS), the standards for vehicle certification are outlined in UN Regulation No. 156 [53]. The rule makes sure that cars stay safe and sound by regulating software updates, such as those that happen over-the-air (OTA).

UNECE, 2021c [54] is the "United Nations Regulation No. 157 on Automated Lane Keeping Systems," which specifies requirements for the certification of ALKS-equipped cars. Among the topics covered by the rule are system safety, fail–safe responses, the ability to engage in a minimal risk maneuver in the event that the driver refuses to regain control when asked, and the human–machine interaction. Additionally, it covers software upgrades, cyber-security, and data storage for autonomous driving.

3.3 European Union

Strict product standards prior to market entry and clearly delineated accountability for safety compliance are examples of how Europe has prioritized consumer protection. A violation of these conditions will result in sanctions [55]. Robust regulation is also required for self-driving vehicles, according to this philosophy. This regulatory task might be approached in different ways. Considerable weight is given to statutory requirements, product quality criteria, and "soft law" processes, such as resolutions and other non-binding measures. Collectively, these steps establish a suitable regulatory structure for autonomous vehicles.

The European Parliament and Council's "EU level Regulation" was approved on November 27, 2019, and it will be in effect starting on July 6, 2022. Words like "automated vehicle" and "highly automated vehicle" appear in this rule [56].

With respect to technical requirements for vehicles and software updates, the "Commission Delegated Regulation (EU) 2022/2236" of 20 June 2022 amended "Annexes I, II, IV and V to Regulation (EU) 2018/858 of the European Parliament and of the Council" and went into effect in 2022.

The Commission Implementing Regulation (EU) 2022/1426, which was created on August 5, 2022, was approved by the European Union in August of that year. In accordance with "Regulation (EU) 2019/2144," which was passed by the European Parliament and the Council, this regulation sets uniform protocols and technical requirements for the type certification of ADSs. In September 2022, this rule went into effect.

The legal framework of ADSs is categorized into two main segments: performance standards for ADSs and adherence evaluation processes for ADSs. The former specifies the critical functionalities required for an autonomous vehicle to achieve type-approval in EU territories, while the latter details the examination, auditing, and testing protocols for the ADS prior to type-approval. Among the required obligations:

- "Documentation": requisite documents for type authorization, which includes a certificate of approval from EU.
- "Monitoring": obligations for manufacturers regarding monitoring inclusive of service and reporting.
- Remote Operation: responsibilities and functions for "on-board and remote operators" of Automated Driving Systems (ADSs)."
- "Software Identification": a distinct identity provider for software related to type-approval of ADSs.
- "Vehicle Modes": dual-mode vehicles that transition between manually operated and automated driving while stationary.

The EU provides for certain safeguards for ethical and cyber-security concerns associated with AI through the "EU Artificial Intelligence Act" [30].

The paper states that the Commission is actively investigating how to incorporate the alternatives and related problems of self-driven cars into their regulations and adapt them to accommodate the new technology.

3.4 United States of America

In 2011, Nevada Assembly Bill No. 511 became the first state in the United States to pass a law allowing ADSs to be used on public roadways. According to Ilkova and Ilka (2017), the majority of US states have passed legislation or issued executive orders pertaining to autonomous vehicles. The federal government's National Highway Traffic Safety Administration has issued recommendations for both businesses and individual states on the topic of autonomous vehicles. Additionally, in 2022, new passenger safety requirements were put into place, displacing older language that assumed a vehicle will always have a human driver and conventional controls.

The National Highway Traffic Safety Administration has mandated that autonomous vehicles meet the same safety standards as their human-driven counterparts [58]. Youngsters are not to be seated in the conventional "driver's seating" position, according to regulations, since this position does not offer enough protection for youngsters in the event of an accident. But that doesn't mean the car has to stop traveling just because a kid is in it [59]. Finally, it's safe to conclude that the US has always been ahead of the curve when it comes to approving ADS testing and use [60].

Cruise and Waymo received complete authorization to offer taxi services in San Francisco in August 2023. Both businesses have received authorization to operate autonomous automobiles for commercial passenger services in San Francisco, permitting them to charge rates for rides at any hour of the day [61] Subsequent to the October 2023 crash, Cruise was compelled to halt its operation due to an unacceptable danger to public safety [62, 63]. Waymo has been offering automated taxi services in Phoenix, Arizona since 2018, with a manual operator present to assume control if necessary. In 2020, human drivers were eliminated, and the service has since operated entirely without drivers.

3.5 United Kingdom

In 2022, "the Road Vehicles (Construction and Use) (Automated Vehicles) Order" was enacted, following the adoption of the Automated and Electric Vehicles Act in 2018. In addition, groups or individuals intending to test out or pilot automated vehicle technology or services should refer to the "Code of Practice: automated vehicle trialing (revised in 2022)". Even more recent developments in UK laws include the 2020 introduction of "Automated Lane Keeping Systems (ALKS)," which enable drivers to securely transfer control to the vehicle. As per Maclean, [64], ALKS regulation was implemented in 2020. No permissions or surety bonds are needed from organizations running trials in the UK.

But to follow the rules, they need to have:

1. Someone who is ready, competent, and inclined to acquire control of the vehicle—either within or outside of the vehicle
2. A vehicle that is roadworthy, or at least partially so, and
3. Adequate insurance coverage.

Despite the status of the code as non-binding guidance, failure to comply with it may result in legal responsibility in any court actions. As it stands, the insurer is held responsible for any damages caused by an accident involving an insured vehicle that happens wherein the vehicle is operating autonomously and results in injury to the person, "(Automated and Electric Vehicles Act, 2018; UK—House of Commons, 2021)."

In the United Kingdom, autonomous vehicles can operate on roadways [65] and other public spaces [66] without a human driver present. It is the owner's responsibility to ensure that their car is insured in the event of an accident. If the insured person has removed or altered the vehicle's operating system files without authorization, or if they have neglected to install necessary software or updates, the insurance may not be held liable.

3.6 China

When it comes to developing legislation to govern autonomous vehicles, China is well ahead of the curve. The country is also one of the global leaders in this transition. Both federal and state governments have passed legislation to ensure the safety of motor vehicles, including AVs. "Regulations on the Administration of Road Testing of Autonomous Vehicles" were drafted by the Ministry of Industry and Information Technology, the Ministry of Public Security, and the Ministry of Transport at the national level, contributing to the advancement of innovation in the nation. Additionally, the investigation discovered that the primary goal of the Chinese government in implementing these regulations is to construct their future based on a sophisticated, technologically driven manufacturing paradigm.

3.7 Australia

An uneven legal framework exists across Australia's states because of the fact that several regulations pertaining to automobile usage and road safety exist on both the state and territory levels. In 2018, the Australian government issued "*Australian Road Rules (or ARRs)*" to address this issue on a national level, following a suggestion from the National Transport Commission. The paper states that the Commission is actively investigating how to incorporate the alternatives and related problems of self-driven cars into their regulations and adapt them to accommodate the new technology.

4 India's Landscape on Automatic Vehicle

India, as a nation in the midst of development, is striving diligently to align itself with contemporary technological advancements. At present, India lacks dedicated legislation to govern autonomous vehicles, particularly self-driven cars. Currently, any occurrences of accidents involving motor vehicles are governed by the "Bharatiya Nyaya Sanhita, 2023" (BNS), previously referred to as the "Indian Penal Code, 1860 (IPC)," alongside the "Motor Vehicle Act, 1988" (MV Act).

The operation of a vehicle is governed by the Motor Vehicle Act. The utilization and licensing of autonomous vehicles in India remain prohibited. In the context of liability concerning death and permanent disablement, the Motor Vehicles Act adheres to the principle of "No Fault" liability as outlined in "Sect. 140."

Here lies the underlying question of whether the principle of "No Fault Liability" dictates that the owner of the car will be held accountable. Should this be implemented in incidents involving autonomous vehicles?

"Sect. 144 of the MV Act, 1988" stipulates a compensation amount of Rs. 50,000 for fatalities and Rs. 25,000 for cases of permanent disability. In the matter of "Haji

Zakaria v. Naoshir Cama," a significant inquiry emerged regarding the potential liability of the defendant for compensation in the absence of any rash or negligent driving. The Supreme Court held that liability cannot be assigned to the vehicle owner in the absence of negligence. Thus, when we assess the incidents involving autonomous vehicles, it becomes evident that the responsibility for negligence lies with the manufacturer rather than the vehicle's owner [67].

The proposed amendment to the MV Act, 2016, remains pending, aiming to exempt specific vehicles from the Act to promote research, development, and innovation.

BNS, formerly recognized as IPC, delineates a range of legal provisions concerning rash driving, negligence resulting in death, infliction of injury, and the causing of grievous harm.

i. Regarding reckless driving, "Sect. 281 [previously Sect. 279 of IPC]"
ii. In the case of causing death through negligence, "Sect. 106(2) [previously Sect. 304A of IPC]" [68]
iii. Regarding Hurt, "Sect. 115 [previously Sect. 319 of IPC]"
iv. Concerning Grievous Hurt, "Sect. 117 [previously Sect. 322 of IPC]"

Regarding the applicability of the IPC and MV Act, the Supreme Court, in its deliberation on the case "State of Arunachal Pradesh v. Ramchandra Ravidas" in 2019 [69], articulated in its judgement that the two statutes are distinct from one another. The IPC adopts a comprehensive perspective on criminal offenses, while MV focuses on upholding road safety standards. Additionally, an individual may be held accountable under both frameworks; however, it is essential to note that neither encompasses self-driving vehicles.

India is projected to become the world's third-largest automotive market by volume by 2020; nevertheless, the advancement and integration of autonomous vehicles are anticipated to go slowly, mostly due to insufficient quality traffic data and inadequate supporting infrastructure.

The advancement of autonomous vehicles in India is being led by prominent domestic manufacturers, like Mahindra & Mahindra and Tata Motors. Government assistance encompasses policies like "the Motor Vehicles (Amendment) Bill 2016", which authorizes the testing of autonomous vehicles.

Significant attempts for autonomous vehicles in India encompass the following:

- Proposed modifications to "the Motor Vehicles (Amendment) Bill 2016" to authorize testing of autonomous vehicles.
- The Automotive Research Association of India and Cognizant Technologies are collaboratively building India's inaugural indigenous autonomous automobile prototype, which is expected to commence testing shortly.
- Tata Motors is building a framework for autonomous vehicles in India and is also working on semi-autonomous cars.
- Minda iConnect and Microsoft India have just established a strategic agreement to locally develop connected vehicle technologies and improve driving experiences.

- A research group of 30 from IIT Kharagpur initiated a start-up for an ambitious project called Auro, with the objective of replacing manually operated carts and buses with autonomous shuttles for usage at university and corporate campuses [70].

5 Challenges Faced by India

The existing legislative system regulating autos in India presents a significant barrier to the deployment of automated vehicles, necessitating a substantial legal and policy shift to facilitate this advancement.

In contrast to its international peers, the Indian Government is wary of employment losses resulting from automation. The government's position on this matter significantly contributes to global automobile companies perceiving India as an unfavorable market for autonomous vehicles. Conversely, it may be inferred that this may indeed result in the creation of additional skilled positions in IT/ITES, engineering, artificial intelligence/robotics, automotive, software development, and associated sectors in India.

A significant obstacle for autonomous vehicles in India is the inadequate highway and transportation system, which contributes to the majority of traffic accidents and fatalities in the country. Nonetheless, these hurdles have not dissuaded Indian automotive companies like Tata, Mahindra, and many tech start-ups from advancing in autonomous vehicle technology.

The advantages of the technology cannot be overlooked. Autonomous vehicles would necessitate enhanced road and transit infrastructure, urban planning, and improved network and wireless connectivity in India. They prefer to improve road safety and decrease road accidents, pollution, carbon emissions, fuel consumption, and traffic congestion, while ensuring transportation accessibility for all, including individuals with disabilities.

6 Recommendations

To prepare for the launch of autonomous vehicles and the regulations that go along with them, it is necessary to conduct a comprehensive review of the current legal framework in India that governs the automotive industry.

1. The issuance of strict regulations offering safety assurance systems and the authorization to issue licenses for autonomous vehicle trials are both necessary.
2. The new law should cover topics including how to determine fault in accident situations, whether there was a programming or design mistake or a malware or cyberattack on the vehicle's algorithm by an outside hacker.

3. Legislators must guarantee that consumers have sufficient privacy and data security protections in the new law, taking into account the various intricacies associated with technological advancement.
4. The crucial issue of how to legally safeguard intellectual property rights is raised with each new breakthrough. It is only fair that the intellectual property rules cover AVs as well.
5. The insurance landscape for motor vehicles will also undergo a metamorphosis as the legal responsibility will now reside with manufacturers, coders, or tech businesses, rather than automobile owners. There will be new and distinct dangers, including software glitches or issues with artificial intelligence that cause insurance claims to arise.

   Additionally, India must concentrate on several critical areas to advance its fully autonomous car development in comparison to other markets.
6. Increased collaboration and funding from government entities, industry associations, and automakers for start-ups focused on developing autonomous solutions.
7. Enhanced governmental policies to facilitate the development, testing, and deployment of electric and autonomous vehicles.

   Increased emphasis and state funding for roadways and traffic infrastructure are essential for the advancement of autonomous vehicles and associated systems. The establishment of domestic intellectual property involves attracting the most talented individuals and experts from advanced markets, including the US and EU, to enhance the capacity and capabilities for the development of batteries, powertrains, and hardware and software integration.
8. 10. Increased collaboration among city developers, infrastructure and solution providers, automakers, and government entities.
9. 11. There has been a rise in acquisitions and partnerships among automakers and companies engaged in the development of software, microchips, GPS, and navigation technologies.
10. 12. Formulation and execution of a policy aimed at phasing out petrol and diesel-powered vehicles to facilitate the adoption of electric and autonomous platforms.
11. 13. Establishing a network of experts from industry, academia, and government to enhance research and development capabilities.

7 Recommendations for Policymakers

Autonomous vehicle technology advances require an adaptive regulatory framework to maintain a balance between innovation and safety. It is necessary to implement technical solutions that ensure responsibility, safety, and transparency when dealing with accountability issues in autonomous vehicles.

Safety protocols can be enhanced by incorporating some sophisticated incident recording systems that include tamper-proof event data recorders and on-board cameras. Event data recorders (EVR) have recently used the most effective technical solution in the context of the advancement of AVs technology. EVR is capable of capturing vital data from sensors, cameras, and GPS which can provide valuable insights into accidental circumstances. Black-box technology aids AVs technology by providing a facility for capturing a secure, tamperproof record of AV operations and it can also help in logging software states, system messages, and sensor data.

This advancement of technology may capture vital pieces of information like vehicle speed, sensor output, system decisions, etc., before and during incidents. This information can be used to investigate any underlying causes of incidents and help to develop AI-based solutions to predict the chances of occurring accidents. To accelerate this decision-making process, artificial intelligence, and machine learning methods must be integrated with autonomous vehicles.

In this context, blockchain technology may take a vital role in maintaining the security of this enormous data that is very required for manufacturers and regulatory bodies. This technological advancement also may help to transfer these comprehensive diagnostic data to cloud-based platforms that can empower autonomous vehicles by providing the facility of real-time monitoring of vehicle performance. With this facility of real-time monitoring, possible faults can be easily detected and reported before those situations intensify into serious problems.

Moreover, real-time monitoring enables communication between adjacent automobiles and immediate infrastructure, such as traffic signals and road signs, thus helping in developing and enhancing situational awareness to avoid accidents.

The operation of AVs can be enhanced by testing through simulations and these tests can be evaluated on various data sets to justify the performance in different topographies conditions, with different environmental settings, and traffic flow patterns. This extensive analysis may help in developing efficient AV by incorporating a robust AI model. Another important aspect is cyber-security measures that embrace data encryption and secure communication, and it also has a large impact on secured data handling and storage. Adopting these different technical solutions may help AVs technology uncover its efficiency and societal impact in the future.

8 Conclusion

Legislators and technological experts need to work together in an evolving and collaborative manner to keep up with the ever-changing field of automated vehicle technology. It is also necessary to have a strong legislative framework to ensure the safe operation of these vehicles. While the European Union and individual states have taken some beginning efforts, more thorough investigation into outstanding concerns is required in any future rules.

In most cases, defining key terms is the first step in creating thorough policies and efficient regulations. Uniform standards for safety, accountability norms, data security methods, and human oversight requirements are extremely desirable, as is a standardized description and arrangement of ADS features. Establishing common language, vehicle categorization, and regulations—ideally on an EU-wide or perhaps global scale—is vital given the cross-border consequences of this issue.

The primary objective of any modern technology is the welfare and safety of the public. No breakthrough is entirely devoid of errors, and autonomous vehicles are no exception. Consequently, comprehensive testing is essential to verify their security criteria for human users, and forthcoming legislation must be rigorous about liability determinations. Furthermore, the government's concerns on loss of employment and unemployment following the deployment of autonomous vehicles in India are not entirely unfounded. Nonetheless, it is undeniable that the technology will enable the emergence of several new and diverse opportunities in sectors such as artificial intelligence, information technology, engineering, robotics, automotive, and software development in India. These industries require attention, as the future significantly relies on them.

References

1. European Commission D-G for the I.S. and M (2003) Final report of the eSafetyWorking Group on road safety. Publications Office of the European Union
2. Wu X, Cao J, Douma F (2021) The impacts of vehicle automation on transport disadvantaged people. Transp Res Interdiscip Perspect 11:100447. https://doi.org/10.1016/J.TRIP.2021.100447
3. Bertolini A, Riccaboni M (2021) Grounding the case for a European approach to the regulation of automated driving: the technology-selection effect of liability rules. Eur J Law Econ 51:243–284. https://doi.org/10.1007/S10657-020-09671-5/METRICS
4. Gurney J (2013) Sue My Car Not Me: products liability and accidents involving autonomous vehicles. U. Ill. J.L. Tech. Pol'y 247
5. Surden H, Williams M-A (2016) Technological opacity, predictability, and self-driving cars. Cardozo Law Rev 38. https://doi.org/10.2139/SSRN.2747491
6. Hansson L (2020) Regulatory governance in emerging technologies: the case of autonomous vehicles in Sweden and Norway. Res Transp Econ 83:100967. https://doi.org/10.1016/J.RETREC.2020.100967
7. National Conference of State Legislatures (2020) Autonomous vehicles | self-driving vehicles enacted legislation [WWW Document]. https://www.ncsl.org/transportation/autonomous-vehicles. Accessed 15 Aug 2024
8. Sever T, Contissa G (2024) Automated driving regulations—where are we now? Transportation Research Interdisciplinary Perspectives, Vol 24, p 101033, ISSN 2590-1982. https://doi.org/10.1016/j.trip.2024.101033
9. Taeihagh A, Lim HSM (2019) Governing autonomous vehicles: emerging responses for safety, liability, privacy, cybersecurity, and industry risks. Transp Rev 39:103–128. https://doi.org/10.1080/01441647.2018.1494640
10. Nu˜nez Velasco JP, Mun Lee Y, Uttley J, Solernou A, Farah H, van Arem B, Hagenzieker M, Merat N (2021) Will pedestrians cross the road before an automated vehicle? The effect of drivers' attentiveness and presence on pedestrians' road crossing behavior. Transp Res Interdiscip Perspect 12:100466. https://doi.org/10.1016/J.TRIP.2021.100466
11. Stange V, Steimle M, Maurer M, Vollrath M (2022) Is the automated vehicle "aware" of the pedestrian? Examining driving behavior adaptation as a cue to inform the passenger of a potential hazard. Transp Res Interdiscip Perspect 16:100701. https://doi.org/10.1016/J.TRIP.2022.100701

12. Surden H, Williams M-A (2016) Technological opacity, predictability, and self driving cars. Cardozo Law Rev 38. https://doi.org/10.2139/SSRN.2747491
13. Teffer P (2018) Society too complex for EU's self-driving cars dream [WWW Document]. Accessed 15 Jan 2023 Euobserver. https://euobserver.com/eu-political/143762
14. Cahlan S. (2018). What's big, orange and covered in LEDs? This start-up's new approach to self-driving cars. Verkkoaineisto. NBC News. https://www.nbcnews.com/mach/science/what-s-big-orange-covered-leds-start-s-new-approach-ncna897151. Päivitetty March 2018
15. Sheridan TB, Parasuraman R (2005) Human-automation interaction. Rev Hum Factors Ergon 1:89–129. https://doi.org/10.1518/155723405783703082
16. Parasuraman R, Sheridan TB, Wickens CD (2000) A model for types and levels of human interaction with automation. IEEE Trans Syst Man Cybern Syst Humans 30:286–297. https://doi.org/10.1109/3468.844354
17. Rai S, Chauhan DK, Kumawat DK (2023) Wheels of justice: analysing the legal status of autonomous vehicles in contemporary legal realm. ShodhKosh J Visual Perform Arts 4(2):843–852. https://doi.org/10.29121/shodhkosh.v4.i2.2023.1754
18. Bradshaw JM, Hoffman RR, Woods DD, Johnson M (2013) The seven deadly myths of "autonomous systems." IEEE Intell Syst 28:54–61. https://doi.org/10.1109/MIS.2013.70
19. SAE International, 2021. J3016_202104: taxonomy and definitions for terms related to driving automation systems for on-road motor vehicles
20. Altunyaldiz Z (2020) Legal aspects of "autonomous" vehicles. Council of Europe, Committee on Legal Affairs and Human Rights, pp 1–14
21. Banks VA, Eriksson A, O'Donoghue J, Stanton NA (2018) Is partially automated driving a bad idea? Observations from an on-road study. Appl Ergon 68:138–145. https://doi.org/10.1016/J.APERGO.2017.11.010
22. Bonnefon JF, Shariff A, Rahwan I (2016) The social dilemma of autonomous vehicles. Science 352:1573–1576. https://doi.org/10.1126/SCIENCE.AAF2654/SUPPL_FILE/BONNEFON.SM.PDF
23. Jeffs J (2022) Autonomous cars. Robotaxis Sensors 2042
24. Karakaya B, Bengler K (2023) Minimal risk maneuvers of automated vehicles: effects of a contact analog head-up display supporting driver decisions and actions in transition phases. Safety 9:7. https://doi.org/10.3390/SAFETY9010007
25. Zhang Y, Angell L, Bao S (2021) A fallback mechanism or a commander? A discussion about the role and skill needs of future drivers within partially automated vehicles. Transp Res Interdiscip Perspect 9:100337. https://doi.org/10.1016/J.TRIP.2021.100337
26. European Parliament (2016) Briefing. January 2016. Automated vehicles in the EU
27. Anderson JM, Kalra N, Stanley KD, Sorensen P, Samaras C, Oluwatola OA (2016) Autonomous vehicle technology—a guide for policymakers, California, RAND, pp 12–16, 59–60
28. Zohn J.R. (2015), When robots attack: How should the law handle self-driving cars that cause damages. University of Illinois Journal of Law, Technology & Policy 2015, (2), pp. 461–485, at p. 471
29. Bradshaw-Martin H, Easton C (2014) Autonomous or 'driverless' cars and disability: a legal and ethical analysis. Eur J Current Legal Issues 20(3). Retrieved 28 June 2024 from webjcli.org/article/view/344
30. De Bruyne J, Tanghe J (2018) Liability for damage caused by autonomous vehicles: a Belgian perspective. J Eur Tort Law 8(3):324–371
31. Surden H, Williams MA (2016) Technological opacity, predictability, and self-driving cars. Cardozo Law Review 38:121–181 at p 136
32. Tesla's Blog. A Tragic Loss, 30 June 2016. Retrieved 28 Sept 2024 from www.teslamotors.com/blog/tragic-loss
33. Bowles N (2024) Google self-driving car collides with bus in California, accident report says. The Guardian, 1 March 2016. Retrieved 28 Sept 2024 from www.theguardian.com/technology/2016/feb/29/google-self-driving-car-accident-california
34. Zohn JR (2015) When robots attack: How should the law handle self-driving cars that cause damages. University of Illinois Journal of Law, Technology & Policy 2015(2):461–485, at pp 469–470

35. Yeomans G (2014) *Autonomous* vehicles—handing over control: opportunities and risks for insurance, Lloyd's, March 2014, pp 1–27, p 6. Retrieved 28 Sept 2024 from www.lloyds.com/news-and-insight/risk-insight/library/technology/autonomous-vehicles
36. Clements LM, Kockelman KM (2017) Economic effects of automated vehicles, presented at the 96th 18 annual meeting of the transportation research board, January 2017, p 7. Retrieved 28 June 2024 from www.caee.utexas.edu/prof/kockelman/public_html/TRB17EconomicEffectsofAVs.pdf
37. Bloom C, Tan J, Ramjohn J, Bauer L (2017) Self-driving cars and data collection: privacy perceptions of networked autonomous vehicles. In: USENIX Association, Thirteenth Symposium on Usable Privacy and Security, 12–14 July, pp 357–375
38. Article 8.1. Koninklijk besluit van 1 december 1975 houdende algemeen reglement op de politie van het wegverkeer en van het gebruik van de openbare weg, published on 9 December 1975 (Highway Code)
39. Article 8.3. Koninklijk besluit van 1 december 1975 houdende algemeen reglement op de politie van het wegverkeer en van het gebruik van de openbare weg, published on 9 December 1975. See in this regard also the decision by the Belgian Court of Cassation, 17 January 1989 as reported in Arresten Hof van Cassatie 1988, p. 599 and in Verkeersrecht—Jurisprudentie 1989, p 181
40. Bloom C, Tan J, Ramjohn J, Bauer L (2012) Self-driving cars and data collection: privacy perceptions of networked autonomous vehicles, USENIX Association. In: Thirteenth symposium on usable privacy and security, 12–14 July, 2017, pp 357–375
41. Glancy DJ (2012) Privacy in autonomous vehicles, Santa Clara Law Rev 52(4):1171–1239
42. Article 2 defines a product as "all movables, with the exception of primary agricultural products and game, even though incorporated into another movable or into an immovable" (Council Directive 85/374/EEC of 25 July 1985 on the approximation of the laws, regulations and administrative provisions of the Member States concerning liability for defective products, *OJ* L 210)
43. According to article 7, the producer will not be held liable if he proves that, having regard to the circumstances, it is probable that the defect that caused the damage did not exist at the time when the product was put into circulation by him or that this defect came into being afterwards
44. A "trolley problem" is a series of thought experiments in ethics and psychology, involving stylized ethical dilemmas of whether to sacrifice one person to save a larger number. The typical example describes a situation where a runaway trolley with five passengers is going to crash, killing all five of them. However, a bystander can intercede and redirect the vehicle to kill just one person on a different track
45. Kadner Graziano T (2016) Cross-border traffic accidents in the EU—the potential impact of driverless cars, European parliament—directorate-general for internal policies of the union. Brussels 2016:1–66
46. Contissa G, Laukyte M, Sartor G, Schebesta H, Masutti A, Lanzi P, Marti P, Tomasello P (2013) Liability and automation: issues and challenges for socio-technical systems. J Aerosp Oper 2:79–98. https://doi.org/10.3233/AOP-130033
47. Contissa G (2017) Automation and liability: an analysis in the context of socio-technical systems. I-LEX 11(1):17–45
48. United Nations (1949) Convention on Road Traffic. Geneva
49. United Nations (1968) Convention on road traffic. Charter of the United Nations, Charter of the United Nations
50. Solum LB (1992) Legal personhood for artificial intelligences. North Carolina Law Review, 1231–1233
51. Wang J, Huang H, Li K, Li J (2021) Towards the unified principles for level 5 autonomous vehicles. Engineering 7(9):1313–1325
52. This is limited to the territory of the contracting party where the relevant national technical and legal regulations apply
53. UNECE (2021a) UN Regulation No. 155—Cyber security and cyber security management system

54. UNECE (2021b) UN Regulation No. 156—Software update and software update management system
55. UNECE (2021c) UN Regulation No. 157—Automated lane keeping systems (ALKS)
56. Engstrom DF, Ho DE, Sharkey CM, Cuéllar MF (2020) Government by algorithm: artificial intelligence in federal administrative agencies. NYU School of Law, Public Law Research Paper, pp 20–54
57. EUR-Lex—4434255—EN—EUR-Lex Type-approval requirements to ensure the general safety of vehicles and the protection of vulnerable road users, 2019. Brussels
58. Lagioia F, Sartor G (2020) AI systems under criminal law: a legal analysis and a regulatory perspective. Philos Technol 33:433–465. https://doi.org/10.1007/S13347-01900362-X/METRICS
59. Wansley MT (2023) Regulating driving automation safety. Emory LJ 73:505
60. Shepardson SUS (2022) Eliminates human controls requirement for fully automated vehicles
61. Shepardson D (2022) U.S. eliminates human controls requirement for fully automated vehicles. U.S News World Rep
62. Rubin I, Baiocchi A, Sunyoto Y, Turcanu I (2019) Traffic management and networking for autonomous vehicular highway systems. Ad Hoc Netw 83:125–148. https://doi.org/10.1016/J.ADHOC.2018.08.018
63. California DMV (2023) Statement on the immediate suspension of Cruise LLC's Deployment and Driverless Testing permits
64. California DMV (2024) Autonomous vehicle testing permit holders—California DMV [WWW Document]. Accessed 14 Jan 2024. https://www.dmv.ca.gov/portal/vehicle-industry-services/autonomous-vehicles/autonomous-vehicle-testing-permit-holders/
65. Sever T, Contissa G (2024) Automated driving regulations–where are we now? Transport Res Interdiscip Perspect 24:101033
66. Maclean R (2022) Rules on safe use of automated vehicles on GB roads—GOV.UK. London
67. Law Commission of England and Wales (2022) Legal reforms to allow safe introduction of automated vehicles announced
68. Dang R, Wang J, Li SE, Li K (2015) Coordinated adaptive cruise control system with lane-change assistance. IEEE Trans Intell Transp Syst 16(5):2373–2383
69. Prakash VWCDS (2024) Global Laws on use of AI in Autonomous Vehicles:(With Suggested Changes to India's Motor Vehicles Act for permitting Autonomous Vehicles). Notion Press
70. The Bharatiya Nyaya Sanhita (BNS) is India's new criminal code, which came into effect on July 1, 2024. It replaced the Indian Penal Code (IPC) and was passed by parliament in December 2023
71. State of Arunachal Pradesh v. Ramchandra Ravidas 2019 (10) SCC 75
72. Available at http://www.agv.iitkgp.ac.in/. Accessed 26 Oct 2024

AI-Driven Analysis of Urban Heat Islands for Sustainable City Planning

Rahul Vadisetty, Sathya Kannan, Ravi Kumar Vankayalapati, Harish Kumar Sriram, Chaitran Chakilam, and Murali Malempati

Abstract Due to the impact of urban heat islands (UHIs) on local temperature, air quality, energy consumption, and public health, the issue is severe in the frame of sustainable urban development. This study presents an investigation into artificial intelligence's (AI) potential for analyzing and mitigating the effects of UHIs, with actionable insight into sustainable city planning. AI models, such as convolutional neural networks (CNN) and recurrent neural networks (RNN), using a wide variety of data ranging from satellite images, climate conditions, population density, and land cover classification, have localized UHI hotspots, analyzed various contributing factors, and simulated the efficiency of several mitigation strategies. The most critical correlations of UHI intensity with urban features-building density, vegetation cover, and material use are shown. The AI models effectively predict temperature fluctuations and quantify the effects mitigated by green infrastructure of green roofs and urban parks. These simulations support city planners in decision-making at a prioritization level by cooling potential and informing data-driven sustainable urban policies. This study shows AI's potential in analyzing high-accuracy UHI and informing resilience strategies in urban areas. This paper describes how AI could improve the urban environment by underpinning sustainable urban planning and improving the quality of life of the urban population in the face of climatic changes. In contrast, most modern challenges related to data quality and computational demands are being addressed. The novelty here is provided through high-resolution UHI data analysis for data-driven sustainable urban policy. The results identify significant correlations of UHI intensity with features such as building density, vegetation cover, and material use. That showed the capability of AI to predict the fluctuation of temperature

R. Vadisetty (✉)
Wayne State University, Detroit, USA
e-mail: rahulvy91@gmail.com

S. Kannan · R. K. Vankayalapati · H. K. Sriram · M. Malempati
Global Payments, Alpharetta, GA, USA

C. Chakilam
Sequel Medtech, Manchester, USA

M. Chakraborty et al. (eds.), *Proceedings of Sixth International Ethical Hacking Conference*, Lecture Notes in Networks and Systems 1513,
https://doi.org/10.1007/978-981-96-8632-2_8

and to quantify the cooling effect from green infrastructure. It provides evidence-based recommendations for urban planners, focusing on AI-driven insights that will enhance urban resilience and improve the quality of life in the face of climate change.

Keywords Urban heat islands · AI-driven analysis · Sustainable city planning · Convolutional neural networks · Recurrent neural networks · Green infrastructure · Temperature forecasting · Land cover classification · Urban resilience · Climate adaptation · Environmental sustainability

1 Introduction

1.1 Overview of Urban Heat Islands (UHIs)

The UHI is a condition whereby there is a continuous development of high temperatures in an urban setting as opposed to the surroundings, which are usually rural. This temperature disparity emerged mainly because of urbanization, characterized by replacing natural features with impervious surfaces such as asphalt and concrete [1, 2].

These materials absorb and store heat internally, making cities "islands" of highly raised temperatures. Human activities, such as energy consumption by buildings, vehicles, and industrial processes, further amplify this effect. Moreover, the meager vegetation in urban cities reduces the cooling effect of nature because meager green spots translate to minimal shading and limited evapotranspiration, a phenomenon wherein plants emit moisture and cool the air. Because of the current rise in global temperature, the impacts of UHIs have become more severe: worsened air quality, high energy consumption, and public health risks expose the population to constant heat. This increases the need to understand and mitigate UHI effects, primarily in the most densely populated cities [3–5].

1.2 Importance of Town Planning for Sustainable Cities

In this context, the need for sustainable city planning assumes significance in handling the UHI challenge. Sustainable urban planning is a multidisciplinary approach that underscores the dire need to balance development with environmental responsibility, resource efficiency, and quality of life. Along with green infrastructure and energy-efficient layouts, sustainable planning reduces UHI and enhances the urban area's resilience [4, 6–8].

Examples of green infrastructure such as parks, green roofs, and tree-lined streets can mitigate the effects of UHI, given their contribution to the albedo, allowing more shade and encouraging natural cooling. Planning of this sort helps make cities all the more livable because of the improved air quality that it guarantees, the biodiversity it ensures, and the lessening consumption of energy—sought in this instance for cooling purposes. With the intensification of the UHI phenomenon, embedding these principles within urban planning lets cities adapt to climate change by protecting urban dwellers from health risks related to extreme heat [6–8].

1.3 Artificial Intelligence and Its Role in Urban Planning

Accordingly, artificial intelligence has become one of the powerful tools against such complex challenges in urban planning and analyzing and mitigating UHIs. Large-scale data analysis enabled by AI can process complex data sets, such as satellite imagery, weather patterns, land cover data, and demographic information. AI algorithms can undoubtedly detect the presence of such hidden patterns; therefore, the identification of UHI hotspots, prediction of temperature fluctuations, and examination of different urban features responsible for such a temperature rise are feasible [9, 10].

For example, deep learning convolutional neural networks use high-resolution satellite images to identify places with sparse vegetation, high construction density, or other causes responsible for UHIs. On the other hand, RNNs help in predicting temperature trends by examining time-series data. Simultaneously, AI allows planners to model mitigations for UHI, including increasing green spaces or reflecting building materials. It indicates which approaches would be most productive in cooling the urban temperatures.

1.4 Research Purpose and Objectives

The study uses AI-driven approaches to analyze UHI comprehensively and puts forward data-driven strategies for sustainable urban planning. Three primary objectives are targeted in this research: first, identification of UHI hotspots in cities through the analysis of spatial and temporal data on temperature; second, this identification will reveal the pattern of how UHI forms and persists in certain areas, hence assisting planners to focus on the most affected zones. These will be followed by assessing the contributing factors to UHI intensity: population density, building materials, and urban layout concerning vegetation cover.

Understanding the factors above will provide insight into how discrete urban features drive temperature variations. Third, the study will examine various mitigation strategies, including green infrastructure, based on their effectiveness at cooling urban areas. It uses AI to simulate these approaches and make pragmatic recommendations through sustainable city planning for how AI can contribute to the goal of resilient and climate-adaptive cities.

In light of this, the research will discuss how AI can be applied to examine and mitigate urban heat islands that support sustainable development within urban areas. By using general UHI hotspots, assessing the contributing factors, and simulating the mitigation strategies, this study proposes how a data-driven approach may be designed for urban resilience. The ability of AI to process complex datasets for UHI impact forecasting provides invaluable insights that will inform urban planners and policymakers in creating more sustainable and adaptive urban spaces. This, therefore, hopes to further existing knowledge of AI-driven UHI analysis and contribute to building greener and healthier urban environments through this research.

2 Literature Review

2.1 Summary of Current UHI Research

UHIs are studied extensively due to their significant contribution to urban climate, energy consumption, and public health. Earlier research on UHI was more related to ascertaining the causes and effects of increased temperatures within city limits than their rural counterparts. According to many studies, the intensities of UHIs are dominantly controlled by urban density, the percentage of impervious surfaces, the type of building material, vegetation cover, and anthropogenic heat contributions from vehicles and industrial activity. These studies indicated that high UHI intensities correspond with urban geometry: highly populated and industrialized areas have experienced an exceptionally high-temperature rise [3, 5–7].

The conventional methods of UHI analysis generally rely on temperature measurements from meteorological stations, satellite imagery, and field surveys. Thermal imagery and remote sensing were widely adopted for mapping surface temperature, whereas statistical models were adopted to identify UHI distribution patterns. Traditional mitigation strategies explored in the UHI studies include increasing green spaces, implementing reflective surfaces, and changing urban geometry to improve airflow. Yet, these techniques need more targeting because they do not utilize the most considerable and complex datasets [11].

2.2 AI in Environmental and Urban Studies

AI has evolved increasingly over the last few years as a vital driving tool for environmental and urban studies, using advanced data processing, analysis, and prediction methods. AI is a distinctively good fit for UHI analysis, which requires high-dimensional variables in large data sets, including important geospatial and temporal information. Applications of AI range from land cover classification, temperature prediction, and air quality monitoring to natural disaster prediction within environmental studies. Only some model types, including neural networks, decision trees, and ensemble methods, have empowered researchers to extract insights from complex data sets well beyond the analytical reach of conventional methods. In urban studies, AI is revolutionizing geospatial data processing, mapping land use, and classifying urban features with ever-increasing precision [11–13].

The CNNs can provide efficient insights into the analyses of satellite imagery, which is used in drawing a complete picture regarding vegetation, building density, and land cover—all direct factors concerning UHI analysis. Similarly, RNNs efficiently forecast temperature trends and air quality by analyzing time-series data. AI-driven models give predictive power to city planners to understand what urban development will mean in terms of environmental conditions and sustainability plans in the short term. Based on AI, urban studies will have the potential for analysis and increased predictions that can be useful for addressing challenging urban issues such as UHIs [13].

2.3 Gaps in Current Research

Although significant developments have been attained in the UHI studies, a vast gulf still prevails on the way to the complete application of AI within UHI analysis and prediction. Most traditional UHI studies are suffering in the scope of their data, which usually relies on ground-based temperature recordings or periodic satellite imagery without integrating various data sources such as socio-economic data, detailed land cover classification, or real-time weather conditions. Besides, conventional statistical models, though serving functional purposes in mere trend analyses, often fail to predict UHI dynamics under different conditions, limiting simulation into future UHI scenarios. Few currently conduct research that fully exploits the predictive power of machine learning in simulating various UHI mitigation strategies [12–15].

It follows that urban planners currently need more specific, data-driven recommendations that are tailored to particular urban contexts. Additionally, while numerous studies emphasize how crucial green infrastructure is in mitigating the UHI effect, only a few quantify the cooling effect of different types of green infrastructures, for instance, green roofs versus urban parks. Only a few comprehensive, AI-driven simulations exist, so it is impossible to effectively test the interventions at scale. Consequently, this leaves a gap in targeted evidence-based recommendations for sustainable urban planning [15].

2.4 *Need for AI-Driven UHI Analysis*

The application of AI in UHI analysis has considerable advantages compared to the more conventional approaches; this is mainly because it can analyze substantial heterogeneous datasets and manipulate them to produce high-accuracy forecasting.

AI UHI analysis will use multi-dimensional input from satellite imagery, weather data, population density gauges, and land use information [13, 15]. Whereas traditional models can offer only trend observations, AI models allow scenarios to be simulated to accurately assess how urban planning changes could affect UHI's intensity. For instance, CNNs can identify very slight patterns in land cover. At the same time, RNNs can predict temperature fluctuations informed by historical trends- a holistic approach that can help unravel the UHI behavior both temporally and across diverse urban settings. Besides, AI allows the simulation of mitigation strategies with greater efficiency [15, 16].

For instance, AI can model and predict the impact of interventions, such as increasing green infrastructure or using more reflective building materials to reduce UHI, enabling planners to focus on those that have the most significant potential cooling value. This predictive capacity allows urban planners to proactively make data-driven decisions that optimize their urban resilience and climate adaptation efforts. In all, AI-driven UHI analysis fills identified research gaps. It scales up urban planning capacity to take the UHI challenge head-on for sustainable city development with reduced environmental and public health impacts [16–18].

3 Methodology

3.1 *Data Collection and Dataset Description*

The methodology used in this analysis is based on integrating multi-sourced data such as satellite images, meteorological data, and socioeconomic variables into a strong analytical framework about UHI. The LSTs have been derived from the radiometric equations trailed from Landsat and Sentinel-2 satellite images by emissivity correction of brightness temperature values according to surface types. Satellite imagery from sensors such as Landsat and Sentinel-2 offers land surface temperature and Normalized Difference Vegetation Index (NDVI) information. Land surface temperature (LST) is retrieved by using TIR data through one standard conversion process given by:

$$LST = \frac{T_{\text{sensor}}}{1 + \left(\lambda * \frac{T_{\text{sensor}}}{\rho}\right) \ln \varepsilon}$$

where T_{sensor} is the brightness temperature derived from satellite data, λ is the wavelength of emitted radiance, ρ is the constant (1.438×10^{-2} m·K), and ε is emissivity dependent on the surface type [15, 16].

It ensures the actual acquisition of the temperature difference between urban and rural areas. The meteorological factors considered include temperature, humidity, and wind speed, derived from reliable meteorological databases that maintain consistency in time and space with satellite data. It integrates such variables through preprocessing steps, normalizing values into one scale to enhance the comparability of diverse datasets.

UHI intensity is the temperature difference between the urban center and its corresponding rural surroundings to validate the model's results for ground truth values. In the case of UHI, this quantity gives a coherent, interpretable measure of the effect across samples. Data Harmonization is performed on their spatial resolution concerning georeferencing satellite imagery to meteorological observations. These steps will provide comprehensive datasets representative of the study area.

The training of the AI models, such as CNNs and long short-term memory (LSTM) networks, is conducted by clearly specifying their parameters. Furthermore, the division of the dataset is maintained at 80:20 for the training–testing ratio, which is sufficient for validating the model while simultaneously preventing overfitting. Models trained to 50 epochs with adaptive learning rates optimized the performance in both spatial and temporal analyses. Such methodological details offer a transparent and reproducible framework for analyzing UHI dynamics [15, 16].

Thus, this conversion allows the study to estimate LST across urban and rural areas under consideration for identifying UHI hotspots.

The meteorological datasets cover historical weather variables such as temperature (TTT), humidity (H), and wind speed (W). The density of the socioeconomic data is integrated into population density (Pd), building density (Bd), and green space area (G) to help understand the contribution of urban characteristics to the effect of UHI [15, 17, 18].

3.1.1 Resolution

The resolution of the satellite imagery used in this study is rather cardinal in capturing the spatial heterogeneity of UHIs. In addition, even though the Landsat and Sentinel-2 imagery has a lot of value, the medium spatial resolution enables only a limited analysis of the fine-scale features of the urban landscape, like single buildings or small patches of greenery. For this reason, advanced image-processing techniques have been incorporated into this study to enhance the spatial resolution. Deep learning-powered super-resolution methods can reconstruct high-resolution imagery from low-resolution inputs, capturing finer urban details with no additional satellite data [15].

The incorporation of ground-based observational data complements satellite imagery through local temperature measurements. Weather stations and mobile temperature sensors dispersed over the test bed can record the data at an excellent resolution that is then interpolated and matched with the satellite-derived data sets. Such a multi-scale data integration approach allows small-scale urban features, such as superb pocket parks or building materials that retain heat, to be represented.

High-resolution imagery combined with ground-based observations produces a full-scale spatial representation of the dynamics of urban heating. The proposed method refines not only the precision of identification of hotspots for UHI but also the detailed assessment of the efficiency of mitigation strategies at the level of the neighborhood and, thus, presents the real added value for the end-users of this type of study planners.

3.2 Data Preprocessing and Feature Engineering

Data preprocessing generally involves multiple steps like normalization and feature extraction to keep each dataset compatible with machine learning models. Normalization scales each feature into a comparable range, normalizing the Min–Max:

$$x = \frac{x - x_{\min}}{x_{\max} - x_{\min}}$$

where x is the original value, and $x_{\min}$ and $x_{\max}$ are the minimum and maximum values of the feature.

This normalization will avoid the problem of variable scaling, especially when integrating spatial data and LST data, weather, and socioeconomic data. After normalization, essential features, including NDVI, LST, Pd, and Bd, are extracted to provide a multidimensional dataset that genuinely reflects the correct urban characteristics of UHI intensity [10, 11, 15].

3.3 AI Models and Techniques

Convolutional Neural Network (CNN) Model for Spatial Analysis.

The CNN model considers the satellite imagery to compute any spatial pattern in UHI distribution. Every image I of size $h \times w$ with channels c is convolved by a filter using a standard 2D convolution operation:

$$(I * f)(x, y) = \sum_{i=k}^{k} \sum_{j=-k}^{k} I(x + i, y + j) * f(i, j)$$

where (x, y) are the spatial coordinates, k is the kernel size, $I(x + i, y + j)$ is the value of a pixel at a position of $(x + i, y + j)$, while $f(i, j)$ is the filter value.

Owing to learning from high-resolution images, the CNN extracts the spatial features of building density or vegetation cover, increasing the model's capability of identifying UHI hotspots [10, 11, 18].

3.3.1 Recurrent Neural Network (RNN) for Time-Series Prediction

A type of RNN, the LSTM network, forecasts temperature trends. For a time-series input, $X = \{x_1, x_2,., x_t\}$, the LSTM will run the sequence by keeping track of a hidden state h_t and cell state c_t at any given time:

$$f_t = \sigma\left(W_f * \left[h_{t-1}, x_t\right] + b_f\right)$$

$$i_t = \sigma\left(W_i * \left[h_{t-1}, x_t\right] + b_i\right)$$

$$\tilde{c}_t = \tanh\left(W_c * \left[h_{t-1}, x_t\right] + b_c\right)$$

$$c_t = f_t * c_{t-1} + i_t * \tilde{c}_t$$

$$o_t = \sigma\left(W_o * \left[h_{t-1,} x_t\right] + b_o\right)$$

$$h_t = o_t * \tanh(c_t)$$

where i_t stands for forget gate, it is the input gate, the output gate, $\tilde{c}_t$ is the candidate of cell, h_t, and c_t are hidden and cell states.

This architecture captures the temporal dependencies in temperature data, enabling the model to forecast UHI patterns over time [14, 15].

3.3.2 Evaluation Metrics

Model performance is assessed using metrics like mean absolute error (MAE) and root mean squared error (RMSE), calculated as:

$$MAE = \frac{1}{n}\sum_{i=1}^{n}|y_i - \hat{y}_i|$$

$$RMSE = \sqrt{\frac{1}{n}\sum_{i=1}^{n}(y_i - \hat{y}_i)^2}$$

where y_i is the actual value, $\hat{y}_i$ is the predicted value, and n is the number of observations.

The lower values of MAE and RMSE mean higher estimation accuracy of the model in terms of both UHI intensity and its trends, and they will help to choose the best analysis model [15].

3.4 Simulation and Mitigation Strategy Assessment

These AI models mimic various mitigation strategies of UHI by predicting their cooling effects using historical and spatial data. For example, increasing vegetation cover by some factor can be tested by changing the NDVI model parameter and observing the resultant predictions. Correspondingly, the change in building materials, simulated by adjustment of the emissivity parameter in the LST equation, enables the study of the effects of reflective materials [12].

These simulated values are then compared against the baseline UHI values to determine the effective cooling strategies. This model will enable data-driven recommendations for urban planners to prioritize strategies that provide maximum UHI reduction while meeting the objectives of sustainable city planning.

4 Results and Analysis

This section will introduce the results of AI-powered analysis for UHI problems, such as determining the UHI hotspots, investigating contributing factors, performing selected AI models, and simulation results for different effective mitigation strategies. Serious analysis with manifold data visualization may be performed in this section, which covers the status of how UHIs manifest in the urban environment and how effective the potential interventions could be.

4.1 Identification of UHI Hotspots

Through the analysis, several UHI hotspots were pinpointed within the study area, concentrated in highly dense urban areas with very little vegetation and big impervious surfaces. The study's CNN-based analysis of satellite images provided heat maps showing the areas most affected by high surface temperatures. Commercial and industrial zones and high-density residential areas demonstrated surface temperatures of 4–5 °C higher than the ambient suburban and green areas. These findings, in turn, highlight the contribution of urban density and construction materials to temperature differences, exhibiting a sharp pattern of heat concentration in urban cores, which diminishes as the green spaces increase in the suburban periphery.

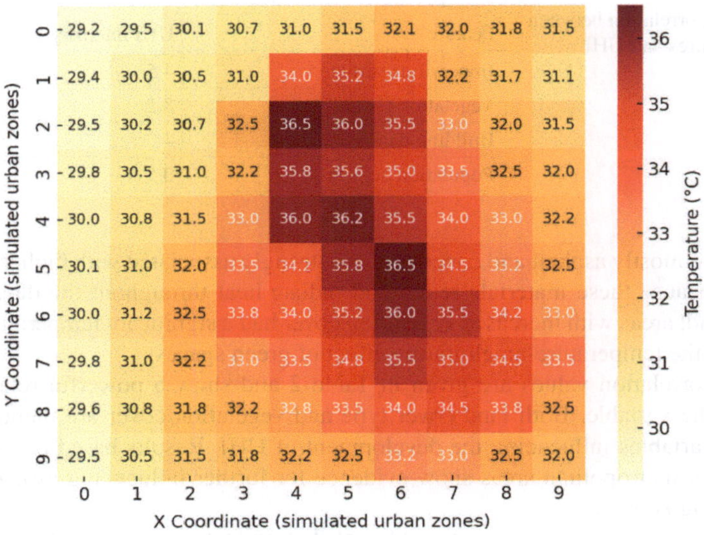

Fig. 1 Heat map of UHI hotspots within the study area

Table 1 Summary of identified UHI hotspots and temperature variations

Area type	Avg. temperature (°C)	UHI intensity
Commercial/business district	36.5	4.5
High-density residential	34.2	3.2
Industrial zone	35.8	4
Suburban residential	30.7	0.5
Green spaces	29.2	–

These variations are captured in the created heat map, as seen in Fig. 1, where the darker shades reflect higher temperatures. The calculated mean temperature and UHI intensity across various classes of urban zones are summarized in Table 1. Further supporting the above visual findings, specific areas where cooling interventions are most urgently needed can be identified.

4.2 Analysis of Contributing Factors

The study also analyzed which characteristic features in an urban setup were best linked with UHI intensity. The significant contributors to the fore were impervious surfaces, vegetation cover, building density, and population density. Impervious

Table 2 Correlation between urban features and UHI intensity

Feature	UHI intensity
Impervious surface	4.5
Vegetation cover	3.2
Building density	4
Population density	0.5

surfaces—mostly asphalt and concrete—are strongly associated with higher temperatures because these materials retain and radiate heat throughout the day. On the other hand, areas with increased vegetation cover had a significant temperature drop, showing the temperature-ameliorating effect of green spaces.

The correlation values are given in Table 2 and show a powerful relationship against the variable. Both land cover type and vegetation cover are mentioned as the top variables influencing the development of UHI. Results based on vegetated cooling of metropolitan areas show evidence for further influencing urban density concerning kept heat.

The graph shows the distribution of contributing factors in different zones of the study area, strengthening the strong association of impervious surfaces with UHI hotspots, especially in the commercial and industrial sectors.

4.3 Model Performance and Evaluation

For instance, the models used in this study have a very high accuracy in predicting UHI patterns. Among these, the CNN model for spatial analysis has been found to have an MAE of 1.25 °C and an RMSE of 1.75 °C, indicating strong performance in identifying variations in temperature on a spatial basis. The temporal analysis likewise presented an MAE of 1.08 °C and an RMSE of 1.58 °C for the LSTM model, an excellent presentation for capturing the trends in temperature. More detailed information on the importance of the urban features given by the Random Forest analysis confirmed that vegetation and building density are two crucial factors determining UHI intensity.

These performance metrics, represented in Table 3, further establish the efficacy of the AI-driven approach to capture UHI dynamics. The low error values across different models validate the dependability of CNNs in spatial data analysis and LSTMs on temporal trends, hence becoming a valuable armamentarium for sustainable urban planning [12, 19].

Table 3 Model performance metrics

Model	MAE (°C)	RMSE (°C)
CNN	1.25	1.75
LSTM	1.08	1.58
Random forest	1.42	1.89

4.4 Simulation Results for Mitigation Strategies

The simulations carried out in this study assessed various UHI mitigation strategies related to green infrastructure additions, the use of reflective materials, and changes to urban configurations. Results of this study indicated that varied increases in vegetation in UHI hotspots through the introduction of green roofs, parks, and tree-lined streets resulted in an average surface temperature reduction of 1.8 °C in areas of high density. The addition of reflective material within the industrial and commercial zones added to a moderate cooling effect, reducing temperatures by as much as 1.2 °C on average without natural shade.

The adjustments in the urban layout, such as green space distribution and building density moderation, created a limited reduction in the surface temperature. Therefore, the changes benefitted the most when combined with the green infrastructure, which suggested that holistic urban design adjustments create more effectiveness than isolated interventions. The summary of the temperature reduction for every strategy is outlined in Table 4 and is visually shown in Fig. 2.

These simulation results confirm that green infrastructure needs to be installed as the primary approach in UHI mitigation. Vegetation was found to have the most significant contribution towards cooling the urban environment. At the same time, reflective materials further support this in cases where the infrastructure is highly dense and green infrastructure is minimal.

Overall, the AI-driven analysis was quite helpful in mapping out UHI hotspots, pointing to contributing factors, and showcasing the comparative effectiveness of various mitigation strategies. The simulations suggest that green infrastructure provides the most significant cooling benefit, followed by reflective materials and strategic layout adjustments. These are solid insights for evidence-based urban planning and valuable recommendations to lessen UHI intensity and enhance urban resilience in a warming climate.

Table 4 Predicted impact of UHI mitigation strategies

Mitigation strategy	Avg. temperature reduction (°C)
Green infrastructure	1.8
Reflective materials	1.2
Urban layout adjustments	1

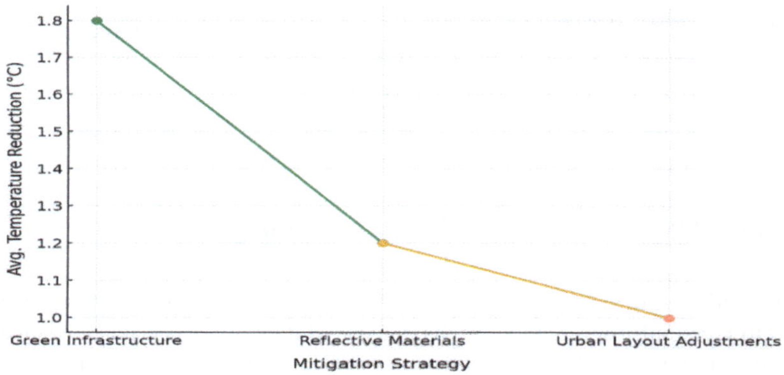

Fig. 2 Temperature reduction by mitigation strategy

5 Discussion and Conclusion

5.1 Interpretation of Results

By showing distinctive patterns in the UHI formation and giving informed knowledge into mitigation strategies, the importance of AI in identifying and analyzing UHIs increases with the findings of this study. The heat maps and model outputs suggested that UHIs are concentrated in high-density urban areas, usually characterized by commercial and industrial zones with extensive impervious surfaces. Therefore, from the analysis of correlations, specific urban features such as land cover type and vegetation density play a proper role in the distribution of temperature variations. At the same time, vegetation proved to be one of the most critical cooling factors, which means incorporating greenery into urban areas can efficiently lower the UHI intensity.

The applications of the AI models, more specifically CNN and LSTM algorithms, lead to the capability of spatial mapping of temperature patterns and their exact temporal trends [19, 20]. While the ability of the CNN model to analyze satellite imagery for UHI hotspots is an example of AI capability in handling big datasets of large spatial extent, an LSTM model captured the seasonal temperature fluctuations with good accuracy and was thus suitable for UHI trends forecasting. These findings suggest that AI can improve urban planning with precise, data-driven evidence of the drivers of UHI effects and the potential cooling consequences of various interventions.

5.2 Implications for Sustainable City Planning

The study results bring about practical implications for sustainable urban planning, as the recommendations put forward active measures a city planner or policymaker interested in mitigating UHI effects can take. First, the results strongly recommend that green infrastructure, such as green roofs, urban parks, and tree canopies, should be included within highly built-up areas. By emphasizing vegetation in high-density regions, planners can capitalize on its natural cooling benefits, depressing temperatures significantly while concurrently improving air quality and lessening energy demands for cooling [21].

Reflective materials from rooftops and pavements can enhance variability in land surface temperature in commercial and industrial areas, serving best when vegetation has very little room. These materials mitigate UHIs by reflecting solar radiation and are most effective in zones with low vegetation cover.

Finally, the study suggested that urban layout adjustments, such as reducing building density in some regions of high temperature and redistributing green spaces, would positively impact temperature regulation. Proper natural air circulation can be encouraged by redesigning city layouts, and heat concentration within the built environment can be reduced. These can serve as valuable guidelines for planners who want to design resilient and climate-adaptive city spaces.

5.3 Limitations and Future Research

Although this study represents the potential of AI in the analyses and mitigation of UHIs, several limiting factors were considered. First, data availability only has the same value in some places, such as satellite imagery and high-resolution weather data. Poorer data in some regions may lower model accuracy and affect generalizability. Long-term datasets, including data from multiple satellite providers and ground-based weather stations, could be exploited in further studies to enhance spatial and temporal coverage.

Another limitation is related to model accuracy and computational constraints. Although the CNN and LSTM outperformed well, their performance depended on fine and high-resolution data, which are computationally expensive. Future studies could further test more advanced machine learning models, such as transformer-based architectures-hybrid models, balance accuracy, and computational efficiency [20–22].

Moreover, although this study revolved around changes in vegetation, building material, and layout, future studies may include socioeconomic patterns, energy consumption, and pollution levels as theoretic variables that will provide a more comprehensive understanding of the UHI dynamics. More investigations into various mitigation measures might be effective over time, studying their longevity and their impact over time, given the continuing pressures in urban environments [23, 24].

5.4 Conclusion

In conclusion, an AI-driven analysis will be crucial to understanding and addressing the issues related to the urban heat islands. The study successfully integrates data from various sources with state-of-the-art machine learning models to understand the identification, contributing factors analysis, and simulation of different mitigation strategies concerning UHI hotspots. The results have shown that green infrastructure plays a significant role in reducing UHI intensity, with complementary solutions using reflective materials and changes in layout.

This work contributes to the cumulation of knowledge within sustainable urban planning, data-driven insights, and actionable recommendations that can be taken to mitigate UHI effects. Applying AI techniques for UHI analysis allows for an increasingly comprehensive understanding of urban temperature dynamics, thereby permitting insights into how planners design adaptation and resilient urban spaces. Equally important is how AI in urban planning could favor sustainable development by reducing the environmental impact and improving living conditions within the urban context to face the climate challenge better.

References

1. Bakir RA, Attia SAM (2023) Advancing urban health assessment through generative AI-driven indicators: GCR case study
2. Bayomi N, Fernandez JE (2023) Eyes in the sky: drones applications in the built environment under climate change challenges. Drones 7(10):637
3. Chauhan S, Jethoo AS (2023) Statistical analysis of diurnal variations in land surface temperature and the UHI effect using aqua and terra MODIS data. Remote Sens Lett 14(5):503–511
4. Chenary K, Soltani A, Sharifi A (2023) Street network patterns for mitigating urban heat islands in arid climates. Int J Digit Earth 16(1):3145–3161
5. Chui KT, Lytras MD, Visvizi A (2018) Energy sustainability in smart cities: Artificial intelligence, smart monitoring, and optimization of energy consumption. Energies 11(11):2869. https://www.mdpi.com/1996-1073/11/11/2869
6. Dyer GM, Khomenko S, Adlakha D, Anenberg S, Behnisch M, Boeing G, Nieuwenhuijsen M (2024) Exploring the nexus of urban form, transport, environment, and health in large-scale urban studies: a state-of-the-art scoping review. Environ Res 119324
7. Elhabodi TS, Yang S, Parker J, Khattak S, He BJ, Attia S (2023) A review on BIPV-induced temperature effects on urban heat islands. Urban Climate 50:101592
8. Ghorbany S, Hu M, Yao S, Wang C (2024) Towards a sustainable urban future: a comprehensive review of urban heat island research technologies and machine learning approaches. Sustainability 16(11):4609
9. Grechina A (2022) Convergence of artificial intelligence and smart city: ethical perspective: case study of Helsingborg City artificial intelligent application for temperature detection. https://www.diva-portal.org/smash/record.jsf?pid=diva2:1704324
10. Hou H, Longyang Q, Su H, Zeng R, Xu T, Wang ZH (2023) Prioritizing environmental determinants of urban heat islands: a machine learning study for major cities in China. Int J Appl Earth Obs Geoinf 122:103411

11. Liu H, He BJ, Gao S, Zhan Q, Yang C (2023) Influence of non-urban reference delineation on trend estimate of surface urban heat island intensity: a comparison of seven methods. Remote Sens Environ 296:113735
12. Makvandi M, Li W, Li Y, Wu H, Khodabakhshi Z, Xu X, Yuan PF (2024) Advancing urban resilience amid rapid urbanization: an integrated interdisciplinary approach for tomorrow's climate-adaptive smart cities—a case study of Wuhan, China. Smart Cities 7(4):2110–2130
13. Mazzeo D, Matera N, Peri G, Scaccianoce G (2023) Forecasting green roofs' potential in improving building thermal performance and mitigating urban heat island in the mediterranean area: an artificial intelligence-based approach. Appl Therm Eng 222:119879
14. Omrany H, Al-Obaidi KM (2024) Application of digital twin technology for Urban Heat Island mitigation: review and conceptual framework. Smart Sustain Built Environ
15. Pan X, Xu Y, Hong T (2024) Surrogate modeling for urban building energy simulation based on the bidirectional extended short-term memory model. J Build Perform Simul 1–19
16. Pugliese Viloria ADJ, Folini A, Carrion D, Brovelli MA (2024) Hazard susceptibility mapping with machine and deep learning: a literature review. Remote Sens 16(18):3374
17. Qiu J, Li X, Qian W (2023) Optimizing the spatial pattern of the cold island to mitigate the urban heat island effect. Ecol Ind 154:110550
18. Sameh S, Zarzoura FH, El-Mewafi M (2024) Spatiotemporal analysis of urban heat island and land use land cover changes using landsat images and CA-ANN machine learning techniques: a case study of Dakahlia government, Egypt. J Spat Sci 69(2):551–572
19. Shang K, Xu L, Liu X, Yin Z, Liu Z, Li X, Zheng W (2023) The study of the urban heat island effect in the Hangzhou metropolitan area was based on the SW-TES algorithm and image dichotomous model. SAGE Open 13(4):21582440231208852
20. Srivastava A, Sharma H (2024) AI-driven environmental monitoring using google earth engine. In IoT sensors, ML, AI, and XAI: empowering a smarter world, pp 375–385. Cham: Springer Nature Switzerland
21. Tabuenca B, Martín S, Greller W, Tillmann A, Uche-Soria M, Castro M, Rodríguez-Artacho M (2024) IoT and generative AI technologies to support urban environmental learning. In 2024 IEEE global engineering education conference (EDUCON), pp 1–4. IEEE
22. Yadav N, Rajendra K, Awasthi A, Singh C, Bhushan B (2023) Systematic exploration of heat wave impact on mortality and urban heat island: a review from 2000 to 2022. Urban Climate 51:101622
23. Yang J, Lu H, Ma Z, Lu Y, Wang W, Zhu J, Sun Y (2024) Predicting CO_2 emissions through partitioned land use change simulations considering urban hierarchy. Int J Digit Earth 17(1):2405541
24. Yin Z, Liu Z, Liu X, Zheng W, Yin L (2023) Urban heat islands and their effects on thermal comfort in the US: New York and New Jersey. Ecol Ind 154:110765

Next-Generation Sequencing of Human Genomes: Ethico-Legal Challenges with Regard to Data Security

Abhisikta Basu and Keya De Mukhopadhyay

Abstract Precision medicine is based on giving the correct treatment to the right patient at the right time. For appropriate treatment, it mainly looks for genomic information. By enabling the precise sequencing of many genes simultaneously, Next-generation sequencing (NGS) technologies are potentially transforming medical practice. Although there is still much debate on the clinical efficacy, this technology is becoming more widespread in oncology. Modern artificial intelligence (AI), especially deep learning, holds promise for improving the accuracy of variant calling, improving variant prediction, and making electronic health record (EHR) systems in NGS-based diagnostics more user-friendly. The influence of NGS on cancer patients is examined in this research based on preliminary data. The present obstacles that prevent all patients with the genetic mutation associated with a particular therapy from receiving that therapy are also highlighted. Additionally, it examines the current level of AI in NGS-based genetics, emphasizing its limitations and potential future approaches. Lastly, the research focusses on the three primary ethical domains: informed consent, privacy, and outcome review.

Keywords Next generation sequencing · Cancer patients · Privacy · Informed consent

A. Basu (✉)
IEM's International Institute of Juridical Sciences (IIIJS), University of Engineering and Management, Kolkata, India
e-mail: abhisikta.basu@uem.edu.in

K. De Mukhopadhyay
Department of Biotechnology, Institute of Engineering and Management, University of Engineering and Management, Kolkata, India
e-mail: keya.demukhopadhyay@uem.edu.in

M. Chakraborty et al. (eds.), *Proceedings of Sixth International Ethical Hacking Conference*, Lecture Notes in Networks and Systems 1513,
https://doi.org/10.1007/978-981-96-8632-2_9

1 Introduction

Next-generation sequencing (NGS) is the outcome of the evolution of sequencing technologies, which revolutionized genomic research and medicine. It all started with Sanger sequencing in the 1970s and remained the most popular one for decades due to the accuracy it had in deciphering DNA [1]. However, its high cost, lengthy time requirement, and lower scalability generated a demand for more efficient approaches. It was the dawn of the early 2000s when the revolution of NGS surfaced as a transformative technology which brought about massively parallel DNA sequencing with drastically reduced cost and turnaround time while achieving an increased throughput. There emerged new technologies such as sequencing-by-synthesis and single-molecule real-time sequencing by platforms like Illumina, Roche 454, and later from Oxford Nanopore and PacBio [2]. These developments made feasible an unprecedented type of big-scale genomics surveys, including transcriptomics, epigenomics, and whole-genome sequencing. Such liberal access that NGS provided for genomic data uncovered areas of scientific progress into personalized medicine, infectious disease, and cancer. Generally, its cost decline has made sequencing an integral part of modern genomics, from millions of dollars per genome during the Human Genome Project era to less than $1,000 today. With each advance in sequencing technology, there is greater sensitivity and accuracy, and this, in turn, will continue to open the door to more translational research and precision treatment [3].

First-generation sequencing technology was based on chemical degradation or enzymatic cleavage of DNA and RNA molecules to generate fragments that would be analyzed. Chain termination-based sequencing method developed by Fredrick Sanger completely revolutionized molecular biology because it allowed the sequencing of DNA and RNA quickly [4]. The first commercial automated sequencing machine, Applied Biosystems ABI 370, came out in 1987, which accelerated and made DNA sequencing much more accurate [5]. The techniques of second-generation sequencing allow the simultaneous sequencing of thousands to millions of DNA fragments, making whole-genome sequencing, transcriptome analysis, and targeted sequencing possible. These technologies have revolutionized genetic variation, disease research, and personalized medicine [6]. Third-generation sequencing technologies allow for long-read sequencing capabilities that enable the sequencing of larger DNA fragments than those of earlier methods. Examples include PacBio sequencing, which uses a single-molecule, real-time (SMRT) approach with fluorescently labeled nucleotides, and Oxford nanopore sequencing, based on nanopore technology [1]. These technologies have really boosted the throughput and speed of DNA sequencing, allowing for breakthroughs in genetics, disease research, and personalized medicine.

1.1 *Overview and Potential Benefits of NGS in Personalized Medicines*

The discovery of cancer drugs has been transformed by next-generation sequencing (NGS), which provides unprecedented insights into the genetic and molecular makeup of cancer. Researchers can find important genetic changes that propel the development of cancer, such as mutations, copy number variations, and structural rearrangements, by sequencing the DNA and RNA of tumour cells. To design focused medicines, this understanding is essential. As an illustration of the significant influence of NGS on precision medicine, the identification of mutations in genes such as EGFR, KRAS, and BRAF using NGS has directly resulted in the development of medications intended to block these particular molecular targets [7]. The finding of biomarkers represents one of NGS's other noteworthy contributions. Biomarkers direct the creation of personalised medicines and aid in predicting a patient's reaction to particular treatments. For example, the development and use of immunotherapies, such as checkpoint inhibitors like pembrolizumab, have depended on the identification of biomarkers like PD-L1 expression or mismatch repair deficit (dMMR). These findings enhance patient outcomes and maximise resource utilisation by lowering needless exposure to poor treatments and increasing treatment efficacy [8]. Additionally, NGS is essential for therapeutic trials because it enables accurate patient categorisation according to the genetic profile of the tumour. NGS data is used in trials like as NCI-MATCH and TAPUR to match patients with treatments that target their particular mutations, enhancing trial results, and speeding up the drug approval process. Furthermore, tracking tumour dynamics and treatment responses is now less intrusive and more successful thanks to the use of liquid biopsies, an NGS application that examines circulating tumour DNA (ctDNA) in blood. The creation of adaptive treatment regimens is further informed by the early detection of resistance and relapse made possible by liquid biopsies [9]. Lastly, by finding neoantigens—unique mutant proteins expressed by tumour cells—NGS has helped to develop immunotherapy. These discoveries have proven crucial in improving immune checkpoint inhibitors and creating customised cancer vaccinations. Furthermore, NGS's capacity to examine epigenetic changes like DNA methylation has created new opportunities for cancer treatment that target epigenetic regulators. These developments highlight how NGS can be used to address a variety of cancer biology and treatment-related issues.

2 Data Security Concerns in NGS

The extremely sensitive data generated by next-generation sequencing (NGS) for personalized treatment raise serious security, privacy, and complexity issues. Individual genomes, exomes, or transcriptomes can be analyzed comprehensively with NGS, which produces a huge amount of detailed data that may be used in precision medicine. These data, comprising single unique genetic variants, single-nucleotide polymorphisms, structural variants, and even expression patterns, would appear to be more informative concerning a person's risk profile for health, their potential susceptibility to illness, or their response to therapy. Nonetheless, since genomic sequences can be considered unique identifiers by themselves, the level of resolution of NGS also represents a privacy concern. NG is, even after anonymization, susceptible to re-identification because such datasets can sometimes contain rare genetic variation or familial patterns traceable to a particular person or his relatives [10].

These challenges become compounded by the complexity of NGS data. In each dataset, there lies billions of base pairs to be stored, analyzed, and interpreted using sophisticated computational tools and bioinformatics pipelines. Integration of genomic data with clinical and phenotypic information adds additional layers of complexity, as the data must be harmonized across disparate formats and standards. The multifaceted nature demands robust frameworks for data governance that not only preserve integrity and the potential for reproducibility but also utility in a clinical context. Moreover, risks are not only an individual's issue but may have familial and population level risks and, therefore raise concerns with regard to issues related to sharing data and consent.

Re-identification is one of the prime challenges faced by genomic data privacy. Rare variants or unique genetic markers can become the individual's identifiers. These markers become very likely to re-identify individuals, as they are linked to public databases or metadata related to geographic origin, age, or ethnicity. The example is demonstrated by genomic data studies that can identify specific individuals using publicly accessible genealogy resources [10]. Such risks require stringent measures related to privacy preservation, especially through advanced cryptographic techniques and secure computation methods and even policies enforcing data access control and informed consent.

Precision medicine relies heavily on next-generation sequencing (NGS), which provides a methodical way to comprehend and cure conditions like cancer. The above-mentioned Fig. 1 depicts the procedure of NGS Testing and the relative ethical and legal implications in various stages. It begins with sample collection, where patients' blood or tumour tissue is collected. Informed consent at this stage is essential, and it means that patients should be informed of the objectives irrespective of risks and consequences of genetic testing. It is important to respect privacy laws such as Digital personal Data Protection Act (India), General Data Protection Regulation (EU) and Health Insurance Protection and Affordability Act (USA). After collection, DNA/RNA extraction is done to separate genetic material. Proper procedures ensure that these biological samples are handled in a secure manner to prevent misuse or unauthorized access. In the sequencing phase, NGS technology is used to identify structural variations, biomarkers, or genetic mutations. In this stage, the data created must be anonymized and encrypted to protect patient privacy. In the next stage, data analysis, the sequencing data must be processed using bioinformatics tools to find actionable mutations. Only authorized people have access to genetic data, and extensive cybersecurity safeguards are to be implemented to protect patient information.

After analysis, the data are used in clinical decision-making to customize treatment regimens according to the tumor's genetic profile. This can involve selecting immunotherapies, personalized therapies, or enrolling the patient in precision-based clinical trials. Patients are made aware of how their data influence treatment, and the results are responsibly communicated to healthcare providers. Finally, monitoring and feedback employ follow-up testing or liquid biopsies to assess therapy response and identify resistance mutations. The process protects the information of patients so that ethical norms and privacy rules are followed. NGS supports legal and ethical obligations, maintains patient trust, and enhances the effectiveness of precision medicine by bringing privacy and consent issues into each step.

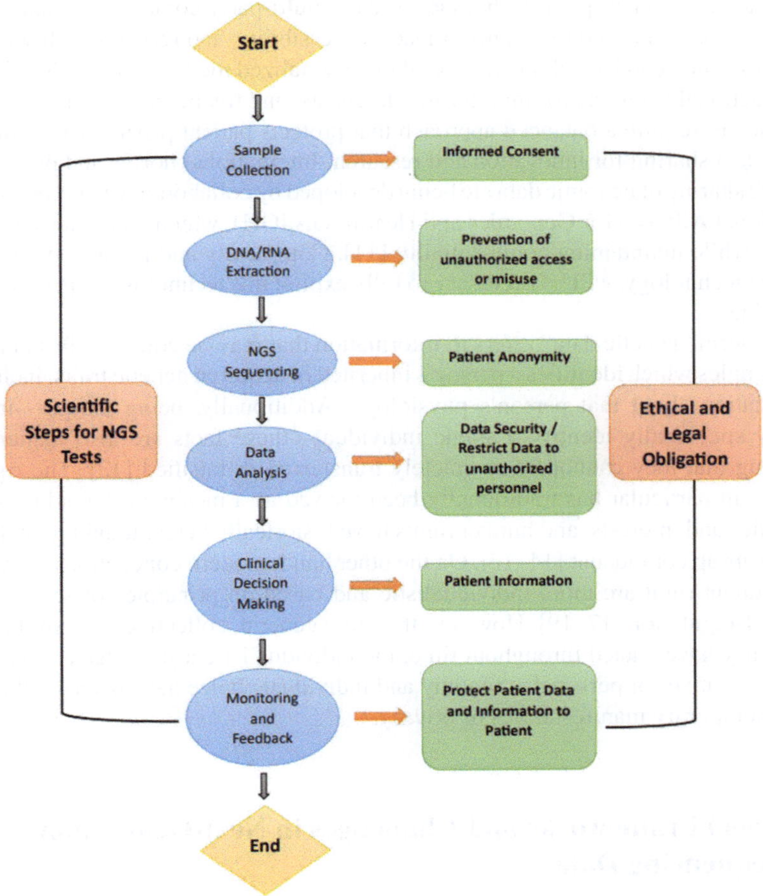

Fig. 1 Procedure of NGS testing and the relative ethical and legal implications in various stages

2.1 Threats and Challenges in Securing Human Genomic Data

The reason behind many threats and challenges the human genetic data holds lies in its sensitive and irreversible nature. Genomic database cyberattacks can lead to breaches that expose individuals to risks such as genetic discrimination or identity theft. Although an anonymized dataset can hide information, it is still susceptible to re-identification through data triangulation with publicly available resources [10]. Poor encryption and data-sharing protocols enhance vulnerability to unauthorized access in storage or transfer. The association of genetic data with EHRs complicates security, since it demands interoperable yet robust protection measures [11]. Advanced encryption techniques, secure multi-party computation, and policy frameworks are needed to balance privacy, accessibility, and research utility [12].

NGS data are indeed clinically useful in personalized medicine and aids in identifying actionable targets for monitoring, diagnosis, and treatment. Its sensitivity and complexity require a balanced approach that protects patient privacy while encouraging data sharing for innovation and research. International norms and procedures for safe sharing of genomic data are being developed by collaborative ventures such as the Global Alliance for Genomics and Health (GA4GH), which guarantee equitable access while maintaining confidentiality [11]. Complexity and privacy, twin issues in NGS technology, will be necessary to fully exploit this technology in personalized medicine.

The term "genetic data" refers to information that may be collected from biological samples which identifies a person's inherited or acquired genetic traits, including information about that person's physiology. Additionally, being unique—in that it may specifically identify a single individual—these facts are also inalienable, meaning that they cannot be completely transferred or falsified [13]. The right to privacy in particular has traditionally been viewed as a means of defending group identities and interests, and human rights have historically been thought to include a strong group component [14–16]. On the other hand, modern conceptions of privacy as a human right are more individualistic and based on principles of self-reliance and self-regulation [17–19]. However, the more general, collective parts of the right to privacy have waned throughout time, the individualistic aspects have flourished. Thus, narratives of personal autonomy and individual choice have been emphasized in contemporary manifestations of privacy.

3 Legal Frameworks and Challenges in Next-Generation Sequencing Data

As stated by Chakrabarty et al., "*Privacy is the genus, whereas data protection is a specie*" where privacy encompasses freedom of thought, autonomy, and control over one's body, reputation, etc. It is a fundamental right, but may be restricted in national security or public safety cases. Data protection is a specific aspect of privacy, encompassing how third parties handle personal information [20].

Through global collaboration in education, research, culture and communication, UNESCO seeks to advance peace and security. In furtherance to promote scientific research, UNESCO formulated two global Declarations specifically concerned with the governance of genetic technologies: the Universal Declaration on the Human Genome and Human Rights in 1997 (UDHG) and the International Declaration on Human Genetic Data, 2003 (IDHGD).

The objective of the UDHG, which is the first intergovernmental agreement specifically designed to protect human rights in the realm of genetics, is to avoid the exploitation of genetic data that is incompatible with human rights and dignity. It declares that private genetic information is confidential and must not be transmitted to third parties without the owner's consent. Articles 5(b) and 7 of the Declaration further prohibits discrimination due to genetic differences but, as observed, Article 6 allows access to a copy of genetic information of any person under Article 5, indicating rights subject to an individual's acceptance and will. But the UDHG also has a long-term objective: protecting the human genome from inappropriate modifications that would compromise the integrity and identity of future generations. Keeping in view this objective, the Declaration under Article 1 clearly states that the human genome as a part of the "heritage of humanity," a phrase that tries to highlight two things: first, that genetic data research involves the responsibility of all of humanity and that the results should benefit both present and future generations; and second, that the human genome as such is not appropriate for appropriation by any state or corporation. Even though this definition in Article 1 is merely symbolic and non-enforceable, it points to more profound conflicts between personal interpretations of rights and obligations and the collective effects that the use and abuse of particular resources can have on social groups—and even the entire world—beyond the immediate rights-holder. Nevertheless, the UDHG is peculiar for its unrelenting concern with individual rights and its inability to realize that the collective impacts on humanity it focuses attention on would not be sufficiently redressed by individualized measures alone. In certain aspects, the International Declaration of Human Genetic Data is an expansion of the UDHG by providing a thorough "framework of principles and procedures to guide States in the formulation of their legislation, policies, and other instruments in the field of bioethics," as well as more specific guidelines for the gathering, use, and storage of genetic data under Article 2(a). The Declaration under Article 7 prohibits genetic discrimination, guarantees the right to know one's genetic composition under Article 10. However, the Declaration under Articles 18 and 19 promotes global cooperation in genetic research and benefit-sharing, and further under Articles 8 and 9 contains requirements for informed consent to genetic testing and confidentiality of genetic data. Furthermore, though, the IDHGD considers genetic data rights to be individually owned, establishing limits for individual consent to handle genetic data and providing individual rights to genetic data within its language. In this regard, the UDHG and the IDHG both uphold the notion that privacy is an individual right whose larger contribution to society welfare is best realized through proceedings for infringement and personalized rights. Therefore, neither text provides a way to acknowledge or even reconcile the privacy consequences of genetic data, or to bring together traditional ideas of privacy as a form of community protection with contemporary notions of private rights.

The Oviedo Convention, commonly known as the Council of Europe Convention on Human Rights and Biomedicine, covers genetic data and human rights. Similar to the IDHGD, the Convention also prohibits the use of predictive genetic tests, requires informed consent for genetic testing, and protects the right to secrecy and information about one's genetic status. However, again, the Convention makes individual agreement the primary mechanism for regulating genetic data and explicitly uses individualistic language to grant rights and safeguards to individuals on an individual basis. The preamble of the Oviedo Convention declares that advances in biology and medicine should be used for the good of present and future generations, and the Explanatory Report to the Convention acknowledges that genetic developments raise new issues for both individuals and the human species. Nonetheless, Article 1 expressly states that the duties of Convention parties are due to individuals and to uphold the "dignity and identity of all human beings." The language of Article 2 states "[t]he interests and welfare of the human being shall prevail over the sole interest of society or science." which strengthens this overtly individualistic goal and reduces the possibility of interpreting the Convention in a way that is considerate of a more expansive definition of privacy. Unlike the UN Declarations, the Convention's preamble expressly rejects the notion that the right to privacy is more than just an openly individualistic right, as opposed to merely failing to recognize or acknowledge it. Article 2 seems quite an anomaly considering the ECHR's history of conception of privacy as always specifically because of concerns about the kinds of harm to groups that diminutions of, or omissions of, privacy rights could lead to, even though it is probably intended to prohibit the placing of corporate, or other, interests over individual interests.

The individualistic language of the ECHR and Convention 108 (also known as Convention for the Protection of Individuals with Regard to the Automatic Processing of Individual Data), the second principal privacy instrument of the Council of Europe, are consistent with Article 2 notwithstanding this discrepancy. Genetic information is particularly mentioned as a "special category" of personal data in Article 6 of Convention 108. It's interesting that the Explanatory Report to the Convention goes into detail about the importance of genetic data in relation to third parties, pointing out that processing one person's genetic data can provide information about that person's health or filiation as well as that of third parties. Thus, Convention 108 recognizes the fact that the nature of genetic data contradicts merely individualistic conceptions of privacy, as claimed by the Article 29 Working Party. However, Convention 108 thus references the wider potential damages that genetic data represent to people other than the immediately consenting individual, just as it does in the formulation of Article 1 of the UDHG, while also failing to provide a way to remedy or alleviate these group privacy consequences. Such declarations provide privacy rights to individuals about their genetic data, which are unyielding in upholding the belief that privacy is an individualistic right whose assurance is based first and foremost on the exercise of individual consent, even though these texts are aware, to varying degrees, of the collective promise—and perils—of genetic data.

Human genetic information is a very sensitive and distinct type of personal data that presents unique challenges for legal frameworks such as the general data protection regulation (GDPR) of the European Union, the Health Insurance Portability and Accountability Act (HIPAA) of the United States, and the Information Technology Act and Digital Personal Data Protection (DPDP) Act of India.

3.1 Human Genetic Data Protection Under Various Jurisdictions

The GDPR, which acknowledges the sensitivity and potential for abuse of genetic data, provides a strong foundation for its regulation. Genetic, biometric, and health data, under Articles 4(13), 4(14) and 4(15) are considered special categories of personal data and are accordingly afforded additional special protection. Save for explicit consent or specific circumstances such as to the purpose of scientific research, or for compelling public interest under Article 9(1) processing genetic data is prohibited. The GDPR's basic principles, listed in Article 5, also include data minimization and storage limitation, which stipulate that genetic data should be collected only for specified purposes and kept for as little time as necessary. This approach aims to find a balance between the privacy protections and the utility of genetic data. Article 6(1) states the legal grounds for the processing of genetic data, like the requirement for explicit consent or the fulfillment of specific legislative provisions. However, Article 22 affords further protection as it guards individuals against decisions that are made exclusively by automated processing, including genetic data profiling, which could be detrimental to a few. Article 89 allows exemptions from several GDPR provisions for scientific research if measures such as pseudonymization are taken to ensure data security and privacy. In addition, Recital 33 allows broad consent for scientific research in cases where specific objectives cannot be defined at the outset, provided that transparency is maintained to inform data subjects about the use of their data.

In the United States, genetic data are somewhat protected by the Health Insurance Portability and Accountability Act (HIPAA), although its reach and comprehensiveness are significantly constrained. Genetic information is considered protected health information (PHI) under 45 CFR §160.103, but only if it is owned by covered organizations, such as healthcare providers, health plans, or their business partners. This narrow scope leaves important data outside of HIPAA's regulatory framework by excluding genetic data gathered by direct-to-consumer (DTC) genetic testing businesses. Although 45 CFR §164.502(d) of HIPAA allows the use of de-identified data for study, the inherent identifiability of genetic information raises worries about potential re-identification, especially with the development of new technologies. As stated in 45 CFR §164.506, HIPAA's provisions permit the sharing of genetic information for purposes of treatment, payment, and healthcare operations without the need for express patient agreement. Although particular consent is not required for such applications, this method may jeopardize patient privacy while facilitating healthcare services. An additional protection is provided by 45 CFR §164.508, which requires explicit authorization for non-healthcare uses like study. Nevertheless, neither dynamic consent nor the changing nature of the use of genetic data in research and innovation are covered by this provision. The Genetic Information Non-discrimination Act (GINA) of 2008, a supplement to HIPAA, prohibits genetic discrimination in employment and health insurance, but does not cover life or long-term care insurance. Amendments are needed to expand coverage and strengthen genetic information safeguards.

Digital technology emphasizes data preservation for sensitive information, rooted in Article 21 of the Constitution, ensuring citizens' right to privacy and data protection. Although Indian legislation, such as the Digital Personal Data Protection (DPDP) Act of 2023 and the Information Technology (IT) Act of 2000, offer a fundamental framework for safeguarding personal data, they do not contain any particular provisions that address the special traits of genetic data. The IT Act's Section 43A requires the protection of sensitive personal data, which may include genetic information if specifically disclosed. Similar to this, the IT (Reasonable Security Practices and Procedures and Sensitive Personal Data or Information) Rules, 2011 mandate consent before collecting and utilizing sensitive personal data. Although it isn't stated specifically, "health conditions" may include genetic data. Contemporary data protection principles are introduced by the DPDP Act, 2023, although the complexity of genetic data is not sufficiently addressed. Genetic data are not specifically classified, which leaves its regulation unclear, even while Section 2(11) provides a broad definition of personal data and Section (12) recognizes sensitive personal data, such as health data. Section 17 regulates cross-border data transfers, permitting them only in cases where consent is given and equal protections are available outside. Section 7 requires express consent before processing personal data. Although these clauses provide some control, they don't address issues like re-identification, dynamic permission for continuing study, or the long-term worth of genetic data in scientific advancement. The principles of data minimization and purpose limitation enforced by Section 9 may be incompatible with the use of genetic data in research that necessitates long-term storage and a variety of applications. Furthermore, the DPDP Act and the IT Act do not specifically address the risks posed by new technologies such as artificial intelligence, which can increase ethical issues and re-identification threats. These discrepancies show how particular legal rules are required to clearly classify genetic data, put protections against re-identification in place, and take into account the particular needs of genetic research and AI-driven analysis.

3.2 Challenges in Cross Border Data Transfers

The question of whether regulatory approval is necessary for the transfer of genetic data from India arises from the country's significant development in medical research.

India's health laws, though not silent, do not fully outline the procedure needed for the transfer of genomic data, and the country's data protection framework is silent as well. Recently, it appears that India's data regulations largely consider data protection from the perspective of information technology and services made possible by it. In any case, the comprehensive law encompasses genomic data. According to Rule 3 of the Information Technology (Reasonable Security Practices and Procedures and Sensitive Personal Data or Information) Rules, 2011 (SPDI Rules), electronic data and information about people that pertains to, among other things, biometric information, physical and mental health issues, and medical records and history are referred to as sensitive personal data or information, or "SPDI." Therefore, electronic genomic data would be considered SPDI, and organizations handling such data would be subject to the SPDI Rules. Rule 7 SPDI may only be transferred if the data subject has given permission or if the transfer is required to fulfil a legitimate agreement between the data subject and the transferring organization. Therefore, Rule 5(1) states that no further regulatory permits are needed as long as the data subject gives their consent.

Additional limitations are enforced under the Digital Personal Data Protection Act of 2023. Section 16 states that sensitive personal information, such as genetic and health data, may be sent outside of India as long as the data subjects have given their express consent and a copy of the information is kept on file in India. Additionally, section 17 of the Act states that the transfer must be made to a nation that the Indian government has determined offers sufficient levels of data security, or it must be made in accordance with a contract or plan authorized by the data protection authority.

The National Ethical Guidelines for Biomedical and Health Research involving Human Participants, 2017 (the "Biomedical Guidelines") of the Indian Council of Medical Research (the "ICMR") govern the testing of biological material and data derived from it. Only with the explicit written informed consent of human subjects may any research involving human participation be conducted. Three elements are involved in this type of consent: giving participants enough information, making sure they can understand it, and confirming that they are competent to provide their consent [21]. The National Guidelines for Gene Therapy Product Development and Clinical Trials, 2019 were released by the ICMR in the specification to research and development of gene therapy products, in addition to the Biomedical guidelines, which offer general guidelines to be followed for conducting various types of research and clinical trials involving human participants. All organizations conducting these kinds of clinical trials must apply for approval from the Review Committee on Gene Manipulation and register with the Clinical Trials Registry of India [22]. Additionally, before a proposal is submitted to the Central Drug Standard Control Organization for product approval, the Gene Therapy Advisory and Evaluation Committee is intended to offer suggestions and examine the clinical trial application for such products. As stated in the Biomedical rules, the rules permit cross-border transfers of data under comparable circumstances. The approval of ethical and biosafety committees, which must be set up by the Indian institutions or companies participating in the transfer, is required before they permit cross-border data transmission in line with the conditions of a material transfer agreement.

When question arises regarding cross border data transfer in the "developed countries," as of December 2024, the US and the EU do not have a bilateral agreement that is explicitly focused on genetic data sharing. The wider sharing of personal data, including genetic data, is made possible by a number of frameworks and procedures. Once governing transatlantic data exchanges for business reasons, the now-defunct EU–US Privacy Shield permitted U.S businesses to obtain personal data from EU firms in accordance with EU privacy regulations. However, because to worries about U.S. surveillance tactics and inadequate security for the data of EU persons, the European Court of Justice declared the privacy shield to be unconstitutional in 2020. Organizations have resorted to Standard Contractual Clauses (SCCs) as a legal means of moving genetic information and other personal data from the EU to the US in the absence of the Privacy Shield. SCCs are contracts that guarantee data protection standards are respected during international transfers and have been certified by the European Commission. Furthermore, an agreement in principle on a new Transatlantic Data Privacy Framework was announced by the US and the EU in March 2022. With an emphasis on strengthening privacy and civil rights protections in U.S surveillance techniques, this framework seeks to address concerns expressed by the European Court of Justice. As of December 2024, the framework is still being developed, despite its promise.

The exchange of genetic information between the US and the EU is further supported via research partnerships and institutional agreements. In order to ensure the ethical and lawful management of sensitive data, many agreements are designed to conform to GDPR and U.S. requirements. The National Institutes of Health (NIH), for instance, encourages the sharing of genomic data to further biomedical discoveries while maintaining participant privacy. In a similar vein, the EU's "1 + Million Genomes" project seeks to improve illness prevention, diagnosis, and treatment by connecting genomic datasets across member states. This project establishes a standard for global genomic data-sharing partnerships, despite being mainly within the EU.

There are difficult legal and ethical challenges when genetic data is shared between the US and the EU. Ensuring that genetic data are processed and transferred legally requires adherence to federal and state legislation in the US as well as data protection requirements such as the GDPR. Given the sensitivity of genetic information, getting people's informed consent is crucial. Strong security measures must also be put in place to shield data from breaches and illegal access, promoting confidence and fulfilling legal requirements.

Table 1 highlights the comparative analysis of the legal framework relating to genetic data protection in the jurisdiction of India, US and EU. The protection of human genetic data also raises ethical issues due to its high impact on biomedical researches.

Table 1 Comparative Analysis of the Legal Frameworks in India USA and EU

Aspect	India	USA	EU
Scope	General data protection laws apply	Health focused	Broad, cross-sectoral
Consent	Sector specific	Authorization required	Explicit, detailed consent required
Genetic data definition	Not clearly defined	PHI when identifiable	Special category data
Enforcement	Fragmented	Office for civil rights	Data protection authorities
Cross border transfers	Limited restriction	No specific rules	Stringent safeguards

4 Ethical Implication of Data Security in NGS

The utilization of human genetic data in biomedical research investigations has increased significantly with the development of genome sequencing technologies. Improvements in biological science, informatics, and bioethics are being assisted by the availability of such data and the increased ability to process it. Aggregate human DNA data, particularly the allele frequencies of hundreds to millions of SNPs across human populations with specific diseases or medical problems, can be conveniently accessed by researchers. However, privacy and confidentiality issues must be addressed. Though there are some distinctions that should be highlighted, data privacy and confidentiality are both important. Data confidentiality is concerned with protecting the data from unwanted access and maintaining the data's integrity while it is being sent or stored. Data privacy pertains to the proper utilization of data. In other words, data should be utilized for appropriate purposes without prejudicing the patient. Undoubtedly, data confidentiality and data privacy are related. Generally speaking, a proper data confidentiality protection policy is necessary for good data privacy. Data privacy, however, may not always be ensured by data confidentiality measures.

For effective data sharing model to work smoothly, certain principles are to be followed. Figure 2 highlights four responsible data sharing principles. Information shared must be free of any personally identifiable information and must have been gathered with informed consent, which includes consent to share information following sufficient anonymization or de-identification. It is forbidden to try to re-identify after anonymization or de-identification without a court order. Furthermore, caution must be exercised to prevent the data resource from being exploited to exclude any ethnic, religious, geographic, or other populations. The data submitter must first get the appropriate ethical authorization. The data must meet strict requirements for quality, objectivity, and verifiability. Ensuring the high quality and validity of given data is the responsibility of the data submitter. Storage needs to be done in a way that facilitates long-term maintenance, searchability, and convenience of access while safeguarding privacy and confidentiality. To prevent a breach of data trust, the storage

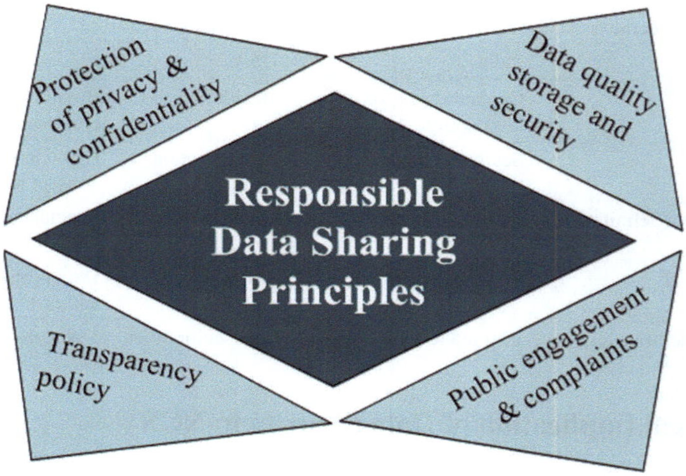

Fig. 2 Principles of ethical data sharing

and access structure must have the proper security features. There may be built-in features that allow tracking the chain of data access. The length of time that data is stored and made accessible must also be covered by the storage and security policy. In order to enhance data quality, access, integrity, and interoperability, mechanisms for gathering input from resource users must be established. Data-sharing policies should be transparent and accessible to the public. Citizens should be involved in the development of data-sharing policies, resulting in improved policies and a formal mechanism for handling complaints of data misuse.

4.1 Navigating Privacy Challenges in Genomic Data: Balancing Individual Rights and Societal Benefits

Genetic data are inherently relational; it tells one about other family members who share genetic characteristics besides the individual. Because genetic information can inadvertently disclose private information about a person's relatives, this relationship aggravates privacy concerns. The court's acknowledgment in *Van der Velden and Gaughran* of the need to protect data that impinge on third-party privacy through associative relationships underscores this. The European Court of Justice has declared Directive 2006/24 on data retention for electronic communications invalid on grounds of disproportionate interference with privacy rights In *Digital Rights Ireland* (C-594/12). This is very important since retained data would allow determining the identity of persons or third parties that communicated in the case and their connections. *Scarlet* v. *SABAM* (C-293/12) and *Breyer* v. *Germany* (C-594/12) settled the principle that dynamic IP addresses could also constitute personal data if tied to further information which makes a person identifiable. The European Court of Human Rights recognized, in *Benedik* v. *Slovenia*, (Application No 63257/14) the importance of strict oversight and proportionality in data processing. In this regard, as the sharing of genetic data is increased for research and personalized medicine, regulatory frameworks have to integrate these precedents in order to balance innovation with privacy.

When the privacy issues related to genetic data are compared with the CJEU and ECtHR judgments, it can be determined that two things come into light. Firstly, the European Court of Human Rights has concluded that collection of genetic data is intrinsically an invasion of privacy because such individually sensitive traits as well as family ties become traceable and mappable. Although the idea is similar, although in the context of genetic information, the CJEU still appears to be following the line of thinking that degrading the private sphere of one person has the effect of revealing not only parts of the personal data of others but also its link to the concerned rights holder. This recognition can easily be taken as a reference to the socio-historic foundations of human rights after World War II and the special circumstances that the privacy guarantees set forth in the Convention (and more subsequently the Charter) support. However, the interpretation of the ECtHR may also just be a realistic consideration of the nature of genetic data and the spread of the effects of a violation of genetic privacy for the network of people genetically related that emanates from the appellant consequently. Not the least, however, do the decisions of the two courts indicate an awareness of the obvious risks for groups that contemporary data collecting and processing brings with it in conflicting with individualistic conceptions of the right and whether the groupings should be related on social grounds or on genetic lines. These are obviously so, when put in perspective of the broader scenario of human rights. In the case of generic data in particular, this damage may, as the dicta of the ECtHR and the Explanatory Report to Convention 108 note, be restricted to family members (though this again widens out the scope of such damage several steps from the person concerned), but it can also, as Article 1 of the UDHG suggests, cause widespread harm.

With the General Data Protection Regulation, the EU has established a powerful data protection framework that creates what can be considered the zenith of personalized privacy principles. According to GDPR Article 4, personal data are any information pertaining to an identified or identifiable natural person who can be identified directly or indirectly, specifically by reference to a datapoint, like a name, location information, an online identifier, or one or more characteristics unique to the natural person's physical, physiological, genetic, mental, economic, cultural, or social identity. Thus, it is clear that genetic information falls within the ambit of personal data under the regulation, but is still very much a product of an individualistic free choice model. Personal data can only be dealt with under the regulation if there is an appropriate justification for doing so, as specified in Article 6 GDPR. This is somewhat different for genetic data, which Article 9.75 classifies as a particular type of personal data. While processing data under Article 9 is presumed to be prohibited, it may be done with the express consent of the data subject (as permitted by Article 9(a)) or on the basis of one of the other grounds specified in Articles 9(b)–(j). When evaluating the feasibility of these alternate bases, it is crucial to read Article 9 in the context of Article 10 GDPR, which states that processing of certain types of personal data, such as genetic and biometric data, is only allowed when it is absolutely required, subject to suitable safeguards [23–25], and authorized by Union or Member State law to protect the vital interests of the data subject or another natural person, or when the processing relates to data that the data subject has expressly made public. These conditions are individually droughted, putting the emphasis on people as data subjects who make individualistic decisions about how to handle their data. Also, neither of the implicit understanding of genetic data as a shared resource, as in the case of the UDHG, nor the explicit understanding, as in Convention 108, that genetic data challenges individually orientated understandings of privacy is put into the text of the GDPR, which also is missing in the Charter.

Rethinking privacy rights as either individual or group rights is not necessary. This is the second point that is implied in the opinions of both courts. Instead, the way the courts discuss the right to privacy in the contexts that are being examined shows a clear belief that the right as it currently exists is strong enough—or perhaps robust enough—to encompass both the widely acknowledged individual focus and a more expansive, "relational" component—which allows within the parameters of the existing right a recognition of how infringements directed at one person diminish privacy for many others.

Since the term "relational" has been used to denote a variety of privacy models or characteristics, it is obviously required to define what is intended by this concept. Other uses of the term "relational privacy" include: the requirement that members of family groups give their informed consent in certain medical situations [26], the specific expectation interests that exist between individuals under the Fourth Amendment of the US Constitution [27], and broader ideas of privacy that deviate from purely individualistic, autonomy-based understandings [28]. The conceptualization presented here finds a home within the wide heading of relational privacy, which provides an alternate account in this last iteration. The conceptualization presented here finds an environment within the wide heading of relational privacy, which provides an alternate account in this last iteration. Such relational explanations of privacy aim to expand the concept of autonomy to incorporate a view of that value as embedded within a larger social environment [29], rather than eschewing the idea of autonomy on which individualistic accounts are predicated [30, 31]. Thus, the fundamental tenet of relational accounts is that the desired outcome—privacy—can only be achieved in "socially supportive" circumstances and by safeguarding the social contexts that make the right in question possible [29]. This expansive view of relational privacy has been used to promote various ideas about the right to privacy and its associated parameters. Concepts of relational privacy can apply to models that aim to preserve a variety of relationships and the environments that allow them to grow, as mentioned by Rachels and Shoeman, among others [32, 33]. In the light of the jurisprudence of the CJEU or ECtHR, that has long held an approach sensitive to the growth of individual personality in a manner encompassing spaces for such development, such notions of privacy as have been expounded, far from being unexpected, would barely be radical enough within a US context. According to what Hargreaves calls "a proper relational account," which "attempts to gauge the severity of any claimed privacy violation by analyzing the harm it does to the web of relationships any individual finds herself in," relational privacy is understood in this article to encompass more than just the growth of this zone of personal development [29, 33, 34]. Although Hargreaves account focuses on the assessment of damages in US tort law, the judgments of the ECtHR and CJEU can be understood as part of a broader conceptualization of privacy when relational privacy is understood as referring to a "web of relationships" and the privacy harm that can spread through that web. This interpretation recognizes that privacy has been established as an individual right rather than a collective one, and it is not inconsistent with the individualistic tone of current privacy protections or the Courts' larger jurisprudence. But it does define privacy as a socially situated right whose invasion and the accompanying harm are reduced or increased by the extent to which the damage is shifted from the appellant to the internet upon the persons with whom he or she is connected. Relational notions of privacy must, however, face challenges to their justiciability in a human rights framework, just as individualistic conceptions do. They have to go over a long-standing argument against collective rights in particular. According to Hargreaves' perspective, individualistic choice-based understandings of privacy that rely on courts identifying the a priori border that can restrict privacy claims in order to make them justiciable as rights claims are defective.

The idea of collective or group rights is not new to the legal system. While Miller sees collective rights as jointly owned rights to communal goods justified (in part) by membership in a social group [35], Finnis has maintained that collective rights exist on the basis of the "general welfare." Raz provides the most well-known and significant explanation of collective rights [36], tying together the strands of different understandings [37]. Raz contends that collective rights exist "because an aspect of human beings' interest justifies holding some person(s) to be subject to a duty" as part of his interest theory of rights [37]. Essentially what distinguishes Raz's understanding of collective rights from the understanding of relational privacy is that collective rights as understood by Raz exist where the interest of no single member of the group in the public good at issue is sufficient by itself to justify holding another person to be subject to duty [37]. Finally and partly on under individualized interest, Raz has concluded that the right to collective goods is inconceivable simply because the collective interest burdens disproportionate numbers of persons to the sought ends. At this point, Raz's attack actually mirrors Waldron's argument. According to Waldron, a common or public good's fundamental quality is that it cannot be enjoyed without its communal component, and as the advantages of the common good cannot be individually realized, collective rights cannot exist [38, 39]. In response to Raz and Waldron, Réaume has maintained that their explanations are predicated on an unduly simplified view of the social interest at play [39]. Réaume specifically acknowledges that there is no right to some public goods, but she argues that this is the case because public goods are participatory in that they require the participation of many people to produce, but their value comes only from this collective involvement. As a result, they must be held by groups rather than individuals, which makes them unsustainable. For a relational view of privacy based on the rulings of the CJEU and ECtHR, this is the most important conflict that must be resolved. There is no way to make a collective right predictably justiciable, according to Raz and others' explanations of collective rights, which assume that a right must be either individually actionable or collective. This is especially true when these perspectives disregard rights, which, like privacy, can include both individually held rights and goals and effects that are focused on the group as a whole.

This raises a key doctrinal question which a relational theory of privacy must answer: How (and if) to incorporate a collective "relational" dimension of privacy within the corpus of rights jurisprudence established by the CJEU and ECtHR. This would most easily be achieved within the framework of the ECHR through a formalized analytical approach, similar to the one currently in use, that takes into account the relational impacts of an alleged privacy infringement when determining the harm caused, and (specifically) the proportionality of a measure that violates privacy under Article 8(2). Because recognition of the rights and interests whose infringement can thus justify justifiable regulation necessarily subsumes individual's enforced right that is held and upheld as a component of ensuring a more all-encompassing, relational right to privacy across and within communities, this would actually address what could be conceived of as some criticism leveled at Raz et al. with regards to the appropriateness threshold established for justifiability based upon Hohfeldian identification of corresponding rights and duties.

In a Hohfeldian schema, the primary right holder in this case is the prospective applicant, who is better placed than the duty bearers, who are represented by state actors. This primary right holder is also a duty bearer in respect to the privacy rights of people in their relational group, according to a relational theory of privacy. However, since the respondent's action, not the applicant's, caused the privacy violation that adversely affects the applicant's relational group, the members of the applicant's relational group's privacy claim is evaluated as causally originating from the respondent rather than the applicant. This ability of an infringement to result in privacy reductions for other members of the applicant's relational group is used to determine how serious the infringement of the applicant's right to privacy is under this relational view of privacy. Thus, the applicant is the causal node through which the violation of privacy reaches the members of his or her relational group—a duty bearer "in trust"—an actor from whom the violation does not originate but through whom it is transmitted. While repeating Hargreaves' approach in a tortious situation, this approach also acknowledges and formalizes the paradigm of analysis used by the ECtHR in respect of genetic privacy and supported, however somewhat, by the CJEU in communications privacy issues.

It would appear that a relational conception of privacy best addresses the problem of how genetic data, as well as the networked effects of violation of privacy, are managed in the twenty-first century, when it comes to determining the extent of data protection breaches under the GDPR and whose rights under the regulation have been violated. Further, this relational model of privacy acknowledges and brings balance to these established individualist mechanisms to their achievement, which include the social role privacy develops within modern and networked society and historical establishment within the ECHR of such privacy as group-identity-affirming, as distinguished in rights from liberalistic protection as conceived within western models of natural rights [40–42]. However, this method still fails to solve two important and useful issues brought up by genetic data in a private setting. The first issue is that it is similarly difficult to get agreement from the members of the group of people that are indirectly identified by the examination of one person's genetic data because they cannot be fully accounted for. This presents a significant barrier to the feasibility of using permission as justification for non-state entities to process genetic data in accordance with the GDPR. Such an issue cannot be eliminated or resolved by a relational understanding of privacy rights. Nonetheless, it can provide a conceptual understanding of privacy in the context of human rights law, which can then provide a fresh framework for structuring and considering consent—as well as the applications and procedures that can be used to extremely sensitive material, like genetic data. Outside the strict confines of a human rights context, an overly cautious view of the relational model could go some way towards supporting the thesis that there must be considered on damages to the relational sphere within tort or data protection claims—the case of Hargreaves cited at. It seems yet again, however, merely to redress the harm that has already occurred, thereby displacing on the shoulders of whose privacy was breached the onus of doing everything to ensure this was not infringed—an inappropriate outcome in such circumstances.

As a consequence, the second problem arises: Whereas the rights and responsibility bearer relationship described above works well when state-led violations occur, it breaks down when the "applicant's" activity causes the privacy violation. Therefore, in many respects, the "applicant" is responsible for the breach when a person gives genetic information to a for-profit genomics firm without the permission of their family members. In these circumstances, the applicant's conduct has diminished the privacy of the biological group members, and a model based on a "duty bearer in trust" no longer seems entirely appropriate. In such cases, it may be necessary to admit that a relational idea of privacy as applied in private contexts assumes a necessarily modified form in addition to the relational conception regulating claims under Articles 7 and 8 of the Charter and Article 8 ECHR. The use of the relational model in support of current data protection law supports reading the GDPR such that members of the relational group are, for the purpose of Article 4, regarded as identified data subjects subject to a threshold in relation established by law.

4.2 Equity and Accessibility

Genomic science and practice have enormous potential to enhance population and individual health for both present and future generations. Enabling the gathering, access, use, and sharing of human genome data both within and between various health and research sectors is necessary to fulfil this potential. Proactively addressing the moral, legal, social, and cultural concerns is necessary to accomplish this. This would be necessary for acknowledging the risks of both use and non-use in human genome data. Balancing and minimizing those risks will need to safeguard interests of people, families, communities, while health and wellbeing interests of current as well as future generations advance. To facilitate research and the integration of genomics into health systems, and in line with the WHO Science Council 2022 report Accelerated Access to Genomics for Global Health [43], which recommended the promotion of ethical, legal, equitable, and responsible use and sharing of human genome data, WHO has developed this document, which outlines principles for human genome data collection, access, use, and sharing. This means that there is a need for the application of genomics into health systems to have a data life-cycle approach. WHO has released guidelines for the ethical collection, use, and sharing of human genetic information. These principles, developed under the guidance of the WHO Technical Advisory Group on Genomics (TAG-G) and other international experts, outline a global framework for the protection of individual rights, furthering equity, and promoting responsible cooperation in genomic research. The Guidance for Human Genome Data collection, access, use, and sharing is a global document promoting human health and well-being through interconnected principles. It builds on top of existing legislation, policies, and frameworks to bring about social and cultural inclusiveness, equity, trustworthiness, integrity, communal benefits, and the utilization of common principles in laws, policies, and guidelines. Moreover, it helps strengthen individual and community capacity over control of genome data.

An assurance to uphold and respect the rights and interests of people who are capable of making knowledgeable decisions regarding their human genomic data at every stage of the data life cycle. Furthermore, a dedication to upholding the rights of and providing assistance to those who lack the mental capacity to make their own decisions. Human genome data utilization affects more than just the individual; hence, it is important to consider family members' and communities' pertinent opinions about data collection, access, use, and sharing at every stage of the data life cycle. It has been recommended by WHO that Human genome data collection, access, use, and sharing should be based on the needs, preferences, and values of individuals, families, and communities throughout the data life cycle. Informed consent is essential for ethical use, with limitations based on public sharing. The consent should be specific, granular, and tailored to social and cultural contexts. The most appropriate informed consent model will depend on the individual/local context and should be supported by governance frameworks and processes. In situations where a particular purpose cannot be identified, informed consent to broad categories of data collection, access, use, and sharing may be permissible provided appropriate safeguards are in place. Individuals and their communities should be involved in the governance and decision-making process regarding data collection, access, use, and sharing. Children should be given the opportunity to affirm or withdraw consent, and the right to an open future should be taken into account when collecting, accessing, using, and sharing data from children.

In order to promote social justice in the human family, it has been recommended by WHO that Human genome data collection, access, use, and sharing should consider local health needs and burdens, considering the interconnectedness of local, national, and international health ecosystems. Return of results to individuals should be clinically relevant, feasible within local health settings, and legally and ethically permissible. Policies and procedures to shield individuals, families, and communities from stigmatization and discrimination should be created in collaboration with communities. The rights and interests of people, families, and communities who would provide human genomic data must also be protected, though. Consideration of both risks and advantages must be used in decision making, and people should have full access to their benefits. Collection of data through commercial interests can't be justifiably fettered. Governance procedures ought to specify roles and consequences for noncompliance in a clear and concise manner.

A commitment to fair data collection, access, usage, and exchange of the human genome and associated benefits. It includes strongly dealing with power inequities and imbalance among diverse stakeholders that will prevent such ventures. Commitment to improving inclusion and diversity among datasets and the decision-makers governing the collection, acquisition of, usage, and dissemination of human genome data that do not come at the costs of the existing or future generation. A commitment to ensuring that the individuals, families, and communities from whose human genomic data is collected, accessed, utilized, and disseminated can also benefit in equitable ways. Further to this end, WHO advises that adequate representation of datasets among diverse populations is necessary, but representation in and of itself is not sufficient to achieve equity. Equitable involvement of people, families, and communities needs to be secured to provide full meaning in decisions-making and governance. It must be taken into consideration that before data collection and sharing, the risks and benefits should be shared equitably. The commitment of promoting mutually beneficial collaboration and partnerships, including public–private partnership, in human genome data collection, access, use, and sharing is underscored. This should include rebalancing power and representation among the stakeholders. The decisions for the governance processes shall involve discussion about possible risks and benefits, and the policies should assure confidentiality and ethical safeguards. Platforms must be interoperable between institutions and sectors. There is a call to strengthen capacity and enhance health literacy among stakeholders, which would be addressed through targeted educational initiatives to increase public awareness of the importance of human genome data.

It has been further recommended that criteria for collecting, accessing, using, and sharing human genome data, decision-making procedures, data protection measures, and the development of such policies should all be covered in publicly accessible policies. These guidelines ought to outline how the right to privacy is safeguarded and who is in charge of making sure that it is upheld at every stage of the data life cycle. Such systems and procedures need to be put in place so that there can be proper communication with people, families, and communities about how their human genome data is used and related research findings. This ought to be publicly accessible to everyone and contain concise summaries of important discoveries and instructional resources. People, families, and communities should be aware of the ways in which they can exercise their rights concerning.

The ethical implications of human genomic data, especially on equity and accessibility, cross with legal gaps in the protection of data in India, the USA, and the EU. This raises a threat of deepening inequalities in the sense that these marginalized groups would have less access to the fruits of genomic progress but have their data exploited. Moreover, the lacuna in international framework that further complicates the determination of fair and worldwide harmonized standards for genomic data protection and ethical use.

5 Conclusion and Recommendation

Human genetic data are cornerstones of modern medicine and research, offering insights into disease mechanisms, treatment strategies, and personalized care. However, its highly sensitive nature, potential for misuse, and long-term implications necessitate comprehensive legal protections. The existing frameworks—the European Union's GDPR, HIPAA, the Indian IT Act and DPDP Act, and international declarations of the Universal Declaration on the Human Genome and Human Rights (UDHGHR) and the International Declaration on Human Genetic Data (IDHGD)—present such gaps in privacy, security, and equitable use in genetic data.

The amendments for genetic data in the GDPR need to consider its unique attributes. In the Recital 26, it should also clearly reflect the fact that anonymization does not quite work, since genetic data can sometimes still be re-identified after anonymization attempts. Articles 6(1)(a) and 9(2)(a) on explicit consent will have to ensure that a dynamic consent model is used so that it is subject to change during the time course of such longitudinal genetic research. Article 4(1) must be revised to address the complexities of co-ownership in familial and multi-party contexts, recognizing that genetic data is inherently shared among relatives. Similarly, Articles 5(1)(c) and 5(1)(e) on data minimization and storage limitation need exceptions for genetic data, permitting its long-term retention for research purposes under strict safeguards. Articles 17(3)(d) and 21(6) should clearly indicate the exceptions to the right to erasure and objection by balancing individual rights with the continuity of critical genetic research. Articles 44–50, dealing with cross-border data transfers, must include specific protections for genetic data to ensure security and compliance with regional standards in cross border data exchanges. Article 22, on automated decision-making, must be expanded to include implications for AI-driven genomic analysis. This should include transparency, mitigation of bias, and safeguards against discriminatory outcomes.

HIPAA, which is essentially a health data protection act, needs significant updating to address the complexities of genetic data. Section 45 CFR § 160.103 shall expand the definition of covered entities to include direct-to-consumer genetic testing companies and research institutions. Section 45 CFR § 164.508 on patient authorization should provide for explicit, informed consent tailored to genetic data and introduce mechanisms of dynamic consent. Section 45 CFR § 164.502, permitting the use of de-identified data, must include very strong safeguards for genetic data against risks of re-identification. There must be new provisions of new technologies such as AI-driven genetic profiling for accountability and anti-bias measures. HIPAA should have protections against discrimination with additional support for GINA, with provisions that make misuse of genetic information in fields such as life insurance and employment illegal. Furthermore, Section 45 CFR § 164.312 on technical safeguards must be strengthened to include genomic data encryption, access control, and breach monitoring.

India's regulatory framework, under the IT Act, 2000, and the DPDP Act, 2023, also needs significant amendments. Section 43A of the IT Act should categorize genetic data as sensitive personal data with higher security requirements like genomic encryption and advanced access controls. Section 72A must introduce specific standards for genetic data anonymization, emphasizing periodic reviews to address technological advancements. Sections 2(11) and 2(12) should be amended to unmistakably categorize genetic data as a separate category of sensitive personal data, while Section 7 should emphasize that consent mechanisms must always be dynamic. Section 17 on cross-border transfer of data should have some conditions such that the standards must be equivalent in terms of protection in the host countries and genetic data locally stored in India. Creating a Genetic Data Protection Board under a new section would provide for compliance, oversight, and periodic assessments for the large-scale processing of genetic data activities.

The UDHG and IDHGD set international standards in ethics but are non-enforceable and lacking in specificity to deal with modern issues such as AI and cross-border data sharing. Both declarations need to evolve to fill in these gaps to stay relevant in an increasingly fast-changing technological environment. There is a need to harmonize national laws with robust international frameworks for the protection of genetic data while promoting innovation and cooperation. Governments and organizations, by addressing these regulatory shortcomings, can safeguard individual rights, promote ethical practices, and unlock the full potential of genetic research for the betterment of society.

References

1. Satam H, Joshi K, Mangrolia U, Waghoo S, Zaidi G, Rawool S, Thakare RP, Banday S, Mishra AK, Das G, Malonia SK (2023) Next-generation sequencing technology: current trends and advancements. Biology 12(7):997. https://doi.org/10.3390/biology12070997
2. Goodwin S, McPherson JD, McCombie WR (2016) Coming of age: ten years of next-generation sequencing technologies. Nat Rev Genet 17(6):333–351. https://doi.org/10.1038/nrg.2016.49
3. Mardis ER (2017) DNA sequencing technologies: 2006–2016. Nat Protoc 12(2):213–218. https://doi.org/10.1038/nprot.2016.182
4. Sanger F, Nicklen S, Coulson AR (1977) DNA sequencing with chain-terminating inhibitors. Proc Natl Acad Sci 74(12):5463–5467. https://doi.org/10.1073/pnas.74.12.5463
5. Schuster SC (2007) Next-generation sequencing transforms today's biology. Nat Methods 5(1):16–18. https://doi.org/10.1038/nmeth1156

6. Pervez MT, Hasnain MJU, Abbas SH, Moustafa MF, Aslam N, Shah SSM (2022) A comprehensive review of performance of next-generation sequencing platforms. Biomed Res Int 2022:1–12. https://doi.org/10.1155/2022/3457806
7. Gulilat M, Lamb T, Teft WA, Wang J, Dron JS, Robinson JF, Tirona RG, Hegele RA, Kim RB, Schwarz UI (2019) Targeted next generation sequencing as a tool for precision medicine. BMC Med Genomics 12(1). https://doi.org/10.1186/s12920-019-0527-2
8. Liu D (2019) Cancer biomarkers for targeted therapy. Biomark Res 7(1). https://doi.org/10.1186/s40364-019-0178-7
9. Lei Z, Tian Q, Teng Q, Wurpel JND, Zeng L, Pan Y, Chen Z (2023) Understanding and targeting resistance mechanisms in cancer. MedComm 4(3). https://doi.org/10.1002/mco2.265
10. Gymrek M, McGuire AL, Golan D, Halperin E, Erlich Y (2013) Identifying personal genomes by surname inference. Science 339(6117):321–324. https://doi.org/10.1126/science.1229566
11. Phillips M, Molnár-Gábor F, Korbel JO, Thorogood A, Joly Y, Chalmers D, Townend D, Knoppers BM (2020) Genomics: data sharing needs an international code of conduct. Nature 578(7793):31–33. https://doi.org/10.1038/d41586-020-00082-9
12. Shabani M, Borry P (2015) Challenges of web-based personal genomic data sharing. Life Sci Soc Policy 11(1). https://doi.org/10.1186/s40504-014-0022-7
13. Griffiths P, Stotz K (2013) Genetics and philosophy an introduction. Cambridge University Press
14. Lerner N (2003) Group rights and discrimination in international law. Martinus Nijhoff Publishers. https://pdfs.semanticscholar.org/5ada/6456ad61a2fb5f085a4f26fadebe2be3b0c2.pdf
15. Floridi L (2017) Group privacy—a defense and an interpretation. SSRN Electron J. https://doi.org/10.2139/ssrn.3854483
16. Mantelero A (2016) From group privacy to collective privacy: towards a new dimension of privacy and data protection in the big data era. In: Springer eBooks, pp 139–158. https://doi.org/10.1007/978-3-319-46608-8_8
17. Hayamizu Y, Kawabe T, Yanase S, Gonda T, Morita S, Ohtsuka S, Yamamoto K (2015) A micromixer using the taylor-dean flow: effects of aspect ratio and inflow condition on the mixing. Open J Fluid Dyn 05(03):256–264. https://doi.org/10.4236/ojfd.2015.53027
18. Rouvroy A, Poullet Y (2009) The right to informational self-determination and the value of self-development: reassessing the importance of privacy for democracy. The Research Portal—University of Namur. https://researchportal.unamur.be/en/publications/the-right-to-informational-self-determination-and-the-value-of-se
19. Zysset A (2015) "Personal Autonomy" and "Democratic Society" at the European court of human rights: friends or foes? J Law Jurisprud. https://student-journals.ucl.ac.uk/laj/article/id/738/
20. Chakrabarty SP, Mondal N, Chattopadhyay S (2024) Secure big-data analytics for emerging healthcare in 5G and beyond. In: The nuances of legal deviations in modern computing: a relook into the privacy and data protection laws in India and beyond, pp 359–377. https://doi.org/10.1049/pbhe063e_ch17
21. Indian Council of Medical Research (ICMR) (2017) National ethical guidelines for biomedical and health research involving human participants. ICMR Ethics Guidelines
22. Indian Council of Medical Research (ICMR) (2019) National guidelines for gene therapy product development and clinical trials. ICMR Gene Therapy Guidelines
23. Quinn P, Quinn L (2018) Big genetic data and its big data protection challenges. Comput Law Secur Rev 34(5):1000–1018. https://doi.org/10.1016/j.clsr.2018.05.028
24. Emam KE, Rodgers S, Malin B (2015) Anonymising and sharing individual patient data. BMJ 350:h1139. https://doi.org/10.1136/bmj.h1139
25. Shabani M, Borry P (2017) Rules for processing genetic data for research purposes in view of the new EU general data protection regulation. Eur J Hum Genet 26(2):149–156. https://doi.org/10.1038/s41431-017-0045-7
26. Goldim JR, Gibbon S (2015) Between personal and relational privacy: understanding the work of informed consent in cancer genetics in Brazil. J Community Genet 6(3):287–293. https://doi.org/10.1007/s12687-015-0234-4

27. Sacharoff L (2012) The relational nature of privacy. https://papers.ssrn.com/sol3/papers.cfm?abstract_id=2201903
28. Bannerman S (2019) Relational privacy and the networked governance of the self. SSRN Electron J. https://doi.org/10.2139/ssrn.3941520
29. Hargreaves, 'Relational Privacy and Tort' (2017) 23 William and Mary. J Race Gend Soc Justice 433.
30. Nedelsky J (2015) Reconceiving autonomy: sources, thoughts and possibilities. http://hdl.handle.net/20.500.13051/6827
31. Nedelsky J (2011) Law's relations: a relational theory of self, autonomy, and law
32. Schoeman (1984) Privacy and intimate information. In: Schoeman (ed) Philosophical dimensions of privacy: an anthology
33. Cohen (1997) Rethinking privacy: the abortion controversy. In: Weintraub, Kumar (eds) Public and private in thought and practice: perspectives on a grand dichotomy
34. Anderson, Honneth (2005) Autonomy, vulnerability, recognition, and justice. In: Christman, Anderson (eds) Autonomy and the challenges to liberalism: new essays
35. Finnis JM (n.d.) Natural law and natural rights. NDLScholarship. https://scholarship.law.nd.edu/law_books/99/
36. Miller, 'Collective Rights' (1999) 13 Public Affairs Quarterly 331
37. Graham G, Raz J (1987) The morality of freedom. Philos Q 37(149):481. https://doi.org/10.2307/2219586
38. Waldron J (2017) Can communal goods be human rights? Routledge eBooks, pp 25–51. https://doi.org/10.4324/9781315094427-2
39. Reaume DG (1988) Individuals, groups and rights to public goods. https://papers.ssrn.com/sol3/papers.cfm?abstract_id=1895704
40. Nissenbaum H (2019) A contextual approach to privacy online. American Academy of Arts & Sciences. https://www.amacad.org/publication/daedalus/contextual-approach-privacy-online
41. Nissenbaum H (2019) Contextual integrity up and down the data food chain. Theor Inq Law 20(1):221–256. https://doi.org/10.1515/til-2019-0008
42. Nissenbaum H (n.d.) Privacy in Context|Stanford University Press. Stanford University Press. https://www.sup.org/books/law/privacy-context
43. Accelerating access to genomics for global health: promotion, implementation, collaboration, and ethical, legal, and social issues: a report of the WHO Science Council (2022). https://www.who.int/publications/i/item/9789240052857

Personality Prediction Through Graphology: Leveraging AI for Enhanced Recruitment Screening and Employee Retention

Ansh Chaudhary and Mohuya Chakraborty

Abstract This paper investigates a novel approach to personality prediction through graphology, combined with artificial intelligence (AI), as a tool for improving recruitment processes and tackling high attrition rates in modern organizations. Using a dataset of 5130 handwritten samples from subjects aged 18 to 67, this study focuses on predicting the "Big Five" personality traits: *Agreeableness, Conscientiousness, Extraversion, Neuroticism, and Openness.* A convolutional neural network (CNN) model trained on this data achieved an accuracy of 74%, showcasing the potential for precise personality analysis based on handwriting patterns. This AI-driven graphological analysis can serve as a preliminary screening tool in recruitment, particularly identifying candidates prone to *Neuroticism,* which may correlate with mental health concerns and higher turnover risk. The performance metrics for this class shows moderate precision (0.51), high recall (0.78), and reasonable F1-score of 0.62, which indicates that the model is moderately effective in distinguishing *Neuroticism* from other traits as desired. Candidates who demonstrate positive indicators in traits associated with workplace stability like *Agreeableness, Extraversion* and *Openness* having high precision of 0.94, 0.95, and 0.80, respectively, and high specificity of 4.33, 9.18, and 5.63, respectively, are advanced in the hiring process. By integrating personality prediction into recruitment, organizations can better align new hires with role demands and organizational culture, fostering a workforce with higher engagement and resilience. The proposed approach addresses the critical challenges of employee retention and attrition, offering a proactive strategy for building a more stable, motivated, and cohesive workforce.

Keywords Artificial intelligence · Graphology · Personality traits · Employability · Recruitment · Convolutional neural network · Image processing

---

A. Chaudhary
Saratoga High School, Saratoga, CA, USA

M. Chakraborty (✉)
Cybersecurity Centre of Excellence, Institute of Engineering & Management, University of Engineering and Management, Kolakata, India
e-mail: mohuyacb@iem.edu.in

M. Chakraborty et al. (eds.), *Proceedings of Sixth International Ethical Hacking Conference*, Lecture Notes in Networks and Systems 1513,
https://doi.org/10.1007/978-981-96-8632-2_10

1 Introduction

1.1 Background Study

Employability is a collection of accomplishments, including knowledge, abilities, and personality traits that increase a person's chances of landing a job and succeeding in their chosen fields, which benefits the workforce, the economy, the community, and the individual [1]. Employability encompasses more than just a person's ability to land and keep a job; it also includes a wider range of qualities and abilities that will contribute to a successful career. When people apply for jobs, businesses search for specific employability skills. By cultivating a variety of skills and attributes that make a person a more desirable and productive employee, employability skills aim to improve a person's preparedness for the job market. These abilities, which are intended to improve employability and job success, include problem-solving, communication, teamwork, flexibility, and personal attributes, among others.

The most crucial point that businesses demand is "employee retention." The ability of a company to retain its workforce and lower turnover is known as employee retention. Companies can increase employee retention in a number of ways, such as by providing competitive pay and benefits, fostering a happy work environment, allowing employees some latitude in their schedules, publicly recognizing top performers, and encouraging work-life balance. Every company wants to retain valuable people, but in industries like construction, education, healthcare, and transportation where there is a severe shortage of employees, it is even more important. To remain competitive and, eventually, to remain in business, leaders need to take action to increase employee retention.

However, with the attrition rate increasing at an alarming in these industries in spite of adopting several modern employee retention policies, it is high time employers think about other areas to retain talented pool, which they may have overlooked so far. One such area is the employee personality traits that play a major role in the overall success of any business. The unique qualities and traits that characterize an individual's behavior, feelings, and mental processes are known as personality attributes or traits. Personality traits are crucial in affecting job performance, relationships with coworkers, and the general atmosphere at work. Establishing a productive and positive work environment requires an understanding of the ability to recognize personality traits in the workplace.

1.2 The "Big Five" Personality Traits Model

Determining the precise number of features has long been a goal of trait theories of personality. Different numbers have been proposed by previous theories. For example, Raymond Cattell identified sixteen personality traits [2], Hans Eysenck proposed a three-factor theory [3, 4], and Gordon Allport's list included 4,000 personality qualities [5]. Many experts believed that Eysenck's thesis was too narrowly focused while Cattell's theory was too complex. This led to the development of the "*Big Five Personality Traits*" model that suggests that individual variations can be described using five main characteristics of personality, which are used to characterize the general features that form the foundation of personality. These characteristics include agreeableness, conscientiousness, extraversion, neuroticism, and openness to new experiences [6–8].

1. *Agreeableness*: People high in agreeableness tend to be well-liked, respected, and sensitive to the needs of others. They likely have few enemies and are affectionate to their friends and loved ones, as well as sympathetic to the plights of strangers.
2. *Conscientiousness*: It is a trait that can be described as the tendency to control impulses and act in socially acceptable ways, behaviors that facilitate goal-directed behavior. Conscientious people excel in their ability to delay gratification, work within the rules, and plan and organize effectively.
3. *Extroversion*: People high in extroversion tend to seek out opportunities for social interaction, where they are often the "life of the party." They are comfortable with others, are gregarious, and are prone to action rather than contemplation.
4. *Neuroticism*: Those high in neuroticism are generally prone to anxiety, sadness, worry, and low self-esteem. They may be temperamental or easily angered, and they tend to be self-conscious and unsure of themselves
5. *Openness*: An individual who is high in openness to experience is likely someone who has a love of learning, enjoys the arts, engages in a creative career or hobby, and likes meeting new people.

With the exception of neuroticism, all other personality traits show a positive human attitude, which is the requirement for a healthy workplace environment. So, during the hiring process, it is believed that candidates belonging to Neuroticism be best avoided for overall benefit of the organization. The idea that there are five fundamental personality qualities is supported by a number of researchers. The psychological evidence supporting this idea has been mounting for many years, starting with research of Fiske [9] and later being built upon by others, such as Norman [10], Smith, McCrae and Costa [12], Goldberg [13], and Robert L. Kentle [14].

1.3 Introduction to Graphology

Considering, personality traits influence how a person interacts with people and handles their work, they can have an impact on employment decisions and career advancement. As part of their hiring process, employers may use personality tests to find applicants who have the skills needed to succeed in a given position. Additionally, supervisors can promote career advancement by offering customized feedback and growth opportunities when they have a thorough understanding of an employee's personality traits. Building solid relationships with coworkers, enhancing job performance, and succeeding in the long run at work all depend on having a solid understanding of personality traits.

There are several personality tests that reveal the personality of a given person. Among the most popular ones that are conducted during the hiring process are psychometric: *measures cognitive ability*, Myers-Briggs Type Inventory (MBTI): *measures affective mind*, Dominance, Influence, Steadiness, and Conscientiousness (DiSC): *measures ability to work, learn, and communicate*, 16 Personality Factor Questionnaire (16PF): *measures how one reacts in real situations at work*, Occupational Personality Questionnaire (OPQ32): *measures work style and preferences*, KOLBE Assessment: *measures the conative mind*, [15].

However, all these tests are time consuming and require lots of questionnaire. The current style of personality prediction is through graphology, the analysis of handwriting in an attempt to determine the writer's personality traits in non-invasive manner [16, 17]. Graphology, the psychological study of handwriting to ascertain the writer's personality, psychological states, temperament, or behavior, is still a contentious field because it lacks a standard and the majority of handwriting interpretations are made subjectively by qualified graphologists [18]. Nevertheless, there are a few studies that back up graphology as a method of predicting human personality. These studies demonstrate that the brain creates characters based on the habits of writers and that each neurological brain pattern results in a unique neuromuscular movement that is similar for people with the same personality type [19, 20].

1.4 Leveraging Artificial Intelligence to Graphology

With the advent of artificial intelligence (AI) showing its potential in predictive analysis of various classification scenarios encompassing various sectors such as leadership, management, finance, healthcare, education, law, advertising etc., research has shown how to incorporate AI in graphology in predicting personality parameters by analyzing offline handwriting [21]. In [22], the authors have used a hidden Markov model and neural network (MLP) to perform classification in order to identify a person's personality based on their handwriting. The latter was used to determine which properties were unrelated to the writer, and former was used to classify the properties that were related to the target writer. Along with back propagation neural network, Xiaoyuang et al. [23] uses fuzzy theory to determine the structure of hidden layer and dynamically adjust additional momentum factor for improvement in the performance. The widely used dataset for conducting these types of analysis is MNIST [24]. In line with the related works, this paper has put forth a convolutional neural network approach to graphology for determining the five personality traits of a candidate during the hiring process and segregating the neuroticism personality.

1.5 Organization

The rest of the paper is organized as follows. Section 2 presents related work followed by the proposed work detailing the architecture, algorithmic steps, dataset, simulation environment and result in Sect. 3. Section 4 is devoted to performance analysis and comparison with a few related works. The paper concludes in Sect. 5 with some highlights on future scope and enhancement.

2 Related Works

This section presents the literature survey of the related works. Pandey et al. introduced the concept of graphology, which combines AI with graphology to explore applications such as stress management, personality prediction, and mental health assessments. By leveraging smart devices and their cameras, graphology enables real-time monitoring to ensure timely interventions. However, the study highlights the challenge of creating accurate, labeled datasets essential for effective implementation [25]. Devanagari script analysis eliminates language barriers in handwriting analysis by using Devanagari scripts instead of English. The approach focuses on extracting structural handwriting features such as slants, letter size, and word spacing. It categorizes personality into five distinct types: introvert, extrovert, optimistic, pessimistic, and stable mindset. Using an artificial neural network, the system achieves an impressive accuracy of 94.75% [26]. Letter feature analysis focuses on analyzing isolated lowercase letters, examining features like contours, intersecting points, and concavities, along with horizontal lines in letters like "t." [27]. This analysis evaluates personal, positive social, and negative social behaviors. Users interact with the system by writing directly on a pad or uploading photos for analysis. Reference

Reference [28] maps handwritten features to personality traits by using an FVM. Employing the polygonization method, it approximates baseline shapes of handwriting and links slant directions to specific personality traits, providing insights into individual characteristics. Parkinson's disease detection applies handwriting analysis to examine bradykinesia, micrographia, and rest tremor in writing samples [29]. Measurements of writing size, such as the width and height of letters, help detect micrographia, offering a non-invasive and quantitative approach to Parkinson's screening, addressing the limitations of subjective clinical assessments [30] traces its history to McIntosh in 1961 and defines the discipline as analyzing aspects like spelling and punctuation. The study discusses the theoretical background and suggests future research into the role of graphology in multimodal communication, stylistics, and digital media. Reference [28], uses grayscale samples, which are then converted to binary images for improved contrast. This method matches handwritten characters with pre-trained templates to identify patterns. The study highlights the potential for incorporating natural language processing in future developments to enhance accuracy and analysis. Reference [31] employs deep convolutional neural networks (CNNs) to generalize across users. The approach achieves an equal error rate of 2.74% by visualizing feature spaces that capture unique writing patterns. This method effectively distinguishes between genuine and random signatures. Reference [32] investigates handwriting size abnormalities like micrographia and macrographia to differentiate Parkinson's disease subtypes from SWEDD. The study finds that micrographia is a stronger indicator of Parkinson's disease than SWEDD, providing valuable insights into these conditions. Reference [33] links static features like size and shape, as well as dynamic features like speed and movement, to personality traits. The study finds impulsivity correlates with fast pen movement, while conscientiousness is associated with fewer space gaps. Reference [34] scans characters to form characteristic vectors based on boundaries and centers. These vectors are matched to a predefined database for personality correlation. However, the study acknowledges the challenge of insufficient predefined data, which impacts accuracy. Reference [35] applies handwriting analysis to assess traits like leadership, motivation, and teamwork. Using backpropagation neural networks, the system classifies profiles with 89% accuracy, aiding human resources in job applicant evaluation. Reference [36] analyzes handwriting individuality through repetitive grapheme patterns called writer's invariants. The study achieves 97.7% accuracy with compressed samples and introduces compressed grapheme representations to reduce computational complexity.

Reference [37] examines handwriting traits across three generations in 100 families. Traits like speed, slant, spacing, and line quality show 60%-75% similarity, with father–son handwriting traits exhibiting the highest resemblance. Reference [38] focuses on the grapheme "th," chosen for its frequency in English and discriminative power. Using genetic algorithms and neural networks, the study achieves up to 94% classification accuracy. Reference [39] uses Arrow R-CNN to recognize diagrams, symbols, and structures. Building on Faster R-CNN with Feature Pyramid Networks, it improves recognition rates for flowcharts from 37.7% to 78.6%, demonstrating effectiveness in diagram recognition. Reference [40] evaluates 140 handwriting samples from students and prisoners to assess job suitability. The system achieves 82.5% accuracy when compared with MMPI results, demonstrating its application in personality evaluation. Reference [41] reviews developments in automatic signature verification, emphasizing the rise of deep learning approaches like CNNs and Siamese networks. The study also evaluates limitations in public and synthetic databases, such as size, diversity, and realism.

3 Proposed Method

3.1 Model Architecture

The proposed model architecture follows a convolutional neural network approach to graphology for personality traits prediction and analysis. It provides a balance of convolutional layers for feature extraction and dense layers for classification, suitable for predicting personality traits from handwriting images as shown in Fig. 1. The different components that are used in this architecture are briefly explained.

1. *Convolutional and MaxPooling Layers*: Extract features from handwriting images by capturing patterns like strokes, curves, and spatial relationships that could be related to personality traits. The model begins with two to three convolutional layers to extract features from the handwriting images. These layers will capture spatial hierarchies in the data, such as edges and contours, which can be useful for distinguishing personality traits based on handwriting.
   - Layer 1: Conv2D with 32 filters, kernel size of (3, 3), activation function ReLU, and input shape set to (224, 224, 1) (grayscale, resized to 224 × 224).
   - Layer 2: Conv2D with 64 filters, kernel size of (3, 3), followed by a MaxPooling2D layer to reduce spatial dimensions.
   - Layer 3: Conv2D with 128 filters, kernel size of (3, 3), followed by a MaxPooling2D layer.
2. *Batch Normalization and Dropout*: The former helps stabilize and speed up the training process, whereas the latter mitigates overfitting by randomly deactivating a portion of neurons during each iteration.
   - Batch normalization is applied after each convolutional block to stabilize learning and improve convergence.
   - Dropout layers are used after each MaxPooling2D layer with a dropout rate of 0.25 to prevent overfitting.

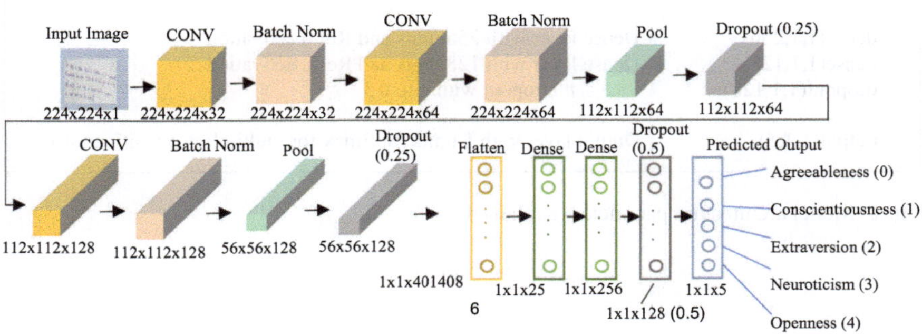

Fig. 1 Model architecture

3. *Flattening Layer*: Converts the 3D output of the convolutional layers to a 1D vector for the dense layers.
    - After the convolutional layers, a flatten layer is used to convert the 2D matrices into a 1D vector before passing it to the dense layers.
4. *Fully Connected (Dense) Layers*: Learn more complex representations of the extracted features.
    - The flattened output is fed into one or two dense layers to learn higher level representations.
    - Dense Layer 1: Contains 256 units with ReLU activation.
    - Dense Layer 2: Contains 128 units with ReLU activation, followed by a Dropout layer with a rate of 0.5.
5. *Output Layer*: Provides probabilities for each personality trait class.
    - The final output layer has 5 units (one for each personality trait) with a SoftMax activation function for multi-class classification. This layer outputs the probability of each personality trait class.

The model architecture is explained in Fig. 2.

```
input(224, 224, 1)           # Grayscale image, resized to 224x224
conv(224,224,32)             # Conv2D with 32 filters and ReLu activation
batch_norm(224,224,32)       # Batch Normalization

conv(224,224,64)             # Conv2D with 64 filters
batch_norm(224,224,64)       # Batch Normalization
pool(112,112,64)             # MaxPooling2D to reduce spatial dimensions
dropout(112,112,64)     # Dropout with rate 0.25

conv(112,112,128)       # Conv2D with 128 filters
batch_norm(112,112,128)      # Batch Normalization
pool(56,56,128)              # MaxPooling2D to reduce spatial dimensions
dropout(56,56,128)      # Dropout with rate 0.25

flatten(1,1,401408)     # Flatten layer to convert 3D to 1D vector

dense(1,1,256)          # Dense layer with 256 units and ReLu activation
dense(1,1,128)          # Dense layer with 128 units and ReLu activation
dropout(1,1,128)             # Dropout with rate 0.5

output(1,1,5)           # Output layer with 5 units (Softmax for multi-class classification)
```

Fig. 2 Model architecture parameters explained

3.2 The Algorithm: Training and Evaluating CNN Model for Personality Classification

The step-by-step algorithm for personality classification and prediction is shown below.

- **Input:**

  Dataset of handwriting images with labelled personality traits.
  Base model architecture (MobileNetV2).
  Training and validation configurations.

- **Output:**

  Trained model.
  Performance metrics (accuracy, confusion matrix).

Detailed Steps

1. **Data Preparation**

   1.1. Define dataset paths for training and validation sets.
   1.2. Resize images to a uniform dimension (150 × 150 × 3) for input consistency.

2. **Data Augmentation**

   2.1. Apply data augmentation techniques to enhance the diversity of training data:

   - Rescale pixel values to [0, 1].
   - Apply transformations: rotations, width/height shifts, shear, zoom, horizontal flips, and nearest fill mode.

   2.2. Create separate data generators for training and validation.

3. **Class Imbalance Handling**

   3.1. Compute class weights to balance training contributions across imbalanced classes.

4. **Model Design**

   4.1. Load MobileNetV2 as the base model with pre-trained weights (imagenet) and exclude the top layers.
   4.2. Add custom layers for feature extraction:

   - **GlobalAveragePooling2D** to reduce spatial dimensions.
   - Fully connected (dense) layers:

     – Dense layer with 128 units, ReLU activation.
     – Output layer with a softmax activation for multi-class classification.

   4.3. Freeze all layers in the base model to preserve pre-trained feature representations.

5. **Model Compilation**
   5.1. Configure the optimizer as **Adam** for efficient training.
   5.2. Set the loss function to categorical_crossentropy for multi-class classification.
   5.3. Define accuracy as the evaluation metric.

6. **Early Stopping Configuration**
   6.1. Monitor validation loss during training.
   6.2. Stop training if validation loss does not improve for 5 consecutive epochs.

7. **Model Training**
   7.1. Train the model using the training generator and validate using the validation generator.
   7.2. Use class weights to address class imbalance.
   7.3. Set the number of epochs to 50 and include the early stopping callback.

8. **Model Evaluation**
   8.1. Predict personality classes for validation data.
   8.2. Convert predicted probabilities into class labels using the argmax function.

9. **Performance Analysis**
   9.1. Compute the confusion matrix to evaluate classification accuracy for each class.
   9.2. Display the confusion matrix graphically for better interpretation.

10. **Training and Validation Trends**
    10.1. Plot training and validation accuracy across epochs.
    10.2. Plot training and validation loss across epochs to assess model convergence.

End Output

- Trained CNN model optimized for personality classification.
- Evaluation metrics including confusion matrix and accuracy trends for validation.

3.3 Simulation Environment

The simulation environment is Google compute engine (GCE) backend running Python 3, which is a virtual machine provided by Google cloud platform (GCP). It has been used for executing the code for utilizing scalable compute resources. 1.21 GB is the amount of random access memory (RAM) currently in use by the code or session. 12.67 GB is the total available RAM allocated to the virtual machine. RAM is essential for temporarily storing data during computation, and these values tell us how much of this resource is being used versus how much is available. 32.47 GB is the amount of disk space currently being used (e.g., for files, libraries, or temporary data generated during the session). 107.72 GB is the total disk space available on the virtual machine. Disk space is a persistent storage resource used for saving files and data.

3.4 Dataset

Sixty-two primary (P) handwriting samples were collected through LinkedIn and Facebook surveys, friends, family, and peers. Sixty-eight samples were collected from secondary (S) dataset available at various open-source databases like Kaggle, MNIST, NIST, and ICDAR as per the venture. Five thousand samples were derived (D) samples that were synthetically generated by augmentation-based synthetic generation (ASG) method, which is a broadly used procedure in data preprocessing and enrichment, predominantly in circumstances where the dataset is inadequate. This process comprises applying several alterations to prevailing images or data samples to generate new, assorted variants while retaining the fundamental features of the original data. Common augmentation methods that are used include rotation, flipping, scaling, cropping, brightness adjustment, and the addition of noise thereby enriching the dataset and improving model generalization. In our study, synthetic handwriting images were generated using a Python-based pipeline that applied augmentations like rotation (within $\pm$ 15 degrees), horizontal flipping, and random cropping with resizing. This approach proficiently extended the dataset without the need for computationally expensive deep learning models, making it a real-world choice for tasks where the goal is to increase data diversity without generating entirely new content.

The 5130 samples were further categorized into the five big personality classes such as "*Agreeableness,*" "*Conscientiousness,*" "*Extraversion,*" "*Neuroticism,*" *and* "*Openness.*" For the secondary dataset, the personality types were measured through [42] that gave the personality type result based on a few questionnaires. The subjects were then asked to write a paragraph (Fig. 2) on a blank sheet and upload the scanned copy along with their personality type. These samples were further analyzed for correctness by a psychologist and a graphologist for generating the final dataset before feeding it to the CNN model. The samples were collected from subjects belonging to the age ranging from 18–67, both male and female. Table 1 shows the details of the dataset used for the work, whereas the Appendix presents some handwriting samples, their categories, and three most important features.

Table 1 Dataset

Personality trait	Encoded (Labelled) value	Number of samples			
		P	S	D	P + S + D
Agreeableness	0	10	16	1000	1026
Conscientiousness	1	18	14	1000	1032
Extraversion	2	12	12	1000	1024
Neuroticism	3	11	14	1000	1025
Openness	4	11	12	1000	1023
Total		**62**	**68**	**5000**	**5130**

Dataset Type: Primary (P); Secondary (S); Derived (D)

3.5 Simulation Result

The performance metrics used for simulation result for all the five classes are accuracy, precision, recall, F1-score and specificity. The details of the simulation result metrics are depicted in Fig. 3. Figure 4 shows the confusion matrix followed by model accuracy and loss in Fig. 5.

	precision	recall	f1-score	support
Agreeableness	0.94	0.85	0.89	1026
Conscientiousness	0.50	0.73	0.60	1032
Extraversion	0.95	0.41	0.57	1024
Neuroticism	0.51	0.78	0.62	1025
Openness	0.80	0.55	0.65	1023
accuracy			0.67	5130
macro avg	0.74	0.67	0.67	5130
weighted avg	0.74	0.67	0.67	5130

Specificity for each class:
Agreeableness: 4.33
Conscientiousness: 2.25
Extraversion: 9.18
Neuroticism: 2.15
Openness: 5.63

Fig. 3 Performance metrics

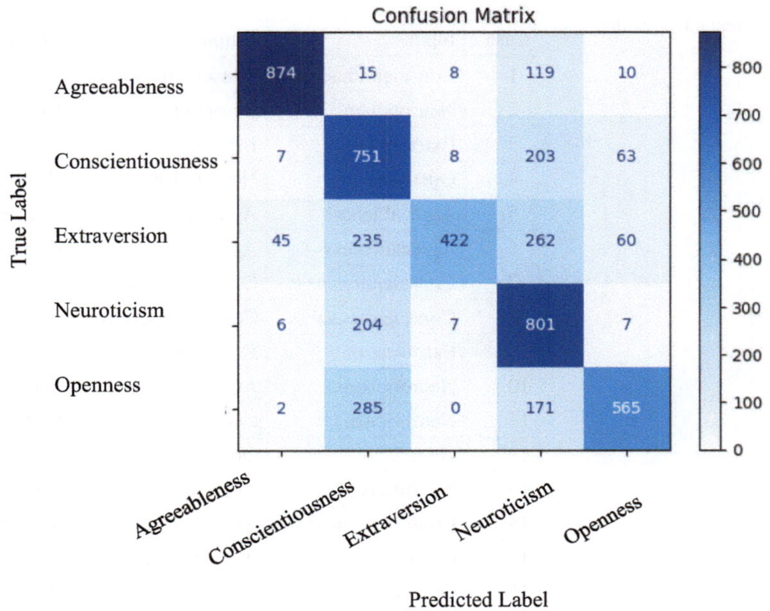

Fig. 4 Confusion matrix

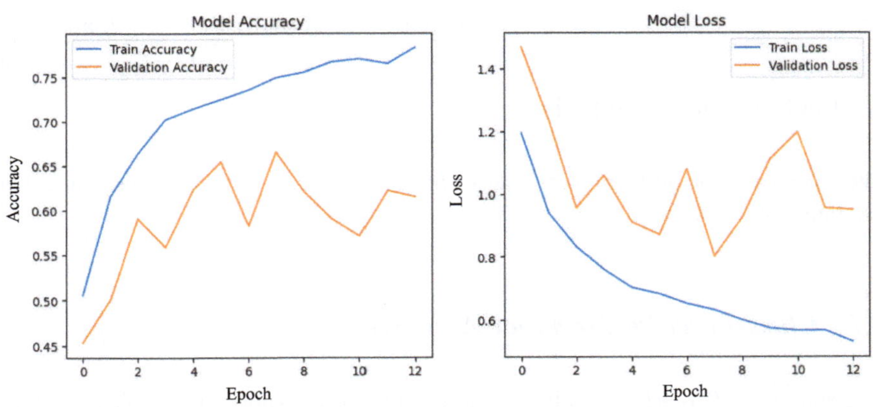

Fig. 5 Model accuracy and loss with epoch

3.6 *Test Cases*

The test case results depicted in Table 2 shows that out of 20 simulated test cases, 15 furnished exact output (75% accuracy) almost close to the performance metrics in Fig. 3.

Table 2 Test case results

Case	Input	Output	Prediction
1	Conscientiousness	Conscientiousness	Correct
2	Neuroticism	Neuroticism	Correct
3	Extraversion	Extraversion	Correct
4	Openness	Neuroticism	Wrong
5	Agreeableness	Agreeableness	Correct
6	Agreeableness	Agreeableness	Correct
7	Conscientiousness	Conscientiousness	Correct
8	Conscientiousness	Conscientiousness	Correct
9	Extraversion	Extraversion	Correct
10	Neuroticism	Agreeableness	Wrong
11	Neuroticism	Neuroticism	Correct
12	Openness	Openness	Correct
13	Neuroticism	Neuroticism	Correct
14	Conscientiousness	Neuroticism	Wrong
15	Conscientiousness	Agreeableness	Wrong
16	Openness	Agreeableness	Wrong
17	Neuroticism	Neuroticism	Correct
18	Agreeableness	Agreeableness	Correct
19	Conscientiousness	Conscientiousness	Correct
20	Extraversion	Extraversion	Correct

4 Performance Analysis

The analysis of the performance in terms of each metric as shown in Figs. 3 and 4 is presented in this section.

4.1 Class-Level Performance Analysis

Agreeableness (0)—High recall suggests that the model is good at identifying *Agreeableness* instances. Very high precision indicates that predictions for this class are mostly correct. This class has the highest F1 score, suggesting a good balance between precision and recall. Specificity value of 4.33 implies that the model can correctly avoid falsely labeling non-*Agreeableness* instances as *Agreeableness* in most cases. However, it is slightly lower than other two classes (viz., *Extraversion* and *Openness*), meaning there is some overlap where other personality traits may be misclassified as *Agreeableness*, affecting the model's ability to isolate this class precisely. The confusion matrix indicates that the model correctly predicted 874 instances as

Agreeableness but misclassified a very few as other classes: 15 as *Conscientiousness*, 8 as *Extraversion,* 119 as *Neuroticism* and 10 *Openness.*

Conscientiousness (1)—The confusion matrix indicates that the model correctly classified 751 instances but misclassified a few as other classes: 7 as *Agreeableness*, 8 as *Extraversion*, 203 as *Neuroticism*, and 63 as *Openness*. The relatively lower precision and recall show the model's struggle with accurately identifying *Conscientiousness.* The relatively low specificity value indicates the model sometimes confuses other personality traits as *Conscientiousness.* The misclassification might be due to similar features shared by *Conscientiousness* with other traits like *Neuroticism* and *Openness*, as seen in the confusion matrix.

Extraversion (2)—The confusion matrix indicates that the model correctly classified 422 instances, but some were misclassified as Agreeableness (45), *Conscientiousness* (235), *Neuroticism* (262), and *Openness* (60). Recall is slightly low, suggesting the model is missing some true *Extraversion* instances. The highest specificity score for *Extraversion* indicates the model is good at recognizing instances that are not *Extraversion,* rarely misclassifying other traits as *Extraversion.* This suggests the model has learned features unique enough to differentiate Extraversion from other classes.

Neuroticism (3)—The confusion matrix indicates that the model classified 801 instances correctly, with misclassifications as *Agreeableness* (6), *Conscientiousness* (204), *Extraversion* (7), and *Openness* (7). A balanced F1-score indicates the model is fairly reliable at identifying *Neuroticism,* but still prone to confusion with other classes. The specificity here shows the model is moderately effective in distinguishing *Neuroticism* from other traits. The model is relatively accurate in identifying instances that are not *Neuroticism,* but there is still some room for improvement, as indicated by occasional misclassifications with traits like *Conscientiousness.*

Openness (4)—The confusion matrix shows that the model correctly classified 565 instances, but some were misclassified as *Agreeableness* (2), *Conscientiousness* (285), and *Neuroticism* (171). The lower recall suggests the model struggles to capture all *Openness* instances, which also impacts precision. The relatively high specificity for *Openness* means the model can avoid confusing non-*Openness* instances as *Openness* in most cases. This value reflects the model's effectiveness in isolating *Openness,* though a small number of other classes might still be misclassified as *Openness.*

As shown in Fig. 5, the performance of the convolutional neural network (CNN) model used in this study was evaluated based on training and validation accuracy and loss across 12 epochs. The results, as depicted in the charts, indicate the model's learning behavior over time. The accuracy plot shows an upward trend for both training and validation accuracy, suggesting that the model improves its ability to correctly classify personality traits as training progresses. While the training accuracy increases steadily, the validation accuracy exhibits some fluctuations, which is common due to the variations in the validation dataset. By the final epoch, both accuracies converge around similar values (0.78), indicating that the model generalizes well to unseen data without significant overfitting. The loss plot demonstrates a consistent decline in training but a zig-zag pattern in validation loss, signaling that the model is effectively minimizing errors during training, however has difficulty removing errors during validation. The validation loss aligns closely with the training loss at the final epoch with a value of 0.2, further reinforcing the absence of overfitting. The steady reduction in loss reflects the model's ability to learn meaningful patterns from the data during training.

The convergence of training and validation metrics near the final epochs suggests that the model is well-tuned and not overfit. Fluctuations in validation accuracy and loss could be attributed to the model's difficulty in generalizing to unseen data, potentially caused by insufficient regularization, which may amplify the effect of data variance. The overall trends confirm the model's capability to effectively predict personality traits using handwriting data. The validation set may have a smaller size, class imbalance, or differing characteristics from the training set, leading to inconsistent performance. Stochastic factors like mini-batch gradient descent and batch normalization can also contribute to fluctuations during validation. Additionally, high learning rates or mismatches caused by data augmentation can create instability in validation metrics. To address this, increasing the validation set size, ensuring data balance, applying regularization techniques (e.g., dropout or early stopping), fine-tuning hyperparameters, or using cross-validation can help stabilize validation performance and reduce the generalization gap.

4.2 Overall Performance and Averages

Accuracy of 0.74, with macro- and weighted averages for precision, and for recall, and F1-score at 0.67, i.e., an approximate balance suggests that the model's performance across all classes is fairly consistent, but there is room for improvement to avoid class imbalance effects.

4.3 Confusion Matrix Patterns

Agreeableness (874), Conscientiousness (751), Neuroticism (801), and Openness (565) have relatively strong true positive counts, indicating that the model correctly classified many samples for these classes while Extraversion has only 422 true positives, showing weaker performance in identifying this class accurately as compared to the others. There are significant off-diagonal entries (e.g., 235 samples of Extraversion were classified as Conscientiousness and 285 samples of Openness were classified as Conscientiousness), which indicates confusion between certain personality classes. Extraversion and Conscientiousness seem to overlap significantly with other classes, highlighting potential feature similarity or insufficient distinction in the data.

4.4 Discussion

To address the misclassifications and further refine the model for future scope, some strategies may be adopted. Using extended data augmentation techniques, such as slight color variations, could help increase data variety and improve generalization. For underrepresented classes or classes with overlapping features, more synthetic data generation using deep learning techniques such as generative adversarial networks (GAN) could help improve classification accuracy. Adding dropout layers, L2 regularization, or increasing dropout rates might reduce overfitting and improve generalization for closely related classes. Examining features within the training data to highlight differences between classes like *Agreeableness* and *Extraversion* or *Conscientiousness* and *Neuroticism.* Feature extraction methods (e.g., PCA, t-SNE) could also help reduce dimensionality and highlight distinct traits. Experimenting with batch sizes, learning rates, and optimizers (like switching between Adam and SGD) could yield slight improvements. Cross-validation could be used more extensively to confirm the model's robustness across different subsets. Considering other models with strong transfer learning capabilities, like EfficientNet or ResNet, which may handle subtle inter-class variations more effectively. Ensemble methods, combining predictions from multiple models, could also improve class differentiation.

Implementing some of the above strategies could enhance the model's ability to better distinguish these classes and improve overall performance, especially for the more challenging traits. The use of synthetic derived data using ASG resulted in a significant improvement in model accuracy, demonstrating the effectiveness of our approach. Future work will focus on further refinement of our model using GAN architecture and exploring additional methods for data augmentation.

5 Conclusion

This study explored the integration of graphology and artificial AI as an innovative method for predicting personality traits to improve recruitment processes and reduce attrition rates. By utilizing a CNN trained on handwritten samples, the model demonstrated promising training accuracy of 78% and effectiveness, particularly in identifying traits like *Neuroticism*, which are linked to higher turnover risk, as well as traits such as *Conscientiousness* and *Agreeableness*, which align with workplace stability. The findings underscore the potential of AI-driven personality analysis as a preliminary screening tool to better match candidates with organizational roles, fostering engagement and resilience among employees. However, the study has certain limitations, including the relatively small dataset and potential cultural and contextual variations in handwriting. Future research could focus on expanding the dataset, exploring diverse demographic groups, and refining the model's accuracy and applicability across industries. This approach offers a proactive and data-driven strategy for enhancing recruitment outcomes and addressing critical challenges in workforce management, paving the way for a more cohesive and motivated organizational culture.

Appendix

Personality type	Handwriting sample	Three most important features
Agreeableness	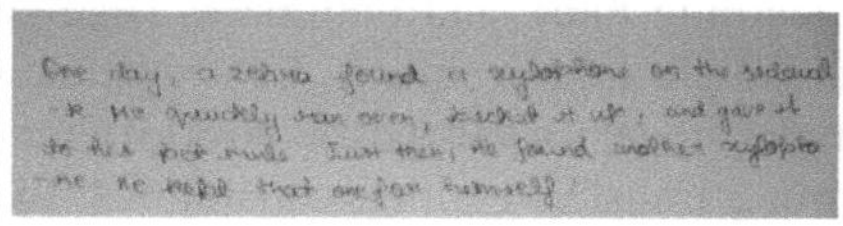	*Connecting strokes*: strongly connected, *Word stant*: extreme left slant, *Lowercase letter "t"*: very low
Conscientiousness	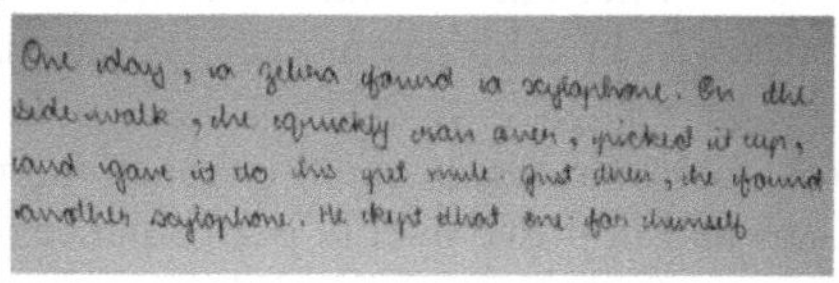	*Word slant*: extreme left stant, *Baseline*: descending, *Lowercase letter "f"*: cross-like
Extraversion	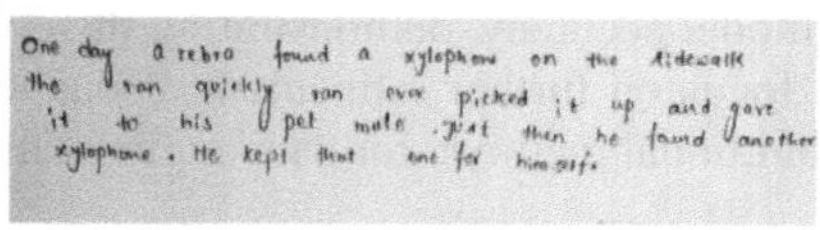	*Connecting strokes*: weak connectivity, *Baseline*: ascending, *Word slant*: extreme right stant
Neuroticism	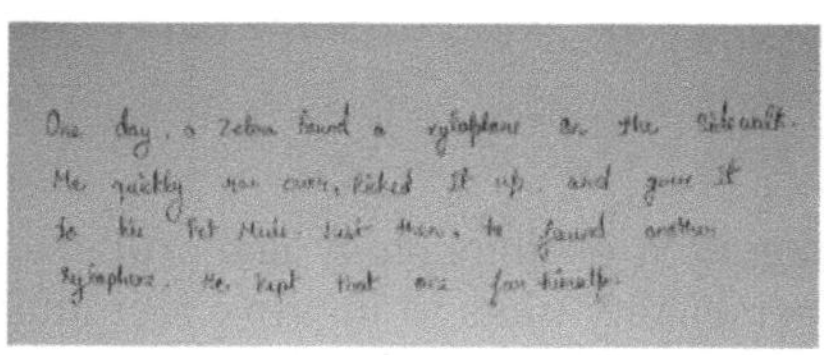	*Baseline*: leveled, *Word slant*: moderate right stant, *Lowercase letter "f"*: angular point
Openness	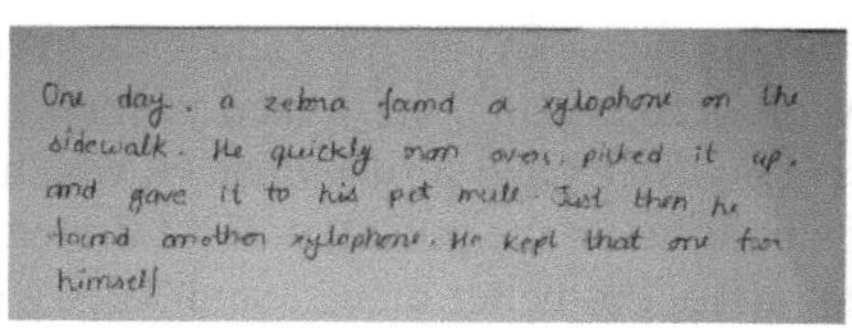	*Lowercase letter "f"*: balanced, *Connecting strokes*: medium connectivity, *Word slant*: moderate right stant

References

1. Yorke M (2004) Employability in higher education: what it is—what it is not. The Higher Education Academy/ESECT
2. Cattell Heather EP, Mead A (2008) The sixteen personality factor questionnaire (16PF)
3. Revelle W (2016). Hans Eysenck: personality theorist. https://doi.org/10.1016/j.paid.2016.04.007
4. Matthews G, Gilliland K (1999) The personality theories of H. J Eysenck and J A Gray: A comparative review. https://doi.org/10.1016/S0191-8869(98)00158-5
5. Elms AC (1993) Allport's personality and Allport's personality. In: Craik KH, Hogan R, Wolfe RN (eds) Fifty years of personality psychology. Perspectives on Individual Differences. Springer, Boston, MA. https://doi.org/10.1007/978-1-4899-2311-0_3
6. https://www.businessinsider.in/the-big-5-personality-traits-could-predict-who-will-and-wont-become-a-leader/articleshow/55863269.cms
7. Lebowitz Matthew S (2016) Stigmatization of ADHD. J Atten Disord 20:199–205
8. John Oliver P, Srivastava S (1999) The big five trait taxonomy: history, measurement, and theoretical perspectives
9. Fiske DW (1949) Consistency of the factorial structures of personality ratings from different sour sources. J Abnorm Psychol 44(3):329–44. https://doi.org/10.1037/h0057198
10. Norman WT (1967) 2800 personality trait descriptors: Normative operating characteristics for a university population. Department of Psychology, University of Michigan, Ann Arbor, MI
11. Smith Gene M (1967) Usefulness of peer ratings of personality in educational research. Educ Psychol Measur 27:967–984
12. McCrae RR, Costa PT (1987) Validation of the five-factor model of personality across instruments and observers. J Pers Soc Psychol 52(1):81–90
13. Goldberg LR (1990) An alternative "description of personality": the big-five factor structure. J Pers Soc Psychol 59(6):1216–1229
14. Kentle RL (1994). An examination of five personality factors. https://doi.org/10.1016/0191-8869(94)90221-6
15. https://www.kolbe.com/kolbe-wisdom
16. Xu Y, Tang Y, Suen CY (2022) Review of handwriting analysis for predicting personality traits. In: Krzyzak A, Suen CY, Torsello A, Nobile N (eds) Structural, syntactic, and statistical pattern recognition. S+SSPR 2022. Lecture Notes in Computer Science, vol 13813. Springer, Cham. https://doi.org/10.1007/978-3-031-23028-8_6
17. Kedar S, Nair V, Kulkarni S (2015) Personality identification through handwriting analysis: a review. Int J Adv Res Comput Sci Softw Eng 5(1):548–556
18. Gavrilescu and Vizireanu (2018) Predicting the Big Five personality traits from handwriting. EURASIP J Image Video Process 2018:57. https://doi.org/10.1186/s13640-018-0297-3
19. Plamondon R (2010) Neuromuscular studies of handwriting generation and representation. Int Conf Front Handwrit Recognit (ICFHR) 261. Kolkata, November 2010
20. Naghibolhosseini M, Bahrami F (2008) A behavioral model of writing. In: International conference on electrical and computer engineering (ICECE), pp 970–973
21. Fallah B, Khotanlou H (2016) Identify human personality parameters based on handwriting using neural networks. Artif Intell Robot (IRANOPEN)
22. Epshtein B, Ofek E, Wexler Y (2010) Detecting text in natural scenes with stroke width transform. Computer vision and pattern recognition (CVPR), IEEE Conference
23. Xiaoyuang L, Bin Q, Lu W (2009) A new improved BP neural network algorithm. In: Second international conference on intelligent computation technology and automation, pp 19–22
24. Deng L (2012) The MNIST database of handwritten digit images for machine learning research [best of the web]. IEEE Signal Process Mag 29(6):141–142
25. https://www.researchgate.net/publication/374396693_GrAI-phology_Integration_of_Graphology_With_Artificial_Intelligence
26. https://library.acadlore.com/ATAIML/2022/1/2/ATAIML_01.02_04.pdf

27. https://ietresearch.onlinelibrary.wiley.com. https://doi.org/10.1049/trit.2019.0051
28. https://www.researchgate.net/publication/284205700_Handwriting_Analysis_for_Detection_of_Personality_Traits_using_Machine_Learning_Approach
29. https://pmc.ncbi.nlm.nih.gov/articles/PMC4031150/
30. https://www.researchgate.net/publication/292159988_An_introduction_to_graphology_Definition_theoretical_background_and_levels_of_analysis
31. https://arxiv.org/pdf/1607.04573
32. https://pubmed.ncbi.nlm.nih.gov/22532169/
33. https://www.researchgate.net/publication/266620744_The_relationship_between_handwritten_signature_production_and_personality_traits
34. https://www.cs.uic.edu/~srizvi/BIT_Thesis.pdf
35. https://www.researchgate.net/publication/220861557_Graphological_Analysis_of_Handwritten_Text_Documents_for_Human_Resources_Recruitment
36. https://www.researchgate.net/publication/3966345_Writer_identification_by_writer's_invariants
37. https://www.researchgate.net/publication/360091471_A_comparative_study_of_handwriting_characteristics_in_three_generation_of_a_family_in_Haryanvi_population
38. https://www.researchgate.net/publication/220932994_Extraction_and_Analysis_of_Document_Examiner_Features_from_Vector_Skeletons_of_Grapheme_'th'
39. https://www.researchgate.net/publication/348974392_Arrow_R-CNN_for_handwritten_diagram_recognition
40. https://www.researchgate.net/publication/373029504_Personality_Recognition_based_on_Handwriting_Types_using_Fuzzy_Inference
41. https://www.researchgate.net/publication/380820435_A_Perspective_Analysis_of_Handwritten_Signature_Technology
42. https://www.personalityassessor.com/bigfive/

AI and Cybersecurity

NFT Ecosystem Challenges and Opportunities: Systematic Literature Review

Abhilash Kayyidavazhiyil, Raja Muthalagu, and Mohuya Chakraborty

Abstract A non-fungible token (NFT) is a distinct digital identity registered on a blockchain to verify ownership and legitimacy. Its inability is to be duplicated, replaced, or divided makes NFT unique. In this study, the Systematic Literature Review (SLR) by Preferred Reporting Items for Systematic Reviews and Meta-Analysis (PRISMA) model of the NFT ecosystem, its challenges and opportunities are elaborately discussed. The novelty of the paper lies in the conduction of the SLR of the NFT ecosystem using the PRISMA model, whereby the benefits and drawbacks of the stakeholders of the system were analyzed and synthesized. The SLR started with 216 papers from various databases and ultimately selected 31 papers published during 2020–2024 that satisfied the selection criteria of the present study. The nine research questions and gaps in the current research trends have been identified with a focus on ethical analysis, market value prediction, security, smart contract, and use cases of NFT ecosystem. The study would enable future researchers to find many open issues of this new technology and to develop novel solutions to enable the NFT market to proceed smoothly.

Keywords SLR · Blockchain · NFT · PUF · Smart contract · SCM · Tokenomics

A. Kayyidavazhiyil
Birla Institute of Technology and Science Pilani, Dubai Campus, United Arab Emirates
e-mail: p20230909@dubai.bits-pilani.ac.in

R. Muthalagu
Department of Computer Science, Birla Institute of Technology and Science Pilani, Dubai Campus, United Arab Emirates
e-mail: raja@dubai.bits.pilani.ac.in

M. Chakraborty (✉)
Cybersecurity Centre of Excellence, Institute of Engineering & Management, University of Engineering and Management, Kolkata, India
e-mail: mohuyacb@gmail.com

M. Chakraborty et al. (eds.), *Proceedings of Sixth International Ethical Hacking Conference*, Lecture Notes in Networks and Systems 1513,
https://doi.org/10.1007/978-981-96-8632-2_11

1 Introduction

Non-fungible token (NFT) is the digitization of an asset that is kept on a blockchain, much like other cryptocurrencies like Bitcoin. However, unlike Bitcoin or other cryptocurrencies, each NFT is unique and cannot be exchanged for another asset on a one-to-one basis. This is because each NFT represents a specific item, such as a digital artwork, video clip, music, or even tweets. The NFT ecosystem is a network of platforms, marketplaces, and tools that enable the creation, trading, and ownership of NFTs. The key components of the NFT ecosystem are originator, platform, blockchain, wallet, investors, curators, and regulatory authority. The originators are individuals or organizations that create NFTs such as digital artists, musicians, game developers and collectors. The OpenSea, SuperRare, and Nifty Gateway are NFT Platforms that offer a transparent and safe market place to confirm asset ownership and transfer where developers may sell their NFTs and customers can purchase and exchange them. NFTs are kept on the blockchain, which includes Ethereum, Binance, and other platforms.

Users are given access to NFT marketplaces allowing them to transmit, receive, store, and swap NFTs and other digital assets through an NFT digital wallet such as MetaMask, MyEtherWallet, and Trust Wallet. Capital for NFTs as a longstanding undertaking is provided by an investor. Conversely, an NFT curator is a person or institution that focuses on promoting NFT collections or artists. Regulations pertaining to tax reporting and consumer protection can be established by an organization that has regulatory jurisdiction over the use of NFTs.

Advantage of NFTs lies in their competence to permit the originators to sell their digital work in innovative ways directly to the collectors, avoiding distributers and maintaining transparency and trust as they are stored on non-permissioned blockchains. This allows collectors to authorize the validity and possession of their digital assets, plummeting the danger of deception and rising the asset's worth.

However, NFTs have their disadvantages as well. To mine and process cryptocurrencies like Ethereum, for the manufacture of numerous NFTs, considerable amount of energy is consumed causing issues of environmental sustainability. Since NFTs are exclusively obtained in the cryptocurrency market, the potential market for digital originators may also be limited. There is also a lack of regulation of the NFT market, which can result in deception, cheats, and additional exploitations. As many countries have not yet regulated NFTs, therefore it can be challenging for NFT stakeholders to trade and protect their digital assets [1].

The NFT ecosystem has been widely studied in this Systematic Literature Review (SLR) using Preferred Reporting Items for Systematic reviews and Meta-Analysis (PRISMA), with an emphasis on several application areas, including games, art, music, and opportunities as well as difficulties. The gap analysis has been done to find the scope of research in this field for the benefit of society at large.

The remainder of the paper is structured as follows. Section 2 provides explanations of the ideas and terminologies while Sect. 3 provides details on the study process. Section 4 highlights the existing related works and performance analysis of existing frameworks. Section 5 explains the research gaps and findings with an analysis of the research questions. Section 6 concludes the paper with highlights of prospects.

2 Concepts and Terminologies

In this section, we elucidate the various technical components connected with NFT events. For a completely functional NFT ecosystem, these components play an important role in laying its structural foundations.

2.1 *Blockchain*

A distributed digital ledger that keeps track of transactions on a computer network is known as a blockchain. Data may be safely stored and shared using this immutable, transparent database. As it provides the foundational infrastructure for developing, managing, and trading these distinctive digital assets, blockchain technology is closely tied to NFTs. An NFT being a digital asset representing ownership of a unique object or piece of content are created using blockchain technology thereby maintaining secured and decentralized ownership record. NFTs are invaluable resources due to their distinct and non-duplicable nature. Blockchain gives security, decentralization, transparency, and interoperability to the NFT transactions and so it forms a vital constituent of NFT ecosystem. There are several token standards of unique digital assets viz., ERC-20, ERC-721, ERC-998, and ERC-1155 that can be possessed, traded, and controlled by means of smart contracts on the Ethereum blockchain.

2.2 *Tokenomics*

Tokenomics is a combination of two words "token" and "economics" that deals with the monetary standards for the development, manufacturing, circulation, and supervision of digital tokens, or cryptocurrencies. The elements of tokenomics are Distribution, Supply, Demand, Governance Structure, Token Utility, and Inflation Rate. These elements are destined to offer value to users and investors, inspire the adoption of the token, and build a robust and lasting ecosystem around it. There are various types of tokens, such as security tokens, utility tokens, and governance tokens, each having distinct features and purposes that can be created by tokenomics.

2.3 Smart Contract

Smart contracts give the software design logic for determining the terms of conditions for ownership, and transfer of NFTs. They are self-executing piece of software code stored in blockchain that expresses the terms and conditions between a seller and a buyer. As they are used by blockchain, a safe, decentralized platform for managing the ownership and transfer of digital assets is maintained by these smart contracts. A smart contract might stipulate, for example, that the original developer of the asset is liable to get a certain percentage of the sale price or that only the current owner of the asset may transfer ownership. So, they form a crucial part of the NFT ecosystem. NFT smart contracts are created by Solidity, a computer programming language [2]. Figure 1 depicts the Solidity code of the NFT smart contract "TestNFT" [3]. The source of this code is OpenZappelin [4], a library for developing secure smart contracts with ERC-721 implementation.

```
// SPDX-License-Identifier: MIT
pragma solidity ^0.8.20;

// Import OpenZeppelin ERC721 and Ownable
import "@openzeppelin/contracts/token/ERC721/ERC721.sol";
import "@openzeppelin/contracts/access/Ownable.sol";

contract TestNFT is ERC721, Ownable {
  uint256 private _tokenCounter;

  // Constructor sets the collection name and symbol
  constructor() ERC721("TestNFT", "TNFT") Ownable(msg.sender) {
    _tokenCounter = 0;
  }

  /// @notice Mint a new NFT to the specified address
  /// @param to The address that will receive the NFT
  function mintNFT(address to) external onlyOwner returns (uint256) {
    uint256 newTokenId = _tokenCounter;
    _safeMint(to, newTokenId);
    _tokenCounter += 1;
    return newTokenId;
  }

  /// @dev Optional: provide a base URI for metadata
  function _baseURI() internal pure override returns (string memory) {
    return "https://your-metadata-endpoint.example/api/token/";
  }

  /// @notice Optional burn function (owner or approved)
  function burn(uint256 tokenId) external {
    require(
      _isApprovedOrOwner(msg.sender, tokenId),
      "Not owner nor approved"
    );
    _burn(tokenId);
  }

  /// @notice View current token counter (next token ID)
  function currentCounter() external view returns (uint256) {
    return _tokenCounter;
  }
}
```

Fig. 1 Smart contract code using solidity for the contract "TestNFT"

2.4 Blockchain Address and Transaction

Transactions on a blockchain network are directed and retrieved using special codes for identification in the form of a string of characters and numbers that corresponds to a particular public key of a user. These are known as blockchain addresses. A user needs to provide their blockchain address to a sender in order to receive cryptocurrencies or other digital assets on a blockchain network. The sender can then use this address to start a transaction, which will be entered into a public ledger and documented on the blockchain. A transaction on a blockchain network is broadcast to every node in the network for verification and confirmation before appending it to the blockchain ledger. Once a transaction has been accepted and appended to the blockchain, there is no way it can be altered or deleted. Hence, blockchain addresses and transactions are crucial components of a blockchain network.

2.5 Token Standards

This section provides a comparative study of NFT token standards, such as ERC-20 [5], ERC-721 [6], ERC-998 [7], and ERC-1155 [8]. Fungible tokens are created on the Ethereum network using the ERC-20 standard. ERC-721 and ERC-998 standards are used to create non-transferable unique NFTs on the blockchain. On the other hand, ERC-1155 is used to create both fungible and non-fungible tokens. ERC-20 is used for financial matters, ERC-721 and ERC-998 are used for virtual or real estate, gaming and artwork matters. ERC-1155 is used purely for gaming. Table 1 depicts a comparative analysis of token standards.

2.6 Figure Formatting

An NFT protocol is a set of rules and specifications that regulates the formation, proprietorship, and exchange of NFTs on a blockchain network. Two design approaches exist for the NFT paradigm. These are top-down and bottom-up approaches.

(1) *Top-down approach*: This approach contains two additional roles: NFT owner and NFT buyer. The steps involved are:

- *Step 1*: Digitization of NFT—The NFT owner verifies that the file, title, and descriptors are correct. The raw data is then digitized and converted into the suitable format.
- *Step 2*: Storage of NFT—The NFT owner stores the raw data outside of the blockchain in an external database. They are also permitted to keep raw data within a blockchain, even though this activity consumes gas.

Table 1 Comparative analysis of token standards

Token standard	Token type	Interoperability	Token attributes	Use cases	Complexity
ERC-20	Fungible	Interoperable with different wallets and exchanges	Basic (name, symbol, and total supply etc.)	Financial	Low
ERC-721	Non-fungible	Exchangeable with other ERC-721 and ERC-20 tokens	Additional metadata (image, description etc.)	Virtual real estate, gaming items, and artwork	High
ERC-998	Non-fungible	Interoperable with other NFTs and ERC-20 tokens	Composable NFT that can contain other NFTs as "sub-tokens" or "child tokens"	Complex structures in gaming, art, or real estate	High
ERC-1155	Both fungible and non-fungible	Interoperable with other token standards	Basic and additional metadata	Game item (fungible), rare and unique skin for a game character (non-fungible)	Low

- *Step 3*: Signing of NFT—In this step, the NFT owner signs a transaction, providing the hash of the transaction to a smart contract.
- *Step 4*: Mint and Trade of NFT—The minting and trading procedure begins shortly after the smart contracts are created. Smart contracts are simply programs stored on a blockchain that run when predetermined conditions are met. It gets the transaction that includes the NFT data. The logic of the Token Standards is the primary mechanism underlying NFTs.
- *Step 5*: Confirmation of NFT—Once the transaction is confirmed, the minting procedure is complete. As a result of this strategy, NFTs will always be linked to a unique blockchain address as proof of persistence.

(2) *Bottom-up approach*: This approach involves two additional roles: NFT creator and NFT buyer. The steps involved are:

- *Step 1*: Template Formation—Using the smart contract, the project founder launches a template to set up numerous basic rules, such as distinct elements (character style, weaponry, or accessories) in the game.
- *Step 2*: Randomization of NFT—When a customer bids on an NFT, they can customize it with a set of extra features on top of the basic lines. These extra features are chosen at random from a database that was predefined at the start.

- *Step 3*: Mint and Trade of NFT—When the related smart contract is triggered, the minting and trading procedure begins.
- *Step 4*: NFT Confirmation—The procedures are carried out using smart contracts. When the consensus mechanism is done, the created NFT will be persistently stored on-chain.

Every block in the blockchain system is restricted in size. Other transactions will enter a subsequent block connected to the first data block when a block's capacity is achieved. Finally, all linked blocks have produced a long-term history that will last forever. A new transaction must be completed to activate the smart contract each time an NFT is created or sold. The ownership information and NFT metadata are added to a new block when the transaction is confirmed, preserving the NFT's history and ownership. The NFT ecosystem work flow comprising 11 steps is depicted in Fig. 2.

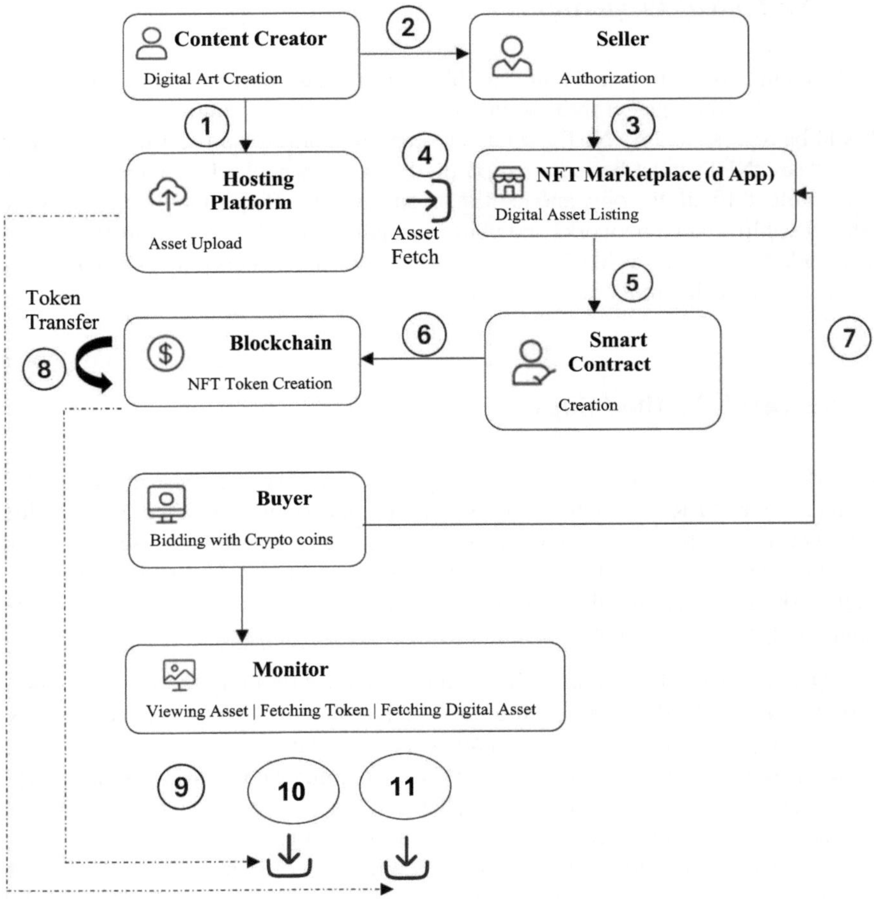

Fig. 2 NFT ecosystem workflow and market participants' interactions

The content creator creates the digital art and uploads the asset to the hosting platform in step 1. In Step 2, the content creator authorizes the seller. The seller then lists the digital asset in the NFT marketplace through the decentralized application (dApp) in Step 3. Thereafter, from the hosting platform the digital asset is fetched by the NFT marketplace in Step 4. In Step 5, the creation of the smart contract takes place through the trusted exchange interface of the website of NFT. NFT tokens are then created through the minting process in the blockchain in step 6. The buyer does the bidding (sending an offer to an NFT holder) using CryptoCoins in Step 7. In Step 8, token transfer takes place in the blockchain. The buyer views the digital asset in the monitor, the monitor fetches the token from the blockchain and the monitor fetches the digital asset from the hosting platform in Steps 9, 10, and 11, respectively.

2.7 NFT Ideal Properties

NFTs ought to be distinct and non-fungible, which means they cannot be traded one-to-one like conventional cryptocurrencies. The ownership or legitimacy of an object should be represented by NFTs, which are kept on a blockchain address linked to a private key. NFTs must be recorded on the blockchain network as they need to be unalterable, difficult to copy, and legitimate and unique. They should be compatible with other blockchain networks and platforms. NFTs must be programmed to incorporate additional features like incentives for creators and owners or royalty payments to enhance their flexibility and customization.

3 Research Methodology

SLR of the current body of research is presented using the PRISMA model [9, 10] in this section. SLR is a methodical procedure concerning a detailed examination of numerous databases and literature sources, screening pertinent research based on recognized criteria, extracting data from selected studies, and synthesizing the results. The four steps involved in SLR thereby completing the whole process with a clear understanding of the research objectives are as follows.

- *Step 1*: Research Questions Formulation—includes defining the research question, followed by developing a thorough search plan to discover pertinent papers, databases, keywords, and inclusion/exclusion standards.
- *Step 2*: Resource Identification—comprises screening studies using inclusion and exclusion criteria, examining titles, abstracts, and complete texts to determine which ones meet requirements and assess applicability.
- *Step 3*: Data Collection and Analysis—encompasses extracting information such as participant specifics, interventions, outcomes, sample size, and other relevant particulars from selected studies.

- *Step 4*: Data Synthesis—embraces synthesizing the results of studies, either through statistical analysis or qualitative synthesis, to guide practice, policy, or future research.

SLRs are commonly used in various fields of study to provide evidence-based answers to research questions, identify research gaps, and guide future research. They are considered the gold standard in research synthesis because they use a rigorous and transparent approach to identify and evaluate evidence, minimize bias, and ensure the reliability and validity of the findings.

3.1 Research Questions Formulation

The goal of this study is to review the existing literature on NFT applications, datasets, wallets, users, marketplace, and environmental impact to pinpoint any gaps in the literature. The project began with a deep review of NFT literature, preparing an NFT framework that produced a list of 182 questions to be surveyed by blockchain technology and NFT leaders, users, curators, and investors, and analyzing the answers from market surveys. Nine key Research Questions (RQs) vis-à-vis Research Areas (RA) were identified as given by the research flow diagram (Fig. 3). Table 2 presents the snapshot of the research table correlating the Research Objectives (ROs), data collection methods, and contribution to research.

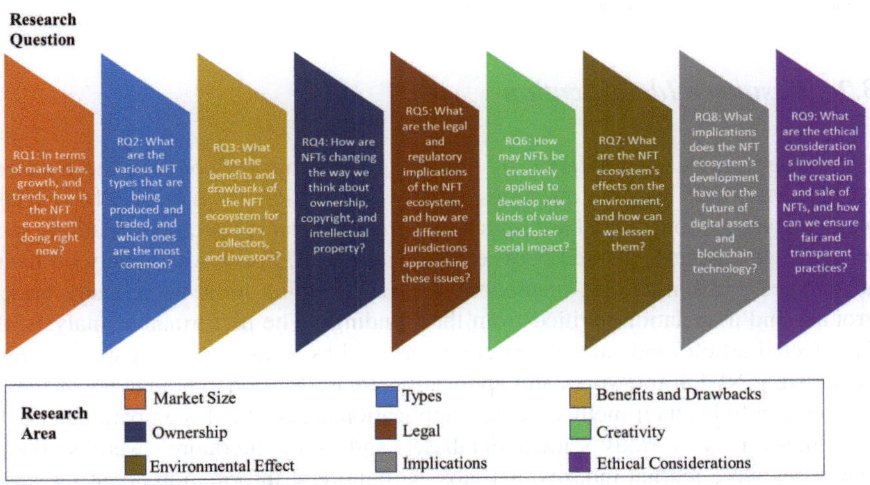

Fig. 3 Research flow

Table 2 Research table

Research Objectives (ROs)	Data collection/analysis	Contribution to research
RO1: To investigate the existing research works on the NFT ecosystem during 2020–2024	Market survey/statistical analysis	NFT research
RO2: To investigate types of current NFTs	Market survey/statistical analysis	NFT collectibles
RO3: To investigate the challenges and opportunities of the NFT ecosystem	Market survey/statistical analysis	NFT ecosystem
RO4: To investigate IPR issues of NFTs	Market survey/statistical analysis	IPR
RO5: To investigate the legal implications of the NFT ecosystem	Market survey/statistical analysis	Legal
RO6: To investigate create-level use cases of NFT	Market survey/analytical reasoning and statistical analysis	Applications
RO7: To investigate the environmental impacts of NFT	Literature review/analytical modeling	Environmental impact mitigation
RO8: To investigate the implication of digital assets and blockchain technology in connection with the NFT ecosystem	Literature review/analytical modeling	Blockchain
RO9: To investigate the values and ethics of NFT creation and trade	Market survey/qualitative analysis	Values and ethics

3.2 Resource Identification

In this step article identification by searching pertinent keywords in the field of NFT based on the research questions identified in section 3A is done. 216 articles were analyzed through database searching across several research publications and industry whitepapers released between 2020 and 2024. Focus was mainly on their advantages and limitations, methodologies, investigation tools used, study background, and association verified from their findings. The performance analysis of 31 selected articles indicates the strengths as well as weaknesses or gaps between the existing NFT frameworks and application areas. An in-depth analysis of these articles highlights their methodologies, application areas, and design parameters.

Articles from various known databases and 'grey literature' were studied. Databases were journal indexes that gave us important information about the articles under consideration for study including its publisher, author, topic, etc. These databases have hundreds and thousands of searchable articles. To perform an SLR, the databases used are IEEE Xplore, Directory of Open Access Journals (DOAJ), ScienceDirect, Google Scholar, PubMed, Scopus, Web of Science, ResearchGate, Science Citation Index, arXiv and SpringerLink, and DBLP. To get articles based on 'grey literature' Google Scholar, whitepapers, and the university's library list were searched in order not to omit a single article relevant to the research. Initially, solely the titles, abstracts, and keywords during 2020 through 2024 were screened. The keywords for search are 'Non-Fungible Token', 'NFT Marketplace', 'Token Standard', 'NFT Digital Collectibles', 'NFT Challenges', 'NFT Opportunities', 'NFT Smart Contracts', 'NFT Blockchain', 'Security in NFT Ecosystem', 'Crypto Collectibles', 'Risk and Return of NFT', 'NFT Transactions', 'NFT Trading', etc. 216 publications were nominated for study in this phase. The studies included a selection of searchable keywords over the above-mentioned eleven databases excluding articles based on headings, keywords, summaries, and complete manuscripts. The non-English articles having missing keywords/published outside the defined year range/5 to 10 citations per year were eliminated.

3.3 Data Collection and Analysis

The research was narrowed down through the inclusion and selection criteria: (a) papers published during 2020 and 2024, (b) peer-reviewed papers, and (c) publications in well-known English-speaking journals. Most importantly papers published in reputed journals were considered to maintain the standard.

3.4 Data Synthesis

The Preferred Reporting Items for Systematic Reviews and Meta-Analyses (PRISMA) [11] model was used as the basis for the data analysis flow diagram (Fig. 4). The diagram depicts that a total of 216 papers were recognized based on keywords mentioned in section 3B. In the initial examination procedure, 125 papers were recognized to meet the eligibility criteria. Duplication was removed by online software. Now, 94 papers were nominated. After passing the articles through the initial screening procedure for inspection of the titles and abstracts, unrelated articles were removed and 57 publications endured. At the last phase, 31 articles were designated for study after removing those that were not directly connected with this research. All of the numerical values are represented by the variable 'var'. To summarize, 91 articles were disqualified based on initial phase noncompliance with the exclusion criteria listed in section 3B for the title and abstract. Thereafter at the screening phase, 31 articles were excluded as we found duplication in the related titles and abstract keywords. In the third phase, 37 articles were excluded after scanning through the full articles as they were found to be not directly focused on the research question. After investigating and exploring the articles in their entirety, 26 publications were discarded and 31 publications were included in the final inclusion stage as shown in Fig. 4 in the PRISMA flow diagram.

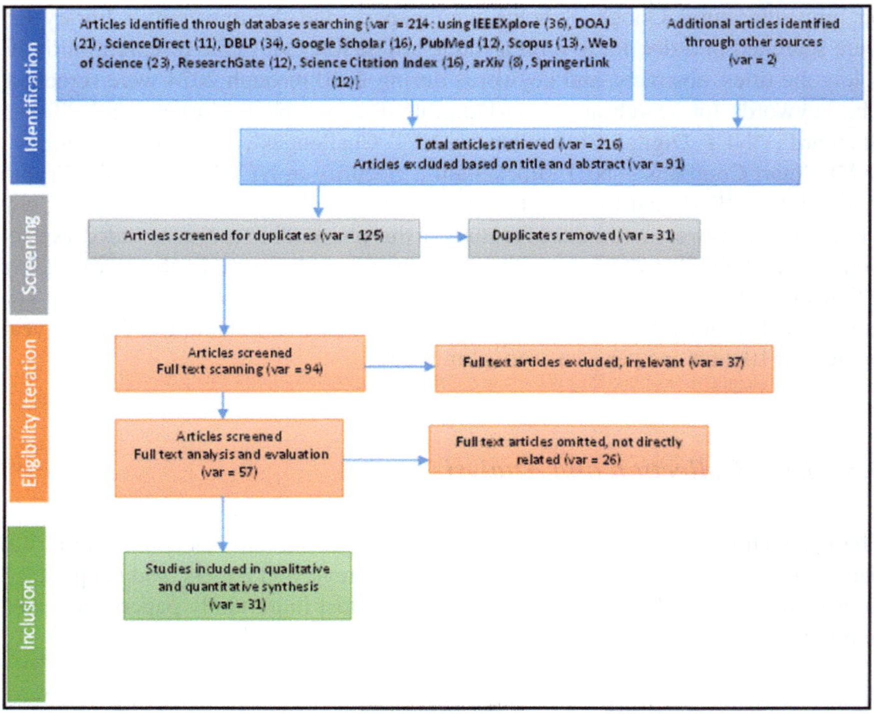

Fig. 4 PRISMA flow diagram for systematic literature review payments to enhance their flexibility and customization

4 Related Works

In this section, we present the existing related works of NFT. The 31 study articles included at the last stage of the PRISMA model were segregated into several categories (C) such as ethical analysis, market value prediction, security, smart contracts, and use cases.

C1: *Ethical Analysis*

For an ethical analysis of NFT, an exhaustive inspection of the consequences of NFT on the environment, intellectual property, artistic inheritance, and monetary permanency is essential. One of the ethical concerns is the environmental influence of NFT as it demands excessive energy usage making it unsustainable and interfering with slowdown of climate change. Apart from this, likelihood of dishonesty and exploitation linked with NFTs is another ethical apprehension. Yet another ethical concern is the sale of artists' NFT creation without their accord rising anxieties about the legitimacy of possession and copyright. NFTs have the potential to monetize cultural artifacts and may further cement existing imbalances within the art world by conferring benefits upon those who already easily gain access to channels of power in order to promote their work. The most recent Code of Ethics and Professional Conduct by the Association for Computing Machinery (ACM) can be referred to conduct an ethics analysis of NFT along with deployment, implementation and sustainability [1].

C2: *Market Value Prediction*

The market value of NFTs is affected by elements that include the cultural significance or reputation as well as popularity of its underlying assets one creation might have. Each NFT itself may also vary in value based on its rarity and unique qualities, with things like limited editions or even interactive arts components fueling the prices for those that go beyond need-to-have collectibles to objects flaunting expensive good taste from abroad. Overall demand for NFTs and their values are likewise influenced by more general movements within the blockchain and cryptocurrency markets. But it is crucial to understand that, much like any asset, the value of NFTs can change dramatically, and investing in them always carries a risk of losing money. When making any investment decisions, like with any other, it is crucial to carry out extensive research, weigh the advantages and disadvantages, and consult a financial advisor. With a primary focus on primary and secondary sale price prediction analysis, a comprehensive statistical analysis of NFT market trade and networks of interaction was carried out using datasets from the Ethereum and WAX blockchains, encompassing over 6 million trades of roughly 5 million NFTs between 2017 and 2021 [12]. There is always a chance that the NFT market will experience another price bubble, as they have happened in the past in a number of areas, such as the cryptocurrency and art markets.

A significant spike in demand without a corresponding increase in NFT supply, possibly due to hype or speculation, could be one of the factors contributing to a price bubble in the NFT market. Lack of control or guideline in the NFT market may result in huge price variations and market instability [13]. Feature vectors of previous NFT collections may be extracted through unsupervised learning by applying techniques such as dimensionality reduction or clustering on the transaction data to classify these collections based on factors including the type of NFTs, the originator or artist, the rarity or uniqueness of the NFTs, and the total worth of the collection [14]. This aids in calculating the potential market value of future newly minted transactions of NFT collectibles. Polychain, NEKO, EtherThings, Hashmasks, and BAYC, were the five market-traded NFT collections that were used for training using a generative contextual approach to understand the market capitalization reflecting the total value of the outstanding tokens owned by NFT holders at a given point in time. Testing, result analysis, and evaluation on four such NFT collections such as Chubbies, CryptoTrunks, VeeFriends, and MAYC depicted a combination of good and poor performance. The authors came to the conclusion that their method was linear as opposed to the adaptive strategy used in the production of NFT transactions. In [15], the risk and return features of NFT startups that are traded on the cryptocurrency market were examined.

C3: *Security*

NFT security is a complex and multifaceted issue that requires careful consideration and attention to detail [16]. By taking appropriate security measures, such as using reputable marketplaces, secure storage solutions, and verified transfer channels, it is possible to minimize the risk of security breaches and protect NFT investments. In the NFT ecosystem issues such as digital content ownership, low-quality NFTs, frauds, possible money laundering, and laws are important considerations while dealing with NFTs. Various design considerations for improving the NFT ecosystem should be kept in mind. Works on potential security challenges and the corresponding solutions of NFTs along with opportunities and challenges of NFT applications have been studied in [17, 18]. The effectiveness of multimodality in market value prediction of NFT using both pictorial and non-pictorial evidence by using machine learning and end-to-end deep learning approaches have been studied in [19]. To track summary data on pricing and sales activity across several NFT collections, intelligence-based tools have evolved for market research [20]. Various models have been developed to detect issues leading to substantial financial loss of several NFT marketplaces having high volume of transactions. These are related to fraudulent user behavior like counterfeiting NFT creation, and trading malpractices including wash trading, shill bidding, and bid shielding [21]. The typical asset-based portfolios in the NFT market have varied consequences [22, 23].

C4: *Smart Contracts*

Software abstraction for generating deployable smart contracts for NFT on the Ethereum blockchain using ERC-721 standard and a specification language having the advantage of code extensibility and reusability with quality metrics of Solidity smart contracts have been developed [24]. The design of a smart contract-based Federated Learning (FL) framework for incentivizing the participation of devices in the Federated Learning process (FLP) and claiming ownership rights to a global model is a possibility [25]. One of the proposed solutions introduces a blockchain-based solution toward registration, reputation, and incentivization for the NFT ecosystem by incorporating a decentralized Inter-Planetary File System (IPFS) and storing its hash on the chain to enforce trust in the NFT marketplaces [26]. In short, blockchain technology enables a transaction to be held in a decentralized manner.

C5: *Use Cases*

NFTs have the potential to revolutionize many industries and provide new opportunities for creators, collectors, and investors. However, it is important to note that NFTs are a relatively new and rapidly evolving technology, and their use cases and value propositions may continue to evolve over time. NFTs have the possibility to transform the event ticketing industry, providing new opportunities for secure and personalized ticketing, as well as creating new revenue streams and markets for event organizers and attendees, tokenize digital products, reduce fraud, and better manage secondary market trades. [27].

Sentiment analysis is a technique used to identify and extract the emotions and opinions expressed in a piece of text, such as social media posts, news articles, or customer reviews. In [28], the authors have used Bidirectional Encoder Representation from Transformers (BERT) model to do sentiment analysis of NFT on social media such as Twitter. They utilized data crawling and pre-processing with 86% accuracy and 87% precision. Because each NFT has several traits, there must be a means to combine the rarity of each attribute into a single value for each NFT to rank them. An interpretable rarity meter for crypto collectibles is a valuable tool for collectors and investors who are looking to identify the most rare and valuable NFTs and crypto collectibles and to make informed decisions about buying and selling in the market. The authors in [29] developed a pipeline for designing the best rarity meter for NFT and articulated the optimal rarity meter problem, which will help collection designers get the desired diversity in a single collection as well as when comparing different collections. Data collection, algorithmic analysis, interpretable metrics, transparency, and auditability are some of the features that could be introduced in the designed tool. Smart contracts on Ethereum have been written in [30].

The problems of the real estate system are certainly obvious and there is an option of the NFT being used as a likely solution [31]. In [32], the authors proposed a blockchain-based electronic registry framework while the authors of [33] developed a working prototype. Another use case is the use of NFT certificates and delivery evidence for priceless jewelry and precious stones, which are highly valued and cherished forms of personal adornment that are often associated with luxury and high-end fashion. Traditional paper-based certificates have issues such as loss, theft, and counterfeiting that are solved with NFT-based certificates to provide superior verifiability, traceability, immutability, security, auctioning, delivery, and ownership management [34]. NFT-based traceability and ownership management may also be used for medical devices [35].

In a different use scenario, the researchers used a soliton communications-based strategy to raise the effectiveness of NFT-based distribution systems in optical links with aggregated EDFA amplification [36]. By employing uninterrupted contracts as NFT in maintaining inter-relationships of components for Decentralized Finance is presented in [37]. The reliability of both hardware and software of end-to-end IoT devices by binding these devices to smart NFT using Physical Unclonable Functions (PUF) is discussed in [38]. NFT has been used for transactions between producers, consumers, rechargeable automobiles, power corporations, and storage providers in yet another application [39]. Anonymous NFT trading by hiding the address of NFT owner during trading is achieved in [40]. Concerns for using NFT usage on the social media platform Twitter has been studied in [41].

5 Research Gaps and Findings

Although the NFT ecosystem is a relatively new and rapidly evolving field, there are still several gaps in research that need to be addressed. In this section, we present the Research Gaps (RGs) and findings after careful investigation and analysis of the study articles under consideration [42]. The key areas where research is lacking are depicted in Fig. 5.

Below is a presentation of the findings and analysis of RQs after a vigilant analysis of RGs.

RQ1. In terms of market size, growth, and trends, how is the NFT ecosystem doing right now?

Analysis—RQ1

The NFT ecosystem has extended and transformed intensely during the last few years as a result of several factors, including improved consciousness and acceptance among collectors, the use of NFTs as a new source of income for producers, and the rising interest in blockchain technology and its conceivable applications. As such the current trends in the NFT ecosystem includes ongoing expansion and diversification, wider acceptance by the general public, integration with other technologies, and synergetic apprehensions posing questions about sustainability and solutions to lessen carbon footprints.

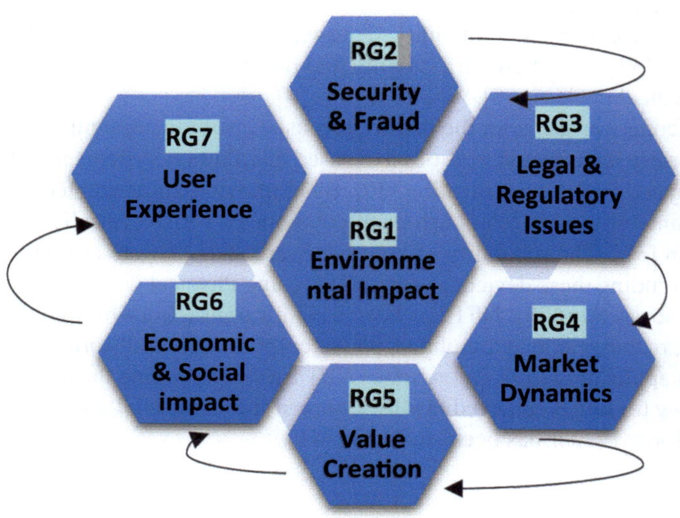

Fig. 5 nResearch gaps

RQ2. What are the various NFT types that are being produced and traded, and which ones are the most common?

Analysis—RQ2

There are many NFTs being created and distributed, covering a wide range of subjects and classifications such as art, fashion, meme, collectibles, gaming, music, sports, real-world assets, etc. The most popular and best to buy NFTs as of May 2023 are AiDoge (AI-Driven NFT Memes Platform), Launchpad XYZ, yPredict (all-in-one AI ecosystem for NFT traders), Robotera (project based on and revolves around a robot-based theme), Love Hate Inu (a memecoin featuring a unique voting mechanism), Tamadoge, Azuki NFTs, Invisible Friends (Ethereum based NFTs from the Random Character Collective), Stepn App Sneakers NFTs, Moonbirds NFT (digital art of cartoon owls), CryptoPunks, Axies, Bored Ape Yacht Club (online avatars), Pudgy Penguins (whose focal point is 8,888 Pudgy Penguins), VeeFriends, Decentraland (first fully decentralized world), Meebits, Loot for Adventurers, Town Star, Matrix NFTs (avatars based on characters living in the Matrix) [43]. These types of NFTs are by no means exhaustive, and more types are constantly being created. Depending on elements like market demand and societal trends, the popularity of certain NFT categories can change over time.

RQ3. What are the benefits and drawbacks of the NFT ecosystem for creators, collectors, and investors?

Analysis—RQ3

For its participants, including producers, collectors, and investors, the NFT ecosystem offers a number of advantages and disadvantages. This is depicted in Table 3.

- *Benefits for creators*: New methods of revenue generation are available to creators, including the sale of NFTs. NFTs can be programmed by creators to automatically include royalties, ensuring continued income from secondary market sales.

Table 3 Benefits and Drawbacks of the NFT Ecosystem

NFT stakeholders	Benefit	Drawback
Creators	New revenue streams	High barriers to entry
	New revenue streams	Environmental concerns
	Direct engagement with fans	
Collectors	Ownership and provenance	Volatility
	Investment potential	Limited utility
	Transparency	
Investors	New asset class	Volatility
	High growth potential	Lack of regulation

Through NFTs, creators may interact with fans directly, opening up a new channel for community development and fan interaction.

- *Drawbacks for creators*: For some creators, the costs and complexity of producing and selling NFTs create substantial entry barriers. NFTs' high energy consumption has sparked environmental worries, which may influence some developers to steer clear of them.
- *Benefits for collectors*: NFTs' price could escalate over time, offering collectors investment opportunities. NFTs offer ownership and provenance transparency, thereby reducing the likelihood of scam and increasing assurance of buyer and seller.
- *Drawbacks for collectors*: NFT values can swing significantly, which poses risks to collectors. Some limited- use NFTs might lessen collectors' income.
- *Benefits for investors*: Investors have a new asset class, which may offer prospects for diversification and substantial returns on investment.
- *Drawbacks for investors*: Investors may face risk due to the potential for considerable volatility in the value of NFTs. Due to unregulated nature of NFT industry, investors may face the perils of fraud and scam.

RQ4. How are NFTs Changing the way we think about ownership, copyright, and intellectual property?

Analysis—RQ4

NFTs can give authorization to creation of digital ownership, permitting proprietary agreement to digital assets of creators for monetization, be used to represent ownership of copyrighted works, be configured to automatically pay royalties to creators on secondary marketplaces. NFTs can further be used to represent allocated ownership of physical properties and assets for investment in formerly inaccessible possessions by delivering transparency in ownership and attribution of digital assets to reduce deception and improve trust between buyers and sellers. Since NFTs offer new methods for representing and making money from digital goods, they are upending conventional ideas of ownership, copyright, and intellectual property.

RQ5. What are the legal and regulatory implications of the NFT ecosystem, and how are different jurisdictions approaching these issues?

Analysis—RQ5

The NFT ecosystem has complex legal and regulatory consequences that alter depending on the jurisdiction. Intellectual property rights, consumer protection, anti-money laundering, and securities laws are a few of the major legal and regulatory challenges pertaining to NFTs. Diverse governing bodies are undertaking these difficulties in various ways. For example, the Securities and Exchange Commission (SEC) of the United States has cautioned that a few of the NFTs may be viewed as securities and as such be ruled by securities regulations. Similarly, the European Union is examining ways to regulate NFTs in agreement with its regulatory outline for digital assets. These legal and regulatory complications of the NFT ecosystem must be understood by creators, collectors, and investors for complying with local laws and regulations.

RQ6. How may NFTs be creatively applied to develop new kinds of value and foster social impact?

Analysis—RQ6

NFTs can in a variety of ways foster social impact and provide new types of value. They can be used to verify the authenticity of historical artefacts, works of art, and collectibles thereby raising the product worth and avoiding deception. They can offer additional sources of income for artists and creators. NFTs can create and sell distinctive digital assets to raise money for philanthropic causes. Decentralized and distributed social networks can be made to give ownership to common people who can have control over their data by using NFTs. Tokenization of assets from the real world, including equities, bonds, and real estate can be made through NFTs making them more accessible to a wider spectrum. NFTs can be used to authorize the ownership and identity of tangible items such as certificates, passports, and land titles thereby reducing fraud and raising transparency.

RQ7. What are the NFT ecosystem's effects on the environment, and how can we lessen them?

Analysis—RQ7

The energy-intensive blockchain technology that underpins the creation and trading of NFTs has a detrimental environmental impact. A large amount of computing power is needed to create an NFT, which uses energy and emits carbon dioxide. Several strategies exist to alleviate the NFT ecosystem's detrimental impact on the environment. One option to reduce the carbon footprint of the NFT ecosystem is to use renewable energy sources, such as solar, wind, or hydroelectric power, to power the computing infrastructure. Another strategy to lessen the negative environmental consequences of the NFT ecosystem is to encourage NFT producers and collectors to adopt more eco-friendly practices. Investigating alternative technologies, such as non-blockchain digital ownership solutions, can also aid in reducing the environmental impact of the NFT ecosystem.

RQ8. What implications does the NFT ecosystem's development have for the future of digital assets and blockchain technology?

Analysis—RQ8

Among the most significant changes are (i) divergence of NFT use cases, (ii) incorporation with other technologies, (iii) challenging out new blockchain platforms, (iv) consideration on sustainability. These propensities have important consequences for the development of digital assets and blockchain technology. As the NFT ecosystem grows, it will be vital for stakeholders to work in a composite manner to address the challenges faced by the NFT industry and ensure that NFTs are used in a responsible and sustainable manner.

RQ9. What are the ethical considerations involved in the creation and sale of NFTs, and how can we ensure fair and transparent practices?

Analysis—RQ9

NFTs that are suitable for cultural artifacts or designs without prior permission and those that infringe on the intellectual property rights of others can be seen as ethically challenging. Ethical concerns are in place due to the excessive use of energy and emissions of carbon dioxide by the NFTs. High manufacturing cost of NFTs prevent many people from using and obtaining them, creating and inequality of access. To mitigate these issues, we must ensure that there is clear statement about licensing terms and ownership of NFTs, use renewable energy sources to produce NFTs for lessening the carbon footprint, think of strategies to reduce cost of NFT creation, and enforce industry associations and organizations to develop ethical guidelines for the creation and sale of NFTs, and creators and buyers to adhere to these guidelines [44, 45].

6 Conclusion

This paper has provided an SLR of the NFT ecosystem challenges and opportunities. The related works exploring various categories of NFT research, such as ethical analysis, market value prediction, security, smart contract, and use cases, have revealed seven RGs or problems that have not yet been answered or solved by existing literature. These are closely related to environmental, economic, and social impact, safety, security and scam, legal and regulatory issues, market undercurrents, worth creation, and user experience. The findings of the study in the form of nine RQs have been critically analyzed for an overall impression of the future scope and prospect of the NFT ecosystem.

References

1. Flick C (2022) A critical professional ethical analysis of non-fungible tokens (NFTs). J Responsible Technol 12:100054. ISSN 2666-6596. https://doi.org/10.1016/j.jrt.2022.100054
2. Solidity Programming Language|The Solidity language portal is a comprehensive information page for the Solidity programming language. It features documentation, binaries, blog, resources & more
3. How to Write & Deploy an NFT (Part 1/3 of NFT Tutorial Series) | ethereum.org
4. https://www.openzeppelin.com/
5. ERC-20 Token Standard | ethereum.org
6. ERC-721 Non-Fungible Token Standard | ethereum.org

7. ERC-998: Composable Non-Fungible Token (ethereum.org)
8. ERC-1155 Multi-Token Standard | ethereum.org
9. Kitchenham B (2004) Procedures for performing systematic reviews. Keele, UK, Keele University 33:1–26
10. Kitchenham B, Charters S (2007) Guidelines for performing systematic literature reviews in software engineering
11. Moher D, Shamseer L, Clarke M, Ghersi D, Liberati A, Petticrew M, Shekelle P, Stewart LA (2015) Preferred reporting items for systematic review and meta-analysis protocols (PRISMA-P) 2015 statement. Syst Rev 4:1–9
12. Matthieu N (2021) Mapping the NFT revolution: market trends, trade networks, and visual features. arXiv:2106.00647v4 [q-fin.ST]
13. Kensuke ITO et al (2022) Bubble prediction of non-fungible tokens (NFTs): an empirical investigation. arXiv:2203.12587v2 [q-fin.ST]
14. Wesley Joon-Wie T et al (2023) Projecting non-fungible token (NFT) collections: a contextual generative approach. arXiv:2210.15493v2 [q-fin.CP]
15. Mieszko M (2021) Non-fungible tokens (NFT). The analysis of risk and return. https://ssrn.com/abstract=3953535. https://doi.org/10.2139/ssrn.3953535
16. Lavrova A, Iakushkin O (2022). NFT Performance and Security Review. In: Gervasi, O., Murgante, B., Misra, S., Rocha, A.M.A.C., Garau, C. (eds) Computational Science and Its Applications—ICCSA Workshops. Lecture Notes in Computer Science, vol 13377. Springer, Cham. https://doi.org/10.1007/978-3-031-10536-4_15.
17. Sharma T et al (2022) It's a blessing and a curse: unpacking creators' practices with non-fungible tokens (NFTs) and their communities. arXiv:2201.13233v1 [cs.CY]
18. Wang Q et al. (2021). Non-Fungible Token (NFT): Overview, Evaluation, Opportunities and Challenges (Tech ReportV2). arXiv:2105.07447v3 [cs.CR]
19. Zheng X (2022) Multimodal learning for improved NFT price prediction. In: IEEE international conference on e-business engineering (ICEBE), Bournemouth, United Kingdom, pp 74–79. https://doi.org/10.1109/ICEBE55470.2022.00022
20. Pavel K (2022) NFT marketplace design and market intelligence (January 6). INSEAD Working Paper No. 2022/03/MKT. https://ssrn.com/abstract=4002303. https://doi.org/10.2139/ssrn.4002303
21. Das D et al (2022) Understanding security issues in the NFT ecosystem. In: Proceedings of the 2022 ACM SIGSAC conference on computer and communications security (CCS'22), November 7–11, Los Angeles, CA, USA. ACM. ISBN978-1-4503-9450-5/22/11. https://doi.org/10.1145/3548606.3559342
22. Ko H et al (2022) The economic value of NFT: Evidence from a portfolio analysis using mean-variance framework. Financ Res Lett 47(Part A):102784. ISSN 1544-6123. https://doi.org/10.1016/j.frl.2022.102784
23. Sharma K, Bhalla R (2022) Stock market prediction techniques: a review paper. In: Luhach AK, Poonia RC, Gao XZ, Singh Jat D (eds) Second international conference on sustainable technologies for computational intelligence. Advances in Intelligent Systems and Computing, vol 1235. Springer, Singapore. https://doi.org/10.1007/978-981-16-4641-6_15
24. Chirtoaca D et al (2020) A framework for creating deployable smart contracts for non-fungible tokens on the Ethereum blockchain. In: Proceedings of IEEE international conference on decentralized applications and infrastructures (DAPPS). https://doi.org/10.1109/DAPPS49028.2020.00012
25. Majeed U et al (2023) FL-Incentivizer: FL-NFT and FL-tokens for federated learning model trading and training. IEEEAccess 11:4381–4399
26. Hasan HR et al (2022) Incorporating registration, reputation, and incentivization into the NFT ecosystem. IEEEAccess 10:76416–77643
27. Regner F et al (2019) NFTs in practice—non-fungible tokens as core component of a blockchain-based event ticketing application. ICIS Proc 1. https://aisel.aisnet.org/icis2019/blockchain_fintech/blockchain_fintech/1

28. Widianto MH et al (2023) Sentiment analysis towards cryptocurrency and NFT in Bahasa Indonesia for twitter large amount data using BERT. Int J Intell Syst Appl Eng IJISAE 11(1):303–309. ISSN:2147-6799
29. Krasnoselskii M et al (2023) KRAMER: interpretable rarity meter for crypto collectibles. IEEE Access 11:4283–4290
30. Oku D (2023) Non-fungible tokens (NFT): exploring the possibilities of NFTs—non-fungible tokens (NFT): exploring the possibilities of NFT. https://www.researchgate.net/publication/368755601.
31. Moringiello JM, Odinet, CK (2023) Blockchain real estate and NFTs (August 3, 2022). William Mary Law Rev 64(4):1131-1191. U Iowa Legal Studies Research Paper No. 2022-16, Widener Law Commonwealth Research Paper, Texas A&M University School of Law Legal Studies Research Paper No. 24-62, Available at SSRN: https://ssrn.com/abstract=4180876
32. Yadav AS, Kushwaha DS (2021) Blockchain-based digitization of land record through trust value-based consensus algorithm. Peer Peer Netw Appl 14:3540–3558. https://doi.org/10.1007/s12083-021-01207-1
33. Serrano W (2022) Real estate tokenisation via non fungible tokens. In: ICBCT'22: the 2022 4th international conference on blockchain technology, pp 81–87. https://doi.org/10.1145/3532640.3532651
34. Alnuaimi N et al (2022) NFT certificates and proof of delivery for fine jewelry and gemstones. IEEE Access 10:101263–101275
35. Senay AG et al (2022) NFT-based traceability and ownership management of medical devices. IEEE Access 10:126394–126411
36. Kamalian M et al (2017) On the design of NFT-based communication systems with lumped amplification. J Light Technol 35(24):5464–5472
37. Kim H et al (2022) Perpetual contract NFT as collateral for DeFi composability. IEEE Access 10:126802–126814
38. Arcenegui J et al (2021) Secure combination of IoT and blockchain by physically binding IoT devices to smart non-fungible tokens using PUFs. Sensors 21(9):3119. https://doi.org/10.3390/s21093119
39. Karandikar N et al (2021) Blockchain based transaction system with fungible and non-fungible tokens for a community-based energy infrastructure. Sensors 21(11):3822. https://doi.org/10.3390/s21113822
40. Chen Z et al (2022) Toward achieving anonymous NFT trading. IEEE Access 10:130166–130176
41. Meyns SC et al (2022) What users tweet on NFTs: mining twitter to understand NFT-related concerns using a topic modeling approach. IEEE Access 10:117658–117680
42. Hofstetter R, de Bellis E, Brandes L et al (2022) Crypto-marketing: how non-fungible tokens (NFTs) challenge traditional marketing. Mark Lett 33:705–711. https://doi.org/10.1007/s11002-022-09639-2
43. Best NFTs to Buy in 2023 (Top NFT Tokens Reviewed) (insidebitcoins.com)
44. Calvo P (2024) Cryptoart: ethical challenges of the NFT revolution. Humanit Soc Sci Commun 11:370. https://doi.org/10.1057/s41599-024-02872-2
45. https://medium.com/coinmonks/nft-token-development-in-2024-a-comprehensive-guide-to-legal-compliance-and-sustainable-business-19b1c6f0b885

Meta-Transactions as a Scalable and Optimal Solution in Public Blockchains

Mohuya Chakraborty and Sudip Kumar Palit

Abstract Public blockchain networks typically require users to create accounts and pay transaction fees using cryptocurrencies to access services. These requirements can present challenges, especially for new users, due to the complexities of setting up wallets and managing cryptographic keys. Additionally, scalability concerns in public blockchains often result in network congestion and slower transaction times during peak usage. This paper introduces a solution designed using solidity, enabling users to interact with the blockchain via meta-transactions without the need to create accounts or use cryptocurrencies for fees. Instead, users can pay fees in their local fiat currencies, which make blockchain services more accessible and scalable. By allowing multiple users to engage with the blockchain using fiat, this approach optimizes the system's scalability, facilitating a higher volume of transactions. We have implemented this solution on a blockchain platform to showcase its practical feasibility and effectiveness.

Keywords Public blockchain networks · Solidity · Meta-transactions · Cryptocurrencies · Fiat currencies

1 Introduction

Blockchain is an innovative technology, used in many businesses. This is a decentralization of the business data stored across multiple nodes. The decentralized nature and inbuilt cryptographic techniques make it more secure, reliable, and transparent. Although this technology is used by many sectors, it still possesses several challenges, especially in public blockchains. Each user in this blockchain manages wallet

M. Chakraborty · S. K. Palit (✉)
Cybersecurity Centre of Excellence, Institute of Engineering and Management (IEM), University of Engineering and Management, Kolkata, India
e-mail: sudipkumar.palit@uem.edu.in

M. Chakraborty
e-mail: mohuyacb@iem.edu.in

M. Chakraborty et al. (eds.), *Proceedings of Sixth International Ethical Hacking Conference*, Lecture Notes in Networks and Systems 1513,
https://doi.org/10.1007/978-981-96-8632-2_12

addresses like metamask, xdcpay, etc., and private keys for those addresses. They have to maintain the cryptocurrency balances on that user address, and these balances are used as transaction fees when they interact with the blockchain, especially when they write something in the blockchain [1]. These mechanisms may create overhead for non-technical users, as they have to deal with cryptocurrencies, and manage wallets. Additionally, transaction throughput and latency depend on their consensus mechanism causing scalability issues.

Scalability is a big problem for public blockchain. For instance, Bitcoin processes approximately 7 transactions per second (TPS), and Ethereum handles about 15–30 TPS, which is far below the capacity of centralized systems like Visa, which processes thousands of TPS [2, 3]. Although blockchain-based systems are more secure and transparent than centralized systems, this transaction delay is the overhead for blockchain-based solutions, as it leads to network congestion and high transaction fees during peak demand, making blockchain less practical for mass adoption. Such scalability constraints are intrinsic to the architecture of most blockchains, where decentralization and security often come at the expense of performance.

Few works have been seen in the literature that highlights the recent works on meta-transactions, blockchain usability, and scalability challenges. Blum et al. [4] present a structured approach to designing blockchain-based applications, enhancing attributes like privacy, operational cost, transparency, risk, and user experience. In this approach, authors use the meta-transaction concept. Seres [5] discusses the potential benefits of meta-transactions in permission less blockchain, suggesting that they can improve user experience without compromising the security of cryptocurrency platforms. Dorri et al. [6] introduce the concept of atomic meta-transactions, requiring consumer and producer signatures. This scheme is resilient against attacks and implementation results show it reduces monetary cost and delay compared to existing solutions.

However, blockchain technology, known for its unchangeability, faces scalability issues like low processing transactions per second [7]. The scalability issues of blockchain's effects on its performance are discussed by Yang et al. [8]. They suggest solutions for blockchain scalability-related problems, including sharding mechanisms, DAG-based systems, off-chain payment networks, and cross-chain technology, to improve efficiency and extend functionality.

The proposed technique tried to minimize these gaps by introducing meta-transactions, enabling users to interact with blockchain services without managing wallet addresses, private keys or cryptocurrency balances. Instead, transaction fees are paid by the users in fiat money, and the execution process is offloaded to relayers [9]. Non-technical users can pay the fiat money to the relayers, and the relayers give their wallet addresses to the users so they can run specific functions on the blockchain. This reduces the technical and financial barriers for new or non-technical users. Moreover, this approach gives a good solution to the problem of scalability by reducing direct blockchain interactions and minimizing network congestion.

The contribution of this work is as follows:

1. This paper introduces the concept of meta-transactions in blockchain to enable users to interact with the blockchain network with the help of fiat currencies instead of cryptocurrencies.
2. The proposed model minimizes network congestion by allowing the relayers to work as middlemen in the execution of blockchain applications on behalf of the user.
3. Non-technical users do not need to use cryptocurrency to get the service of blockchain, which minimizes the system's complexity.
4. This paper shows how meta-transactions help in improving blockchain scalability and optimizing performance.

The rest of the paper is organized as follows: Sect. 2 gives an overview of the prerequisites. Section 3 demonstrates the proposed framework. The implementation of the framework is described in Sect. 4. Performance analysis is shown in Sect. 5. Security analysis is defined in Sect. 6. Section 7 discusses the use cases and applications of the work, and finally, Sect. 8 ends the paper with a conclusion and future work.

2 Prerequisites

2.1 Blockchain Basics

Blockchain is a decentralized, distributed ledger which makes it a system that securely records transactions across a network of nodes. Each transaction that executes in the blockchain requires verification by the validator or miners, after that these transactions can be included in the block. Consensus mechanisms like Proof of Work (PoW) or Proof of Stake (PoS) are used for the inclusion of these transactions in a block [10]. The miners or validators received incentives for this verification process. Gas fees are required to execute each transaction and cryptocurrencies are charged for this process. Cryptocurrencies are mandatory for transactions and become a barrier for users unfamiliar with this system [11].

A smart contract is a program on a blockchain that automatically carries out agreements between parties when specific conditions are met [12]. It removes middlemen by carrying out transactions directly. This ensures that all actions are clear, secure, and permanent. To create smart contracts, it's important to understand basic blockchain ideas and know programming languages like solidity, which is used for Ethereum and similar blockchains [13].

2.2 Meta-Transactions

Meta-transactions are a procedure in blockchain, where users can initiate a transaction without having cryptocurrency. The relayer works as a middleman who has cryptocurrency. Users pay fiat money to the relayer, sign their transaction data cryptographically and pass it to the middle layer or backend. Then the relayer handles these transactions and pays gas fees on behalf of the users. This kind of abstraction minimizes the technical and financial obstacles for users. Thus, enabling blockchain applications to reach a wider range of users. Additionally, meta-transactions always ensure the integrity of signed data and enhance the scalability of the network. Therefore, not only maintaining trust, and performance but also enhancing user convenience.

3 Proposed Framework

In the proposed framework, the blockchain interactions of the user become simple due to the integration of a fiat payment mechanism and relayer service to handle transaction fees and streamline the user experience. Therefore, users in this framework do not need to maintain cryptocurrency wallets. Instead, they submit requests via a Graphical User Interface (GUI) in the user interaction layer, including the wallet address of the relayer, the name of the transaction with the required parameters want to run in the blockchain, and the fiat payment proof. This information is transferred to the backend layer where the user inputs are validated. Then a signed transaction is submitted to the blockchain, the smart contract layer. Finally, the blockchain verifies this signed data with the help of the inputted parameters. Signing the transaction and verification is done with the help of cryptographic functions, hence ensuring security and trust in the system.

The proposed framework has the following layers or components as part of the architecture. These are as follows:

1. **User Interaction Layer**: This layer is responsible for handling user inputs via a GUI.
2. **Backend Layer**: This layer takes input from the user interaction layer, and then validates the input. After successful validation, this layer interacts with the smart contract layer.
3. **Smart Contract Layer**: After getting input from the backend layer, this layer verifies transaction authenticity and executes blockchain operations.

3.1 User Interaction

The users of this framework interact with the system through a user-friendly GUI in the user interface layer to fill in the following inputs: (1). the blockchain function that needs to be executed (e.g., transferring tokens or updating data) with required parameters, (2). the wallet address of the relayer, and (3). fiat payment proof (e.g., receipt or confirmation). After getting all the information, this information is sent to the backend layer for processing.

3.2 Backend Processing

- Input Validation: The backend layer verifies the inputted wallet address, transaction details, and payment details.
- Payment Validation: The backend layer takes input of payment proof, and then sends the payment proof information to the payment gateways to verify the fiat payment. The payment gateway, on the other hand, verifies the payment of the user. If it is genuine, then control proceeds to the next step, otherwise give the appropriate error message to the user.
- Transaction Construction: After successful payment verification, the backend constructs the message hash, which includes information like the user address, and function signature. Function signature contains the encrypted blockchain-based functions with the parameter that the user wants to run. Then sign the message hash using the private key of the relayer. In the end, backed calls meta-transactions by inputting the function signature, wallet address, and cryptographic parameters which are required to validate the signed message hash.
- Blockchain Submission: This information is then passed to the deployed smart contract.

3.3 Smart Contract Execution

- Signature Verification: After getting the call for meta-transaction execution, the smart contract generates the message hash from the inputted function signature and wallet address. Thereafter, it generates the signer's address from the generated message hash, and other inputted cryptographic parameters, and finally verifies whether the provided wallet address and the signer address are the same or not. If the above-mentioned addresses are the same then the process proceeds to the next step. Otherwise, the process stops here by sending the backend layer the appropriate error message. This error message is forwarded to the user interaction layer by the backend layer.
- Function Execution: After successful signature verification, the smart contract runs the specified function on the blockchain that the user sends to execute.
- Response Handling: The result of the successful transaction in the blockchain is then returned to the backend layer, which is finally sent to the user interaction layer.

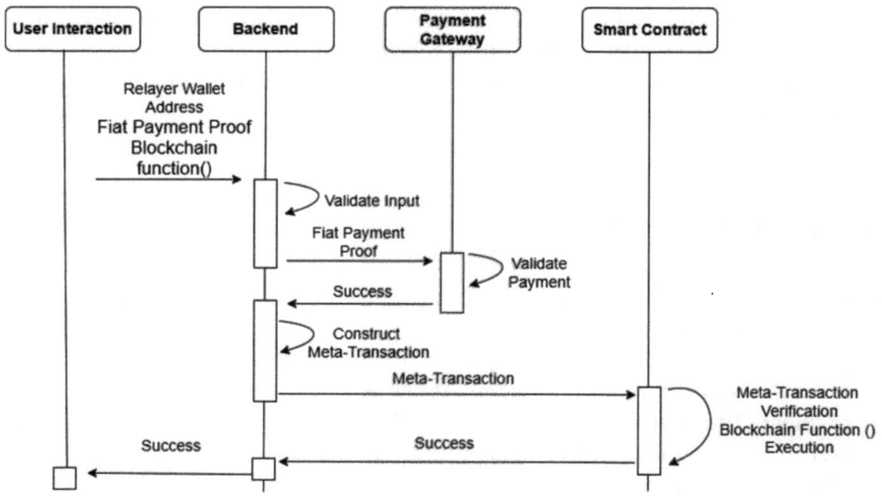

Fig. 1 Interaction sequences of the dApp

Figure 1 shows the interaction sequences of the proposed framework.

4 Implementation

4.1 Technologies Used

To enable an efficient blockchain interaction and user-friendly dApp, the proposed framework uses several technologies. The technologies used for this are as follows:

1. Solidity: The smart contract logic of the proposed framework is written in solidity. Solidity is a very popular programming language for designing smart contracts for dApp on Ethereum and compatible public blockchain-based applications. In the smart contract layer, using solidity, the logic for executing meta-transactions, signature verifications, and user function execution is defined [14].
2. Hardhat: Hardhat is used as the development environment for deploying the smart contract written in solidity, and testing the smart contract. Hardhat provides a convenient environment to compile, deploy and test the dApp in localhost, and live network. It integrates with ether.js to interact with the blockchain.
3. Node.js: Node.js is used in this dApp to implement the backend layer. It handles user interactions, validates inputs, interacts with the payment gateway, and communicates with smart contracts. It acts as an intermediary between the GUI and the smart contract and ensures that the meta-transaction is constructed correctly, signed by the relayer, and then submitted in the blockchain.

4. Square APIs: Square API is used in this dApp as a payment gateway. This API is integrated with the backend to handle the fiat payment mechanism. It allows users to make payments in fiat currency (such as USD, EUR, and INR), thus eliminating the burden for users to have the cryptocurrency. During the execution of the dApp, it actively verifies the genuine transactions made through this API. Therefore, meta-transactions are called after getting payment confirmation from this API.
5. Other APIs, Languages, and Libraries:
   - Ether.js: For generating message hash, and signing of the meta-transaction function, the backend uses the ethers.js library to interact with the blockchain.
   - HTML and JavaScript: The GUI for the dApp is designed with the help of these technologies. Together, these technologies are used to create an intuitive graphical user interface (GUI) for users. HTML is used to design the form JavaScript is used to interact with the backend.

4.2 *Meta-Transaction Execution in Backend*

The code snippet for the execution of the meta-transaction is given in Fig. 2.

```
// Generate the function signature
const functionSelector = ethers.utils.id("exampleFunction(string)").slice(0, 10);
const encodedParameters = ethers.utils.defaultAbiCoder.encode(["string"], [message]).slice(2);
const functionSignature = functionSelector + encodedParameters;

// Create the message hash
const messageHash = ethers.utils.solidityKeccak256(
    ["address", "bytes"],
    [userAddress, functionSignature]
);

// Sign the message hash
const signingKey = new ethers.utils.SigningKey(wallet.privateKey);
const signature = signingKey.signDigest(messageHash);

// Extract v, r, s from the signature
const { v, r, s } = signature;

// Execute meta-transaction
const tx = await contract.executeMetaTransaction(userAddress, functionSignature, v, r, s, {
    gasPrice: ethers.utils.parseUnits('20', 'gwei'),
    gasLimit: 100000
});

await tx.wait();
```

Fig. 2 Code snippet for the execution of meta-transaction

1. Generate Function Signature: The backend first creates the function selector by taking the function's name to be executed and encodes the function parameters given as input by the user. The function signature is then created by merging the function selector and encoded parameter. The function signature is then created by combining both parts.
2. Create Message Hash: Then a hash of the relayer's address (wallet address) and the function signature is generated using the function solidityKeccak256(). This hash value ensures the uniqueness of the message.
3. Sign the Message: The generated message hash is then signed by the relayer's private key to prove the authenticity of the signer. The signature then forms the components v, r, and s to verify the signature later.
4. Execute Meta-Transaction: At the end, the parameters: wallet address, function signature, and the signature components v, r, and s are sent to the smart contract's executeMetaTransaction() function. This function is then executed by the relayer with the help of his/her wallet address, and private key, on behalf of the user. The transaction is then confirmed on the blockchain.

4.3 Smart Contract Details

A smart contract is the main logic that executes in the blockchain. Below, the key snippets of the smart contract are given that describe how the meta-transaction is handled.

1. Generate message hash: The function getMessageHash() generates a message hash based on the inputted parameters: user address, and function signature. During the message hash generation, this function used the keccak256() library function. Figure 3 gives this code snippet.
2. Execute meta-transaction: The function executeMetaTransaction() is responsible for executing the actual meta transaction. It inputs the wallet address, function signature, v, r, and s as input. Then use the function signature and user address as input to call the getMessageHash() function. getMessageHash function returns the message hash. This message hash is then used along with input parameters v, r, and s to get the address of the signer (relayer) of the function. The library function ecrecover() is used to recover the address of the signer. Then, the inputted user address and this generated signer address are matched. If the relayer is genuine, then this test passes and then calls the function specified in the function signature.

```
// Helper function to get the message hash
function getMessageHash(address userAddress, bytes memory functionSignature) internal pure returns (bytes32) {
    return keccak256(abi.encodePacked(userAddress, functionSignature));
}
```

Fig. 3 Code snippet of generate message hash

```
function executeMetaTransaction(
    address userAddress,
    bytes memory functionSignature,
    uint8 v,
    bytes32 r,
    bytes32 s
) public payable returns (bytes memory) {
    // Construct the message hash for verification
    bytes32 messageHash = getMessageHash(userAddress, functionSignature);

    // Verify the signature
    address signer = ecrecover(messageHash, v, r, s);
    require(signer == userAddress, "Invalid signature");

    // Execute the function call
    (bool success, bytes memory result) = address(this).call(functionSignature);
    require(success, "Function call failed");

    emit MetaTransactionExecuted(userAddress, msg.sender, functionSignature);
    return result;
}
```

Fig. 4 Code snippet of execute meta-transaction

The execution is done on behalf of the user by the relayer, without the user needing to pay gas fees directly. Figure 4 gives the code snippet of this function.

3. Example of a transaction function: The exampleFunction() is a sample function that can be called through the meta-transaction. The function executeMetaTransaction() receives the function signature of this function. Therefore this function is being called by the executeMetaTransaction() function in the line:

    (bool success, bytes memory result) = address(this).call(functionSignature);

    In this call, the user can interact with the blockchain. Figure 5 gives the code snippet of this function.

Figure 2 shows the backend layer logic to prepare and call the meta-transaction function, whereas Figs. 3, 4 and 5 show how the smart contract layer handles this function call and executes the meta-transaction. By combining all these three functions, meta-transactions can be executed that enable users to directly interact with blockchain without directly dealing with cryptocurrencies.

```
// Example function that can be called via meta-transaction
function exampleFunction(string memory message) public pure returns (string memory) {
    return message;
}
```

Fig. 5 Code snippet of transaction function

5 Performance Evaluation

5.1 Scalability

In the proposed framework, any number of users can interact with the blockchain without having a wallet address of their own. Instead, they are required to input the wallet address of the relayer, and users pay these relayers the fiat charges to interact with the blockchain. This concept helps in maintaining a small number of wallet addresses with cryptocurrency in blockchain, so performance is not down as happens in normal blockchain due to the huge number of wallets. A higher number of wallets make the system more complex due to the management of each wallet's transactions and associated states. Therefore, the proposed framework increased scalability.

5.2 Stress Testing

Stress testing is essential to test the robustness of a system. The popular stress testing tool Artillery is employed to perform stress tests of the proposed framework [15]. Artillery can simulate a high range of transaction volumes to assess how a system can perform under a load. In this stress test, many concurrent users are simulated to show the resilience of the proposed scheme. During testing, a local blockchain is used to assess the stability of the proposed framework, excluding real-time transaction processing and the real-time blockchain environment. Otherwise, the stress test results would be influenced by other network architectures, loads for payment and blockchain. In this case, Artillery is used to send multiple meta-transactions to the backend layer of the proposed model.

5.2.1 Summary of Performance Results

To investigate the performance, the proposed system underwent a stress test with 5 virtual users generating requests at a rate of 5 requests per second. The total test duration set was 60 s. The test is performed to imitate meta-transactions of the backend layer. Table 1 shows the performance of the stress test, emphasizing the system's performance.

5.2.2 Performance Observations

- Response Time Consistency: The response time is throughout consistent with very slight variation at the 95th and 99th percentiles.
- Scalability: The system shows good scalability as it can manage 5 requests per second per virtual user.

Table 1 Key metrics

Metric	Value
Total requests	300
Successful responses (200)	300
Total downloaded bytes	10,200 bytes
Request rate	5/sec
Mean response time	20.8 ms
95th percentile response time	30.9 ms
99th percentile response time	34.8 ms
Virtual users completed	300
Virtual users failed	0
Mean session length	22.6 s

- No Failures: The efficiency of the backend can be realized as there are no failed requests.
- Session Lengths: The average session length for virtual users was 22.6, indicating a stable user session.

5.3 Optimality

This framework is optimal as it efficiently handles meta-transactions by significantly reducing the number of wallet users, thus minimizing overall system complexity. The system is more accessible and cost-effective because users can interact with the blockchain without directly paying gas costs. The backend layer effectively handles transaction processing, and signing reduces delays whereas, the use of signature verification and smart contract execution ensure transaction security and integrity. Furthermore, from the stress test, we can see that the system can handle many concurrent users virtually without any error, hence the system is scalable. These features ensure that the proposed framework utilizes each resource effectively while maintaining a high level of security and user accessibility, thus offering an optimal solution.

6 Security Analysis

6.1 Payment Verification

In the proposed dApp, only the genuine users, who have paid the fiat money to the relayers, can only run the blockchain features. The proposed system, after getting input from the users, first checks whether the transaction fees are genuine or not by calling a payment gateway. The payment gateway, in its server, verifies the payment and then gives feedback to the backend layer. So unauthorized transactions are easily filtered out.

6.2 Signature Verification

In any blockchain-based application, it is crucial to verify the genuineness of each transaction. The cryptographic digital signature generation and verification methods in the backend layer and smart contract layer respectively ensure the genuineness of the meta-transaction. The backend layer creates the function signature, message hash, and signature components using the relayer's private key and wallet address. The keccak256() method is used to create the message hash. Alternatively, the smart contract layer generates the message hash using the same function keccak256(), recovers the address of the signer using the method ecrecover(), verifies the wallet address and the generated signer addresses are same, then sends the transaction function in blockchain to execute for the users. The signed message is tamper proof due to use of Ethereum's signature scheme. These processes make the proposed framework extremely difficult for the attacker to forge the signature or manipulate the transactions.

6.3 Attack Vectors

Although the proposed framework provides robust security measures, it is still important to analyze potential attack vectors that could compromise the security of the proposed system.

- Man-in-the-Middle Attack: In this attack, an attacker can alter the transaction data. But, in the proposed framework, due to the use of the cryptographic signature (by relayer), this attack is not feasible by the attackers.
- Server Compromised Attack: As a blockchain-based system is distributed in nature, all transactions are recorded, and stored in several nodes. This decentralized nature of the network protects the proposed system from compromised server attacks, which occur in centralized server-based systems when a single server is compromised or fails.
- Sybil Attacks: In this attack, an attacker can create multiple identities to flood the system or to get control over the network [16]. In the proposed scheme, each transaction is verified under a private key of the relayer, therefore if the attacker creates multiple identities, still this attack is not feasible as the private keys are unavailable to the attackers.
- Identity Spoofing: Identity spoofing, where an attacker tries to impersonate a relayer, is mitigated in the proposed scheme through signature validation. The

use of the function ecrecover() ensures that only the legitimate relayer's private key can produce a valid signature for a transaction. This cryptographic check prevents spoofing attempts.
- Data Tampering: Secure hash-based functions are utilized to generate the message hash. Therefore, if this hashed data is generated, no one can alter this without invalidating the hash. Therefore, attackers cannot change the executed transactions.
- Malicious Relayer Activity: Relayers in this framework may attempt to alter transaction data or misuse their position. However, the use of their cryptographic signature for transaction verification ensures that they cannot modify the transactions without invalidating the signatures. Additionally, user-signed messages include the intended relayer fee, ensuring transparency.

7 Use Cases and Applications

This proposed framework combines individual users with enterprise adoption of blockchain technology. For individual users, it removes the barrier of having cryptocurrency or wallets. This allows people of low-income groups, or unskilled to get the services of blockchain. Users can easily participate without upfront costs since transaction fees are covered by relayers, promoting fair access to digital services. For businesses, the system is easy to scale and allows for quick onboarding of users. This makes it simple to use blockchain in areas like supply chain management, loyalty programs, and international payments. By supporting both personal and business uses, this framework promotes the wider adoption of blockchain technology.

8 Conclusion and Future Work

In this paper, a meta-transaction-based framework is proposed that enhances both the scalability and optimality of the public blockchain network. This novel approach significantly reduces the complexity of the users in accessing the blockchain-based application, because they can engage with the applications without having wallet addresses or cryptocurrencies. Furthermore, the proposed framework allows a fiat-based payment mechanism that optimizes scalability by promoting a higher rate of transactions without affecting network congestion or slower processing times. This solution is scalable and makes efficient use of resources, helping to lower transaction costs and reduce execution times. The proposed approach is secure and can effectively grow and enhance blockchain interactions. This helps more people adopt the technology across different user groups.

Future improvements for this system include supporting multiple blockchains for better cross-chain interaction, automating the conversion from fiat to crypto for a smoother user experience, and optimizing transaction processing for greater scalability.

References

1. Böhme R, Christin N, Edelman B, Moore T (2015) Bitcoin: economics, technology, and governance. J Econ Perspect 29(2):213–238
2. Xie J, Yu FR, Huang T, Xie R, Liu J, Liu Y (2019) A survey on the scalability of blockchain systems. IEEE Network 33(5):166–173
3. https://archlending.com/glossary/transactions-per-second
4. Blum F, Severin B, Hettmer M, Hückinghaus P, Gruhn V (2020) Building hybrid DApps using blockchain tactics-the meta-transaction example. In: 2020 IEEE international conference on blockchain and cryptocurrency (ICBC), pp 1–5. IEEE
5. Seres IA (2020) On blockchain metatransactions. In: 2020 IEEE international conference on blockchain (Blockchain), pp 178–187. IEEE
6. Dorri A, Luo F, Kanhere SS, Jurdak R, Dong ZY (2019) SPB: a secure private blockchain-based solution for distributed energy trading. IEEE Commun Mag 57(7):120–126
7. Kim S, Kwon Y, Cho S (2018) A survey of scalability solutions on blockchain. In: 2018 international conference on information and communication technology convergence (ICTC), pp 1204–1207. IEEE
8. Yang D, Long C, Xu H, Peng S (2020) A review on scalability of blockchain. In: Proceedings of the 2020 2nd international conference on blockchain technology, pp 1–6
9. NETWORK X. XDC Gasless Subnet: Gasless Subnet Staking dApp
10. Arya J, Kumar A, Singh AP, Mishra TK (2021) Chong PH blockchain: basics, applications, challenges and opportunities
11. Drescher D (2020) Blockchain basics. Ascent Audio
12. Mohanta BK, Panda SS, Jena D (2018) An overview of smart contract and use cases in blockchain technology. In: 2018 9th international conference on computing, communication and networking technologies (ICCCNT), pp. 1–4. IEEE
13. Khan SN, Loukil F, Ghedira-Guegan C, Benkhelifa E, Bani-Hani A (2021) Blockchain smart contracts: applications, challenges, and future trends. Peer Peer Netw Appl 14:2901–2925
14. Dannen C, Dannen C (2017) Solidity programming. Introducing ethereum and solidity: foundations of cryptocurrency and blockchain programming for beginners, pp 69–88
15. Artillery docs retrieved from https://www.artillery.io/docs
16. Aggarwal S, Kumar N (2021) Attacks on blockchain. In: Advances in computers, vol 121, pp 399–410. Elsevier

A Decentralized Application for Secure Authentication and Communication in Industry 4.0 Networks

Sudip Kumar Palit, Subhojit Chakraborty, Sarthak Srivastava, and Mohuya Chakraborty

Abstract Industry 4.0 networks are crucial to maximize profitability, optimize processes, and accelerate demand. Internet of Production (IoP) is the networked environment within Industry 4.0 paradigm, focusing on connecting production assets through advanced communication networks. In IoP, stakeholders can communicate with each other when required to share information about the products they send and receive in the supply chain. A secure authentication mechanism is required in this new concept to make safe communication, where stakeholders of a separate supply chain communicate with each other. This research paper presents a decentralized application (DApp) for stakeholders of the IoP. The proposed protocol provides tamper-resistant records, secured data transfer, and peer verification between stakeholders. The performance analysis of the proposed protocol shows its supremacy over three other related protocols under varied situations with a superiority score range of 50.84 to 454. The security analysis using the well-known tool Automated Validation of Internet Security Protocols and Applications (AVISPA) demonstrates that the offered scheme is free from replay and man-in-the-middle attacks. Performance analysis shows that it outperforms other related schemes.

Keywords Decentralized Application (DAPP) · Industry 4.0 · Secure authentication · Supply chain networks · Blockchain · Ethereum · Internet of production

---

S. K. Palit (✉) · S. Chakraborty · M. Chakraborty
Cybersecurity Centre of Excellence, Institute of Engineering and Management (IEM), University of Engineering and Management, Kolkata, India
e-mail: sudipkumar.palit@uem.edu.in

S. Srivastava
Babu Banarasi Das University, Lucknow, Uttar Pradesh, India

M. Chakraborty et al. (eds.), *Proceedings of Sixth International Ethical Hacking Conference*, Lecture Notes in Networks and Systems 1513,
https://doi.org/10.1007/978-981-96-8632-2_13

1 Introduction

Blockchain, a decentralized-based technology, is widely used in Industry 4.0 networks, including the Internet of Production (IoP), Supply Chain Management (SCM), and the Industrial Internet of Things (IIoT), because of its inbuilt features like transparency, security, and efficiency. In the global economies, blockchain is demanding as it requires secure, transparent, and verifiable interconnections among the stakeholders. Through this paper, an initiation is taken to propose a decentralized authentication protocol for the IoP that leverages the immutability of blockchain combined with cryptographic methods to provide integrity, confidentiality and authenticity of data. The protocol is designed to enhance the security of the interaction of the stakeholders, improve transaction efficiency, and promote supply chain efficiency through the use of cryptographic primitives like hash-based function and exclusive or (XOR) operation, timestamp-based session key verification, and blockchain storage.

Industry 4.0 and IIoT work in fields like machine learning, automation, data, autonomy, and interconnectivity [1]. The main motive of these sectors is to automate the supply chain processes. However, these sectors are unable to provide services like cross-domain collaboration among the stakeholders. Moreover, cybercriminals are targeting IIoT due to the technology paradigm transformation [2].

However, IoP is a better alternative to Industry 4.0, and IIoT, which aims to connect different supply chain stakeholders, enhancing cooperation and communication, increasing productivity, and reducing time to market [3]. However, implementation of IoP is challenging as it needs reliable infrastructure, security and privacy of production data, and effective interest alignment among stakeholders [4].

A numerous number of authentication protocols have been offered in IIoT. Authentication protocol for machine-to-machine (M2M) communication in IIoT networks was offered by Esfahani et al. in 2017 [5]. In this protocol, cryptographic primitives like Exclusive OR and hashing operations were used for authentication, confidentiality, and cyber-attack resistance. Li et al., in the same year, suggested a three-factor user authentication approach for IoT security [6]. Again in 2017, an authentication protocol utilizing cryptographic primitive elliptic curve cryptography (ECC) was offered for wireless sensor networks (WSN) and the Internet of Things (IoT) by Li et al. [7]. A secure data transmission solution for heterogeneous Cyber-Physical Systems (CPS) was offered by Harishma et al. in 2018 [8]; A privacy-preserving strategy for IIoT contexts was presented by Das et al. in 2018 [9]. Eldefrawy et al. [10] introduced a user authentication method for IIoT systems in 2019. However, mutual authentication between users and sensor nodes was not present in this offered method.

Authentication protocol is crucial for a complex system like IoP where lots of stakeholders put their personal information. A huge number of authentication schemes have been offered to discuss security aspects of diverse environments like global mobility networks, 5G networks, wireless IP-based networks [11–15]. Blockchain is considered an essential tool to enhance the security and reliability of a network and is heavily used in designing authentication schemes. The efficiency of an authentication protocol depends on the security features offered and on the fast execution of the scheme. Proper use of lightweight cryptographic primitives like simple one-way hash function, and XOR operation gives both the security features and fast execution time.

However, some blockchain-based schemes for SCM systems were created for betterment. In 2020, Dwivedi et al. developed a blockchain-based information exchange and authentication approach for an SCM platform with a key allocation mechanism [16]. All security requirements, including confidentiality, integrity, and authentication, are met by the proposed plan. In 2022, Hasan et al. introduced a blockchain-based authentication solution that leveraged IoT in SCM [17]. The parties in this scheme, the producer, distributor, store, and customer, authenticated the product and each other before processing it in the chain. In 2023, Mishra et al. offered a protocol based on blockchain to resist vaccine record forgery [18].

In order to improve the security and dependability of IoP, we envisioned to design a secure framework in this study. Participants in the network can share their data with one another under this framework. Manufacturing organizations are more productive and efficient when they share data among IoP participants. In this framework, an algorithm is created to ensure that participants communicate with each other in a trustworthy manner. The suggested framework's algorithm design incorporates blockchain technology to manage the massive volumes of data produced and exchanged by the parties.

This research concept is a big innovative step in putting all of these pieces into one framework: the IoP-DApp framework. This is where blockchain, IoP protocols, and decentralized applications act together to assure security, transparency, and efficiency in the communications and authentication processes throughout supply chains. Other notable deficiencies concerning scalability and applicability of the previous works are addressed here, to upgrade the preceding systems, thus enhancing the security, efficiency, and usability.

The main contributions of the proposed paper are as follows:

1. A new framework is proposed for IoP for interconnection among the stakeholders.
2. An authentication mechanism is proposed for secure communication in the IoP.
3. A decentralized application for IoP (IoP-DApp) is proposed which is developed on the Ethereum blockchain.
4. Lightweight cryptographic primitives like hash methods and XoR operators are utilized to keep execution time very low.
5. The proposed scheme leverages the power of blockchain, smart contracts, and the IoP to improve transparency, security and efficiency. The framework improves supply chain transaction security and efficiency by reducing the chance of tampering, counterfeiting, and border access.
6. The proposed IoP-DApp framework showcases its practicality for real-world supply chain scenarios.

This is how the remaining part of the paper is structured. The preliminary knowledge about blockchain, smart contracts is given in Sect. 2. The suggested framework and the security algorithm are discussed in Sect. 3. The security analysis is covered in Sect. 4, and the performance analysis is covered in Sect. 5. Section 6 concludes the paper.

2 Preliminary Knowledge

2.1 Blockchain

Blockchain technology is a cutting-edge data management method that enables open and honest information exchange across a network. A blockchain database stores the data in blocks that are joined together to form a chain. Without network consensus, this chain cannot be broken or changed, hence the data in this network stays consistent. Therefore, users can create an unchangeable or immutable ledger to manage accounts, payments, orders, and other transactions using blockchain technology. The built-in safeguards in the system prevent unauthorized transaction entries and guarantee a consistent shared picture of these transactions.

2.2 Ethereum Blockchain and Smart Contract

Generally regarded as the second most popular cryptocurrency after Bitcoin, Ethereum is a second-generation blockchain. Ether is the name of the currency produced by the Ethereum blockchain. Ethereum serves as an operating system for creating smart contracts and decentralized apps (DApps).

On the Ethereum blockchain, smart contracts are computer programs. They start up when a user transaction initiates them. They greatly expand Ethereum's capabilities. These applications serve as the foundation for decentralized organizations and applications. For as long as Ethereum is in existence, a smart contract that has been published on Ethereum will be accessible and functional. Even the developer of the blockchain program is unable to remove it. Smart contracts are always available for usage and do not discriminate against any user because they are automated. Smart contracts are commonly used in insurance, quadratic funding, decentralized trade exchanges, and lending applications.

3 Proposed Framework and Algorithm

The IoP-DApp framework is presented in this part, followed by an authentication protocol. This framework can be applied to any production sector, such as automotive, pharmaceutical, etc. The raw material suppliers are the industry's stakeholders. Many mid-tier manufacturers use the supplier's inputs to make the specified parts, which they then supply to mid-tier manufacturers at a higher level. Ultimately, the mid-tier manufacturer sends the completed pieces to the final manufacturer, who then produces the finished goods. These stakeholders come together to form the Internet of People (IoP) in our suggested structure, where they can exchange information and communicate with one another to optimize output. Any stakeholder can communicate with its predecessor and subsequent stakeholders directly, but if they wish to communicate with other people, they must use this IoP. Since there are many stakeholders in this network and IoP provides safe and easy contact with all of them, it is evident that IoP is necessary in this scenario.

Figure 1 illustrates how goods are transformed, beginning with suppliers and moving through various mid-tier manufacturers before final manufacturers produce the final products. It also shows how information is disseminated among stakeholders, both directly and indirectly with the aid of the IoP.

Information sharing directly or indirectly is beneficial for enhancing production rates. However, it also poses risks because the trustworthiness of the information recipient may be unknown. Trust among partners is crucial in such situations, and indirect partners are particularly vulnerable. Therefore, a robust security architecture needs to be designed for this framework to ensure that any stakeholder can trust others when sharing information.

In IoP, data can be exchanged via lookup services, distributed networks, trustworthy third parties, or direct communication [4]. Participants in the suggested concept exchange data over a distributed network, using blockchain technology.

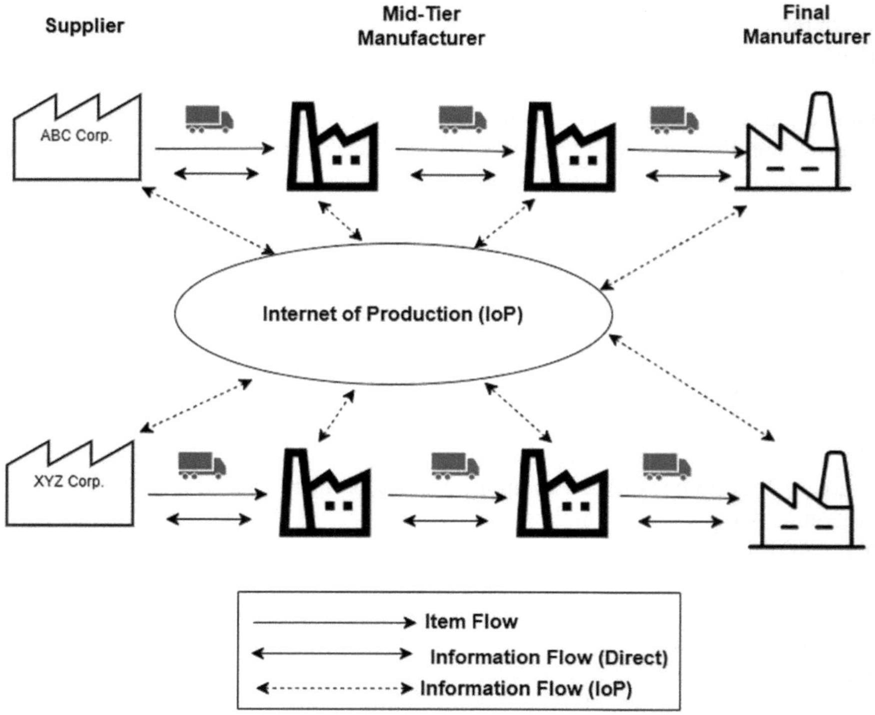

Fig. 1 Communication among the participants in the IoP

3.1 Proposed IoP-DApp Framework

The proposed IoP-DApp framework integrates the blockchain, decentralized application (DApp), and Internet of production (IoP) protocol to provide secure, transparent, and efficient industrial production. Four main components are part of the framework: The blockchain network utilizes smart contracts to securely store and share data, ensuring that transactions are tamper-proof and traceable. Real-time monitoring and tracking made possible by the DApp gives all involved parties access to production data, empowering decision-making. Figure 2 shows the proper framework and its working principle.

3.2 Proposed Model and Algorithm

The suggested model makes use of a blockchain network to record and exchange data for all kinds of participant communications. Suppliers, mid-tier manufacturers, and final manufacturers are all involved in this model. The service provider is also

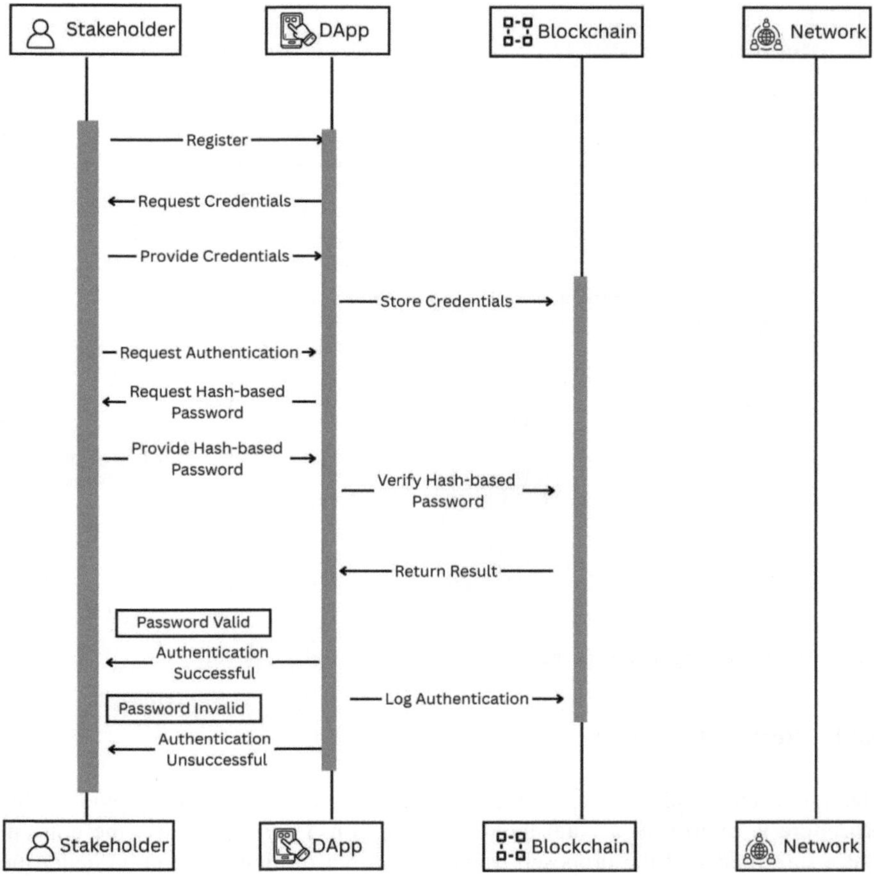

Fig. 2 Proposed framework and working principle

presented as an important stakeholder who serves as a go-between for participants' data sharing. The service provider is a reputable and registered organization that makes services available on the IoP. Upon joining the network, each participant registers with the service provider. Participants can interact with one another within the same supply chain or with participants in other supply chains after registering. The blockchain network is linked to both participants and service providers.

Under the suggested model, mid-tier manufacturers receive items from suppliers. The mid-tier manufacturers prepare the products after receiving the items, and they can then deliver them to the final manufacturers or other mid-tier manufacturers. Any stakeholder within the same supply chain or across supply networks can communicate with registered stakeholders when necessary. As a result, a security method with two separate stages—the registration phase and the authentication phase—is suggested in order to execute the suggested architecture.

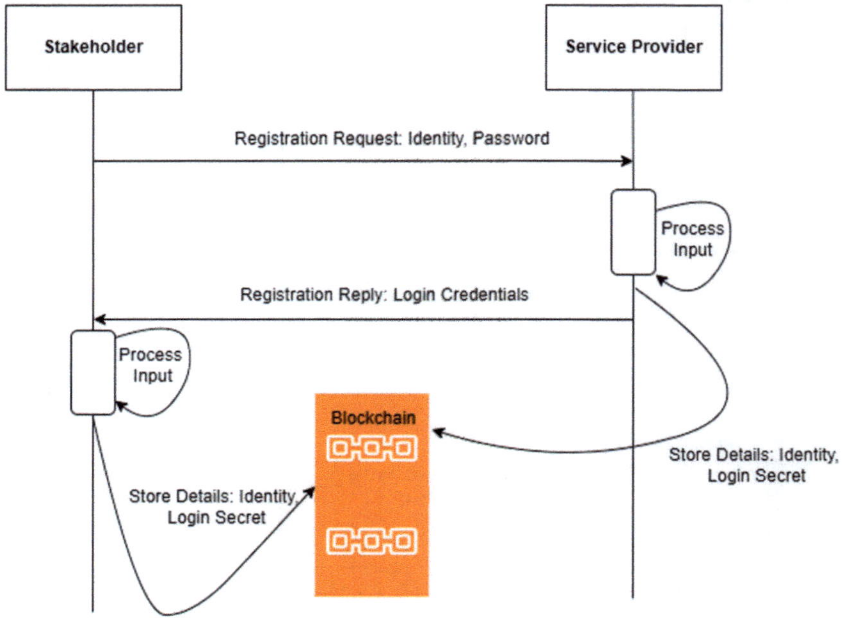

Fig. 3 Stakeholder registration phase

3.2.1 Registration Phase

The initial step for all IoP stakeholders is registering with the service provider. This procedure is referred to as stakeholder registration. Figure 3 depicts the concept that a stakeholder of the proposed framework is performing registration under the service provider. During this phase, the stakeholders select identity, and password, and send registration requests to the service provider. The service provider uses the secret key and acknowledges the stakeholders. Service providers use blockchain to store personal information shared by the stakeholders for safekeeping. Stakeholders also store received information from stakeholders in the blockchain. All stakeholder completes registration under service provider to get services of IoP.

3.2.2 Authentication Phase

All parties involved can exchange information and ask queries with one another at this phase. Information regarding the parties involved in different supply chains is shared by the service provider. A list of the participants in the same supply chain as well as other supply chains is available to any stakeholder upon request from the service provider. The stakeholders' names are on the list. Each member of the blockchain is identified by their own identity, and everyone on the blockchain can be

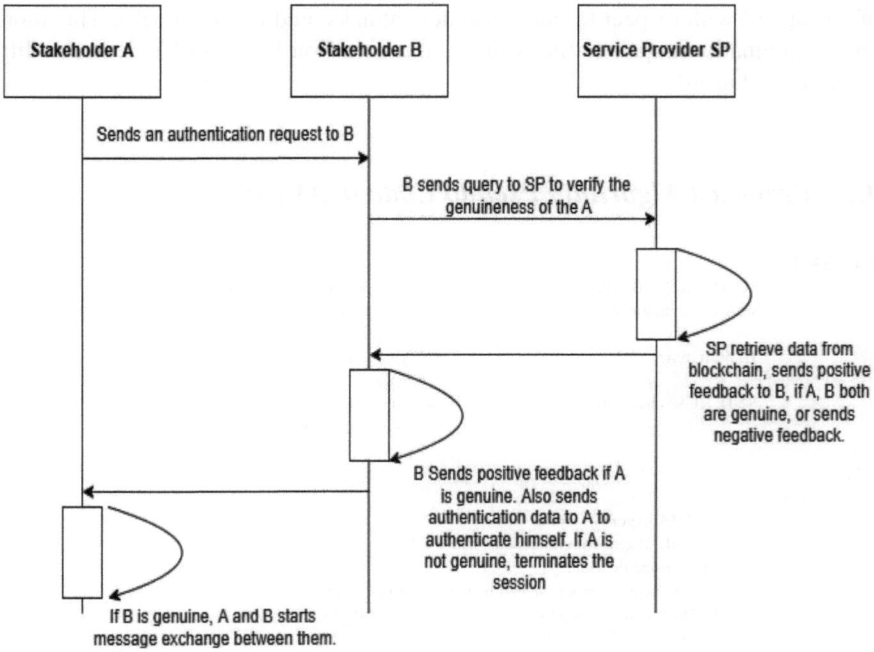

Fig. 4 Stakeholder authentication phase

referred to by their identity. As a result, before beginning an information exchange, any stakeholder, A, should authenticate one another before asking a question or sharing information with another stakeholder, like B. The steps for this phase are shown in Fig. 4.

Following the successful authentication, stakeholders A and B are able to interact and share information. Any two stakeholders can securely exchange information and communicate with one another in this fashion.

4 Security Analysis Using AVISPA

In the proposed model, the stakeholders communicate with each other securely. A security analysis is necessary to ensure this claim. Security analysis of the proposed protocol is done through the well-known security analysis tool, Automated Validation of Internet Security Protocols and Applications (AVISPA) [19]. The simulation of the proposed protocol in AVISPA is accomplished using the High-Level Protocol Specification Language (HLPSL) [20]. The protocol execution is held using the SPAN, which allows us to visualize the protocol's operation, a security protocol animator designed for HLPSL [21]. The AVISPA simulation tool is used to identify the security of a protocol with respect to man-in-middle attacks, and replay attacks. This thorough examination improves the secure communication framework's dependability and reduces hazards.

4.1 Proposed Algorithm Pseudo Code in AVISPA

```
Role STKHA (
        STKHA, STKHB, SP: agent,          # Agents involved in the session
        SKAS: symmetric_key,              # Shared symmetric key for
                                          # encryption
        H: hash_func,                     # Cryptographic hash
                                          # function
        SND, RCV: channel(dy)             # Communication channels
                                          # for sending and receiving
        )
 Init:
     Input:
            - STKHA: Agent initiating the role
            - STKHB: Agent communicating with STKHA
            - SP: Service Provider
            - SKAS: Symmetric key shared between STKHA and SP
            - H: Hash function to perform cryptographic operations
            - SND: Channel for sending data
            - RCV: Channel for receiving data
       State = 0
       IDSTKHA: Identity of Stakeholder A
       PWSTKHA: Password of Stakeholder A

 Transitions:
   1. If State == 0 and Received(start):
     - Generate nonce M
     - Compute PIDSTKHA = H (IDSTKHA || M)
```

```
        - Compute PSA = H (PIDSTKHA ⊕ PWSTKHA)
        - Send PSA encrypted with SKAS to SP
        - State = 3

    2. If State == 3 and Received (SKSA encrypted with SKAS):
        - Generate nonce MA1 and TMA
        - Compute MC1 = H (SKSA || MA1 || TMA)
        - Compute SSTA = PSA ⊕ TMA
        - Compute PIDSTKHA_N = H (IDSTKHA || MA1)
        - Compute PSA_N = H (PIDSTKHA_N ⊕ PWSTKHA)
        - Compute SRA = (MA1 || PSA_N) ⊕ H (SKSA || TMA)
        - Send {MC1, SSTA, TMA, SRA} to STKHB
        - State = 5

    3. If State == 5 and Received (MC3, SSB, SSKA, TS):
        #Validate MC3 using SKSA, MA1, TS
        - Compute SKSA_N = H (PSA_N ⊕ SKSP)
        - Compute SSKA = SKSA_N ⊕ H (TS || SKSA)
        - State = 12

End Role

Role STKHB (
            STKHA, STKHB, SP: agent, # Agents involved in the session
            SKBS: symmetric_key, # Shared symmetric key for
                                              # encryption
            H: hash_func,                     # Cryptographic hash function
            SND, RCV: channel(dy)             # Communication channels for
                                              #sending and receiving
           )
  Init:
       Input:
                - STKHA: Agent initiating the role
                - STKHB: Agent communicating with STKHA
                - SP: Service Provider
                - SKBS: Symmetric key shared between STKHB and SP
                - H: Hash function to perform cryptographic operations
                - SND: Channel for sending data
                - RCV: Channel for receiving data
          State = 2
          IDSTKHB: Identity of Stakeholder B
          PWSTKHB: Password of Stakeholder B

  Transitions:
    1. If State == 2 and Received(start):
        - Generate nonce M1
        - Compute PIDSTKHB = H (IDSTKHB || M1)
        - Compute PSB = H (PIDSTKHB ⊕ PWSTKHB)
        - Send PSB encrypted with SKBS to SP
        - State = 6

    2. If State == 6 and Received (SKSB encrypted with SKBS):
        - State = 8

    3. If State == 8 and Received (MC1, SSTA, TMA, SRA) from STKHA:
        - Generate nonce MB1 and TMB
        - Compute MC2 = H (SKSB || MC1 || MB1 || TMB)
        - Compute SSTB = PSB ⊕ TMB
```

```
        - Compute SRB = (MB1||PSB_N) ⊕ H (SKSB || TMB)
        - Send {MC1, MC2, SSTA, SSTB, TMA, TMB, SRA, SRB} to SP
        - State = 9

    4. If State == 9 and Received (MC3, MC4, SSA, SSB, SSKA, SSKB, TS) from SP:
        # Validate MC4
        - Send {MC3, SSB, SSKA, TS} to STKHA
        - State = 11
End Role

Role SP (
        STKHA, STKHB, SP: agent,      # Agents
        SKAS, SKBS: symmetric_key,            # Shared symmetric keys for SP
        H: hash_func,                         # Hash function
        SND, RCV: channel(dy),                # Communication channels
       )

Init:
        Input:
                - STKHA, STKHB, SP: Agents involved
                - SKAS, SKBS: Symmetric keys for secure communication
                - H: Cryptographic hash function
                - SND, RCV: Communication channels for interaction
        State = 1
        SKSP: Secret key of SP

  Transitions:
    1. If State == 1 and Received (PSA encrypted with SKAS):
        - Extract PSA
        - Compute SKSA = H (PSA ⊕ SKSP)
        - Send SKSA encrypted with SKAS to STKHA
        - State = 4

    2. If State == 4 and Received (PSB encrypted with SKBS):
        - Similar process as PSA handling
        - Compute SKSB
        - Send SKSB encrypted with SKBS to STKHB
        - State = 7

    3. If State == 7 and Received ({MC1, MC2, SSTA, SSTB, TMA, TMB, SRA, SRB})
                    from STKHB:
        #Validate MC1 and SSTA for STKHA
        #Validate MC2 and SSTB for STKHB
        - Generate nonce TS
        - Compute MC3 = H(MC1||MA1||SKSA) ⊕ TS
        -Compute MC4 = H(MC2||MB1|SKSB) ⊕ TS
        - Compute SSA = MA1 ⊕ H(SKSB||TS)
        - Compute SSB = MB1 ⊕ H(SKSA||TS)
        - Compute  SKSA_N=H (PSA_N ⊕ SkSP)
        - Compute  SKSB_N=H (PSB_N ⊕ SkSP)
        -Compute SSKA_N= SKSA_N⊕ H (TS|| SKSA),
        -Compute SSKB_N= SKSB_N ⊕ H(TS||SKSB).
        - Send {MC3, MC4, SSA, SSB, SSKA_N, SSKB_N, TS} to STKHB
        - State = 10

End Role
```

```
Role Session (STKHA, STKHB, SP: Agent, SKAS, SKBS: Symmetric_Key, H:
Hash_Function):
  Variables:
    SN1, SN2, SN3: Communication_Channel
    RV1, RV2, RV3: Communication_Channel_Response

  Composition:
    - Initialize Stakeholder A:
      StakeholderA(STKHA, STKHB, SP, SKAS, H, SN1, RV1)

    - Initialize Stakeholder B:
      StakeholderB(STKHA, STKHB, SP, SKBS, H, SN2, RV2)

    - Initialize Service Provider:
      ServiceProvider(STKHA, STKHB, SP, SKAS, SKBS, H, SN3, RV3)
End Role

Role Environment ():
  Constants:
    stkha, stkhb, sp: Agent
    skas, skbs: Symmetric_Key
    h: Hash_Function
    IntruderKnowledge:
    KnownEntities = {stkha, stkhb, sp, h} # Entities initially known to the
                                          # intruder

  Composition:
    - Start a valid session between legitimate participants:
      Session (stkha, stkhb, sp, skas, skbs, h)

    - Start a session where the intruder impersonates Stakeholder A:
      Session (i, stkhb, sp, skas, skbs, h)

    - Start a session where the intruder impersonates Stakeholder B:
      Session (stkha, i, sp, skas, skbs, h)

    - Start a session where the intruder impersonates the Service Provider:
      Session (stkha, stkhb, i, skas, skbs, h)

  End Role
```

4.2 Security Analysis

The proposed protocol is written in HLPSL and then executed in SPAN simulation tool. The CL-AtSe backend of AVISPA is used to verify the output of the protocol. The output of this test ensures that the proposed protocol can resist the replay and man-in-the-middle attacks. Figure 5 shows the output of the experiment conducted.

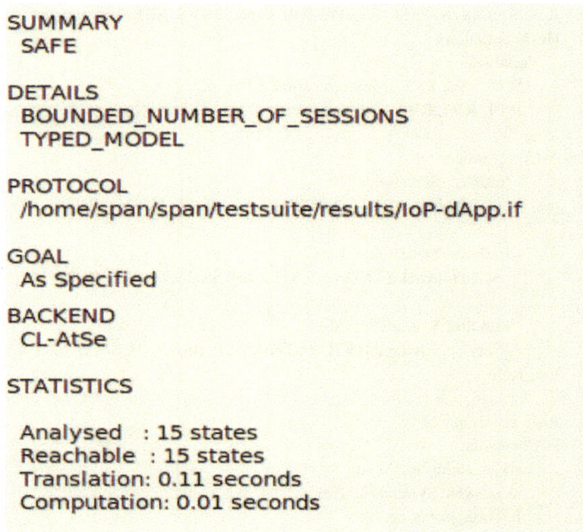

Fig. 5 CL-AtSe backend output from the AVISPA simulation

5 Performance Analysis

A performance analysis of the suggested model is carried out in this section. Execution time and communication time are the two characteristics that are assessed. These parameters compare the suggested model with three existing systems [16–18]. Focus is placed on the authentication phase of the suggested scheme for the analysis and comparison of execution time and communication time, leaving out the registration phase because the authentication phase is carried out repeatedly as needed by the participants. In contrast, the registration phase is completed only once for each participant.

The total time required by the several cryptographic operations employed during the proposed scheme's authentication phase's implementation is used to calculate execution time. Consequently, the suggested system uses the execution durations of cryptographic primitives including asymmetric key operations, elliptic curve cryptography (ECC) point multiplication, and one-way hash functions, while accounting for the other techniques [16–18]. The execution times of these cryptographic primitives are calculated using the data documented in [22]. There are just 36 one-way hash functions used in the suggested design. On the other hand, Hasan et al. [17] employs 2 asymmetric key operations, 12 one-way hash functions, and 12 ECC point multiplications, whereas [16] uses 8 one-way hash functions and 7 asymmetric key operations. Mishra et al. [18] uses 21 exponential operations, and 11 ECC point multiplication. The ECC point multiplication, asymmetric key operation, exponential operation and one-way hash function execution times are 63.075 ms, 522 ms, 522 ms, and 0.5 ms, respectively. Consequently, the suggested scheme outperforms the other schemes [16–18] in terms of execution time, taking just 18 ms to complete

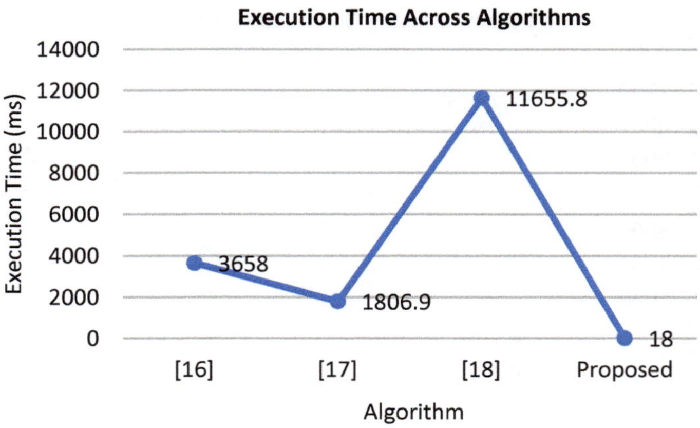

Fig. 6 Comparison of execution time across algorithms

since the protocol design uses a simpler one-way hash function. Figure 6 provides a summary of the findings.

Calculating the total size of all parameters transferred during the authentication process yields the communication overhead. The following are the sizes of various computation-related parameters: There are 320 bits of output from an ECC point multiplication, 2048 bits from an asymmetric key operation, 256 bits from a MAC function, 256 bits from a hash function, 160 bits for the participant's identity, 128 bits for a random number, 256 bits for the participant's password, 128 bits for the transaction identity, 32 bits for the timestamp, 32 bits for a random difficult digit to reach a selected target value, and 16 bits for a Boolean flag. The suggested method exchanges data with a total size of 5152 bits. The communication overhead for the suggested strategy and the other strategies [16–18] is displayed in Fig. 7.

Let us analyze the superiority score of the proposed protocol with respect to the other protocols under consideration.

- *Execution Time Comparison (Speedup Factor):* The Speedup Factor (S) may be defined as the ratio of execution time of a protocol to that of the proposed protocol.

$$S = \frac{\text{Execution Time of Protocol}}{\text{Execution Time of Proposed Protocol}} \quad (1)$$

From Eq. (1) and Fig. 6, we can calculate S for the related protocols under consideration as given in Table 1.

So, the proposed protocol is significantly faster than the other three, with the biggest speedup over [18].

- *Communication Overhead Comparison (Communication Efficiency Factor)*: The Communication Efficiency Factor (C) is the ratio of communication overhead of a protocol to that of the proposed protocol.

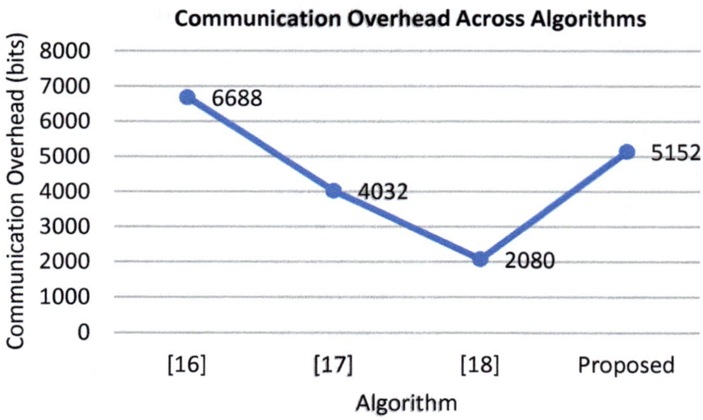

Fig. 7 Comparison of communication overhead across algorithms

Table 1 Speedup factor

Protocol (P)	$S_{[P]}$	Remark
[16]	$S_{[16]} = \frac{3658}{18} = 203.22 \sim 203.2$	203.2 times faster
[17]	$S_{[17]} = \frac{1806.9}{18} = 100.383 \sim 100.4$	100.4 times faster
[18]	$S_{[18]} = \frac{11655.8}{18} = 647.544 \sim 647.5$	647.5 times faster

$$C = \frac{\text{Communication Overhead of Protocol}}{\text{Communication Overhead of Proposed Protocol}} \tag{2}$$

From Eq. (2) and Fig. 7, we can calculate C for the related protocols under consideration (Table 2).

While the proposed protocol has higher communication overhead than [17] and [18], it is still more efficient than [16].

Let us define an overall superiority score (S_{total}), which combine both parameters execution time and communication overhead. This gives the result based on the weighted sum of speedup factor S and communication efficiency factor C.

$$S_{\text{total}} = \alpha \times S + \beta \times \frac{1}{C} \tag{3}$$

Table 2 Communication efficiency factor

Protocol (P)	$C_{[P]}$	Remark
[16]	$C_{[16]} = \frac{6688}{5152} = 1.35 \sim 1.3$	1.3 times more overhead
[17]	$C_{[17]} = \frac{4032}{5152} = 0.785 \sim 0.78$	Lower overhead, 22% less
[18]	$C_{[18]} = \frac{2080}{5152} = 0.451 \sim 0.4$	Lower overhead, 60% less

Table 3 Numerical superiority score

Protocol (P)	$S_{\text{total}}[P]$	
	$\alpha = 0.7$, and $\beta = 0.3$	$\alpha = 0.5$, and $\beta = 0.5$
[16]	$S_{\text{total}}[16] = 0.7 \times 203.2 + 0.3 \times \frac{1}{1.3} = 142.47$	$S_{\text{total}}[16] = 0.5 \times 203.2 + 0.5 \times \frac{1}{1.3} = 101.98$
[17]	$S_{\text{total}}[17] = 0.7 \times 100.4 + 0.3 \times \frac{1}{0.78} = 70.66$	$S_{\text{total}}[17] = 0.5 \times 100.4 + 0.5 \times \frac{1}{0.78} = 50.84$
[18]	$S_{\text{total}}[18] = 0.7 \times 647.5 + 0.3 \times \frac{1}{0.4} = 454$	$S_{\text{total}}[18] = 0.5 \times 647.5 + 0.5 \times \frac{1}{0.4} = 325$

Here, α, β are weights assigned to the factor's S and C, respectively. As execution time plays vital role for ensuring real time performance, reduce security risks, lowers energy consumption, and scales efficiently in large networks, we may consider $\alpha = 0.7$, and $\beta = 0.3$. Then, the superiority score of the proposed protocol against other protocols as shown in Table 3 can be calculated from Eq. (3), taking S and C values taken from Tables 1 and 2, respectively. Considering $\alpha = 0.5$ and $\beta = 0.5$, the results depict the supremacy of the proposed protocol.

The proposed protocol significantly outperforms all existing protocols in both execution time and overall efficiency. When $\alpha = 0.7$ and $\beta = 0.3$ (giving more weight to execution time), the superiority gap widens, making the proposed scheme the best choice for real-time applications. When $\alpha = 0.5$ and $\beta = 0.5$ (equal weighting for execution time and communication overhead), the proposed protocol remains superior, confirming its efficiency in both aspects. Existing protocols have either high execution time (e.g., [18]) or higher communication overhead (e.g., [16]), making them less practical compared to the proposed solution.

6 Conclusion

This paper discusses the necessity of information sharing in Industry 4.0 networks concerning IoP and then suggests a safe framework that makes use of a blockchain network. Through direct and indirect communication inside the IoP, this framework enables stakeholders to exchange or request information from any other stakeholder. The registration phase and the authentication phase are the two stages of the proposed protocol. The authentication phase, which verifies the mutual authentication of stakeholders prior to any contact, strengthens the security of the suggested framework. Furthermore, the framework's defense against replay and man-in-the-middle attacks is confirmed by the AVISPA simulation. According to performance studies, the suggested method achieves an exceptionally low execution time of about 18 ms, outperforming other current schemes. Furthermore, a decentralized application called IoP-DApp is created for the Ethereum blockchain, which is where the paper is implemented. To improve blockchain's use in the Internet of Things, we intend to release a new consensus method in the future. We will also create a specific application to put this consensus technique into practice.

References

1. Munirathinam S (2020) Industry 4.0: Industrial internet of things (IIOT). Adv Comput 117(1):129–164. Elsevier
2. Hussain Z, Akhunzada A, Iqbal J, Bibi I, Gani A (2021) Secure IIoT-enabled industry 4.0. Sustainability 13(22):12384
3. Pennekamp J, Glebke R, Henze M, Meisen T, Quix C, Hai R, Gleim L, Niemietz P, Rudack M, Knape S, Epple A (2019) Towards an infrastructure enabling the internet of production. In: 2019 IEEE international conference on industrial cyber physical systems (ICPS) 2019 May 6, pp 31–37. IEEE
4. Pennekamp J, Henze M, Schmidt S, Niemietz P, Fey M, Trauth D, Bergs T, Brecher C, Wehrle K (2019) Dataflow challenges in an internet of production: a security & privacy perspective. In: Proceedings of the ACM workshop on cyber-physical systems security & privacy 2019 Nov 11, pp 27–38
5. Esfahani A, Mantas G, Matischek R, Saghezchi FB, Rodriguez J, Bicaku A, Maksuti S, Tauber MG, Schmittner C, Bastos J (2017) A lightweight authentication mechanism for M2M communications in industrial IoT environment. IEEE Internet Things J 6(1):288–296
6. Li X, Peng J, Niu J, Wu F, Liao J, Choo KK (2017) A robust and energy efficient authentication protocol for industrial internet of things. IEEE Internet Things J 5(3):1606–1615
7. Li X, Niu J, Bhuiyan MZ, Wu F, Karuppiah M, Kumari S (2017) A robust ECC-based provable secure authentication protocol with privacy preserving for industrial internet of things. IEEE Trans Industr Inf 14(8):3599–3609
8. Harishma B, Patranabis S, Chatterjee U, Mukhopadhyay D (2018) POSTER: authenticated key-exchange protocol for heterogeneous CPS. In: Proceedings of the 2018 on Asia conference on computer and communications security 2018 May 29, pp 849–851
9. Das AK, Wazid M, Kumar N, Vasilakos AV, Rodrigues JJ (2018) Biometrics-based privacy-preserving user authentication scheme for cloud-based industrial internet of things deployment. IEEE Internet Things J 5(6):4900–4913
10. Eldefrawy MH, Ferrari N, Gidlund M (2019) Dynamic user authentication protocol for industrial IoT without timestamping. In: 2019 15th IEEE international workshop on factory communication systems (WFCS) 2019 May 27, pp 1–7. IEEE
11. Palit SK, Chakraborty M, Chakraborty S (2022) MASKA: mutual authentication and session key agreement protocol in global mobility networks. In: Applications of machine intelligence in engineering 2022 Apr 21, pp 309–321. CRC Press
12. Palit SK, Chakraborty M, Chakraborty S (2022) AUGChain: blockchain-based mobile user authentication scheme in global mobility network. J Supercomput 78(5):6788–6816
13. Palit SK, Chakraborty M, Chakraborty S (2023) Performance analysis of 5GMAKA: lightweight mutual authentication and key agreement scheme for 5G network. J Supercomput 79(4):3902–3935
14. Palit SK, Chakraborty M (2024) TLMMChain: blockchain-based hierarchical security management scheme for three-level mobility model in next-generation networks
15. Palit SK, Chakraborty M (2020) A mobile user authentication technique in global mobility network. In: Proceedings of international ethical hacking conference 2019: eHaCON 2019, Kolkata, India 2020, pp 171–190. Springer Singapore
16. Dwivedi SK, Amin R, Vollala S (2020) Blockchain based secured information sharing protocol in supply chain management system with key distribution mechanism. J Inf Secur Appl 54:102554. https://doi.org/10.1016/j.jisa.2020.102554

17. Hasan AT, Sabah S, Haque RU, Daria A, Rasool A, Jiang Q (2022) Towards convergence of IoT and blockchain for secure supply chain transaction. Symmetry 14(1) https://doi.org/10.3390/sym14010064
18. Mishra R, Ramesh D, Edla DR, Qi L (2023) VaccineChain: a checkpoint assisted scalable blockchain based secure vaccine supply chain with selective revocation. J Ind Inf Integr 1(34):100485
19. AVISPA Automated validation of internet security protocols and applications. http://www.avispa-project.org/
20. Oheimb DV (2005) The high-level protocol specification language HLPSL developed in the EU project AVISPA. In: Proceedings of 3rd APPSEM, Frauenchiemsee, Germany, pp 1–17
21. AVISPA SPAN, the security protocol animator for AVISPA. http://www.avispa-project.org/
22. Banerjee S, Odelu V, Das AK, Chattopadhyay S, Kumar N, Park Y, Tanwar S (2018) Design of an anonymity-preserving group formation based authentication protocol in global mobility networks. IEEE Access 6:20673–20693. https://doi.org/10.1109/ACCESS.2018.2827027

DEChain: Private Blockchain-Based Digital Evidence Management System

Mohuya Chakraborty, Sudip Kumar Palit, Debanjan Nag, Abhirup Bag, Pallabi Mondal, and Anupama Bain

Abstract This paper proposes a Private Blockchain-enabled Digital Evidence Management System using Hyperledger Fabric and Composer to aid the activities of Law and Enforcement department. Applying the principles of decentralization and non-tamperability, the blockchain system guarantees proper registration, transfer, and verification of all digital evidence at all stages of their usage. These processes enhance the transparency and security largely eliminating the vulnerabilities of the evidence succumbing to fiddling before the legal processes. By using smart contracts, the system captures activities like evidence registration, custody transfer, among others, plus integrity verification. It also enforces exclusive controls of participation for example, only police, investigators, technicians, or attorneys are allowed to engage with the evidence. This makes efforts of trying to manage the digital evidence require minimal intervention from human beings thus increasing efficiency. Performance trials show good system performance under 'live' mode. In general, registration of the digital evidence lasts 23.8 ms, transfer of the evidence takes 15.1 ms, verification requires 9.4 ms and generation of the reports requires 0.8 ms.

Keywords Digital evidence · Blockchain · Hyperledger fabric · Hyperledger composer · Smart contracts

1 Introduction

The judicial system relies very much on the accuracy of the digital evidence to uphold justice and confidence in the court case. However, the following challenges are well-known about the traditional evidence management systems: tampering, unauthorized access, and rather inefficient storage and retrieval. Since more and more often such evidences get included in court cases it is crucial to sort out these problems to maintain

M. Chakraborty · S. K. Palit (✉) · D. Nag · A. Bag · P. Mondal · A. Bain
Cybersecurity Centre of Excellence, Institute of Engineering and Management (IEM), University of Engineering and Management, Kolkata, India
e-mail: sudipkumar.palit@uem.edu.in

M. Chakraborty et al. (eds.), *Proceedings of Sixth International Ethical Hacking Conference*, Lecture Notes in Networks and Systems 1513,
https://doi.org/10.1007/978-981-96-8632-2_14

the integrity of justice. Similar to the research by Shanmugam et al., making a strong case for development of more creative solutions that offers indicates that conventional systems struggle with issues of data integrity and chain of custody [1].

Blockchain technology offers an innovative solution to these challenges due to the inherent properties of decentralization, transparency and non-interference. Evaluations on conventional divided systems reveal that blockchain technology offers safe data sharing as well as storage, cryptographic solidity, and shielding from alteration. Hyperledger Fabric is one of the most suitable blockchain frameworks for performing sensitive legal works and is an enterprise level platform specifically designed for enabling the deployment of dApps safely and at scale. Hyperledger Fabric is a highly popular enterprise blockchain solution that can provide safely scalable and high resilience implementing a decentralized application intended for delicate jobs such as managing legal proofs. According to Androulaki et al., Hyperledger Fabric has features that make it suitable to process evidence needed in courts [2].

Several studies have explored blockchain-based solutions for secure evidence management. To tackle issues like fraud and data abuse in wireless networks, for example, IBM engineers unveiled a concept that makes use of Hyperledger Fabric. Comparably, new developments in blockchain technology have shown that it can improve data traceability, withstand manipulation, and guarantee strong access control in a variety of applications. However, current methods frequently neither fully solve scalability and performance issues nor have domain-specific modifications for judicial needs. In contrast to conventional evidence management systems, Nakamura and Chen showed how blockchain-based platforms might resolve chain-of-custody problems in the legal field while enhancing performance and scalability [3].

Matthias Gottlieb [4] proposed an innovative blockchain implementation to enhance the techniques of credential authentication, data exchange and academic administration alongside issues such as standardization. In 2024, Shiyong Huang et al. [5] put forward TMAS, a graph-based framework to discover and categorize transaction anomalies to enhance the efficiency and accuracy of blockchain fraud detection. In 2024, Shiyong Huang et al. [5] proposed TMAS, a graph-based framework to detect and categorize transaction anomalies, improving blockchain fraud detection accuracy and scalability. In the same year, Shalini Dhara et al. [6] invented a hybrid blockchain with quantum cryptography to build a reliable and safer mechanism for IoT than the normal cryptographic system.

The stablecoins were discussed in the empirical analysis undertaken by Lennart Ante et al. [7] in 2022 reviewing the financial performance of stablecoins discussing adoption issues and regulatory approaches using examples related to Tether and DAI among others. Nevertheless, Shi Dong et al. [8] provided only the particular aspects of the discovered conceptual, practical applications, and technical development of blockchain in various fields.

In the year 2021, Uzma Jafar et al. [9] presented a review of e-voting systems with blockchain solutions and pointed out the benefits of tamper-proofing and sharing the issues like scalability, voter privacy and adoption of the e-voting system. Based on the research conducted in 2023, Shrabani Sutradhar et al. [10] put forward a blockchain managerial identity model that uses Hyperledger Fabric Framework and OAuth 2.0, aiming at bringing about improved scalability and enhanced security for the healthcare sector. In 2019, researchers Harihara Gopalan et al. [11] explored usage of blockchain for digital forensics concerning the reliability of evidence preservation and non-tampering proofs during the investigation.

Back in 2021, McSeth Antwi et al. had discussed analyzing [12] the capability of Hyperledger Fabric for health care including patient data sharing with high security, private and efficient health care delivery systems. Sakshi et al. [13] suggested the surveys on blockchain IoT forensic solutions in 2023 with the trends in evidence which include the preservation of the solution and the issues such as scalability and real-time implementation.

The traditional systems for digital evidence management face several challenges, including vulnerability to tampering, high computation costs, limited scalability, and risks associated with centralized data management. These systems often struggle with inefficiencies in handling large volumes of transactions and ensuring secure and transparent access to evidence. The absence of robust audit trails and the susceptibility to data breaches compromise their reliability in sensitive legal and forensic contexts. Moreover, in the traditional forensic system, memory limitations of collection devices lead to evidence data being overwritten, causing delays in reporting crucial evidence and inefficiencies in investigations [14]. Addressing these gaps requires a solution that combines scalability, security, and transparency while maintaining affordability and efficiency in managing digital evidence.

To address these gaps, this paper presents DEChain, a private blockchain-based architecture for digital evidence management based on Hyperledger Fabric and Composer. By enhancing scalability, lowering the risks connected with centralized management, and ensuring data integrity, the suggested architecture seeks to address the shortcomings of traditional systems.

The contributions of this paper are as follows:

1. A novel blockchain-based framework, DEChain, tailored for digital evidence management.
2. Using Composer and Hyperledger Fabric to develop and implement safe procedures that allow for the tracking of the reliability of the evidence.
3. A comparison of several performance metrics, including the built framework's computational cost, latency, and finished throughput.

The remaining sections of this paper are organized as follows: Composer, blockchain, digital proof, and Hyperledger Fabric are among the ideas discussed in Sect. 2. The authors provide a thorough explanation of the processes, flow diagrams, and algorithms of the proposed DEChain architecture in Sect. 3. Additionally, Sect. 4 provides illustrations of the simulation environment and performance analysis. Section 5 concludes this study with a few observations and recommendations for additional research.

2 Preliminaries

2.1 Digital Evidence

Digital evidence refers to any form of information in digital form that can be used in investigative processes as well as in legal proceedings. Due to its ever-growing dependence on digital technologies, such evidence has become crucial in forensic investigations. Digital evidence takes many forms, including computer files, emails, or system logs from personal devices or servers. Network data, such as communication logs, packet traces, and firewall logs, also serve as valuable evidence.

Messages, call history, and app information from smartphones and tablets are used under mobile device evidence. Today, Internet of Things (IoT) devices have emerged, which add another layer of complexity because information from smart assistants, cameras, and even wearable devices is actively used in investigations. Other tools include cloud service which stores back-up, documents and shared resources, and these are also good sources of evidence. Social media platforms further contribute with posts, interactions, and private messages. The numerous sorts of digital evidence include multimedia files such as photographs, movies, and audio recordings, as well as location data from GPS-enabled devices.

The issues arising from handling of digital evidence are as follows: Digital data is less stable since it can be quickly modified or deleted. The vast amount of data gathered every single day might intimidate the investigators and yet the issue of evidence authenticity and legal admissibility is critical. Authentication is necessary to eliminate any possibility of tampering with the evidence and meeting the legal standards for admissibility in legal proceedings requires careful handling [15].

2.2 Blockchain Technology

Blockchain refers to a distributed ledger technology (DLT) that records transactions in a very secure system that cannot be modified. Unlike most conventional systems reliant on centralized control, blockchain is a decentralized network which creates trust among its participants. This capability is especially useful where the data needs to be permanently and irretrievably stored.

Blockchain offers several key features. Immutability ensures that once data is recorded, it cannot be changed. Decentralization distributes data across multiple nodes, eliminating single points of failure. Transparency enables open access to view transactions for authorized viewers, thus, promotes trust. Cryptographic security ensures the data remains secure and the user's privacy is well safeguarded by strong algorithms. Additionally, smart contracts, where terms are coded, allow agreements to self-execute without dishonesty.

In the context of digital evidence, blockchain is more than helpful. It confirms that all documents tendered are protected and cannot be edited; further, presents an efficient way of accessing the electronic material through time stamps. Of all the processes involved in managing evidence, the chain of custody is specially regulated, and on the blockchain, such changes as unauthorized modifications or gaps in paperwork cannot occur [16].

2.3 Hyperledger Fabric

Hyperledger fabric on the other hand is an open source blockchain solution by Linux foundation. It is specifically designed to help developers create safe, sustainable, and efficiently-balanced blockchain solutions. Fabric is a permissioned system where only those with permission can transact on the system and or add to the chain. Its flexibility and integrated mechanism mean that it can be adjusted and has provisions for modification of other attributes including approval flows and smart contracts. It also supports chain code (Smart Contracts) that enables the use of languages such as JavaScript, Go and Java in developers to develop the business of the firm. Strong features of the application including role access control, privacy security and optimization position it for those applications that entail security and proprietary knowledge. In digital evidence management, the most common application of Hyperledger Fabric is to offer metadata repository, COC, and the isolation of privy information.

2.4 Hyperledger Composer

Hyperledger Composer increases the efficiency of creating blockchain applications with Hyperledger Fabric. By focusing on enterprise networks, Composer offers a range of tools for modeling entities, actors, and groups. Developers can leverage JavaScript to establish business logic and govern access through customizable guidelines. The framework features a user-friendly playground interface for checking out blockchain networks, as well as a CLI device for advanced network administration. With its REST API integration and rapid simulation abilities, Composer allows the advent of virtual proof answers and simplifies the development of packages for handling proof of lifestyles, statistical control, and file sharing [2].

3 Proposed Framework

3.1 Purpose

The goal of the cutting-edge blockchain technology DEChain is to alter the way that digital evidence is gathered and processed. Additionally, it offers a secure, immutable setting that may support all stages associated with evidence, including submission, validation, and review up to the legal body. DEChain ensures the legitimacy, accountability, and audibility of digital evidence by using the core benefits of blockchain technology, which include transparency, tamper-proofing, and diffusion. In order to preserve the chain of custody, guarantee data integrity, and provide clients with timely and reliable evidence, this paper addresses important concerns in maintaining digital evidence.

3.2 Architecture

DEChain framework contains the following components:

3.2.1 Stakeholders

The DEChain system involves multiple stakeholders who play distinct roles:

- Administrative Staff: These users oversee administrative processes of the system, enter stakeholders' information and grant permission by roles.
- Law Enforcement Agencies: Someone accountable for inputting digital evidence in the system, and for authorizing the transfer of documents between users or offices.
- Forensic Experts: Responsible for cryptography, and forensic methodology for the processes that validate the truthfulness of the evidence submitted in court. They also prepare reports of their findings.
- Legal Professionals: Achieve admissibility of evidence in courtroom proceedings by accessing validated evidence and reports.

3.2.2 System Components

The system components comprise.

- Blockchain Network: Serves as the basis for further safe and unalterable registration of all the transactions and interactions associated with the digital evidence.

- Smart Contracts: Other forms of functional recommendations include; Actualizing process automation for instance in registration of evidence, custody transfers and verification procedures.
- Secure Storage: Integrates both the hot and cold storage for the ease for accessing as well as archiving of the evidence and the relevant reports.
- Audit Module: Records all the transactions from the system to make it easy to check compliance of the activities with the existing legal and organizational requirements.

3.3 Workflow

There are six stages of workflow.

1. **Evidence Registration**
   - Participants: Law Enforcement Agency
   - Actions: Initiates RegisterEvidence transaction, inputs all details: evidence ID, description, and cryptographic hash.
   - Outcome: DigitalEvidence asset is created on the blockchain, and EvidenceRegistered event is emitted to inform all participants.
2. **Evidence Transfer**
   - Participants: Law Enforcement Agency, Forensic Expert, or Legal Professional.
   - Actions: Initiates TransferEvidence transaction and also updates new custodian's details.
   - Outcome: CurrentCustodian field in DigitalEvidence is updated; EvidenceTransferred event is emitted with an indication of the transfer.
3. **Evidence Verification**
   - Participants: Forensic Expert
   - Actions: Initiates VerifyEvidence transaction, compares stored hash with the new hash for integrity.
   - Outcome: Integrity status updated as "Verified" or "Compromised". EvidenceVerified event is emitted to state integrity status in the event.
4. **Report Generation**
   - Participants: Forensic Expert
   - Actions: Initiates GenerateReport transaction after verification has been done. Latest findings and metadata are then inserted into the report.
   - Outcome: ReportGenerated event is emitted with the analysis of the provided evidence.
5. **Legal Review**

- Participants: Legal Professionals
- Actions: Queries blockchain for the records of evidence. Only verified evidence is shown.
- Outcome: Relevant DigitalEvidence assets are returned when the query conditions such as its status, or the identity of the custodian has been inserted.

6. **Access Control and Security**

- Enforced Throughout: Only those participants, who have been pre-authorized, can perform certain actions. Example: Only current custodians can transmit evidence, and only law enforcement can register evidence.
- Outcome: Protect information and the system by implementing security on a role basis.

3.4 Algorithm

Step 1: Evidence Registration Process

The purpose of the evidence registration procedure is to reliably and securely generate a digital record for fresh evidence. Upon receiving a request for evidence registration, the system verifies the request and creates a special identification number for the evidence. This stage records important metadata, such as the evidence's description, cryptographic hash, registration timestamp, and details about the original custodian. After that, the evidence is digitally documented and given the initial status "Registered". For future monitoring and auditing reasons, a registration event is emitted, and the original chain of custody is registered to guarantee traceability.

Step 2: Evidence Transfer Mechanism

Transferring evidence in a secure and responsible manner might provide another significant problem in evidence management. Before initiating the transfer procedure, the system verifies that the person making the request is the current custodian, confirms that the new custodian's information is complete, and verifies that the record is present in the system. As part of the transfer procedure, the custodian details are changed, the evidence is marked as "in transfer," and the record of prior custom details is maintained. The chain of custody log is retrieved using various time stamps and transfer record segments. Ultimately, a transfer event is initiated in order to generate an audit trail for the aforementioned transaction.

Step 3: Evidence Integrity Verification Algorithm

The validity of the collected digital evidence is especially questioned. This procedure involves employing popular algorithms like SHA-256 to compare the hash value of the evidence that has been submitted for authentication with the hash value of the evidence that has been saved. "Verified" is the label applied to evidence if its hashes match those stored in the database. The evidence is labeled or designated as "Compromised," which is an alarm, if they disagree. To increase confidence, a timestamp is appended to each verification attempt, and an event of integrity status is raised.

Step 4: Report Generation Protocol

The aim of the organizational report development process is to prepare technical, detailed reports for substantiated facts. In order for a new Report to be created, the document must be "Verified," and the requester must have the appropriate access level. This indicates that, apart from the metadata collection, the client verification status should also be analyzed and authenticated. The evidence is accredited, has identification number, has details of the custody history, and is laser marked with the time the report was produced. A report-generation event is then created to release the report for more scrutiny or prosecution.

Step 5: Overall System Governance Principles

The system is built on principles that support its continued protection, added efficiency and continued observation. On the specific issue of transaction repudiation, a sound COCO is developed with every action or event in the process documented. In order to safe-guard the audit process, protocols that authenticate and authorize participants' identities as well as cryptographic processes that assure the integrity of the data available to participants are incorporated. Other requirements for the next studies, transaction time, logs creation, and audits for dependability and transparency accountabilities are crucial for the system performances.

Key Characteristics

Preservation of the primitiveness and integrity of evidence-related data, enhanced traceability, and cryptographic methods identify them as the hallmark of the system's algorithms. The solution makes the DE architecture straightforward and dependable by incorporating and transferring controls while keeping wholesome events and transactions documented.

The flow diagram of the proposed work is shown in Fig. 1.

4 Simulation and Performance

The graphs (Figs. 2, 3, 4 and 5) show the performance of evidence-related tasks where milliseconds are used to quantify registering, transmitting, and verifying evidence. The effectiveness and adaptability of the system during crucial operations are expressed in terms of these metrics.

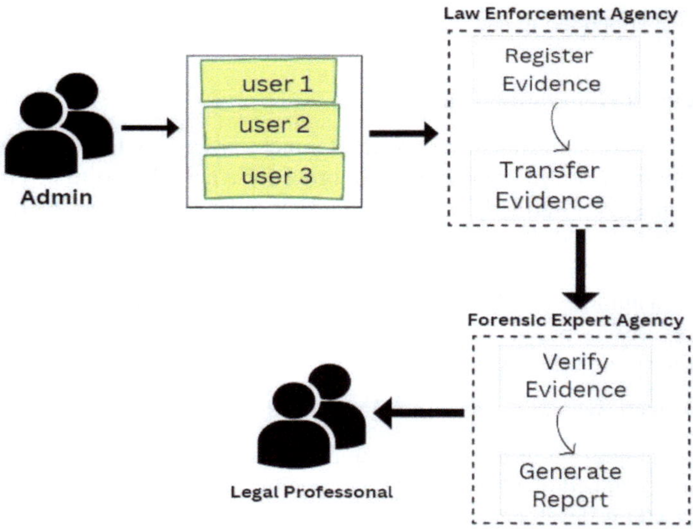

Fig. 1 Flow diagram of DEChain

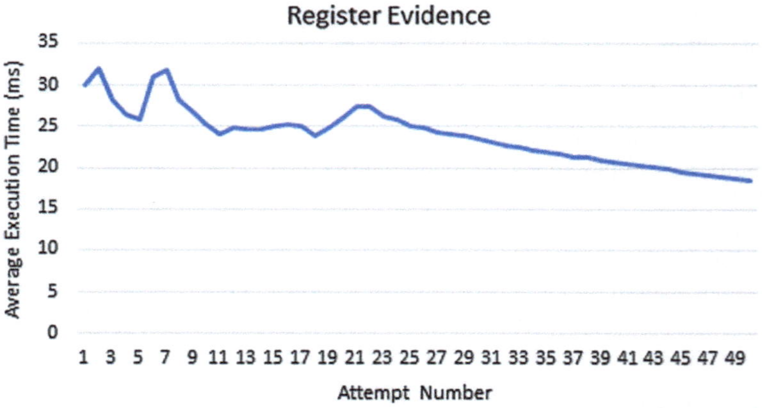

Fig. 2 Register evidence

4.1 Register Evidence

Figure 2, titled "Register Evidence," examines the duration required to register evidence. As the attempt number increases, the average execution time for the event Register Evidence decreases and gradually parallel to the x axis. This indicates the stability and scalability of this event with the increased number of attempts. The average execution time for this event is approximately 18.58 ms for 50 attempts.

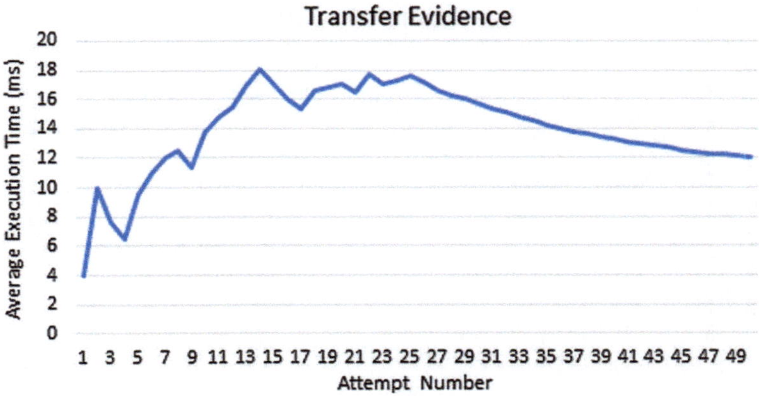

Fig. 3 Transfer evidence

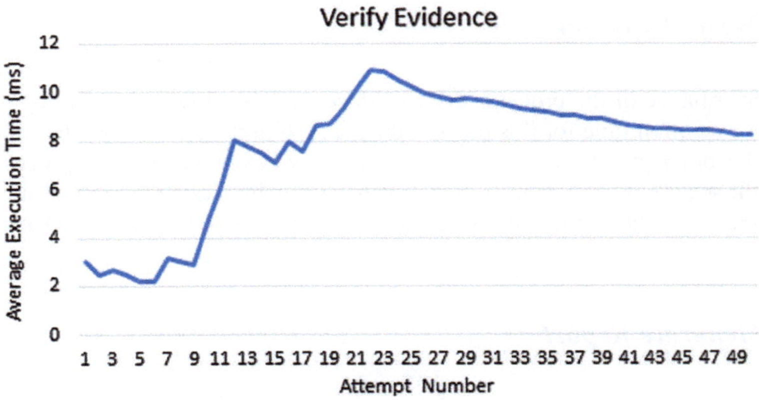

Fig. 4 Verify evidence

4.2 Transfer Evidence

Figure 3 shows the performance of the process transfer evidence. The average execution time for this process initially increases with the attempt number, then gradually decreases and becomes parallel to the x axis as the attempt number increases. This demonstrates that the Transfer Evidence process is stable and scalable to handle large numbers of operations. The average execution time for this event is approximately 12.04 ms for 50 attempts.

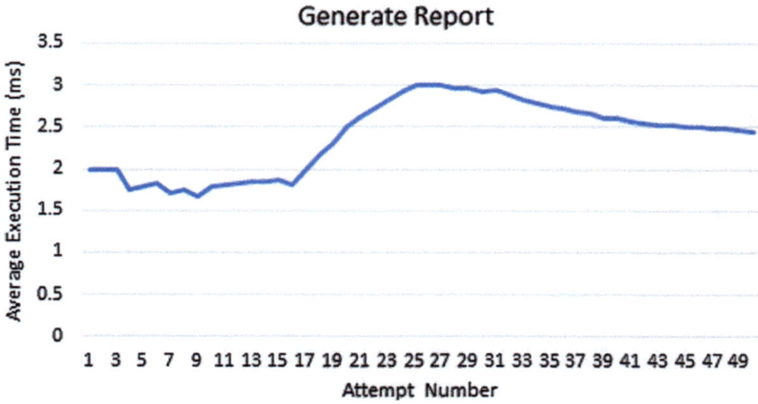

Fig. 5 Generate report

4.3 *Verify Evidence*

The performance of the process verify evidence can be observed from Fig. 4. The average execution time for this process increases with the attempt number initially, then it becomes parallel with the x axis. Like previous processes, this process also shows the scalability and stability of this event with an increasing number of attempts. The average execution time for this event is approximately 8.25 ms for 50 attempts.

4.4 *Generate Report*

Figure 5 depicts the performance of the process generate report. This process also becomes stable with the increasing number of attempts. It gives average execution time approximately 2.48 ms up to 50 attempts.

4.5 *Analysis*

These performance metrics are derived from accumulated times of the system's performance on specific activities under controlled testing conditions that are not representative of real usage. The time before the start of the test's procedures and the time after each test's registration, transmission, verification, and generate report processes are completed are noted. In order to determine the timer interval and the elapsed time, the two times are therefore subtracted. These three results make it abundantly evident that this system is well suited to handling jobs in a robust manner without experiencing excessive delay or throughput loss.

Table 1 System specifications

Processor	AMD Ryzen 5 5600H with Radeon Graphics 3.30 GHz
Installed RAM	8.00 GB (7.34 GB usable)
System type	64-bit operating system, × 64-based processor

Considering all the steps, the evaluation shows that the recommended solution performs better in terms of evidence management, guaranteeing quick and safe processing of digital evidence.

4.6 Environment

System specifications used in the simulation are given in Table 1.

4.7 Performance Analysis and Comparison

This section compares the performance of DEChain with existing evidence management systems based on three key parameters: 1. Processing Efficiency, 2. Security Features, and 3. Computational Cost.

4.7.1 Processing Efficiency Comparison

In crucial activities, DEChain exhibits remarkable processing efficiency with low latency. It takes 18.58 ms for register evidence, 12.04 ms for transfer evidence, 8.25 ms for verifying evidence, and only 2.48 ms for generating reports.

DEChain outperforms the traditional methods by working much faster and responding more quickly while most such systems work slowly and often experience delays for guaranteeing smooth processing of huge activities. In terms of the following two parameters, DEChain supersedes traditional methods.

- Throughput: DEChain works better than conventional systems that are easily clogged by performing a large number of transactions efficiently.
- Latency: Considering other systems with longer transit times resulting from centralization, test results reveal relatively fast response times during the registration, transfer, verification of evidence and their report generation.

4.7.2 Security Features Comparison

Based on the blistering of services, DEChain employs the Hyperledger Fabric version of blockchain to provide unalterable storage, safe communication, and easily reportable transactions. DEChain secures the evidence and revisions from other people's access unlike a normal system that might be eased to hack into thereby causing data breach.

DEChain, leveraging Hyperledger Fabric, ensures unalterable storage, encrypted communication, and role-based access control, making it far more secure than traditional methods. While traditional systems rely on centralized databases prone to hacking, tampering, and single points of failure, DEChain's decentralized nature prevents data breaches and unauthorized access. It provides transparent, immutable transaction logs for easy auditing, unlike traditional systems that often lack robust security measures and are vulnerable to manipulation. This makes DEChain a reliable solution for safeguarding digital evidence.

4.7.3 Cost of Computation Comparison

It is clear that DEChain's Hyperledger Fabric and modularity enable it to be designed for low calculation costs, even under high traffic situations. DEChain maintains its full scalability and affordability without sacrificing security or performance, in contrast to conventional systems that need significant investments in the underlying infrastructure.

DEChain, powered by Hyperledger Fabric's modular design, is optimized for low computation costs, even during high-traffic scenarios. Its efficient consensus mechanisms ensure scalability and performance without requiring extensive computational resources. In contrast, traditional methods often rely on centralized infrastructures that demand substantial investments in hardware and maintenance, leading to higher computation costs and slower processing times under heavy workloads. DEChain's streamlined architecture thus delivers faster, more affordable computation without compromising security or reliability.

5 Conclusion

Blockchain, a very secure, transparent and reliable technology has revolutionized the handling of digital evidence completely. This paper introduces the DEChain, a concept of managing digital evidence through the Hyperledger Fabric and Composer. Criminating the main problems associated with the traditional solutions of evidence handling or dealing with the evidence management systems including but not limited to; ease of manipulation, unauthorized access, and challenges in handling the evidence.

DEChain preserves the verity, genuineness, and auditability of the digital evidence through a decentralized and immutable blockage. Elimination of dependency on the central control in the architecture of the framework also eliminates possible delicate points that might negatively affect the judicial process. Performance evaluation proves DEChain efficient, low-latency, high-throughput, and with reduced computational burden, making it suitable for practical use. DEChain is a holistic approach that assists law enforcement, forensic specialists, and lawyers, ensuring comprehensive digital evidence processing. It is both innovative and secure due to the integration of blockchain as data is very transparent and nearly impossible to tamper with.

These contributions show that the work lays down the foundation of how blockchain technology could bring drastic change to the process in which legal institutions manage evidence. Further research can focus on increasing the efficacy of DEChain to accommodate higher amounts of data and implementing modern cryptographic technologies for transmission of data that promote privacy. If such problems are solved, DEChain may further evolve into a full-service solution that would allow legal systems of the world to meet the growing demand of the digital age.

References

1. Shanmugam V et al (2020) Challenges in digital evidence management systems. Int J Legal Technol 45(3):15–29
2. Androulaki E, et al (2018) Hyperledger fabric: a distributed operating system for permissioned blockchains. In: Proceedings of the 13th EuroSys conference. pp 30–45
3. Nakamura Y, Chen L (2021) Blockchain applications in evidence management systems. Blockchain Innov J 15(2):78–90
4. Gottlieb M, Deutsch C, Hoops F, Pongratz H, Krcmar H (2024) Expedition to the blockchain application potential for higher education institutions
5. Huang S, Hao X, Sun Y, Wu C, Li H, Ren W, Choo KK (2024) TMAS: a transaction misbehavior analysis scheme for blockchain
6. Dhar S, Khare A, Dwivedi AD, Singh R (2024) Securing IoT devices: a novel approach using blockchain and quantum cryptography
7. Ante L, Fiedler I, Willruth JM, Steinmetz F (2022) A systematic literature review of empirical research on stablecoins
8. Dong S, Abbas K, Li M, Kamruzzaman J (2003) Blockchain technology and application: an overview
9. Jafar U, Aziz MJ, Shukur Z (2021) Blockchain for electronic voting system—review and open research challenges
10. Sutradhar S, Karforma S, Bose R, Roy S, Djebali S, Bhattacharyya D (2023) Enhancing identity and access management using Hyperledger Fabric and OAuth 2.0: a block-chain-based approach for security and scalability for healthcare industry
11. Harihara Gopalan S, Akila Suba S, Ashmithashree C, Gayathri A, Jebin Andrews V (2019) Digital forensics using blockchain
12. Antwi M, Adnane A, Ahmad F, Hussain R, ur Rehman MH, Kerrache CA (2021) The case of HyperLedger Fabric as a blockchain solution for healthcare applications
13. Sakshi, Malik A, Sharma AK (2023) A survey on blockchain-based IoT forensic evidence preservation: research trends and current challenges
14. Li M, Lal C, Conti M, Hu D (2021) LEChain: a blockchain-based lawful evidence management scheme for digital forensics. Fut Gener Comp Syst 115:406–420

15. Casey E (2011) Digital evidence and computer crime: forensic science, computers and the internet. Academic Press
16. Mougayar W (2016) The business blockchain: promise, practice, and application of the next internet technology. Wiley

Generative AI for Advanced Recycling Processes in Polyethylene and Polypropylene Manufacturing

Rahul Vadisetty, Sai Teja Nuka, Srinivas Kalisetty, Chandrashekar Pandugula, Jai Kiran Reddy Burugulla, and Venkata Narasareddy Annapareddy

Abstract Linked with environmental impact, rapid growth in the production of PE and PP makes searching for effective recycling solutions a matter of urgency. In traditional recycling, enormous challenges, contamination, degradation, and high-energy input make processes inefficient and not scalable. Generative AI can transform such challenges into optimized data-driven processes that enhance recycling outcomes. The paper reviews generative AI applications in advanced polyethylene (PE) and polypropylene (PP) recycling processes to improve process efficiency, quality control, and chemical recycling. This may allow high-purity sorting, predictive maintenance, and consistent chemical recycling due to generative AI over predictive modeling and real-time decision-making, turning plastic waste into reusable materials. However, generative AI implementation requires a significant amount of data infrastructure and also faces particular challenges because of data privacy concerns and compatibility with existing systems. Further, the paper discusses some environmental and ethical issues related to adopting AI-driven recycling and how such technology is envisaged to potentiate sustainable development. The review

R. Vadisetty (✉)
Wayne State University, Detroit, MI, USA
e-mail: rahulvy91@gmail.com

S. T. Nuka
Sustaining Mechanical Engineer, Argon Medical, Plano, TX, USA

S. Kalisetty
Integration and AI Lead, Miracle Software Systems, Novi, MI, USA

C. Pandugula
Lowes Inc., Mooresville, NC, USA

J. K. R. Burugulla
American Express, Phoenix, AZ, USA

V. N. Annapareddy
Sr Enterprise Application Developer, Madison, WI, USA

M. Chakraborty et al. (eds.), *Proceedings of Sixth International Ethical Hacking Conference*, Lecture Notes in Networks and Systems 1513,
https://doi.org/10.1007/978-981-96-8632-2_15

investigated recent developments and prospects that illustrated the potential of generative AI to be a game-changer in the recycling revolution regarding waste reduction, PE, PP, and circular economy objectives. In the final analysis, from the results obtained in this study, it seems that responsibly developed generative AI has quite a good chance of overhauling recycling regarding sustainable efficiency and the global imperative set by Earth.

Keywords Generative AI · Polyethylene recycling · Polypropylene recycling · Process optimization · Chemical recycling · Predictive maintenance · Sustainable development · Circular economy · Data-driven recycling · Environmental impact

1 Introduction

1.1 Overview of Polyethylene and Polypropylene

Among the different types of plastics, PE and PP are rated the most highly used globally, mainly because of their versatility, toughness, and reasonable prices. The long-chain hydrocarbon structure of PE has been widely used in manufacturing packaging materials, plastic bags, containers, and pipes [1–3]. Because of its vital attributes toward strength and heat resistance, there is a high demand for PP made from propylene monomers, with claims of usefulness in automobile parts, textiles, and appliances. Although in widespread use, PE and PP are severe environmental problems because they are non-biodegradable and create a plastic accumulation burden in landfills and oceans [3, 4].

1.2 Current Recycling Limitations

Traditional PE and PP recycling suffer from severe limitations. For instance, the most adopted mechanical recycling process quickly leads to contamination and quality degradation; the final products are often weaker and less durable than their virgin counterparts. In addition, both segregations of PE and PP from mixed waste streams are cumbersome and energy-intensive, further reducing the economic viability of the recycling processes. Another critical concern is energy consumption since high temperatures must be used to melt plastics. This contributes to higher Carbon emissions and offsets whatever gains have been made by recycling [1–5].

1.3 The Role of Advanced Technology

The challenges listed, therefore, make the redevelopment of recycling processes using advanced technology quite urgent. Here, generative AI has become prominent, with a probable role in enhancing recycling efficiency, the quality of recycled materials, the disposal of energy consumption, and similar situations. Generative AI will enable predictive modeling, process optimization, and real-time decision-making to overcome many intrinsic bottlenecks within current recycling methods.

It could also ensure better scalability of PE and PP recycling, support global sustainability goals, and encourage movement by the industry toward the idea of a 'circular economy. Generative AI has the potential to enhance the quality and yield of recycled plastics significantly, given its data-driven approach that changes the face of plastic waste management [4–7].

2 Generative AI in Manufacturing and Recycling Technologies

2.1 Overview of Generative AI

Generative AI is a class of AI specifically built and designed to generate new content, ideas, or solutions by learning patterns from existing data. Unlike traditional AI, which mainly analyzes and classifies information, generative AI creates new outputs; thus, it is valid, especially in creative and optimization fields. Generative AI includes deep learning, reinforcement learning, and generative adversarial networks, which involve training neural networks to produce results using patterns learned from data [8, 9].

In manufacturing, generative AI is used to optimize product designs to decide on the best production processes by simulating various scenarios and their respective outcomes. The technology finds broad application in the automotive and aerospace industries, whereby design optimization through AI cuts down material waste, reduces production time, and saves costs. Generative modeling enables manufacturers to simulate different materials virtually, understand their performance under various conditions, and bring potential issues to light before they happen. These capabilities allow generative AI to address several challenges in manufacturing, especially in material processing and recycling, where accuracy and quality are of the essence [10, 11].

2.2 Applications in Recycling and Waste Management

Generative AI has massive potential in recycling and waste management, acting on the optimization of sorting processes, prediction of material flow, and enhancement of recycling outcomes. Generative AI models, consideration of big data on waste composition, performance of recycling facilities, and operational parameters, will predict how different types of materials, including plastics, metals, and electronics, can be processed best. These models can be installed on-site to monitor and adjust sorting and processing equipment to give high yields with minimum contamination rates among recycled products [10–12].

When generative AI is applied in real-world applications, it helps improve recovery by identifying optimal sorting paths for different waste streams. For instance, AI-enabled systems installed in recycling plants use sensors and cameras at the incoming waste streams to classify materials with precision.

After that, generative AI models the ideal processing pathways to achieve maximum purity using minimum energy. For instance, it can forecast fuel values and make decisions on handling non-recyclable materials more efficiently in waste-to-energy facilities, thus contributing to a cleaner waste management cycle in general. Given the rising demand for quality recycled materials, these steps are crucial for developing more innovative and efficient recycling facilities [12].

2.3 Benefits of Generative AI for PE and PP Recycling

This generative AI has certain advantages over PE and PP recycling, enabling solving very complicated tasks: process optimization, quality control, and decision-making. It opens a lot of possibilities for optimization of the recycling process:

1. Predictive Modeling: Generative AI can enable predictive modeling, where a system predicts specific outcomes in recycling based on the initial composition of materials and processing conditions. Therefore, the feature assists a recycling facility in predicting its product quality and makes necessary modifications concerning the input materials or parameters associated with processing. For instance, if a batch of recycled PP is contaminated, predictive modeling may suggest suitable pre-processing that could reduce the impurities to ensure a better-quality final product [12, 13].
2. Process Optimization: With the advent of generative AI, processes involved in the recycling of PE and PP would be optimized by their processes of finding various pathways and choosing the most efficient one, for example, arriving at the best temperature, various sorting methods, and the time for processing each lot to yield the highest using the least quantity of energy. The same would extensively apply to a chemical recycling process where generative AI can aid in controlling reaction pathways that break plastics into their raw materials so that such a process is far more efficient and predictable [13, 14].

3. Real-time Decisioning: Thanks to generative AI, real-world insight and real-time decision-making are pivotal in recycling operations. Generative AI can detect variances in material quality or equipment performance by parsing continuous data streams from sensors and monitoring systems. When a particular variation in the feedstock composition of PE or PP arrives, AI-driven systems can immediately adjust their processing parameters to maintain product consistency and avoid waste [12, 14].
4. Quality Control and Consistency: Generative AI helps in quality control by finding patterns in the recycling process that affect the consistency of the recycling product. It could ensure that the PE and PP recycling standards meet the industrial standards by considering impurities and predicting durability in the recycled material. AI-driven quality control is becoming essential to high-value applications as high-purity recycled plastics increasingly find their place in food-grade or medical applications, for instance, where even a minute contamination could affect its usability [12, 14].
5. Energy Efficiency and Environmental Gains: Generative AI opens the door to energy efficiency in recycling, as it optimizes process efficiency and energy utilization. For instance, AI models can simulate energy usage using different processing methods and provide operating conditions resulting in optimal recycling outputs with the lowest carbon emissions. This will save on operational costs and reduce the environmental footprint of recycled PE and PP, further contributing to environmental sustainability goals [15].

Overall, generative AI brings transformational potential into PE and PP recycling, given the predictive modeling, efficiency in the process, real-time monitoring, and quality it establishes in the setting. Generative AI approaches the complexity of recycling with data-driven decision-making and optimization, thus opening ways toward high-quality, greener recycling. The capability for dynamic response, with variation in material input and operational conditions, makes the technology a tool of high value in finding a solution to the challenge of recycling high-volume plastics such as polyethylene and polypropylene.

3 Applications of Generative AI in Advanced Recycling Processes for PE and PP

Generative AI can mark a generational leap in the recycling processes of PE and PP. Because of generative AI-enabled process optimization and predictive maintenance, together with chemical recycling innovations, the recycling plants will be able to work on a new dimension of improvement for productivity and quality with a commitment to sustainability [15–17].

3.1 Process Optimization

Process optimization of recycling facilities implies the assurance that each process, from sorting and shredding to melting, is as efficient as possible with minimal resource use. Generative AI processes data gathered from sensors and other equipment and then makes real-time recommendations for adjustments to achieve ideal conditions during processing. This capability is precious when it comes to PE and PP recycling, where the purity of the material at every step has to be very high [17, 18].

Table 1 illustrates how generative AI addresses traditional recycling challenges: increasing accuracy in sorting, making shredding uniform, and optimizing melting processes through real-time adjustments for improved efficiency and quality.

Sorting is where the generative AI does its magic: taking input from the imaging systems, it can very accurately discern PE, PP, and other materials with high accuracy. This helps reduce contamination because AI models, once trained on a large number of images, can then recognize that material property and adjust sorting paths dynamically. The better the sorting, the cleaner and more consistent the feedstock for recycling will turn out [16–18].

This represents the shredding stage, where generative AI controls shredder parameters to produce uniformly sized particles. Uniformity of the particles is a primary concern down the value chain for downstream processes. Consistent particle size improves melting efficiency because large and small particles may melt at different rates and cause defects in recycled material. The AI-driven shredding systems adjust rotational speed, blade angle, and motor power based on the incoming materials' characteristics, ensuring uniformity and minimizing material loss [16–18].

Generative AI generates predictions for the appropriate temperature in each batch during a melting process, thus optimizing energy consumption. This is an essential issue in recycling PE and PP since the melting profile could affect the homogeneity of the product. By lessening energy demands, AI-driven melting processes contribute to more environmentally friendly operations and help facilities reduce their overall carbon footprint [19].

Table 1 Process optimization through generative AI in PE and PP recycling

Recycling stage	Traditional challenges	Generative AI solutions
Sorting	Difficulty in separating mixed plastics accurately; contamination from other materials	AI-powered imaging and sensor analysis for precise identification and separation of PE and PP from mixed waste streams
Shredding	Inconsistent shredding sizes lead to irregular melting and inefficiencies	Adaptive shredding algorithms that maintain uniform particle size for consistent downstream processing
Melting	High energy requirements and variability in melting profiles, resulting in inconsistent product quality	AI-predicted optimal temperatures and energy inputs, adjusting in real-time for consistent melt quality and reduced energy consumption

3.2 *Predictive Maintenance and Quality Control*

Generative AI lets recycling facilities track the health of their equipment and predict scheduled maintenance times. Predictive maintenance is vital in recycling plants because the failure of any equipment may disrupt continuous operations. Therefore, costly downtime causes injuries to quality consistency. In this respect, generative AI also contributed knowledge to quality control to detect defects, contaminants, and inconsistencies in real-time, facilitating high standards for maintaining quality in recycled products [19, 20].

Table 2 summarizes how generative AI overcomes classical issues in recycling applications with the possibility it offers for predictive maintenance, improving quality assessment thanks to real-time monitoring, and detecting micro-defects thanks to advanced image analysis.

Generative AI can use sensor data from machinery to uncover patterns of wear and stress on parts and warn technicians well before an actual equipment failure. For instance, AI can detect slight vibrations, temperature fluctuations, or performance drifts that indicate upcoming problems. Through the predictive nature of their maintenance needs, recycling facilities can avoid unplanned stops and prolong the service life of machinery [16, 18].

For quality control purposes, AI-driven models constantly follow up on material data so that products manufactured from recycled input meet the requirements of any industry standard. Generative AI can be specially trained in contaminant detection, color variations, and physical inconsistencies in real time, simultaneously adjusting the processing parameters to maintain consistency. This is very important in industries where purity is crucial. Minor problems with contamination can make recycled materials unusable for packaging foodstuffs [14, 16, 18].

The other domain where AI performs exceptionally well is defect detection. Generative AI can spot micro-defects that might be present in the plastic, like tiny air pockets or weak spots, which usually are not visible if inspected physically.

Table 2 Predictive maintenance and quality control applications of generative AI

Application	Traditional issues	Generative AI benefits
Equipment monitoring	Frequent breakdowns due to wear, tear, and heavy usage	Predictive maintenance using anomaly detection to identify potential issues before they cause disruptions
Quality assessment	Manual inspection is prone to errors and variability	AI-based quality monitoring for real-time consistency, detecting contaminants and other quality issues instantly
Defect detection	Limited visibility into micro-defects in recycled materials	AI-enabled image analysis and pattern recognition to detect even minor defects, ensuring recycled plastic quality

By analyzing imaging sensor data, generative AI ensures that defective batches are identified and corrected before quality issues reach the customer [18–20].

3.3 Chemical Recycling Innovations

While mechanical recycling breaks down plastics into their usable parts, chemical recycling reduces them to the original chemical building blocks so that they can be reused as raw materials in new products. Generative AI is leading the way toward perfecting the chemical recycling of PE and PP by ensuring higher reaction efficiency, reduced energy consumption, and product consistency [21, 22]. AI models can simulate various reaction conditions and choose the best conditions concerning temperature, pressure, and catalyst usage for every batch to achieve qualitative and quantitative raw materials.

Table 3 shows that in the reaction pathway, generative AI emulates a breakdown process for PE and PP by optimizing all the chemical reactions in converting plastics into their raw materials, such as monomers. AI models consider millions of pathways and identify those peaking the yield while having purity with minimal partial-breakage processes, hence improving the quality of the material from the recycling processes [21–23].

Conversion efficiency is improved by AI in determining the best reaction conditions, such as temperature and concentration of catalyst, to arrive at maximum output with the least energy input. Most chemical recycling processes are highly energy-intensive; generative AI reduces these costs by predicting how the various conditions affect reaction efficiency and adjusting parameters to improve the process.

Real-time analysis of reaction conditions by AI-driven models ensures that the product is consistent since each batch will be chemically and qualitatively identical. This consideration is of critical importance in chemical recycling since minor alterations in temperature or pressure can lead to changes in the properties of the recycled product. This process keeps the generative AI reaction environment stable. It ensures

Table 3 Chemical recycling innovations enabled by generative AI

Recycling stage	Traditional limitations	Generative AI contributions
Reaction pathways	Variability in reaction outcomes leading to inconsistent product quality	AI-driven optimization of reaction pathways for precise breakdown of PE and PP molecules
Conversion efficiency	High energy costs and complex processing requirements	Reduced energy usage through predictive modeling of optimal reaction conditions
Product consistency	Inconsistencies in recycled raw material properties	Real-time AI control of reaction environments to maintain consistency and uniformity in output

that high-quality material for high-quality applications is ensured, further widening the circle of use for recycled PE and PP [23, 24].

Generative AI is a critical driver in improving the processes of recycling PE and PP via further process optimization, predictive maintenance, and quality control in the efficiency of chemical recycling. Due to these improvements, AI will automatically enable contributions toward a more circular recycling industry: energy use reduction, quality of product increase, and operational continuity. These advantages are complete in the circular economy, where high-quality recyclates can replace virgin plastics, decrease dependence on fossil fuels, and limit any contamination of the environment. Generative AI holds all the potential to make real-time changes and optimize complex recycling processes that promise a sea change in plastic waste management and the promotion of best sustainable practices across industries that depend on PE and PP products [24, 25].

4 Data-Driven Benefits and Challenges Related to the Use of Generative AI

Significant opportunities for improvement in efficiency, quality, and sustainability can be expected from implementing generative AI in recycling PE and PP. A data-centric approach such as the one described requires significant infrastructure and planning. While it can facilitate operations and reduce waste, it also brings many challenges, most implicitly in data management, integration, and security.

4.1 Data Requirements and Infrastructure

Generative AI requires specific data for a recycling facility with related infrastructure. Generative AI models will need real-time sensor data, detailed material composition data, and historic operational records. The most crucial data requirements for generative AI in recycling facilities:

- Real-time plant data sensor-driven metrics can also include values relating to machine speed, temperature, level of vibration, and energy use; these make runtime adjustments in operations such as optimization of the temperature of melting or adjustment of shredder parameters based on material properties through the use of enterprise AI [20, 21].
- Material compositional and quality data: AI models require heavy data on the chemical composition, contaminants included, and incoming plastic waste quality levels. This will be instrumental in ensuring the accurate sorting of materials and that the resultant PE and PP grades meet the stipulated industrial standards. This data can be sourced from various tools like imaging systems, chemical sensors, and spectral analysis [16, 17, 22].

- Environmental and operational data: Several ambient temperature and humidity factors can influence the effectiveness of recycling, especially in chemical recycling. Environmental data lets AI change process conditions to maintain product quality consistently [23].
- Training of AI Models with historical data: AI needs historical data during the training phase, like past performance records, history of maintenance, and records regarding quality control. In this way, AI can identify patterns, simulate different scenarios, and improve prediction capabilities [24].

The data storage and processing capacities must be robust to support such demands. Most use cloud-based storage for secure data handling and deploy high computing infrastructure to process data promptly and sustain real-time AI insights into operations. Real-Time Data Processing Facilities need to process data so that an AI can let it adjust the rate as soon as possible. In contrast, access to historical data helps refine the accuracy of the models.

4.2 Benefits of Data-Driven Insights

There will be many benefits to integrating generative AI into the recycling processes of PE and PP. Still, perhaps the most valuable assets will come from the real-time insights that result from constant data analysis. The efficiency of the recycling process can be significantly enhanced, waste reduced, and operational costs decreased.

AI-powered insights enable facilities to iron out their processes by identifying inefficiencies and recommending real-time changes. For instance, the AI may calculate the optimum conveyor speeds that will provide the most effective sorting or alter the angle of shredder blades depending on the material density to minimize material loss and maximize throughput. Such fine-tunings allow recycling facilities to process volumes of PE and PP with minimal waste [19, 20, 22].

Generative AI also plays a vital role in the reduction of waste. With AI supporting high-precision sorting and quality control, contamination within the recycled material will be reduced, thus increasing its purity and usability. This is simply because, with precision, there will be fewer unusable materials. AI can sort out contaminants to compose materials and meet tight industry standards for better quality final products suitable for various applications [23–25].

Another considerable advantage is energy efficiency: AI dynamically adjusts energy use to specific processing needs, such as suggesting the lowest practical temperature required to melt plastics. This capability reduces energy consumption and contributes to a lower overall carbon footprint for recycling facilities. AI models support predictive maintenance through continuous machine performance monitoring, enabling repairs to be performed proactively and preventing unexpected breakdowns that disrupt operations. Predictive maintenance helps extend the life of the equipment and optimize uptimes, essential features in recycling facilities for high outputs.

Besides, AI continuously monitors material properties and processing conditions so that the quality of recycled products maintains a pretty high level, enabling AI to make real-time adjustments to keep the quality consistent for high-quality applications. AI generation generally reduces one's dependency on manual checks for quality and quality inconsistencies.

4.3 Challenges of Integration with AI

While generative AI seems to offer promises of effective means of enhancing recycling efficiency, integrating them into existing recycling models presents numerous challenges. Solving these challenges will be crucial for successfully implementing AI and maximizing its full potential through data-driven recycling.

- Data privacy and security concerns: In recycling plants, since the plants use cloud-based data storage, problems related to data privacy may come forward. The operational data would be considered insensitive but should be safeguarded from unauthorized access and potential cyber threats. Furthermore, data regulations, including the GDPR, mandate a company to manage data transparently and securely, further complicating data management for AI purposes [17, 18, 23].
- Data quality and availability: The quality of AI is only as good as the quality and integrity of the underlying data. Any missing information, inconsistency, or inaccuracy in the data hampers the effectiveness of models in arriving at optimum recommendations and optimizing processes. All their processes need sensors and data collection systems that ensure consistent and accurate data capture. If facilities have not invested in data tracking, historical data from where to train the models will be challenging [18, 19].
- Training and maintenance of the AI Model: Training generative AI models is complex and requires specialized data science and machine learning skills. Facilities may need to partner with AI experts to develop and refine models for applications specific to the recycling processes. Once deployed, AI models require periodic maintenance to update with new data, which is resource-intensive, especially for facilities lacking technical expertise.
- Integration with existing systems: Most waste management facilities use ancient machinery and software systems, which might not be compatible with new AI platforms. Integrating AI mandates alteration in the prevailing infrastructure, such as sensor upgrades or interlinking machinery with a central data processing system. Integrating this may be costly and time-consuming; operations can be disrupted during the transition.
- Scalability and cost concerns: Implementing AI recycling technology can be expensive, especially in smaller facilities. It could be anything from hardware installation to collect data to model development and its maintenance costs for integrated generative AI, which is expensive. Additionally, scalability might

be limited as small-scale facilities could not create a volume of data that can effectively train an AI model [17, 19, 21].
- Ethics and environment: Although generative AI will reduce environmental impacts associated with recycling, the energy needed for AI training can be pretty high on large-scale machine learning infrastructure. It, therefore, means balancing the ecological benefits of AI-driven recycling intelligently with the energy cost of AI itself, which requires careful planning as facilities strive toward implementing sustainable practices without inadvertently increasing their carbon footprint.

Therefore, while generative AI supports transformational potential in the PE and PP recycling arenas, rigorous planning and investment will replace any challenges driven by data management, integration, and scalability. The timely balancing of technical and ethical considerations in infrastructure to enable robust data collection and the successful implementation of AI will depend on whether AI satisfies such requirements—efficiency, waste reduction, and sustainability.

5 Future Prospects and Ethical Considerations: Generative AI for Recycling

The integration of generative AI into polyethylene and polypropylene recycling provides exciting frontiers, with consideration for future innovations that come with potential ethical implications. Ongoing AI evolution will likely continue shaping the nature of the recycling process via new applications, improved efficiencies, and a sharper focus on issues related to sustainability. These technological advances will be married to responsibilities concerning environmental impact, transparency, and ethical use of AI. A potential and ethical analysis offers a comprehensive understanding of the road ahead regarding AI in recycling.

5.1 Potential Future Innovations

Generative AI for recycling is still in its infancy, and ongoing developments have enhanced processes and the sustainability of polyethylene and polypropylene. The most promising area is self-learning AI models, which have improved even. These would, in practice, be adaptive models that learn to optimize the recycling processes on their own by large datasets and advanced algorithms, making changes in real-time according to changing input material mixes and operational and environmental parameters. Such would enable recycling facilities with minimal human interventions, minimizing errors while maximizing efficiency [11, 13, 15].

Another emerging application of generative AI in this setting is the design of chemical recycling processes, which will be even more sophisticated. The AI-driven models would predict complex chemical reactions in the chain that break down the polymers into their original monomers. Such techniques would unlock infinitely recyclable material flows where polyethylene and polypropylene waste get recycled repeatedly into high-quality raw materials without degradation. Improved designs for chemical recycling would also lead to the better realization of closed-loop systems whereby facilities can produce higher-purity outputs that fit well into high-demand areas such as food-grade or medical packaging, widening the usability scope of recycled plastics [16, 19, 21, 25].

Other fields of innovation involve the integration of AI into IoT technologies, creating interconnected systems that may allow the widest monitoring and data sharing across the phases of the recycling process. Sensors and smart devices embedded within such recycling equipment can record complex data. This provides generative AI models with intricate insight into every aspect of the operation—from sorting and shredding to melting and chemical breakdown. Such granular data analysis by AI allows for detecting even minor anomalies, thus enabling proactive adjustment to maintain continuity of operations with overall improvement in recycling performance.

Further on, generative AI has the potential to introduce active capabilities for waste sorting. Such would be the case where AI, together with robotics and advanced imaging, would identify and sort various types of plastics that show changes in polymer composition independently with high accuracy [25]. The accuracy would also enable facilities to handle mixed plastic wastes far more efficiently, reduce the contamination rate, and produce purer recycling output that meets industrial standards.

5.2 Environmental and Ethical Implications

Some of the profound environmental implications that may be tied to the role of Generative AI consist of waste management, especially regarding plastic waste, resource conservation, and carbon emissions reduction. With greater sorting accuracy, less contamination, and optimization of energy use, AI-powered recycling processes reduce the general amount of plastic waste reaching landfills and oceans. Moreover, with AI's enhancement power, these facilities can transform discarded polyethylene and polypropylene materials into productive raw materials that complement economic circularity by keeping plastic for productive use and never as waste. Beyond that, generative AI is making an actual, positive impact on energy conservation [19–21].

Conventional recycling involves vast amounts of energy, especially for energy-intensive chemical recycling, further adding to the industry's carbon footprint. Generative AI works on the ability to model and predict optimal energy requirements of a process, significantly decreasing such demands and allowing a facility to cut down on energy consumption sans any compromise in output quality. The rationale for this energy efficiency aligns with global sustainability goals, whereby the recycling industry tries to reduce greenhouse gas emissions and encourage cleaner practices.

However, there is also an ethical dimension regarding AI's environmental impact that must be considered.

However, while AI-mediated recycling can reduce plastic waste and energy consumption, training large-scale AI models comes at a computational cost that often offsets the environmental benefit. The development and maintenance of high-performance AI infrastructure call for large amounts of energy, which can lead to substantial carbon emissions if the facilities depend on non-renewable energy sources. Thus, AI-driven recycling is bound to reach a balance that will balance the environmental gains made by AI-optimized recycling against the cost of computations if this AI-driven recycling is to remain genuinely sustainable. Besides, it is relevant that the gains made for the environment are considered within the broad ethical framework of generative AI's societal impact [22, 24].

The more dependence on automated AI systems, the less human oversight will be. It remains essential, therefore, that responsibilities under these systems are transparent and accountable: decision-making in recycling-maybe decisions that involve waste processing and material quality be traceable, wherein the facility operator and other stakeholders can understand the logic of AI recommendations that have taken place. Given that ethical guidelines would demand transparency in AI decisions, this would automatically bring in the responsible use of AI, with limited chances of 'unwanted' consequences, thus meeting the societal expectations regarding accountability.

5.3 Responsible AI and Sustainability

Ethical use of generative AI in recycling goes beyond environmental concerns; it is about the core idea of responsibility in AI development, openness of decision processes, and sustainability compatibility. Responsible AI is designed and assessed according to the best-set principles that protect human interests and serve the common good. For recycling, this means its developers design the AI equipment so that production maximizes sustainability, energy efficiency, and material quality over optimal operational speed or simply cost-cutting [24, 25].

Given that generative AI models have a highly complicated way of functioning, one of the fundamental principles of responsible AI for recycling can be considered transparency. This means, in essence, that the mechanism of such AI recommendations should be known to operators, and the whys should be understandable if the outcome is to influence something as important as waste composition, processing conditions, or final material quality. The AI systems allow developers to ensure that any decision-making process is traceable and understandable. These build in an environment of trust with low chances of bias in algorithms or even operational changes not intended. Generation AI in recycling also needs to fall crucially within sustainability goals.

These recycling facilities are on the go to meet the circular economy principles, and the AI systems need to be designed to enable sustainable material flows, focusing, among other things, on recyclable outputs produced in such a manner as to return to manufacturing cycles repeatedly. The responsible AI frameworks promise to align more by underlining the prime importance of AI designs leaning toward long-term material quality, with low energy use and low environmental impact over short-term gains. Furthermore, AI has to flex into some new metrics and certification on sustainability. It has to enable any recycling plant to go on par with upgraded environmental performance standards and contribute positively to the upgraded standards on the global recycling yardstick. Other ethical matters involve the staff.

While AI offers an unprecedented capability for automation, it could come at the cost of jobs currently filled by manual labor in a recycling facility [13, 14, 17, 20, 22]. To balance automation needs with the workforce and needs for continued social benefit, workers could be placed at facilities to oversee AI systems, quality assurance, and contributions to design as these roles evolve for workers in the industry. Responsible AI usage as an exchange may give the recycling facilities an active opportunity to contribute to an efficient and transparent recycling future that is tuned to the values of society at large. In other words, generative AI hints at a bright future for recycling, likely with significant improvement in process efficiency, reduction of waste, and energy conservation.

On the same breadth, such progress is not devoid of ethical imperatives. AI-driven recycling that has been forwarded above should fall in line with environmental protection goals. It should turn transparent around issues of social sustainability. As the facilities study the use of generative AI, they will be balancing technological capability with moving on to an ethical obligation toward the care of the environment, demand in society, and the workforce.

References

1. Wróbel M, Szymańska S, Kowalkowski T, Hrynkiewicz K (2023) Selection of microorganisms capable of polyethylene (PE) and polypropylene (PP) degradation. Microbiol Res 267:127251
2. Niu L, Chen Y, Li Y, Wang Y, Shen J, Wang L, Zhao B (2023) Diversity, abundance and distribution characteristics of potential polyethylene and polypropylene microplastic degradation bacterial communities in the urban river. Water Res 232:119704
3. Xu Z, Munyaneza NE, Zhang Q, Sun M, Posada C, Venturo P, Liu G (2023) Chemical upcycling of polyethylene, polypropylene, and mixtures to high-value surfactants. Science 381(6658):666–671
4. Munte S, Banjarnahor M, Tanjung DA, Budi RS (2023) MECHANICAL test and thermal stability on thermoplastic sago (Metroxylon saga rotted.) combination of polyethylene and polypropylene. Rasayan J Chem 16(1)
5. Babiker DM, Usha ZR, Wan C, Hassaan MME, Chen X, Li L (2023) Recent progress of composite polyethylene separators for lithium/sodium batteries. J Power Sources 564:232853
6. Dubdub I (2023) Kinetics study of polypropylene pyrolysis by non-isothermal thermogravimetric analysis. Materials 16(2):584
7. Peng Y, Yang J, Deng C, Deng J, Shen L, Fu Y (2023) Acetolysis of waste polyethylene terephthalate for upcycling and life-cycle assessment study. Nat Commun 14(1):3249
8. Rane N, Choudhary S, Rane J (2024) Enhancing sustainable construction materials through the integration of generative Artificial Intelligence, such as ChatGPT or Bard. SSRN 4681678

9. Akhtar P, Ghouri AM, Ashraf A, Lim JJ, Khan NR, Ma S (2024) Brilliant product platforming powered by AI and Generative AI: personalization for the circular economy. Int J Prod Econ 273:109283
10. Boussioux L, Lane JN, Zhang M, Jacimovic V, Lakhani KR (2023) Generative AI and creative problem solving
11. Bashynska I, Prokopenko O (2024) Leveraging Artificial Intelligence for circular economy: transforming resource management, supply chains, and manufacturing practices. Sci J Bielsko-Biala Schl Finan Law 28(2):85–91
12. Mohammed MY, Skibniewski MJ (2023) The role of generative AI in managing industry projects: transforming industry 4.0 into industry 5.0 driven economy. Law Bus 3(1):27–41
13. Rane N (2023) Role of ChatGPT and similar generative artificial intelligence (AI) in the construction industry. SSRN 4598258
14. Rane N, Choudhary S, Rane J (2024) Role and challenges of ChatGPT or similar generative artificial intelligence in reinforced concrete technology. SSRN 4681731
15. Elahi M, Afolaranmi SO, Martinez Lastra JL, Perez Garcia JA (2023) A comprehensive literature review of the applications of AI techniques through the lifecycle of industrial equipment. Discov Artif Intell 3(1):43
16. Hollauer C, Garcelán J, Ragam N, Vaish T, Asensio OI (2023) Generative AI and discovery of preferences for single-use plastics regulations. In: Proceedings of the AAAI symposium series, vol 2(1), pp 68–77
17. Kuusisto M (2024) Artificial intelligence solutions for the recycling and utilization of biomaterials for industrial applications
18. Kumar A, Bakshi BR, Ramteke M, Kodamana H (2023) Recycle-BERT: extracting knowledge about plastic waste recycling by natural language processing. ACS Sustain Chem Eng 11(32):12123–12134
19. Paraye P, Sarviya RM (2024) Advances in polymer composites, manufacturing, recycling, and sustainable practices. Polym-Plast Technol Mater 63(11):1474–1497
20. Mishra A, Jatti VS, Paliwal S Sustainable materials: the role of artificial intelligence. https://doi.org/10.1201/9781003437369
21. Wei Y, Hadigheh SA (2023) Development of an innovative hybrid thermo-chemical recycling method for CFRP waste recovery. Compos B Eng 260:110786
22. Chung C, Kim J, Sovacool BK, Griffiths S, Bazilian M, Yang M (2023) Decarbonizing the chemical industry: a systematic review of sociotechnical systems, technological innovations, and policy options. Energy Res Soc Sci 96:102955
23. Liu Q, Martinez-Villarreal S, Wang S, Tien NNT, Kammoun M, De Roover Q, Richel A (2024) The role of plastic chemical recycling processes in a circular economy context. Chem Eng J 155227
24. Zhang L, Leung MY, Boriskina S, Tao X (2023) Advancing life cycle sustainability of textiles through technological innovations. Nat Sustain 6(3):243–253
25. Ncube A, Mtetwa S, Bukhari M, Fiorentino G, Passaro R (2023) Circular economy and green chemistry: radical, innovative approaches in designing new products are needed. Energies 16(4):1752

Automated AI-Driven Phishing Detection and Countermeasures for Zero-Day Phishing Attacks

Rahul Vadisetty, Anand Polamarasetti, and Dinesh Kalla

Abstract Among the most prevalent cyber threats, phishing attacks are still state-of-the-art and, thus, difficult to detect, such as zero-day phishing attacks. Most of the conventional phishing detection mechanisms rely on signature-based or rule-based systems that cannot identify such new emerging threats and neutralize them in real time. The paper proposes applying automated artificial intelligence (AI)-driven approaches for detecting and responding to zero-day phishing attacks. Therefore, this research suggests developing a robust phishing attempts detection system, identifying phishing through diverse digital platforms, such as emails, websites, and social media, using advanced machine learning techniques that include but are not limited to Natural Language Processing (NLP), image recognition, and anomaly detection. The proposed system will integrate supervised and unsupervised learning models within a framework that will deliver real-time countermeasures, including automated user alerts and blocking mechanisms. Experimental results are presented that evaluate the detection accuracy, false favorable rates, and speed of response by comparing performance with existing methods. These results show in detail how well the AI-driven approach reduces the success rate of zero-day phishing attacks and proves to be more adaptive and scalable than traditional methods. This research forms the basis for understanding the full potential of AI-based systems for cybersecurity defenses and resiliency while setting the ground for future developments in automated phishing countermeasurement research that may impact sensitive digital infrastructures.

R. Vadisetty (✉)
Wayne State University, Detroit, MI, USA
e-mail: rahulvy91@gmail.com

A. Polamarasetti
Andhra University, Visakhapatnam, AP, India

D. Kalla
Microsoft, Raleigh, NC, USA

M. Chakraborty et al. (eds.), *Proceedings of Sixth International Ethical Hacking Conference*, Lecture Notes in Networks and Systems 1513,
https://doi.org/10.1007/978-981-96-8632-2_16

Keywords Zero-day phishing · AI-driven detection · Cybersecurity · Machine learning · Phishing countermeasures · Automated response · NLP · Anomaly detection · Real-time security · Cyber threat mitigation

1 Introduction

1.1 Background and Importance of Phishing Detection

Phishing attacks have now dominated the cybersecurity threats organizations and users face globally. Since the digital ecosystem keeps expanding, phishing methods evolve and grow complex, and most of the old detection techniques find them challenging and less effective. Among many, zero-day phishing attacks are extremely dangerous; new vulnerabilities not known by cybersecurity defense mechanisms get exploited. Since these attacks are so unpredictable, there is an immediate need for the detection process to be automated to save sensitive data and reduce other potential damage effectively. Fast detections and mitigations of phishing threats have become essential in information security [1–3].

1.2 Problem Statement

Zero-day phishing attacks pose significant challenges due to their capability of exploiting unknown vulnerabilities and bypassing predefined security protocols. Traditional modes of phishing detection, such as signature-based and rule-based systems, usually lag in identifying these new emerging threats, leaving users and organizations vulnerable to data breaches and unauthorized access. These limitations of traditional detection strategies create a gap in cybersecurity, demanding an adaptive AI-based solution that could respond to the dynamic nature of zero-day phishing attacks.

1.3 Purpose and Objectives

This work aims to design an advanced AI-driven framework that automatically detects and mitigates zero-day phishing attacks for improved response times and efficiency in general system performance. This research would commit to the design and implementation of a proposed AI-based detection model, which would be specific and customized for detecting zero-day phishing attempts across multiple varied platforms such as emails, social media, and web applications. The research also aims to study the automated countermeasures that can be an adequate reaction and minimize successful phishing attacks. By integrating detection and real-time response, the proposed framework will provide a more effective defense mechanism against evolving phishing threats.

1.4 Research Questions

The following two general research questions will guide the present study:

1. How effectively could AI-driven models detect zero-day phishing threats since these models rely so much on adaptive learning and anomaly detection?
2. What are the countermeasures that can be integrated and automated to grant a real-time response for each phishing attempt detected, thus minimizing the user's exposure to those threats?

These questions will, therefore, govern the investigation and discuss the level at which AI and machine learning algorithms can identify new phishing attempts and apply necessary countermeasures on time.

1.5 Significance of the Study

This research goes a long way in the cybersecurity world, potentially raising the bar when it comes to defending against zero-day phishing attacks. With particular emphasis on the challenges of phishing detection, this research informs needed insights into AI-driven cybersecurity challenges. It will try to include a realistic, scalable, adaptive framework that will evolve with shifting threat landscapes. This will hopefully be a foothold upon which further work can be laid for the research and development of automated cybersecurity measures. The results of this study contribute to an extension of the theoretical understanding of AI applications in phishing detection and have pragmatic value for an industry practitioner who wants to protect his digital infrastructures against increasingly sophisticated attacks.

2 Literature Review

2.1 Phishing and Zero-Day Attack Landscape

Phishing attacks are considered one of the most widespread and destructive types of cyber threats facing almost every organization and individual from all walks of life. Over the years, phishing methods have transformed to a great extent into social engineering, impersonation, and multi-vector means of delivery that mislead users. A particularly threatening genre within this spectrum is zero-day phishing attacks, which find their basis in previously unknown vulnerabilities. Such attacks are anything but dumb and can easily pass through any signature-based intrusion detection system. The subsection below critically analyzes the techniques used in basic and zero-day phishing attacks. It discusses the effect of these attacks on critical industries, meaning it is requisite to emphasize more sophisticated adaptive unwavering attack detection methods [3–5].

2.2 Existing Detection and Prevention Mechanisms

Traditional phishing detection techniques can generally be categorized into heuristic-based, signature-based, and rule-based systems. In signature-based detection, incoming data is matched against known phishing indicators. Heuristic-based approaches depend on predefined rules and patterns that detect soluble phishing-like behavior. A rule-based system is based on logical conditions that qualify or quantify an attempt as phishing or otherwise. Each technique has its inefficiencies while dealing with evolving, unknown threats. Signature-based systems, for instance, cannot work against zero-day phishing since they cannot catch patterns yet to be documented. While various heuristic and rule-based approaches may grant better flexibility, these cannot provide the much-needed adaptive learning to tackle techniques of dynamic zero-day phishing. Following is a discussion on the strengths and weaknesses of these approaches that underline the shortcomings that make necessary the AI-driven solution presented next [6–10].

2.3 Recent Development in AI and Machine Learning in Cybersecurity

Artificial intelligence and machine learning (ML) have contributed significantly to furthering cybersecurity, including phishing detection. AI-driven techniques employ NLP, image recognition, and anomaly detection to find phishing attempts across e-mails, web pages, and social media. The different machine learning models are supervised, unsupervised, and deep learning-based, all finding their applications in cybersecurity [11–13]. With supervised learning, data classification can be allowed by using labeled datasets. On the other hand, unsupervised learning may find anomalies without predefined labels, making it practical for zero-day threat detection. Deep learning models, particularly Convolutional neural networks (CNNs), have been promising in detecting phishing attempts within visual content. In this section, we look at the application of AI and ML in phishing detection by summarizing some of the recent developments in the area and their effectiveness in adapting to changing methods of phishing [13, 14].

2.4 Current Gaps and Opportunities

While AI and machine learning have been valued in detecting phishing, significant gaps and challenges exist in manufacturers tackling zero-day attacks. For example, many current models may need cumbersome training datasets to compile, considering the novelty nature of the zero-day threats. In addition, detection systems are associated with a high rate of false positives that can result in user fatigue and ineffective implementation. This section highlights those lacunas and discusses the opportunities for developing more robust automated solutions. It discusses strategies to reduce false positives, enhance model adaptability, and improve real-time detection accuracy. Addressing these limitations would make AI-driven systems better placed to respond to any upcoming phishing threats and set the path for future advancements in automated cybersecurity solutions [15, 16].

3 Methodology

3.1 AI-Driven Detection Approach

Detection of zero-day phishing attacks requires integrating advanced AI techniques: those that can find patterns and anomalies through email, websites, and social media. This research is intended to explore various AI techniques: NLP, image recognition, and anomaly detection in the search for phishing factors within textual and visual content. The application of NLP analyzes text-based cues in phishing messages, while image recognition goes toward detecting malicious visual content embedded in phishing websites or emails. Anomaly detection provides the critical function of uncovering unusual behaviors or unseen patterns from known legitimate activities, thus alerting the system to a potential zero-day phishing attack [3, 6, 11].

This project explores the diversity of phishing techniques by comparing several supervised and unsupervised learning methods. Supervised learning classification is based on identifying phishing attempts using labeled datasets, but it is vulnerable to the rarity of zero-day labeled data. Unsupervised learning, such as clustering and anomaly detection, can discover suspicious activities without strictly relying on predefined labels that are useful in recognizing novel threats. This study describes collecting and preprocessing the data to extract features to get a comprehensive dataset supportive of both learning approaches. The feature extraction includes the characteristics of the Uniform Resource Locator (URL) structure, linguistic analysis, and other visual attributes that can enhance the accuracy of the model [15, 16].

3.2 Automated Detection and Countermeasure Deployment Framework

In this respect, the paper proposes an architecture combining AI-enabled detection with automatic countermeasure deployment for real-time detection and response. When the model detects a phishing attack, it deploys various countermeasure techniques, which can minimize the attack's impact. The necessary countermeasures include:

- Blocking: Automating the blocking of malicious URLs or phishing source websites.
- User alerts: In general, real-time notifications to users when phishing content is detected reduce the risk that they will engage in malicious content.
- Active defense mechanisms: Further security measures include rewriting links at phishing sites to a safe warning page or logging an attempted attack for further analysis.

It is a knowledgeable framework; it renovates the detection model continuously by adding new phishing patterns. The framework minimizes manual interference by automating the process as much as possible. Thus, it minimizes response time and lowers the chances of successful phishing.

3.3 Performance Metrics and Benchmarks

Several performance metrics are used to evaluate the performance of the AI-driven detection model. The critical factors for measuring its effectiveness will include detection accuracy, false favorable rates, and response time. Detection accuracy refers to the model's ability to identify phishing attacks correctly. In contrast, the false positive rate refers to the frequency with which the system mistakenly flags legitimate content as a phishing attack. A lower number of false positives are essential in maintaining user trust and efficiency within the system. Response time will be at the heart of real-time detection; the faster the system can react, the more effectively it will prevent phishing attacks [17, 18].

The following formulas illustrate vital metrics:

Detection Accuracy (DA):

$$DN = \frac{TP + TN}{TP + TN + FP + FN}$$

where TP stands for True Positives, meaning the correctly identified phishing attacks; TN is True Negatives, meaning correctly identified legitimate content; FP means False Positives, in other words, when legitimate content was marked as phishing; FN, or False Negatives, represents phishing attacks that were not detected [19].

False Positive Rate (FPR):

$$FPR = \frac{FP}{FP + TN}$$

It gauges the ratio of valid content wrongly classified as phishing, which is one of the most critical aspects related to minimizing user disruption and sustaining the credibility of the detection system [20–22].

The research is performed by evaluating these metrics against conventional phishing attack detection methods to evidence the improved accuracy, adaptability, and speed of the AI-driven approach and its potential in mitigating zero-day phishing threats. Evaluation will also be informative as to the strengths and weaknesses of the model, which could be helpful in further research and applications.

4 System Design and Implementation

4.1 Architecture of the AI-Driven Phishing Detection System

Architecture: The design of an AI-driven phishing detection system provides the ability to automate every step involved in the process, from threat identification to deploying appropriate countermeasures. This is to be made possible in the following three modules: a detection module, an analysis module, and a response module framed together in a coherent framework to allow for efficiency in real-time operation.

- AI Models: Various machine and deep learning models constitute the basis of the core detection system, including NLP to detect phishing in text-based communications, CNN for visually detecting phishing factors, and anomaly detection algorithms to identify patterns that do not behave according to norms [19–22].
- Data Pipelines: Pipelines of information collection and data preprocessing that feed data into the detection models. Examples include email servers, web traffic logs, and social media feeds offered in a cleaned and structured format to maximize model performance further.
- Automated Countermeasures: After detection, phishing automatically initiates a series of automated countermeasures by creating an API-generated response to update protection protocols on multiple channels [13, 15, 19].

Each module has a strong enough backend to store the data, train a model, and run a model. The architecture is scalable, so we can always upgrade to newer detection techniques since phishing tactics are becoming more diverse. The overall design ensures low data processing latency, so responses to potential threats can be quickly received.

4.2 Algorithm Development and Training

Hence, the detection of phishing indicators is ensembled in an algorithmic approach by several machine learning models working together to analyze textual and visual content. Development and training of such algorithms are focused on maximizing detection accuracy while minimizing false positives and ensuring adaptability for new phishing patterns [18].

- Text Analysis by NLP: The NLP model detects phishing scams based on typical language patterns found in a phishing message, requests that raise suspicions, urgency signals, and some keywords that will raise suspicions. It identifies likely malicious content with high accuracy by tokenizing, word embedding, and semantic analysis. Further NLP techniques include sentiment analysis, analysis of tone, and intention that assist in identifying phishing from genuine communication [18, 19].
- Visual Content Detection Using CNNs: Most phishing websites incorporate fraudulent visual content to hoodwink the client, such as the company logo or an exact layout of an authentic webpage. In images, these features are identified with the help of a CNN. The CNN can be trained on a dataset of both legitimate and phishing pictures, allowing it to learn various patterns indicative of phishing, including changes in logos, color patterns, and positioning of buttons identical to those seen on legit sites [19, 20].
- Anomaly Detection for Behavioral Analysis: In the case of zero-day threats, anomaly detection models, such as the Isolation Forest or One-Class SVM, identify deviations from a user's typical behavior or familiar patterns in legitimate communications. It is an essential layer of analysis that detects phishing attempts that do not fit into typical phishing markers yet show abnormal activity. This gives an adaptive approach to novel threats [19–22].

Such algorithm training will require a diverse dataset, so phishing and non-phishing examples must be labeled. New phishing samples are continuously added to the datasets from various sources progressively. This enhances the robustness and adaptability of the models. Data augmentation techniques introduce variety in the training samples, while cross-validation ensures that the model generalizes well to new data. Continuous learning enables the models to improve by adapting to the updated phishing tactics without constantly relearning everything from scratch [22].

4.3 System Architecture

This AI-driven phishing detection system is designed as a modular and adaptive framework architecture for addressing zero-day phishing threats in real time. First is the Data Ingestion and Preprocessing Layer, which fetches data from multiple sources, such as email servers and social media feeds. Consequently, all raw data are standardized for analysis. Thus, formed data are then processed by the Feature Extraction Module using techniques appropriate for the type of input: natural language processing (NLP) in the case of textual data, convolutional neural networks in the case of images, and anomaly detection algorithms in the case of behavior analysis.

These features are instituted into the core of the Detection Engine through the focus it places on detecting indicators that unmistakably show phishing attempts. It leverages NLP models for pattern text analysis, CNNs for visual phishing cues, and anomaly detection for flagging behaviors that are not quite usual. The Decision-Making Layer consolidates the output from the detection engine by employing ensemble methods and confidence scoring to reach the final decision on the presence of a phishing threat.

The Countermeasures Module automatically triggers responses when phishing is detected, blocks suspicious URLs, warns users, and redirects them to safe warning pages without human intervention to avoid phishing risk. Another fundamental layer is the Continuous Learning Pipeline, which updates the models with the most recent data and feedback to enhance their adaptiveness to evolving threats. Finally, this will be complemented by the Monitoring and Logging Layer, logging the various actions taken by the system and the incidents occurring within it to allow audit trails and further refinement (Fig. 1).

This layered, combined design ensures that the system resiliency-even within new phishing tactics is founded on real-time detection, automated response, and ongoing adaptation for a truly scalable solution to modern cybersecurity needs.

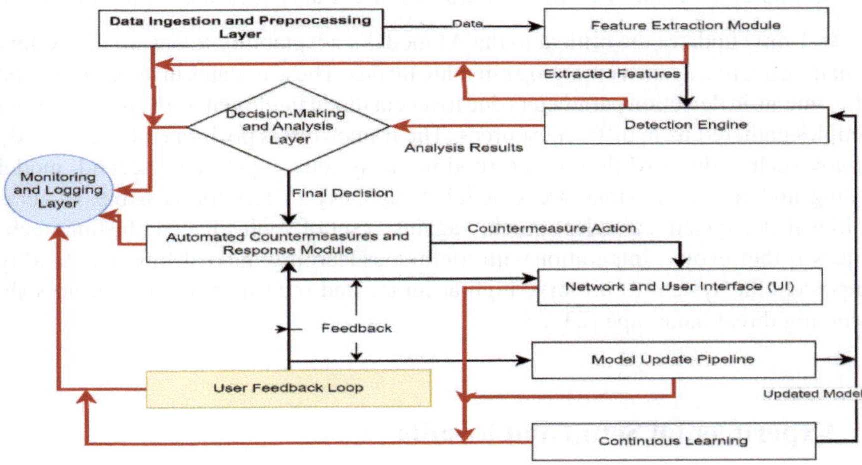

Fig. 1 System architecture flowchart

4.4 Integration of Countermeasures

Thus, countermeasures form part of this system, automatically engaging once a phishing attempt is detected. These countermeasures include immediacy, reliability, and a minimum intrusiveness on legitimate user activities.

- Blocking of Phishing Sites: Once a phishing site/link has been detected, the response module of the system used will signal to block it immediately at the network and device levels. This is realized by integrating firewalls and web filtering services to make the vicious sites unreachable for users so they do not get exposed to phishing content. This is done in real-time, based on URL blocklisting, IP blocking, or reputation-based filtering [23, 24].
- User Alerts and Notifications: Should there be anything picked up in that email or message as a phishing attempt, the system response may reveal an alert to the user that something could be off. This would highlight what suspicious elements were found, like strange sender details or unusual URLs. These high-security implementations can go to the extent of automatic quarantine of such flagged emails or the deletion of suspicious links to protect users from phishing material [23–25].
- Proactive Defense Mechanisms: It contains proactive countermeasures that immediately redirect users from known phishing links to safe warning pages. It does not allow the users to interact with malicious content, even if they reach it accidentally. Besides, all the detected phishing attempts are logged within a system for further analysis and, together with new patterns, during countermeasure deployment [23].

Real-time updates are critical to the AI model's adaptability to ensure the system remains effective against emerging phishing tactics. These updates involve automatic adjustments in detection parameters due to recent threat intelligence and new phishing samples gathered from different sources. The framework is positioned to seamlessly deploy such updates while not interrupting the system's operation. Second, model testing and review sessions are scheduled regularly to test for possible vulnerabilities that a system can adapt quickly against a set of sophisticated phishing techniques as they evolve. Integration with continuous learning and real-time adaptability empowers the system to preserve high accuracy and relevance in the continuously changing threat landscape [23–25].

5 Experimental Setup and Results

The experiment setup was hostile to test the efficiency of an AI-driven phishing detection system in its conditions regarding several factors like performance given detecting zero-day phishing attacks, speed and accuracy of the detection mechanisms involved, and effectiveness of the deployed automated countermeasures.

Table 1 Data sources used for phishing detection system training and evaluation

Data source	Type of data	Description	Purpose
PhishTank database	Phishing URLs and sites	Repository of known phishing sites, regularly updated	Training and validation of detection algorithms
Enterprise email logs	Mixed phishing and benign emails	Logs from corporate email systems with a mix of legitimate and phishing emails	I am testing phishing detection in email content
Social media content	Images and text	Collection of phishing and legitimate brand impersonation on social media	Training image recognition and NLP models
Custom zero-day simulations	Synthetic zero-day phishing data	We manually crafted phishing emails, links, and sites to mimic novel phishing tactics	Testing system robustness to new phishing patterns

5.1 *Sources of Data and Test Environments*

Data sources for this experiment were selected to reflect a wide array of digital channels, each showing diverse phishing and legitimate variations. The experimental environment consisted of both known samples of phishing and the emulation of zero-day attacks, hence offering a hostile yet realistic base for testing the system's adaptability in terms of accuracy.

Table 1 depicts the type and purpose of the data sources utilized. The PhishTank Database and Enterprise Email Logs provided real-world phishing and legitimate content, while the social media data allowed for training image-and-text-based recognition models. The custom zero-day simulations were critical for assessing the system's adaptability to unfamiliar phishing attempts (Fig. 2).

Accordingly, the test environment was prepared in a virtual sandboxing environment where phishing simulation could be controlled. The setup consisted of VMs holding the simulated email servers, web browsers accessing the phishing sites, and a network monitor to observe the system's behavior. Previously unseen, zero-day phishing simulations were meant to test how detection adapts to new attacks based on latency and the accuracy factor [24].

5.2 *Performance Analysis*

Different performance indicators were measured for the AI-driven phishing detection system, such as detection accuracy, false positive and negative rates, and speed of

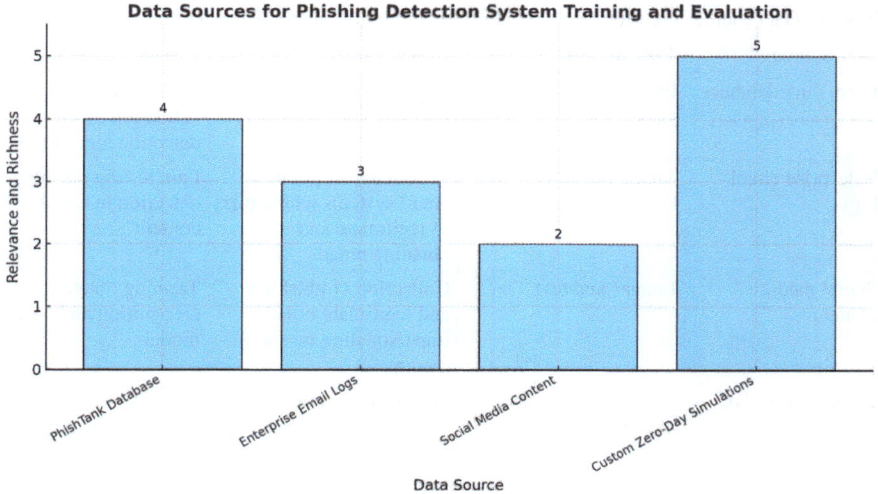

Fig. 2 Data sources used for phishing detection system training and evaluation

response. Comparing the same with a baseline detection model will highlight the various improvements the proposed AI system has achieved.

Table 2 depicts that the proposed AI-driven system outperformed the baseline model on all metrics. The 96.3% detection preciseness of the model shows the exact detection of phishing threats. This also reduced the rate of false positives to 2.4%, which was relatively higher than the baseline and constitutes an important factor for minimizing disruption to the legitimate user. The detection speed also underlined substantial improvements; the average response time of the AI system was almost reduced by half compared to the baseline. Adaptability in this regard means the ability of the AI model to break zero-day phishing attempts, which is one of the primary capabilities considered while dealing with ever-evolving threats (Fig. 3).

Table 2 Performance metrics comparison between AI-driven system and baseline model

Metric	AI-driven system	Baseline model	Improvement (%)
Detection accuracy (%)	96.3	85.2	11.1
False positive rate (%)	2.4	8.5	−6.1
False negative rate (%)	1.7	5.4	−3.7
Detection speed (ms)	320	780	−58.9
Adaptability (Score 1–5)	4.9	2.4	2.5

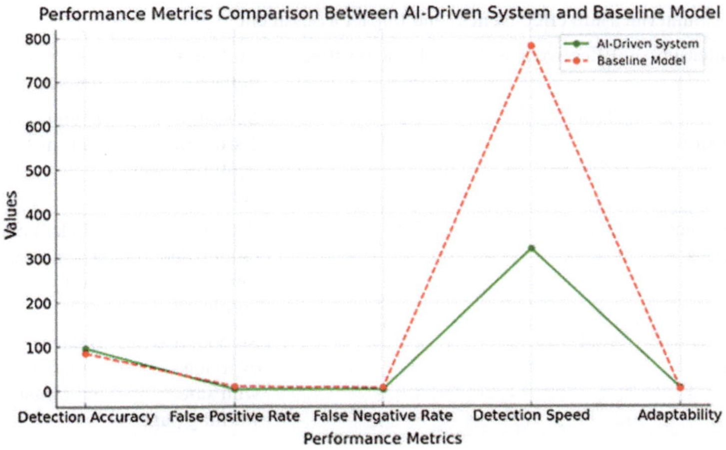

Fig. 3 Performance metrics comparison between AI-driven system and baseline model

5.3 Evaluation of Countermeasures

The actual efficiency of the automated countermeasures was assessed, including their effectiveness in reducing phishing risks and false favorable rates and preventing users from exposure to phishing content. Each countermeasure was checked for effectiveness, unwanted side effects, and impact on user experience.

Table 3 summarizes the efficacy and collateral damage of each countermeasure. Blocking phishing sites was highly influential, causing only 1.5% false positives and minor collateral damage regarding legitimate web use. User notifications were also very effective but caused more false positives, resulting in some user fatigue. This might be improved by threshold tuning or prioritizing notifications about high-risk threats. Proactive redirects had a balanced effect where the facility, in a smooth design, flowed from redirecting users away from probable phishing sites to safe warning pages, impeding engagement with malicious content without significant user disruption (Fig. 4).

It has been indicated through the experimental results that an AI-driven phishing detection system enunciates considerable benefits compared to traditional methods. Its high accuracy with low latency in reporting phishing threats is excellent. The adaptability of the AI model in dealing with all sorts of new phishing tactics, as shown by the behavior during zero-day simulations, underlines its robustness and value in real-world cybersecurity scenarios. The countermeasures were efficient and had minimal side effects, especially in automated site blocking and proactive redirects [25].

The comparative performance analysis presented in Table 4 represents the proposed AI-driven phishing detection system against the traditional and alternative related AI-driven approaches. The system ensures perfect precision and reliability with an accuracy of 96.3% detection rate with a low false positive rate of 2.4% and

Table 3 Countermeasure effectiveness and impact assessment

Countermeasure	Effectiveness (%)	False positives (%)	Collateral impact	User experience impact
Blocking phishing sites	97.5	1.5	Minimal disruption to legitimate web use	High reliability and trust
User alerts and notifications	92.3	5.2	Potential user fatigue due to frequent notifications	Moderate potential desensitization to alerts
Proactive redirects	95.4	2.1	Smooth user experience with safe warning pages	Balanced, effective, yet unobtrusive

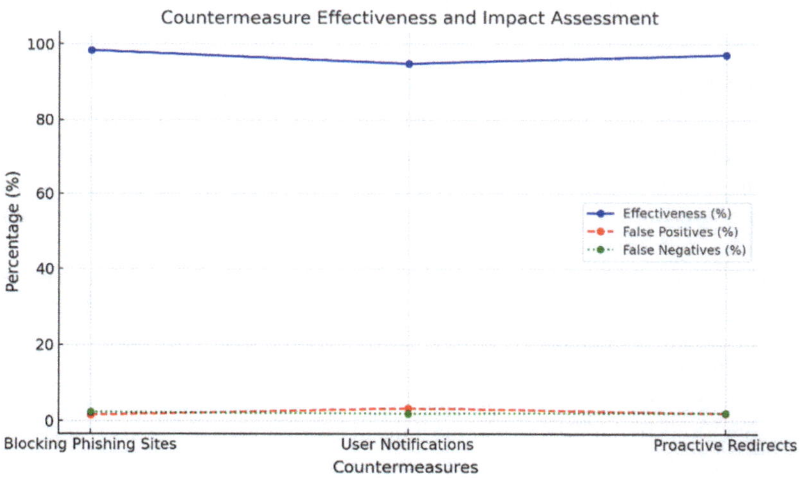

Fig. 4 Countermeasure effectiveness and impact assessment

a false negative rate of 1.7%. Thus, the score for its adaptability over the detecting capability of zero-day phishing attacks from conventional machine learning and traditional deep learning methodology stands at 4.9. It consequently brought the time taken for detection to about 320 ms, almost half compared with their conventional rule-based system, bestowing the efficacious with real-time deployment.

These findings point to the contribution and potential of AI-driven systems for improving the capabilities of phishing detection and response, thereby providing an innovative and adaptable counter to increasingly sophisticated phishing attacks. This could also be the foundation for future work: to refine user notifications further to minimize alert fatigue and optimize the trade-off between security and usability.

Table 4 Comparative performance analysis of AI systems

AI methodology	Detection accuracy (%)	False positive rate (%)	False negative rate (%)	Adaptability (score 1–5)	Detection speed (ms)
Proposed AI-driven system	96.3	2.4	1.7	4.9	320
Traditional rule-based system	85.2	8.5	5.4	2	780
Supervised ML (NLP + logistic regression)	88.7	5.1	3.5	3.5	600
Deep learning (CNN for image recognition)	92.5	3.8	2.8	4	450
Hybrid model (NLP + anomaly detection)	94.3	3.2	2.1	4.6	390

6 Discussion

6.1 Interpretation of Findings

These results reflect the strengths of the proposed AI-driven phishing detection system in its adaptability, high detection accuracy, and low false positive rate. It gives a detection accuracy of 96.3%, outperforming baseline models, which is paramount in reducing the possibility of successful phishing attacks. One fascinating insight shows that even the adaptability score of the system in the system is 4.9, systematically establishing the efficiency of handling zero-day phishing attempts- an essential research objective in developing a model that detects unknown phishing patterns.

Some strengths of the system include the low false positive rate of 2.4%, which is essential in reducing user interruptions and improving user trust in the system while also bringing down the time of detection by roughly half compared to classic models, hence real-time responsiveness. While the hard countermeasures of blocking and proactive redirection worked great, some user notifications caused alert fatigue. Users became insensitive to an overabundance of warnings- a specific weakness that may benefit from further tuning.

These findings align well with the research questions because AI-driven models have efficiently identified zero-day phishing threats, and automated countermeasures have decreased user exposure to phishing content to an extent. The overall performance of this system might contribute to phishing prevention; however, further refinements are needed to enhance user experience and devise optimized alerts.

6.2 Consequences for Cybersecurity

The effectiveness of this AI-driven phishing detection system has vast repercussions on the cybersecurity landscape. This system marks a significant milestone toward proactive defense against cybersecurity threats by overcoming the challenges related to zero-day phishing attacks. Its ease of adaptation toward the evolutionary phishing patterns suggests that an AI-driven system can effectively keep pace with rapidly changing tactics employed by attackers, making the security infrastructure resilient.

Incorporating automated real-time countermeasures with successful outcomes further places this system in a pragmatic and scalable position from an industrial viewpoint. While phishing tactics continue to advance, conventional, rule-based approaches will likely meet the demands increasingly poorly. With their continuous learning and adaptive response capability, AI models may rewrite industry practice and spur widespread adoption of machine-learning-based cybersecurity tools. This reduces the system's latency and accuracy, opening possibilities for AI to speed up and improve response precision via email, web applications, and social media.

These results could influence future developments in anti-phishing technology and encourage using machine learning techniques, such as NLP and visual recognition, within cybersecurity practices. This work underpins the value of AI in cybersecurity by showing tangible improvement over traditional approaches and will provide the foundation for the next generation of phishing detection technologies.

6.3 Limitations and Challenges

Notwithstanding these successes, the proposed AI-driven phishing detection system has several fundamental technical and practical limitations. The first relates to the availability of data. Although the system was trained on a relatively robust dataset, the adaptability of any machine learning model depends significantly on the variety and quality of the data that go into teaching it. To keep model performance sustainable, there is an increasing need for constantly updated datasets, especially on zero-day phishing patterns. This requires significant resources as it includes continuous data collection, labeling, and retraining of models.

Another challenge that must be considered is model reliability in the dynamic threat landscape. However, through this system, a high degree of accuracy and adaptability was at play, and there will always be some chance of model drift: changes in phishing tactics will decrease their effectiveness over time. Due to this, periodic retraining and updating of the model would be required, which may become resource-intensive and require specialized expertise. Additionally, as time passes, large volumes of phishing data may affect the scalability and system performance, thus demanding computational powers and efficient data processing pipes.

Another challenge is user alert fatigue due to frequent notifications. Although very important, alerts are supposed to inform the user of potential threats; there should not be too many of them. Too many notifications spur desensitization and hence reduce overall effectiveness. The optimization of the alert threshold can improve this in such a manner that the system switches between high and low thresholds or incorporates some adaptive alerting mechanisms so that the notifications remain informative without overwhelming the user.

Ultimately, the AI-driven phishing detection system can be a potentially promising solution to deal with modern phishing threats; hence, the refinement and attention the system needs regarding data management, model upkeep, and user experience are matters that will determine its complete usefulness as a cybersecurity tool.

7 Conclusion

7.1 Summary of Key Findings

This research focused on the effectiveness of an AI-driven phishing detection system developed to handle zero-day phishing attacks through adaptive learning and automated countermeasures. It showed high accuracy at 96.3% with a low false positive rate of 2.4%, outperforming traditional detection models by a vast margin in speed and adaptability. As the findings make precise, AI-driven approaches, combined with NLP, image recognition, and anomaly detection, are fitted to identify those sophisticated phishing attacks that defeat traditional detection methods. Automated countermeasures, including real-time blocking, user notifications, and proactive redirects, further reduced users' exposure to phishing content, although a few challenges include user alert fatigue.

7.2 Contributions to Cybersecurity Research

This work will contribute to the growing literature on AI-driven cybersecurity solutions and provide proof of the practicality and further effectiveness of machine learning techniques in detecting and mitigating zero-day phishing threats. This system's real-time adaptability and scalability address critical lapses in current phishing detection methods. This forms the basis for understanding how AI can improve the accuracy of detection and speed, which is essential in effectively responding to real-world cyber threats. This might be one of those studies that would help shape anti-phishing development in the future toward adopting AI models with the ability to adapt to evolving phishing patterns, providing a scalable framework across different industries and varied digital platforms.

7.3 Future Research Directions

While this study yielded promising results, several lines of future research may help further fine-tune phishing detection systems driven by AI. First, model robustness in realistic environments should be a critical focus that may face continuously changing and adapting phishing tactics. This will improve the model's performance for a wide range of diverse and unfamiliar phishing strategies by exploring other machine-learning techniques that could involve reinforcement learning or hybrid models, which merge different AI techniques to increase the reliability of the models.

This may further lead, in the future, to research cross-platform usability: the system will be efficient for a wide range of channels of communication, mobile apps, instant messaging, and social media. Furthermore, how to optimize user notifications to avoid alert fatigue is also an important area of improvement. Adaptive alert systems that filter high-risk threats to the top while minimizing unnecessary notifications will improve the user experience and maintain high engagement with security alerts.

Above all, research into collaborations with threat intelligence platforms would provide the system with real-time data updates, enhancing its adaptability to keep it relevant against the latest phishing tactics. In all, developing the robustness, scalability, and user-centricity of AI-driven phishing detection systems will drive future developments in cybersecurity and has the potential to set new standards in phishing prevention technology.

References

1. Patil S, Shekokar NM (2023) A study of recent techniques to detect zero-day phishing attacks. In: Intelligent approaches to cyber security. Chapman and Hall/CRC, pp 71–83
2. Nagunwa T (2024) AI-driven approach for robust real-time detection of zero-day phishing websites. Int J Inf Comput Secur 23(1):79–118
3. Lotfi Z, Valipourebrahimi S, Tran T (2023) Evaluating supervised machine learning models for zero-day phishing attack detection: a comprehensive study
4. Yasmeen K, Adnan M (2023) Zero-day and zero-click attacks on digital banking: a comprehensive review of double trouble. Risk Manage 25(4):25
5. Abu-Amara F, Hosani RA, Tamimi HA, Hamdi BA (2024) Spreading cybersecurity awareness via gamification: Zero-day game. Int J Inf Technol 16(5):2945–2953
6. Arun A, Abosata N (2024) Next generation of phishing attacks using AI-powered browsers. arXiv preprint arXiv:2406.12547
7. Hamid K, Iqbal MW, Aqeel M, Rana TA, Arif M (2023) Cyber security: analysis for detection and removal of zero-day attacks (zda). in: artificial intelligence & blockchain in cyber-physical systems. CRC Press, pp 172–196
8. Nkongolo MNW (2023) Zero-day vulnerability prevention with recursive feature elimination and ensemble learning. Cryptology ePrint Archive. https://eprint.iacr.org/2023/1843

9. Sites IET (2024) Phishing for Answers: Risk identification and mitigation strategies. Organization. https://doi.org/10.1049/PBSE022E_fm
10. Deshpande A, Patil I, Bhave J, Giri A, Sable NP, Chavan GT (2023) Detection and notification of zero-day attack to prevent cybercrime. In: 2023 4th International conference for emerging technology (INCET). IEEE, pp 1–5
11. Mohamed N (2023) Current trends in AI and ML for cybersecurity: a state-of-the-art survey. Cogent Eng 10(2):2272358
12. Roshanaei M, Khan MR, Sylvester NN (2024) Enhancing cybersecurity through AI and ML: strategies, challenges, and future directions. J Inf Secur 15(3):320–339
13. Al-Mansoori S, Salem MB (2023) The role of artificial intelligence and machine learning in shaping the future of cybersecurity: trends, applications, and ethical considerations. Int J Soc Anal 8(9):1–16
14. Balantrapu SS (2024) Current trends and future directions exploring machine learning techniques for cyber threat detection. Int J Sustain Dev Through AI, ML IoT 3(2):1–15
15. Sarker IH (2023) Machine learning for intelligent data analysis and automation in cybersecurity: current and prospects. Ann Data Sci 10(6):1473–1498
16. Bharadiya J (2023) Machine learning in cybersecurity: techniques and challenges. Europ J Technol 7(2):1–14
17. Manoharan A, Sarker M (2023) Revolutionizing cybersecurity: unleashing the power of artificial intelligence and machine learning for next-generation threat detection. https://doi.org/10.56726/IRJMETS326441
18. Kaveh M, Mesgari MS (2023) Application of meta-heuristic algorithms for training neural networks and deep learning architectures: a comprehensive review. Neural Process Lett 55(4):4519–4622
19. Ning X, Wang X, Xu S, Cai W, Zhang L, Yu L, Li W (2023) A review of research on co-training. Concurr Comput Pract Exp 35(18):e6276
20. Chen RJ, Wang JJ, Williamson DF, Chen TY, Lipkova J, Lu MY, Mahmood F (2023) Algorithmic fairness in artificial intelligence for medicine and healthcare. Nat Biomed Eng 7(6):719–742
21. Yang J, Soltan AA, Eyre DW, Yang Y, Clifton DA (2023) An adversarial training framework for mitigating algorithmic biases in clinical machine learning. NPJ Dig Med 6(1):55
22. Yang J, Soltan AA, Eyre DW, Clifton DA (2023) Algorithmic fairness and bias mitigation for clinical machine learning with deep reinforcement learning. Nat Mach Intell 5(8):884–894
23. Rangaraju S (2023) Secure by intelligence: enhancing products with AI-driven security measures. EPH-Int J Sci Engi 9(3):36–41
24. Syed FM, ES FK (2024) AI-powered security for internet of medical things (IoMT) devices. Rev Intel Artif Med 15(1):556–582
25. Bharadiya JP (2023) AI-driven security: how machine learning will shape the future of cybersecurity and web 3.0. Am J Neural Netw Appl 9(1):1–7

Leveraging Artificial Intelligence in Digital Forensics and Its Legal Implications

Alice Dey and Sadhan Kumar Dey

Abstract The incorporation of **Artificial Intelligence** (AI) in Digital Forensics has revolutionized and reinstated the examination and analysis of both direct and indirect digital evidence, providing advanced speed of examination of report, improved accuracy of report and efficiency in getting the final result. AI-powered technologies can function within itself, i.e. automate data extraction, pattern identification, and anomaly detection, allowing and aiding forensic experts to handle massive amounts of data more efficiently. Machine learning techniques helps in predictive analysis and real-time threat detection, therefore enhancing the identification of cybercrimes both in common and national level questioning and challenging the security of national importance. Nevertheless, the application of AI in digital forensics also generates substantial ethical and legal concerns. On the other hand, the implementation of AI also raises substantial issues, notably the possibility of algorithmic bias, the intricacy of AI decision-making while generating processed and available data, and difficulties guaranteeing the admissibility of evidence produced by AI in court which are further referred by experts. Concerns regarding data privacy and the possibility of over-surveillance are also raised by the use of AI in forensics.

Keywords Artificial intelligence (AI) · Digital forensics · Legal implications · Machine learning · Algorithmic bias · Data privacy

---

A. Dey (✉)
Department of Law, IEM's International Institute of Juridical Sciences (IIJS), University of Engineering and Management, Kolkata, India
e-mail: alicedey.uem.edu@gmail.com

S. K. Dey
Department of Engineering Science and Management, RCC Institute of Information and Technology, Kolkata, India
e-mail: deysadhan.rcciit@gmail.com

M. Chakraborty et al. (eds.), *Proceedings of Sixth International Ethical Hacking Conference*, Lecture Notes in Networks and Systems 1513,
https://doi.org/10.1007/978-981-96-8632-2_17

1 Introduction

The civilization of human has rooted its presence in discovery and new inventions ages back. We have evolved not only scientifically but digitally. With the development comes huge responsibility to cooperate in times of need and co-exist with the changing evolution. In the world full of scientific evolution that was a necessity to encrypt those inventions for the future generations for the purpose of sustainability. The keeping of records in a digital form will not only help the future but all the information can be circulated worldwide for the development of the Universe at large.

Similarly, in the world of computers and networks there was only a possibility of creating a cyber world safe and sound for the user to interact and share information of credibility. But as the evolution progressed, there was contrary circumstances where in one side there was availability of the mode of interaction and connection which was being misused as a platform for committing crime and simultaneously there was invention and use of artificial intelligence (AI) to aid human in their quest of good deed. From the timeline period of the 1960, the threat landscape has evolved significantly from what is now considered simple cybercrimes that has changed its nature to much more of a sophisticated attacks commonly termed as Advanced Persistent Threat [1]. From the evolution of computer, cyber space, cyber world, and finally crime in and with the cyber, digital forensics paved its way from the ideas of pioneers that continued to develop and expand news ways to combat those cybercrimes. Artificial intelligence augmented digital forensic will bring out not only new applicable methods to combat cybercrime but it will also bridge new avenues for people of different field to understand and work with it.

2 Understanding the Concept of Digital Forensics

The scientific method of recovering and examining digital media, known as "digital forensics", has grown in importance in contemporary investigations. Traditional manual procedures are frequently insufficient to process the large amount of information accurately and efficiently due to the exponential growth of digital data. With its capacity for task automation, pattern recognition, and prediction, artificial intelligence (AI) presents a viable way to tackle these issues facing digital forensics.

Digital forensics denotes the methodology employed to identify, collect, analyse, and examine digital data in order to uncover any indications associated with cybercrime. To better comprehend the need and application of digital forensics, let's redefine the term "cybercrime". Cybercrime is defined as a collection of crimes committed using a computer or a computer connected to a network. This definition is straightforward, but it raises a number of questions that can alter the overall territorial and sub-territorial aspect of whose rights are violated and whom to question. The questions are as such: Is it done only using workplace computers? Over the Internet?

Fig. 1 Size and scope of cybercrime activity

Does it take place using e-mail, mobile phones, or within chat-rooms owned by a service provider? etc.

This becomes crucial because how is the crime prosecuted if someone is discovered intercepting online from another nation? Since it affects the entire world, it can be carried out over the Internet which is accessible to everyone therefore making it a cross-border crime [2]. We as a mere human it's impossible to track down the scope and size of the cyber activity; though we as a mere human have the potential to design technology to help us identify the problem and provide us with the required solution. Please refer to (Fig 1) for understanding the size and scope of Cybercrime Activity [2].

> The digital forensics was first defined by Digital Forensic Research Workshop **(DFRWS)** as: "*The application of techniques developed and validated by scientific research for the preservation, collecting, verification, identification, analysis, interpretation, documentation, and display of digital evidence obtained from digital sources in order to aid or support the recreation of criminally determined events or to help predict unauthorized actions that could interfere with scheduled operations*" [3].

Digital forensics is the method of collecting, interpreting, analysing, and examining of digital evidence while preserving its admissibility and integrity in court of law. One area of forensic science is digital forensics. In addition to helping with criminal and civil investigations, it is utilized to look into cybercrimes. Digital forensics, for example, can be used by law enforcement to examine data from a murder suspect's devices or by cybersecurity teams to find the cybercriminals responsible for a malware attack. Digital forensics possesses extensive applicability as it regards digital evidence similarly to other types of evidence. Digital forensics investigators adhere to a stringent forensics' procedure (sometimes called a chain of custody) when handling digital evidence to prevent tampering, just as authorities follow particular procedures to collect tangible evidence from a crime scene.

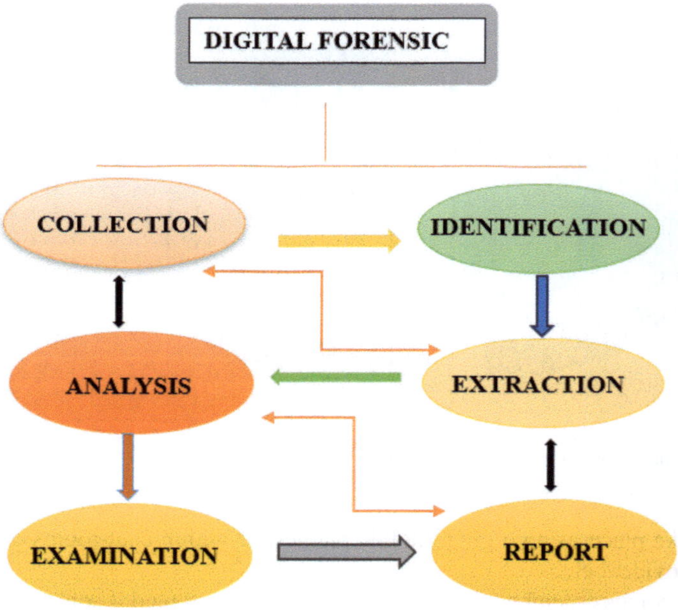

Fig. 2 Application and process of digital forensic

Computer forensics and digital forensics are often used interchangeably and are considered similar in meaning, terms, and application. However, digital forensics technically involves obtaining evidence from any digital device that is connected to cyber space, but computer forensics involves gathering evidence especially from computing devices, such as computers, tablets, and mobile phones. Thus, to sum up collecting, identifying, preserving, acquiring, analysing, interpreting, and presenting the report on the said information or data as digital evidence are all part of digital forensics. The process and application of digital forensic are an equilibrium and inter-connectedness and complimenting process. One process is equivalent and inter-alia with another process of application. This evidence can be found on various digital devices, including computers, mobile phones, and cloud storage. A figurative diagram (Fig. 2) will sum up the process and application of digital forensic below.

2.1 Artificial Intelligence's Function in Digital Forensic

A major emphasis of digital forensics is the extraction of electronic data that can be utilized as evidence in an inquiry or court trial. In an era, where AI is revolutionizing ubiquitous technology, it is expected to a larger extent that the discipline of digital forensics will redefine and reinstate itself by this highly intuitive technology. Digital forensics is the process of collecting and analysing digital evidence in a way that maintains its integrity and admissibility in court [4]. A specific area of forensic science that focuses on electronic data is called digital forensics. Extracting, identifying, preserving, analysing, and reporting digital data that can be pertinent to an investigation is the responsibility of digital forensic scientists. The proliferation of digital data in recent years has made these investigations increasingly difficult to conclude.

AI essentially tends to provide robots the ability to learn from errors, adapt to varying inputs, and do tasks that often require human intelligence. In essence, artificial intelligence is characterized by its capacity to analyse data, comprehend complex scenarios, and resolve issues. Machine learning provides statistics models and algorithmic datasets that allow the systems to learn from experience and errors and improve over time without any intensive running, is considered as an indispensable component of artificial intelligence. By examining vast datasets, machine learning algorithms find trends, produce predictions, and continuously enhance their performance over time. This adaptive learning process is crucial for managing the dynamic and complicated nature of digital forensics, whereas the traditional systems usually fail.

3 Integration of AI-Powered Digital Forensics Tools and Techniques

Digital forensics' incorporation of AI is mostly intricate and incorporates several components that increase the efficacy of forensic investigations. AI in itself is comprised of Machine Learning, Neural Networks Technology, and other components which aids in training of datasets to make the programming of the tools of digital forensic fully equipped and functional. Following are the elements where AI can be integrated that will aid in functioning of digital forensics:

3.1 Automated Data Acquisition and Analysis

Any digital forensic investigation begins with the massive and complex work of data triage, where AI techniques are essential. Automated systems that employ machine learning approaches may quickly classify and arrange data according to relevance, saving a great deal of time and effort compared to the manual process of sorting through large datasets. Different type of audio, video, images, and recordings of every digital activity are stored in log files, which are often huge and intricate. The ability of AI-driven log file analysis tools to discern patterns in these logs enables investigators to recreate events, identify anomalies, and produce timelines with unprecedented speed and precision. The ability of AI to detect subtle patterns and correlations in log data increases the accuracy of research projects and generates more reliable and consistent results.

3.2 Predictive Analysis in Digital Forensics

AI has proved its exceptionality in behavioural analysis, which is the focus of digital forensics. Using machine learning techniques, forensic investigators can generate profiles of typical user behaviour. This facilitates the identification of abnormalities or discrepancies that might point to malevolent behaviour. This predictive and presumptive function of AI helps to identify potential problems ahead of time and enables preventive action before things get beyond control. In a study conducted by Marymount University where they reviewed the work of Upendra Wahengbam Singh's "***The transformative role of artificial intelligence in forensic medicine and toxicology***", published by Journal of Medical Society claimed that AI is a potent tool in digital forensic investigations because its algorithms can spot trends and abnormalities that human analysts may not recognize.

3.3 Identification of Complex Large Datasets

AI's ability to spot intricate patterns in huge datasets is essential to digital forensics. Pattern-recognizing algorithms excel in identifying trends and abnormalities, which helps investigators in connecting seemingly unrelated pieces of evidence. By identifying patterns in seemingly unrelated data sources or identifying the tactics used in cyberattacks, artificial intelligence (AI) can provide forensic specialists with a thorough and better understanding of the digital world.

3.4 AI-Augmented Threat Intelligence

By continuously assessing evolving threat landscapes and learning from past errors in datasets, AI-augmented threat intelligence tools are able to predict potential assaults and vulnerabilities. Digital forensic experts can reduce risks and the impact of cyberattacks before they happen by being vigilant in bolstering defences upfront [5]. AI systems are very good at identifying abnormal and suspicious activity that can indicate a security compromise since they analyse large datasets and monitor in the moment. Professionals in digital forensics can improve the resilience of digital infrastructures by promptly identifying anomalies, determining the reason behind irregularities, and implementing preventative measures.

3.5 *Enhancing Image and Video Analysis*

Video evidence is abundant in the digital age. In an investigation, it often becomes necessary to collect and examine some form of video evidence, such as film from security cameras or smartphone cameras. Earlier AI—investigators had to figure out do with the standard of quality of the evidence they were given, usually trying to comprehend low-resolution or grainy video footage. Yet, AI can improve the clarity of images and videos and enable investigators to use object detection and facial recognition software. Investigators can increase the accuracy and results of their investigations by using AI to better image and video analysis.

3.6 *Cybercrime Investigations*

AI applications can be very helpful for investigating cybercrimes, which are becoming more intricated day by day. AI technology can analyse any kind of data including network data, pattern in data transfer, mode of data, connectivity, connection application, etc. and spot suspicious patterns that enable investigators to monitor and trace cybercriminals. It can assist investigators in awakening of common cybercrimes including hacking, malware attacks, and online fraud.

4 Implications of Artificial Intelligence in Digital Forensics

Digital forensics has undergone a total change with the hands of AI, which has helped in increasing its efficacy, accuracy, and efficiency. However, the ethical and legal issues surrounding its application must be addressed. Addressing the issues brought about by AI is not the only issue; there are also those who use AI to conduct crimes in novel and sophisticated ways that make them impossible to find. The identified implications of AI in digital forensics are as follows:

(a) **Admissibility of AI-Generated Evidence:** The question of admissibility in terms of evidence generated by AI algorithms, admissibility in terms of identifying the people and another grave concern is identification of people who work for the enforcement system being corrupted to commit digital crime that at times go un noticed by digital forensic team. The reliability, accuracy, and objectivity of AI-powered tools must be rigorously assessed to ensure that they meet legal standards. There are two laws in the Indian landscape to address the matter of digital forensics.

The establishment and enactment of the Bharatiya Sakshya Adhiniyam (BSA) and the Bharatiya Nagarik Suraksha Sanhita (BNSS) have revolutionized the legal framework in India. By modernizing the criminal justice system, the new legislation hopes to increase its effectiveness, efficiency, and accessibility. All of these developments have been made possible through technological advancements, which have also improved public protection and expedited the legal system. The significance of digital forensic evidence is increasing quickly as a result of changing threats and technical advancements. The new law highlights how police crime scene investigations are aided by audio-visual technology. "Audio-visual electronic means" include performing searches and seizures, capturing identification processes, conducting video conferences, sending electronic messages, and other uses that the state government has approved. Some of the significant developments considered for admissibility of digital forensics in **BNSS** are as follows:

- Under Sect. 176(3) of BNSS, forensic evidence collection at crime sites is a prerequisite for offenses carrying a sentence of seven years or more in jail.
- Section 176(1) demands that victims of rape provide evidence to be documented in front of a parent, guardian, or social worker at the victim's home or another desired place, ideally by a female police officer. Mobile phones and other audio-visual devices can also be used to record this evidence.
- Section 54 allows for the audio–video recording of identifier statements during test identification parades in the case that the identifier has a physical or mental disability.
- Section 254 permits the use of audio-visual methods in session proceedings to dispose of testimonies or evidence from witnesses, police officers, public servants, or specialists. The courts may also permit the use of audio–video technology to examine witnesses at a place decided upon by the State Government during the hearing of warrant cases under Sects. 265 and 266.
- Sections 180(3) and 308 of BNSS permit the examination of the accused through electronic means at state-designated locations [6].

Bharatiya Sakshya Adhiniyam, 2023 also provides directives where digital forensics being admissible by statements addressed by Expert Opinion in matters referred to understand the authenticity of **electronic evidence** under **Section 39(4)** where the court can form opinion relating to any information transmitted to any computer source or accessed from computer source in reference to contribute or pose any threat in the cyberworld.

(b) **Privacy Concerns**: In order to conduct the analysis of any given data, AI-driven forensic technologies frequently need access to vast amounts of data. It is crucial to ensure the security and privacy of those data, especially when handling private or sensitive data. Data protection laws like the California Consumer Privacy Act (CCPA) in the US and the General Data Protection Regulation (GDPR) in the EU must be followed while using AI in digital forensics. Unauthorized access to any sort of available data on the Internet or improper manipulation of those available data can compromise the validity of forensic investigations and result in privacy violations [7]. The lack of transparency in AI, sometimes known as the "black box", is another major obstacle. AI systems that are transparent and are able to explain their decision-making procedures must be installed in courts and the court should observe those. This openness in the system makes it easier to comprehend how the AI came to its conclusions, which is essential for stakeholders and legal professionals to evaluate the validity and equity of the AI-generated opinions. The privacy protection of system-generated content is still in doubt if it is transposed to other e-platforms [8, 9].

(c) **Bias and Discrimination**: If AI systems are trained on biased data, they may exhibit bias in their results. The development of impartial, transparent, and equitable AI systems is essential. A thorough analysis of the quality of the data utilized to train AI systems is necessary. Inaccurate or biased training data might produce biased or untrustworthy expert views, which can have serious legal repercussions. Examining factors like data collection techniques, data representativeness, and the existence of any known biases or distortions in the training data are all part of evaluating the quality of the data. Courts should also be informed about current initiatives to lessen prejudice in AI systems and make sure that the AI in use conforms with industry best practices in this area [10, 11].

(d) **International Cooperation**: International collaboration and cooperation are crucial in managing the legal ramifications of AI in digital forensics since cybercrime crosses national boundaries. Standardizing legal frameworks and practices can guarantee uniform use of AI-powered products and ease the transfer of digital evidence. The newly enacted **EU E-Evidence Regulation 2023** [12, 13] is a promising game changer in application and bringing in new possibilities for the domain in countering effectively any threat on way to application of digital forensics. This regulation establishes guidelines that permit public agencies and law enforcement to access stored electronic evidence from service providers situated in a different state. A directive that mandates all service providers-based outside the Union but providing services within the EU to appoint a legal representative in an EU member state supplements the legislation. The International Criminal Court also plays an important role in backing up this EU-Regulations in a regularized standard and balancing the legal plethora of justifying complications raised from various Digital Evidences [14].

(e) **Ethio-Legal Implications:** The inherent complexity of AI systems in digital forensics poses a challenge to interpretability and explainability. When AI systems provide insights or make decisions, forensic analysts, attorneys, and the parties involved must understand how those conclusions were reached. The implementation and utilization of AI in the field will be hampered by a lack of transparency, which might lead to scepticism and mistrust if the understanding of the process and functioning of AI in Digital forensics is unclear. One approach in addressing this challenge and assisting practitioners in comprehending and validating the decision-making process is to develop AI models with integrated interpretability features. Working toward explainable AI in the realm of digital forensics not only increases accountability but also facilitates more fruitful collaboration and cooperation between AI systems and human experts which is ethical and legal at the same time [15].

5 Case Examples of AI-Integration in Digital Forensics

5.1 The Ashley Madison Data Breach

Hackers gained access to Ashley Madison's website in 2015, exposing users' private information. To find the attackers and determine the extent of the breach, the investigation into this incident involved a great deal of digital forensics effort. In order to determine the attack's origins, examine the hacker's techniques, and evaluate the effects on the impacted parties, forensic teams had to sort through massive amounts of data. The importance of digital forensics in responding to data breaches and comprehending cyberattacks and restoring cybersecurity is demonstrated by this example [16, 17].

5.2 The Chicago Police Strategic Subject List (SSL)

The Chicago Police Department uses a predictive policing technique commonly known as Predictive Analysis Technique, which in their code operative procedure named as the Strategic Subject List (SSL) to find people who might be involved in violent crimes. Nevertheless, research showed that the SSL algorithm had racial biases and preferentially targeted African Americans. This case's ethical issues highlight how critical it is to remove biases not only in AI-driven systems but also in mind of people who curated these technologies and make sure that forensic technologies and forensic experts should create and implement with accountability and fairness in perspective [17, 18].

6 Future of AI in Digital Forensics

Emerging technology and changing study topics will propel major breakthroughs in artificial intelligence (AI) in digital forensics in the future. AI is anticipated to be crucial in improving forensic procedures as digital forensics continues to adjust to the growing complexity and volume of data. All those working in the field of digital forensics would benefit from a better prospectus. These important areas where AI is anticipated to have an impact include technology advancements and integration with national and international law. The proposed future goals are as follows:

- **Advances in AI Technologies**: The development of AI technology in itself is one of the main areas of progress. The capabilities of AI in digital forensics could be further improved by recent advancements in machine learning techniques, especially in deep learning and neural networks. For example, the accuracy of data processing and pattern recognition can be increased by using more complex neural network structures, such as Transformers and Generative Adversarial Networks. These developments will make it possible for AI systems to recognize and categorize digital evidence more accurately, even in situations that are unclear and intricately grave [19].

- **Improved Data Privacy and Security**: Data security and privacy are becoming more and more crucial as AI technologies are included into digital forensics. Future advancements in AI will need to take up ethical data protection issues into account. Data privacy may be greatly aided by methods like federated learning, which enables AI models to be trained across several dispersed devices without exchanging raw data. Furthermore, secure data processing and analysis will make it possible by developing in cryptographic techniques like homomorphic encryption, guaranteeing that private data is safeguarded during the forensic procedure [19].
- **Improved Multidisciplinary Collaboration and Co-operation**: The future of AI in digital forensics will be enhanced by more multidisciplinary cooperation and collaboration between forensic specialists, legal experts, and AI researchers and engineers. To guarantee that AI technologies are created with a thorough awareness of forensic requirements and legal norms, cooperation between these domains is crucial. AI models used in digital forensics, e.g. there is a need for training a model with unbiased datasets so that it can generate results in real time that are both interpretable and admissible in court of law. In order to combat future digital crime, it is anticipated that the proposed regulations, directives, and other developed laws of the European Union will be integrated into Indian criminal law in addition to concentrating on developing technological advancements [12, 13].
- **Addressing Ethical and Legal Challenges**: Addressing ethical and legal issues will continue to be a top priority as AI develops. To avoid biases that can influence forensic results, it is crucial to guarantee fairness and transparency in AI systems. Additionally, in order to handle emerging issues with data ownership, privacy, and accountability, the legal structure governing the use of AI in forensic investigations must change. Future studies must examine how to incorporate ethical issues into the creation of AI and set rules for its use in forensic settings [19].
- **Continuous Learning and Adaptation**: Continuous learning and adaptability will help AI systems in digital forensics stay up to date with changing technology and dangers. Maintaining the efficacy of forensic tools will require machine learning models that can update themselves in response to fresh data and new trends. AI systems will continue to be useful and able to handle new problems in the quickly evolving field of digital forensics if mechanisms for continual learning are put in place and are updated on a regular scenario [20].

7 Conclusion

The integration of AI into digital forensics has transformed the field of cyber investigations. With its ability to automate tedious tasks and provide complex analytical solutions, artificial intelligence has emerged as a crucial tool for forensic specialists. Its capacity to accelerate investigations and deepen analysis has significantly increased the effectiveness of digital forensic processes. As long as technology continues to advance, it is anticipated that the application of AI in digital forensics will result in groundbreaking advancements. Deep learning, a subset of machine learning, is expected to be essential.

The development of increasingly sophisticated algorithms with contextual analysis and delicate pattern identification capabilities will lead to more accurate and efficient digital forensic investigations. The use of AI in digital forensics is also heavily influenced by ethical issues. Important concerns regarding privacy, accountability, and the possibility of abuse are brought up by the application of AI [21]. Upholding ethical norms in forensic practice and preserving public trust depend on AI systems being developed and deployed ethically. In order to overcome these obstacles and enhance the potential of AI-driven digital forensics, it will be essential to conduct continuous research and development. Moreover, the law that govern the whole social ecosystem for the protection of its people should also bring in new changes from time to time for combating against every predicament of cybercrime and also being an aid to digital forensic experts, legal practitioners, and the people.

References

1. Sachowski J (2016) Understanding digital forensics. In: Implementing digital forensic readiness. Syngress. ISBN 978-0-12-804454-4
2. Shimonski R, Zenir J, Bishop A (2015) Digital reconnaissance and surveillance. In: Cyber reconnaissance, surveillance and defense. Syngress, pp 1–44. ISBN 978-0-12-801308-3
3. Carrier B (2002) Digital forensic examination and analysis tools. In: Digital forensic research conference. Syracuse, New York, USA
4. Badman A What is digital forensics? https://www.ibm.com/think/topics/digital-forensics#:~:text=Digital%20forensics%20is%20the%20process,with%20criminal%20and%20civil%20investigations
5. Li C-T, Grobler M (2014) Digital forensic readiness: are we there yet? JICLT 9(3):173–179. ISSN 1901-8401
6. Revolutionizing digital forensics: India's new legal frontiers; column; bar and bench. Published 27th July, 2024
7. Shen C, Bai S (2020) AI for digital forensics: a review. IEEE Trans Inf Forensics Secur 15:2032–2045
8. Holzinger A, Biemann C, Pattichis CS, Kell DB (2017) What do we need to build explainable AI systems for the medical domain? arXiv preprint arXiv:1712.09923
9. Russel S (2021) Artifical Intelligence: a modern approach, 4th edn. Pearson
10. Surden H (2014) Essay. Machine Learning and Law. Wash L Rev 89:87
11. Goodfellow I, Bengio Y, Courville A, Bengio Y (2016) Deep learning. The MIT Press, p 800. ISBN: 0262035618

12. Regulation (EU) 2023/1543. https://eur-lex.europa.eu/legal-content/EN/TXT/?uri=CELEX%3A32023R1543
13. Directive (EU) 2023/1544. https://eur-lex.europa.eu/legal-content/EN/TXT/?uri=celex:32023L1544
14. Dillon M (2014) Electronic courts and the challenges in managing evidence. a view from inside the international criminal court. Int J Court Admin 6(1). ISSN 2156-7964
15. Monika K, Monga D (2023) Role of AI in digital forensics. IJLPSR 5(4):121–123
16. Casey E (2019) Digital forensics and investigations: people, process, and technologies. Elsevier
17. Yadav RT (2024) AI in digital forensics. IJSRET 10(4). ISSN (Online): 2395-566X
18. Liu Z, Zhang T (2018) Ethical challenges in AI-driven digital forensics. J Dig Foren Pract 10(2):112–120
19. Barocas S, Hardt M, Narayanan A (2019) Fairness and machine learning: limitations and opportunities. MIT Press
20. Li X, Qin Z (2022) Blockchain technology in digital forensics: challenges and opportunities. IEEE Access 10:1–14
21. Yampolskiy RV (2019) Artificial Intelligence safety and security. CRC Press

Analyzing the Role of Artificial Intelligence in Shaping Impulsive Buying Behavior: A Study on Consumer Decision-Making in the Online Fashion Retail Sector

Joysri Datta, Shweta Kishore, and Sanchita De

Abstract Online fashion business is flourishing and the use of various cutting-edge technologies is impacting consumer purchase behavior. Impulse buying as we know involves spontaneous and unplanned purchases which is a common phenomenon across online shopping platforms. Most of the times, it is a result of a complicated mix of several factors. In the context of online fashion retail, our research aims to explore the factors that are driven by Artificial Intelligence (AI), and affects the impulsive buying habits of customers. The current study uses a quantitative method and a structured questionnaire to gather data from a group of different online fashion buyers. A regression analysis is performed to capture the impact of various factors such as personalized recommendations, virtual try-ons, advertising strategies, chatbots, and navigation ease in determining the impulse buying behavior of online fashion shoppers. The study explores the mental processes behind impulsive decisions by analyzing the interaction between consumer purchase decisions and different AI-driven stimuli. Various retail organizations can use the potential of AI technologies to make better marketing strategies and boost sales, while consumers become increasingly aware about how AI-related technologies may affect their options for purchase.

Keywords Impulse buying behavior · Online fashion retail · Consumer decision-making · Artificial Intelligence · Fashion industry · Personalized recommendations · Promotional strategies

---

J. Datta (✉) · S. Kishore
Department of Business Administration, Institute of Engineering & Management, University of Engineering and Management, Kolkata, India
e-mail: joysri.datta@gmail.com

S. De
Department of Basic Science and Humanities, Institute of Engineering & Management, University of Engineering and Management, Kolkata, India

M. Chakraborty et al. (eds.), *Proceedings of Sixth International Ethical Hacking Conference*, Lecture Notes in Networks and Systems 1513,
https://doi.org/10.1007/978-981-96-8632-2_18

1 Introduction

Technology has become a very vital factor across various industries influencing consumer choices, especially in the digital marketplace. Varied emerging technologies have brought huge changes in the online retail landscape, especially in the fashion segment by offering more personalized experiences, increase in operational efficiency in operations, and enhancing consumer engagement. In recent times, emerging technologies like Artificial Intelligence (AI) is being employed in the Indian online fashion retail to forecast consumer preferences, estimate trends, and speed up procedures, all of which has a huge impact on consumer purchase behavior. AI-powered recommendation systems, virtual assistants, and automated marketing tools have transformed how customers interact with online fashion platforms. Various AI enabled systems utilize large amounts of consumer data to provide customized suggestions that match individual tastes and preferences thus creating an environment suitable for impulsive purchases by making consumers available to a wide variety of products they might not have actively thought of buying but are probably going to buy on an impulse.

In the present scenario, change in customer buying preferences, has led many retailers to shift toward consumer-centric operations. Artificial Intelligence (AI) enabled technologies plays a crucial role in enhancing customer modeling by automating the acquisition and analysis of vast amount of customer data, thereby giving rise to the development of personalized customer models (CMs). CMs can provide customized recommendations and improve customer experience that align with evolving needs of customers and support decision-making [1]. AI helps consumers to shop better by providing tailored suggestions based on their likes and dislikes and browsing history. It has the capability to suggest products that match individual tastes, making the shopping experience more efficient. AI-powered chatbots are quite prompt to offer instant assistance, answering questions and guiding users through the purchasing process [2]. The integration of AI enabled strategies to retail e-commerce websites has led to a positive influence on the impulse buying behavior by various customers. Better AI services result in improved user experiences, which can trigger impulsive purchasing behavior. Customers with higher hedonic motivations (those seeking enjoyment) are more affected by system quality, while those with higher utilitarian motivations (focused on practical benefits) respond more to information quality. AI has the potential to facilitate the shopping experience for both the category of customers in different ways. It creates an enjoyable and engaging shopping experience for customers who derive enjoyment and pleasure from shopping by enhancing the system quality thereby leading customers to purchase various items spontaneously. On the other hand, for utilitarian customers, who emphasize more on real-time benefits, AI improves the information quality available on shopping platforms. This allows them to make informed decisions quickly and purchase items on an impulse whenever they find lucrative offers or products that meet their criteria [3].

The Indian online retail fashion segment is growing rapidly, with many e-commerce platforms such as Amazon, Flipkart, and Myntra gaining popularity. These platforms offer an extensive range of trendy fashion products and apparel, meant for a wide range of consumer tastes. Factors such as increased internet access, rising disposable incomes, and a growing desire to follow global fashion trends are responsible for the fame of such online retail platforms. The incorporation of AI to these e-commerce platforms has not only enabled them to run more smoothly, but it has also brought in a change the way people shop for clothes. In a country so diverse like India, the potential of AI to satisfy varied tastes, preferences, and affordability among various customers has got a significant outcome on the consumer purchase behavior. AI helps companies to create customized shopping experiences, therefore facilitating the search for products that fit their preferences.

The essence of Artificial Intelligence (AI)-powered digital assistance (AIPDA) in luxury brand's sector is also significant. It offers an exciting shopping experience to consumers, which has already been proven in various studies [4].

According to research by Jain and Gandhi [5], Artificial Intelligence significantly influences consumer purchase decisions by providing personalized product recommendations, improving interaction with products, and providing detailed product information. The study found that among Indian online fashion shoppers, product recommendations, and interactive experiences are quite pertinent in driving impulsive buying behavior, while the impact of product information was found to be minimal. It also suggested that adopting advanced technologies like smart mirrors and self-checkout systems provide a better shopping experience and boost impulse purchases [5].

A research by Bhagat et al. [6] examined the role of technology in determining consumer purchase intentions in the online retail sector. Their study focuses on the way AI positively impacts customer satisfaction and their buying behavior while identifying key factors that encourage its adoption. Factors like subjective norms, faith, and consciousness have a huge impact on the shopping intentions by various online users, making technology more available and user-friendly. Simplifying processes like product searches and comparisons is another forte of AI which makes the overall shopping experience a hassle-free journey, creating greater convenience and satisfaction for consumers. The positive relationship between AI and consumer behavior leads to improved trust and an increased desire to buy from e-retailing platforms that utilize AI technology [6].

A study conducted by Chauhan et al. [7] describes the change brought in the online purchasing habits specially in the fashion industry due to the pandemic. The study also discovers different elements such as hedonic shopping value and positive emotions, revealing important insights into impulse buying behavior. Here, three key independent variables that influence impulse buying behavior has been taken into consideration namely: Fashion Involvement (FI), Hedonic Shopping Value (HSV), and Sales Promotion (SP). Positive Emotion (PE) plays a crucial facilitating role in the relationships between these independent variables and impulse buying (IB). Specifically, PE enhances the effects of HSV and SP on IB, stating that positive emotions can influence consumers to make impulsive purchases. It was also found that, FI alone was found to be insignificant, its impact on IB becomes significant when mediated by PE, emphasizing the importance of emotional responses in consumer purchasing behavior during the pandemic [7].

A research article highlights numerous critical factors that drives impulsive buying in online shopping environments, notably discounts and promotional offers being primary motivators. The urge for instant satisfaction and the attempt for enhancement of mood profoundly influence consumers' tendency to engage in impulsive buying. It has also been found out that demographic factors like income have a huge influence impulsive buying behavior, as individuals with higher income levels tend to display a greater inclination toward impulsive purchases. Other demographic factors such as age, gender, education, and occupation are also responsible for variations in impulsive buying patterns among customers, suggesting that both behavioral and psychographic influences, along with demographic characteristics, shape consumers' online shopping behaviors [8].

The impact of AI on impulsive purchasing behavior in India's online apparel retail business is apparent. AI delivers highly tailored shopping experiences by utilizing complex algorithms and data analytics, encouraging consumers to make impulsive purchases. While Artificial Intelligence has the potential to improve customer pleasure and streamline the shopping experience, it also raises serious ethical concerns about manipulating consumer behavior. Understanding how AI influences impulsive buying behavior is critical for both merchants and customers as they navigate the increasingly digital retail market.

2 Literature Review

According to a recent study, various modern technologies can be utilized to improve sales for the modern-day retailers. The paper focuses on how several AI-related technologies such as smart mirrors, virtual rooms, one-tap payments, and self-checkout systems can enhance the customer shopping experience. These technologies boost several factors like searching for the required products faster thereby effectively saving time and making the purchasing process hassle free. In today's competitive fashion retail landscape, retailers can encourage impulsive buying behavior of customers by using such modern AI-related innovations, which ultimately results into higher sales [5].

Researcher explores the purchase decisions of Instagram shoppers' using the Engel-Kollat-Blackwell (EKB) theory. It appraises the purchase behavior in the fashion segment and evaluates the impact of perceived eWOM, emotional value, quality, price, and risk. The research applies the Partial Least Squares Structural Equation Modeling technique (PLS-SEM) to evaluate the model proposed, which describes 42.1% of variance in purchase decision. The result indicates that the purchase decision gets positively impacted by perceived eWOM, emotional value, quality, while perceived price and risk do not have any significant impact.

The research also emphasizes the role of AI in enhancing personalized shopping experiences on Instagram [9].

The online retail industry has grown rapidly, necessitating the use of Artificial Intelligence (AI) to identify consumer preferences and demands, enabling businesses to customize products and services to meet unique customer needs. The research assessed the influence of Artificial Intelligence (AI) on consumer buying behaviors and decisions, encompassing preferred payment methods, expenditure values, online shopping frequency, and employment of electronic gadgets [10].

Artificial Intelligence has had a major impact on e-commerce and online consumer happiness, which has raised customer satisfaction and brand trust. The present study examines technological developments that augment client experiences, with a particular emphasis on e-satisfaction and its influence on purchase intentions. The framework of this review can be applied to empirical research. Provides an overview of the several AI technical advancements that enhance the online fashion retail customer experience. AI tools and processes have greatly affected the e-commerce sector and client satisfaction with online buying [11].

According to the industrial life cycle, the e-commerce sector, which is characterized by e-commerce platforms, is presently in its growing stage. Since online shopping is already one of the biggest trends in the contemporary digital economy, it is critical to monitor consumer behavior data and promptly forecast market shifts. Online transactions will be one of the key elements of the digital economy, and Artificial Intelligence will be the primary driver of digital economy's growth in the future [12].

The intention to make an impulse buy is strongly and favorably impacted by attitudes toward technology and perceived utility and simplicity of use in relation to AI [13].

The researcher studied the impact of Artificial Intelligence on the purchase intent of consumers in the area of e-retailing. The result indicates that the AI enables ease of operations, which influences subjective norms, faith, and consciousness in a positive manner, which as a result significantly enhances the purchase intentions. Consumer confidence is improved by AI, creating a seamless virtual experience and encouraging trust in online purchases. The study indicates that AI reduces cost of purchase and search time and at the same time also provides personalized alternatives which improves the complete shopping experience. In recent times, consumers choose AI enabled online retail websites, which shortens product search time, product comparison, decision-making, resulting in increased satisfaction and improved intention for purchases [6].

The study observes the influence of different factors, concentrating on demographic (income), behavioral, and psychographic drivers like mood, discounts, and instant gratification on impulsive buying in online shopping. Though discounts and promotions significantly impact impulsive purchases, they create challenges for sellers as they result in customer churn once the offer ends. The study suggests that impulsive buying should be promoted through product variety and quality to ensure satisfaction of customer, rather than relying on deceptive discount practices.

At the same time, various factors like ease of purchase, product authenticity, and speed of the delivery are identified for future research [8].

The research explores the impact of Artificial Intelligence (AI) on Generation Z's purchasing decisions and brand trust. A survey of 224 respondents revealed that AI exposure and accuracy perception significantly influence brand trust, which in turn affects consumer behavior. The findings underscore the importance of understanding AI's role in customer interactions [14].

The retail industry has been at the forefront of the AI revolution, implementing AI technologies extensively to transform consumer interactions, supply chain management, and business performance. Artificial Intelligence (AI) is revolutionizing the marketing landscape by enabling significant advancements in customer engagement, personalization, and operational efficiency. AI is changing how businesses interact with customers by providing previously unknown insights into consumer preferences and behaviors [15].

The development of e-commerce and Artificial Intelligence's role in enhancing algorithms for increased consumer interaction demonstrate how these technologies have the ability to advance e-commerce by making it more accessible and tailored to each user's demands.

The way AI-powered marketing tools affect customer behavior highlights the intricate relationship between psychology and technology. AI analysis of a consumer's online footprint is used to create personalized marketing techniques that subtly influence their purchase decisions [16].

3 Research Gap

i. Limited research on how different AI-driven marketing strategies impact impulsive buying behavior across various online shopping platforms for fashion products.
ii. Most studies focus on isolated AI elements (e.g., chatbots) without considering their combined impact on consumer behavior.
iii. Limited research on how AI caters to diverse markets like India, with unique cultural and economic contexts.
iv. Lack of research exploring the interaction between AI-induced impulsive buying and demographic/behavioral factors (age, gender, income, etc.).

4 Objective of the Study

The main purpose of this present study is to compute a composite AI-induced impulsive online fashion product purchase index which will quantify the effect of AI-driven determinants on the impulsive buying behavior of the customers in online fashion material shopping. This research targets to identify and measure the impact of 17 AI features, incorporating personalized recommendations, limited-time offers, virtual try-ons, AI chat-bots and retargeting ads on purchasers' tendency to make unintended purchases. By accumulating these AI dimensions into one composite index, the study tries to provide insights into how AI assistance shapes the unplanned buying decisions of the buyers. The empirical results of the study also let the online retailer know the right strategies to enhance customer engagement and sales.

5 Research Questions

i. How do age, gender, and income affect the efficiency of AI-powered marketing tactics in encouraging impulsive fashion product purchases on online retail platforms?
ii. How can behavioral characteristics, like consumer spending habits and the frequency of online purchases made by customers, moderate the effect of AI-generated offers on impulsive fashion product purchases on online retail platforms?

6 Research Methodology

A survey-based research design was adopted to collect primary data with the help of structured questionnaire. Non-probability convenience sampling was employed to collect 127 responses. Closed ended questions were used to measure variables like, age, gender, education, income, and online purchase behavior. Hypothesis were framed to analyze the relationship between variables. Regression analysis was employed to assess the impact of independent variables (gender, age, education, employment status, monthly income, monthly expenditure, frequency of online purchase, education, and gender) on the dependent variable (AI-induced impulsive online fashion purchase behavior).

7 Data Analysis

This section covers the statistical analysis of the gathered data with an emphasis to identify relationship between the variables. The interpretation of the regression analysis provides insight of the significant factors influencing this recent tendency.

7.1 Estimated Results

Table 1 shows that 64.6% of the sample are male, the remaining of the sample comprises of female population (Fig. 1).

Table 2 represents that 45.7% of the sample respondents belong to the age group of (25–34) years. About 18.9 and 18.1% of sample belong to the age groups of (35–44) and (18–24) years, respectively. Around 11% of sample are in age group of (45–54) years. Remaining is aged above 55 years (Fig. 2).

Table 3 projects that 57.5% of the sample hold Master's Degree, while 21.3% have the Bachelor's Degree. About 9.4% of the sample have been awarded PhD degree. Around 0.8% of sample have gathered post-doctorate experience. Only 11% of sample have education up to high school level (Fig. 3).

Table 1 Gender of the respondents

Gender	% of sample respondents
Male	64.6
Female	35.4
Prefer not to say	0

Source Authors' calculation

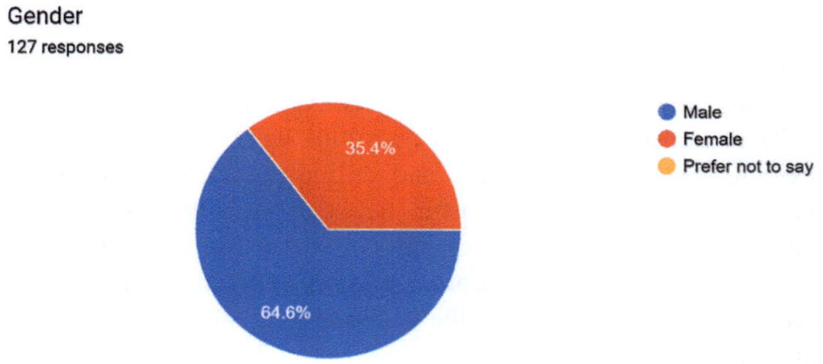

Fig. 1 Gender of the respondents

Table 2 Age of the respondents (in years)

Age (in years)	% of sample respondents (%)
18–24	18.1
25–34	45.7
35–44	18.9
45–54	11
>55	6.3

Source Authors' calculation

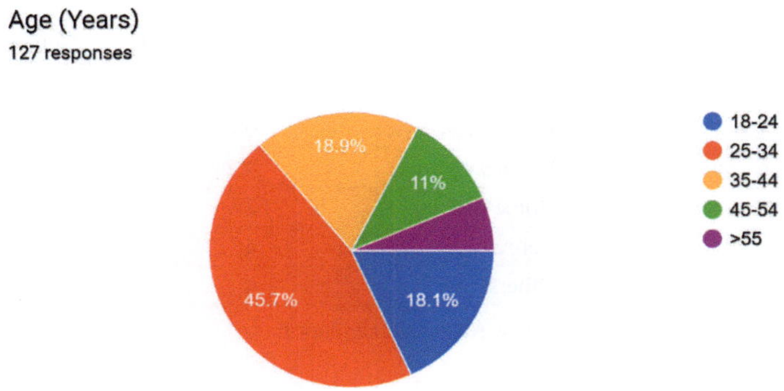

Fig. 2 Age of the respondents

Table 3 Education level of the respondents

Education level of the respondents	% of sample respondents
High school	11
Bachelor's degree	21.3
Master's degree	57.5
Doctorate	9.4
Post-DOCTORATE	0.8

Source Authors' calculation

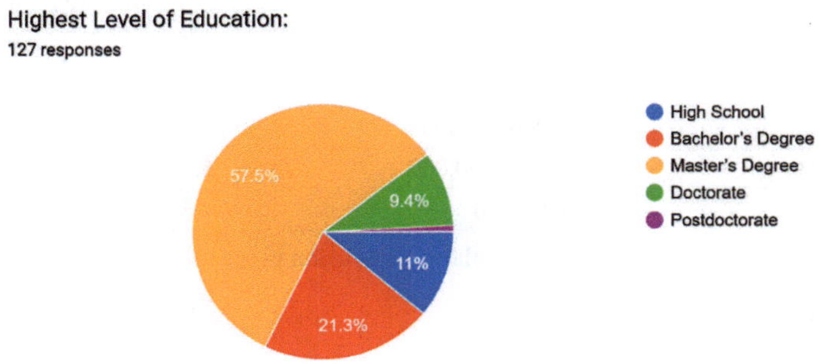

Fig. 3 Education level of the respondents

Table 4 shows that 68.5% of the sample having serviced in formal sector. About 17.3% of sample are students, pursing education. Around 4.2% are running their own business. Rest are either home-maker, retired or self-employed (Fig. 4).

Table 5 represents the monthly income of the respondents. About 27.6% of sample respondents earn above INR 1 lakh per month. 19.7% earn month income below INR 25,000/month. 14.2% of sample fall in the monthly income range of INR (50,000–75,000). Remaining earn monthly income in between INR (75,000–100,000) (Fig. 5).

Table 4 Occupation of the respondents

Occupation	% of sample respondents
Students	17.3
Employed/service	68.5
Business owner	4.2
Home maker	2
Retired	4
Others	4

Source Authors' calculation

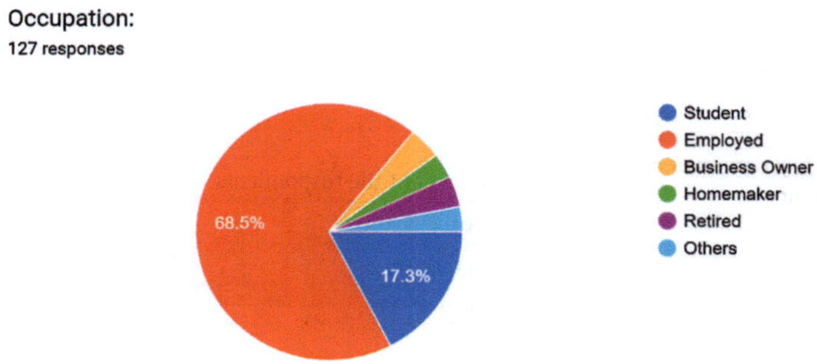

Fig. 4 Occupation of the respondents

Table 5 Monthly income (in INR) of the respondents

Monthly income (in INR)	% of sample respondents
<25,000	19.7
25,000–50,000	14.2
50,000–75,000	22.8
75,000–100,000	15.7
>100,000	27.6

Source Authors' calculation

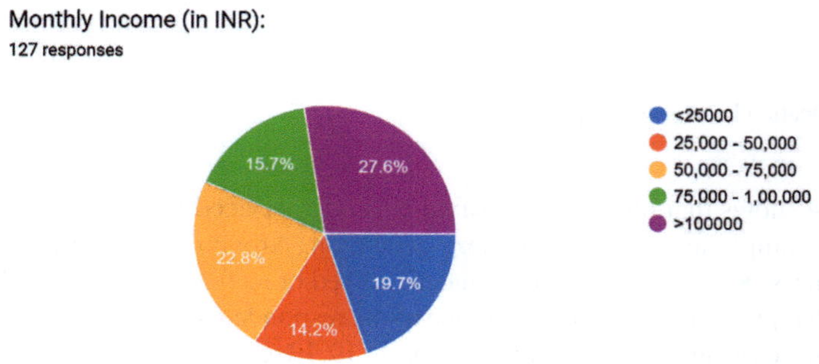

Fig. 5 Monthly income of the respondents

Table 6 Frequency of online purchases by respondents per year

Frequency of online purchases	% of sample respondents
Rarely	21.3
Occasionally	35.4
Frequently	31.5
Very frequently	11.8

Source Authors' calculation

Table 6 presents the overview of frequency of online fashion product purchases per annum. 21.3% of the sample buy online fashion product once or twice in a year. About 35.4% of sample respondents occasionally purchase online fashion product, may be 3–4 times in a year. 31.5% of sample frequently purchase online fashion items. Rest are purchasing online fashion goods very frequently, almost every month in a year (Fig. 6).

Table 7 shows that 40.9% of the sample use to spend INR 2000 per month on an average to purchase fashion product through online platform. Near about 38.6% of sample spend INR 2000–5000 on online fashion purchases. 14.2% have the spending of INR 5000–10,000 on online fashion products through online shopping (Fig. 7).

Figure 8 represents the preferred online fashion retail platform by the respondents.

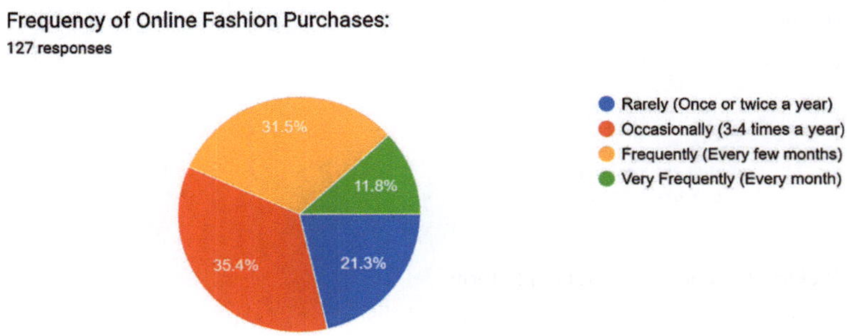

Fig. 6 Frequency of online purchases of the respondents per year

Table 7 Average spending on online fashion purchases per month (in INR) by respondents

Average spending on online fashion purchases per month (in INR)	% of sample respondents
<2000	40.9
2000–5000	38.6
5000–10,000	14.2
10,000–15,000	2.3
>15,000	4

Source Authors' calculation

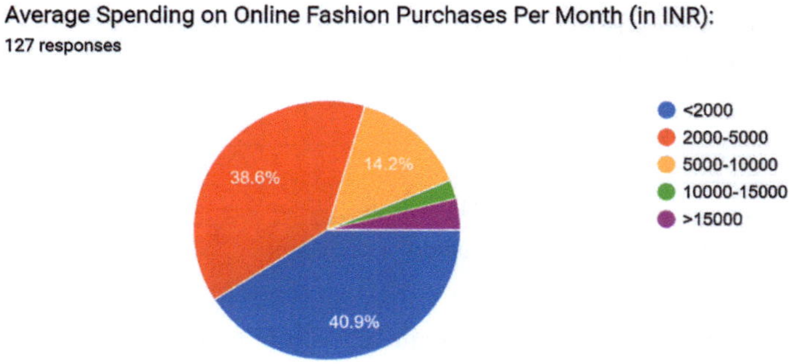

Fig. 7 Average spending on online fashion purchases per month (in INR) by respondents

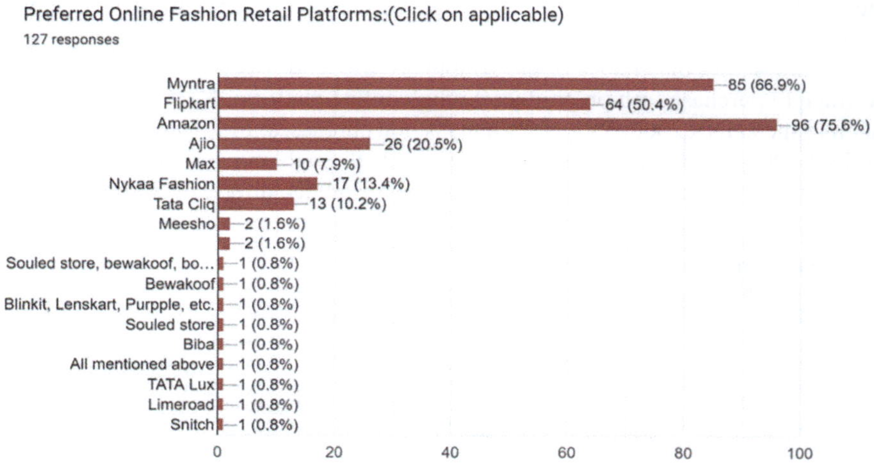

Fig. 8 Preferred online fashion retail platform

7.2 Hypotheses

Hypothesis 1: As employment status gets improved, the AI-induced impulsive online fashion purchase increases.

Hypothesis 2: The AI-induced impulsive online fashion purchase increases with the increase in monthly income of the respondent.

Hypothesis 3: Respondents having higher monthly expenditure are likely to have larger amount of AI-induced online fashion purchases.

Hypothesis 4: Young aged customers are more impulsive to make unplanned buying decisions being triggered by AI stimulation.

Computation of Composite AI-Induced Impulsive Online Fashion Purchase Behavior Index

Composite AI-Induced Impulsive Online Fashion Purchase Behavior Index is computed as the simple average of total 17 dimensions of AI-Induced Impulsive Online Fashion Purchase Behavior of customers. This estimated index satisfies all the properties of composite index. The 17 dimensions of AI-Induced Impulsive Online Fashion Purchase Behavior of customers are:

1. AI-generated personalized recommendations (e.g., "Recommended for you" section) influence the decision to make an impulsive purchase of the buyer.
2. The buyer is more likely to buy a fashion item impulsively when AI shows him/her limited-time discounts or offers.
3. Virtual try-on features (e.g., seeing how clothes fit) provided by AI increase the tendency of the buyer to make unplanned purchases.
4. Product recommendations provided by AI powered chatbots and virtual assistants prompt the buyers to go for purchases which are not previously planned.
5. Marketing techniques such as AI-based advertisements and retargeting (e.g., displaying previously viewed items) happen to influence the buyer to purchase fashion products on an impulse.
6. Creating personalized shopping experiences such as providing customized search results makes the buyer to explore products related to fashion and buy them which they would not have planned to buy under usual circumstances.
7. The potential of AI to predict and display products matching the buyer's preferences is yet another reason that increases the probability of the buyers to purchase fashion related products on an impulse.
8. The recent trend of various AI-powered online fashion sites to provide suggestions based on browsing history or past purchases seems to have a strong influence on buyers to act and buy on an impulse.
9. It is a matter of excitement for the buyers when AI-related marketing techniques highlight the trending fashion items whenever they are opting to buy certain stuffs.
10. It is worth mentioning that AI-generated suggestions make the buyers go for purchasing the interested products according to their preferences and taste.
11. Prompts like "only a few items left," "limited time offer," create a mindset of urgency in buyers to buy fashion related items on an impulse.
12. AI features make the buyer feel more satisfied with his/her impulsive purchases compared to purchases made without AI assistance.
13. The buyer is more likely to make an impulsive purchase when AI shows positive reviews and ratings for a fashion item.
14. The buyer often buys fashion items online without prior planning or intention when exposed to AI-related technologies that prompts them to purchase.
15. The buyer makes more frequent purchases due to AI features in online fashion platforms.

16. The buyer enjoys interacting with AI features such as chatbots and personalized recommendations, which encourages impulsive buying.
17. The buyer feels satisfied with my impulsive purchases when AI plays a role in helping him/her to choose a fashion product.

7.3 Regression Analysis

Composite AI-Induced Impulsive Online Fashion Purchase Behavior Index = f (Gender of the respondent, age of the respondent, education level of the respondent, employment status of the respondent, monthly income of the respondent, monthly expenditure of the respondent, frequency of online purchase per month) (Table 8).

8 Findings

1. The regression analysis shows that AI-induced impulsive online fashion purchase behavior of the customers is highly influenced by age, employment status, monthly income, monthly expenditure of the respondents.
2. Age of the respondent has its positive significant impact on AI-induced online fashion product shopping. People in higher age groups tend to be more exposed to AI applications and more influenced by it to raise their online fashion product shopping in more impulsive manner. Hypothesis 4 has been proven false.
3. Employment status of the respondent is positively affecting the AI-induced impulse online shopping nature of the customers. The possible explanation of this outcome is that the employed persons are likely to hold smart devices in which very frequently the AI shocks via online shopping platform occur which may lead to sudden purchase of fashion products by the customers. Hence, Hypothesis 1 has been proven to be true.
4. Monthly expenditure of the respondents is also having positive significant impact on AI-assisted online purchases of fashion products by customers. All the respondents who are having significantly larger expenditure, their unplanned online fashion material buying induced by AI will be higher. Major portion of their monthly spending will comprise of their impulsive shopping expense encouraged by AI shocks. Hence, Hypothesis 3 has been verified to be true.
5. Another notable result obtained from this regression analysis is the significant impact of monthly income of the respondent on their unintended online shopping decision making. The inverse relation between monthly income of the consumer and their sudden fashion product purchase is obtained, indicating that unexpected online buying induced by AI falls with the rise in the monthly income of the consumer. When income in hand is less, the online fashion product purchasers will be more influenced by AI provided various offers and discount. They may think that it is better to buy the products which are in offer, leading to their

Table 8 Regression table

Source	**SS**	**df**	**MS**	Number of observations = 127		
				F(7, 119) = 3.52		
Model	18.3993056	**7**	2.62847223	Prob > F = 0.001		
Residual	88.9832899	**119**	0.747758739	R-squared = 0.1713		
Total	107.382596	**126**	0.852242822	Adj R-squared = 0.1226		
				Root MSE = 0.86473		
Composite AI-induced impulsive online fashion purchase behavior index	Coefficient	Std. err	t	$P > \lvert t \rvert$	[95% Conf. interval]	
Gender dummy	−0.1546276	0.1822906	−0.85	0.398	−0.5155812	0.2063259
Age of the respondent	0.0149759*	0.0086979	1.72	0.088	−0.0022469	0.0321987
Education level of the respondent	−0.004592	0.0379294	−0.12	0.904	−0.079696	0.0705121
Employment status of the respondent	0.668434***	0.2500212	2.67	0.009	0.173367	1.163501
Monthly income of the respondent	−0.0000149***	3.27e-06	−4.54	0.000	−0.0000214	−8.39e-06
Monthly expenditure of the respondent	0.0000397*	0.0000222	1.79	0.076	−4.29e-06	0.0000837
Frequency of online purchase per month	0.3393209	0.2681426	1.27	0.208	−0.1916282	0.87027
Constant	3.16381***	0.5510971	5.74	0.000	2.072583	4.255037

Note ***, **, *: Significant at 1, 5, and 10% levels, respectively
Source Authors' calculation

impulsive purchase of products of different fashion brands. It may be the situation that those buyers are not having any plan to buy the product, but AI generated offers induce them to buy it immediately. They usually think that if they miss the offer, they may not be able to afford those products in future when they will be in actual need for them. Thus, they make a reserve of those products through their sudden purchase provoked by AI assistance. Hence, Hypothesis 2 has been confirmed to be false.

6. Frequency of online purchase per month also has positive impact on AI-suggested Impulsive Online Fashion Purchase nature of the buyers, though its level of significance is less. More a buyer purchases fashion product through online platform, he/she will be more induced by AI generated offers to have impulsive shopping.
7. Education of the respondent negatively affect unintentional online shopping tendency of the buyer being influenced by AI proposals but this effect is not a significant one. As a buyer is more educated, it is expected that he/she will be less influenced by different AI suggestion to take sudden online purchase decision.
8. The regression does not find out any significant impact of gender of the respondent on impulsive online fashion purchase decision-making, directed by AI.

9 Managerial Implications and Suggestions

i. **Target High-Income, Infrequent Shoppers**: These customer groups are more vulnerable to AI-induced impulse purchases.
ii. **Personalize Offers**: AI should be used to tailor offers on the basis of age, income, employment, and spending habits.
iii. **Create Urgency**: Employ AI to highlight time-limited offers and exclusive deals.
iv. **Optimize Mobile Experience**: Focus on mobile apps for personalized offers and push notifications.
v. **Educate Consumers**: Although AI is a powerful tool for marketers, it is important to educate the customers about the negatives of impulsive buying.
vi. **Monitor and Adjust**: Consumer behavior can be continuously tracked to refine and adjust the future strategies.

10 Conclusion

The present study highlights how AI-driven technologies like personalized recommendations, virtual try-ons, and chatbots, significantly affect impulsive buying behavior on online fashion retail platform. Through analyzing the correlation between consumer decision-making processes and AI stimuli, this current research provides insights into the psychological triggers behind making unplanned purchases by buyers of fashion products on online fashion selling platform.

The regression analysis shows that AI-induced impulsive online fashion purchases are highly influenced by factors like age, employment status, monthly income and expenditure. Older respondents and employed customers are more likely to make unplanned purchases due to their frequent exposure to AI-driven stimuli. Higher monthly expenditure increases AI-induced shopping by the customers, while lower income forces buyers to respond to AI-generated offers by frightening them to miss discounts. On the other hand, education is having negative impact on impulsive buying, although insignificantly, while gender has no impact. Frequent online shoppers are temperately influenced by AI, and the regression results confirm Hypotheses 1 and 3, while disproving Hypotheses 2 and 4.

The main policy suggestion emphasizes that retail companies can employ these AI-technologies to optimize their marketing strategies and boosts up their sales if the consumers are more informed about the impact of AI on their purchasing choices. Finally, the study features the dual role played by AI in providing benefit to the retailers and in empowering consumers with greater awareness in online fashion product shopping.

References

1. Pereira AM, Moura JAB, Costa EDB, Vieira T, Landim AR, Bazaki E, Wanick V (2022) Customer models for artificial intelligence-based decision support in fashion online retail supply chains. Decis Support Syst 158:113795
2. Xu Y, Shieh C-H, Van Esch P, Ling I-L (2020) AI customer service: task complexity, problem-solving ability, and usage intention. Australas Mark J 28(4):189–199. https://doi.org/10.1016/j.ausmj.2020.03.005
3. Zhu Y, Shi H, Hashmi HBA, Wu Q (2023) Bridging artificial intelligence-based services and online impulse buying in E-retailing context. Electron Commer Res Appl 62:101333
4. Rahman MS, Bag S, Hossain MA, Abdel Fattah FAM, Gani MO, Rana NP (2023) The new wave of AI-powered luxury brands online shopping experience: the role of digital multisensory cues and customers' engagement. J Retail Consum Serv 72:103273. https://doi.org/10.1016/j.jretconser.2023.103273
5. Jain S, Gandhi AV (2021) Impact of artificial intelligence on impulse buying behaviour of Indian shoppers in fashion retail outlets. Int J Innov Sci 13(2):193–204. https://doi.org/10.1108/IJIS-10-2020-0181
6. Bhagat R, Chauhan V, Bhagat P (2023) Investigating the impact of artificial intelligence on consumer's purchase intention in e-retailing. Foresight 25(2):249–263. https://doi.org/10.1108/FS-10-2021-0218
7. Chauhan S, Banerjee R, Dagar V (2023) Analysis of impulse buying behaviour of consumer during COVID-19: an empirical study. Millennial Asia 14(2):278–299. https://doi.org/10.1177/09763996211041215
8. Rani MM, Catherine S (2023) A study on impulsive buying behaviour in online shopping. Int J Profess Bus Rev. https://openaccessojs.com/JBReview/article/view/1237
9. Yeo SF, Tan CL, Kumar A, Tan KH, Wong JK (2022) Investigating the impact of AI-powered technologies on Instagrammers' purchase decisions in digitalization era–a study of the fashion and apparel industry. Technol Forecast Soc Chang 177:121551. https://doi.org/10.1016/j.techfore.2022.121551
10. Jangra G, Jangra M (2022) Role of artificial intelligence in online shopping and its impact on consumer purchasing behaviour and decision. In: 2022 Second International conference on

computer science, engineering and applications (ICCSEA). pp 1–7. https://doi.org/10.1109/ICCSEA54677.2022.9936374
11. Pillarisetty R, Mishra P (2022) A review of AI (Artificial Intelligence) tools and customer experience in online fashion retail. Int J E-Bus Res (IJEBR) 18(2):1–12. https://doi.org/10.4018/IJEBR.294111
12. Xiong Y (2022) The impact of artificial intelligence and digital economy consumer online shopping behavior on market changes. Discret Dyn Nat Soc 2022(1):9772416. https://doi.org/10.1155/2022/9772416
13. Arachchi HADM, Samarasinghe GD (2023) Impulse purchase intention in an AI-mediated retail environment: extending the TAM with attitudes towards technology and innovativeness. Glob Bus Rev 09721509231197721. https://doi.org/10.1177/09721509231197721
14. Guerra-Tamez CR, Kraul Flores K, Serna-Mendiburu GM, Chavelas Robles D, Ibarra Cortés J (2024) Decoding Gen Z: AI's influence on brand trust and purchasing behaviour. Front Artif Intell 7. https://doi.org/10.3389/frai.2024.1323512
15. Haque A, Akther N, Khan I, Agarwal K, Uddin N (2024) Artificial intelligence in retail marketing: research agenda based on bibliometric reflection and content analysis (2000–2023). Informatics 11(4):4. https://doi.org/10.3390/informatics11040074
16. Lopes JM, Silva LF, Massano-Cardoso I (2024) AI meets the shopper: psychosocial factors in ease of use and their effect on E-commerce purchase intention. Behav Sci 14(7):7. https://doi.org/10.3390/bs14070616

Deepfakes: Navigating Ethical Concerns and Legal Frameworks

Ankita Roy, Shrabana Chattopadhyay, and Shambhu Prasad Chakrabarty

Abstract The term deepfake has been defined by Cambridge dictionary as "*a video or sound recording that replaces someone's face or voice with that of someone else, in a way that appears real.*" These definitions clearly illustrate that deepfakes refer to synthetic media in which a person's voice, appearance, or behavior can very easily be altered and fabricated. However, deepfakes have become a curse and raised many ethical and societal concerns as it has the capabilities to be a threat in cyber security, forgery, defamation, criminal intimidation, sedition and copyright infringement among many other crimes, their widespread use threatens political integrity, privacy, security, and public trust. While various laws and legislations indirectly address some issues of deepfakes, the difficulty persists as the technology is evolving every day, increasing the potential to harm. As a result, existing laws and regulations fall short in adequately remedying the malice. The present research examines the efficacy of existing legal framework to mitigate the legal concerns arising out of deepfakes. For the purpose of the present research, the existing legal framework of China, Europe Union, United Kingdom, and India in relation to deepfakes will be considered and compared through the doctrinal method. The research expects that by addressing the grey areas in the existing legislations, a scope of enactment of a new legislation on the specific area of deepfakes may be introduced.

Keywords Generative artificial intelligence · Deepfakes · Incompetent laws · Accountability · Enactment of specific law

A. Roy · S. Chattopadhyay (✉) · S. P. Chakrabarty
Department of Law (IIIJS), University of Engineering and Management, Kolkata, West Bengal, India
e-mail: shrabana.uem@gmail.com

S. P. Chakrabarty
e-mail: shambhuprasad.chakrabarty@uem.edu.in

M. Chakraborty et al. (eds.), *Proceedings of Sixth International Ethical Hacking Conference*, Lecture Notes in Networks and Systems 1513,
https://doi.org/10.1007/978-981-96-8632-2_19

1 Introduction

Artificial Intelligence (AI) as a technology is one of the most transformative advancements of this century, redefining what machines are capable of achieving. As in the saying of Stephen Hawking "*Everything that civilization has to offer is a product of human intelligence; we cannot predict what we might achieve when this intelligence is magnified by the tools that AI may provide.*" Further warning "*Success in creating AI would be the biggest event in human history. Unfortunately, it might also be the last,*" [1] quite gives us the impression of the extent the AI is capable of going. Deepfake, a prodigy of Generative AI, which itself is a subset of AI, derives from the combination of "deep learning" and "fake," has invoked a whole new level of debate over the subject of advanced technology [2]. To fully understand deepfakes, it is essential to first examine and consider the purpose behind it—why this technology was developed in the first place and what it aims to achieve. In other words, we must consider the men's rea behind the creation and deployment of deepfake technology [3]. At its core, deepfake technology is designed to deceive and to trick people by providing false information with the intention of creating a false impression or hiding the truth in order to manipulate someone perception, actions, or decisions. It is a tool capable of altering and customize original content and create hyper realistic images, videos and audio recording through original content using deep learning algorithms. It can replace one person's face with another, manipulate facial expression and synthesize speech, blurring the lines between reality and fake. Arguably, the success of this technology is measured by its ability to manipulate perception and influence decisions which can also be said as the wrongful gain of this technology. As a result, the general public falls victim to wrongful loss as they trust something that's not true, in fact, is entirely fabricated [4]. As Bill gates cautioned "*Humans should be worried about the threat posed by artificial intelligence*" [5]. In 2017, the term deepfakes first emerged when a video featuring an actress named Gal Gadot, engaging in a pornographic content went viral. However, it was eventually revealed that the face of the actress used in the video was used by an anonymous user of the social media platform Reddit, who referred himself as deepfakes. This incident served as an eye opener moment in the public awareness of the potential risks associated with this technology [6].

1.1 Evolution of Deepfakes Malice

The concept of machine learning, a segment of AI embodies the remarkable potential of a computer to constantly evolve and improve itself without being explicitly programmed to do so [7]. Deep learning, a subset of machine learning, has made an evolutionary impact over the segment of deepfakes. From the inception of photoshop to recent breakthroughs in Generative Adversarial Networks, the ability to manipulate photos and videos has evolved significantly [8]. As a result, today, the most challenging aspect of deepfakes is to detect authenticity. Earlier, in the past, viewers could easily distinguish between what was real and what was fabricated. However, advancements due to machine learning have made it a thousand times harder for a naked eye to identify deepfakes, blurring the lines between reality and deception [9]. This potential became starkly evident in 2018 in the case of former US President Barack Obama that came into the limelight. A deepfake video, which was created by Jordan Peele, an American actor and comedian went viral, wherein he created a video involving the former president Barack Obama, by manipulating his own facial movements onto Obama's facial characteristics using technology, showing the uncanny capabilities of facial movements deepfakes [10]. In the same year, the human race learned about the concept of synthetic media through the incident of CNN journalist, Jim Acosta. In this incident, a video was circulated showing Acosta to get aggressive to one of the staff of the White House, leading to the decision of revoking Acosta's White House press pass [11]. In the age, which is dominated by digital media, the potential of deepfakes in manipulating through visual content quite raises ethical concerns. Thereafter in 2019, the evolution of deepfakes technology continued to advance through face swapping techniques. A mashup video went viral where the identity of Rashmika Mandanna, a famous Indian actress, was surrealistically merged up into an explicit content without her consent [12]. By 2020, the advent of puppet deepfakes marked a new phase in deepfake advancement. The Industrialist Ratan Tata was showcased in a video ostensibly providing investment advice, manipulating his likeness to promote risk-free investment endeavors that made it difficult to distinguish between fact and fiction [13]. In 2021, the development advanced further with the introduction of audio deepfakes, which was notably included in the Anthony Bourdain documentary. This advancement raised more questions regarding authenticity and manipulation in the visual and aural domains by demonstrating how voices may be synthesized to impersonate actual people [6].

1.2 Contemporary Issues and Challenges

In recent years, the prevalence of deepfakes has increased significantly. Initially, this tool was limited primarily to enthusiasts being exploring a new technology. However, over time, this application has now expanded its use in various sectors including but not limited to political manipulation, phishing scams, revenge, and economic pornography, among other crimes [14]. However, several factors contribute to the rise of deepfakes. The global coverage of the internet has allowed access to vast resources and to virtually perform any activity, enabling its use to a wide range of purposes. Additionally, the widespread availability of high-quality data from social media platforms provides a large dataset to be trained from [15]. This allows AI models to enhance the quantity of the generated results. Advancements in hardware, with the ability to process massive database, have further fueled this trend, allowing manipulation of high-quality data [16]. User interaction also gives valuable insights that further allows to improve the results. Moreover, the affordability of accessing

AI technology has evolved significantly. 1950, which began with the creation of first computer have expanded significantly by 2024 [17]. Lastly, the proliferation of open source deepfake generative tools, allowing more access to the broader audience.

According to the report of *Deeptrace*, a cybersecurity company, the number of deepfakes videos available online surged from 7,964 in 2018 to 14,678 in 2019 [18]. Figure 1 illustrates the websites and apps that provide the general public with easy access to the deepfakes technology. Each of these platforms provides unique functionalities. For instance, *DeepSwap* allows users to generate deepfake videos, photos, and GIFs. Replace creates deepfakes videos by applying one's face with movie characters. Face *Swap Live* swaps one's face with other in no time. *Deepfakes Web* generates high quality deepfake videos [6]. *Facemagic* creates deepfake videos by applying one's face on movies characters. *Avatrify*, a face animation app that brings images to life. *DeepBrain* generates AI videos based on provided scripts. *AnyoneSwap* allows users to replace one face with another [19].

This research paper is structured into five parts. The first part provides an overview over the evolutionary impact of deepfake technology, highlighting the significant advancements and the potential risks. The second part of this paper while exploring the process of creating deepfakes deals with the application of deepfake technology, exploring both the positive and negative effect across various sectors. The third part of the paper addresses the legal issues surrounding deepfake technology, focusing on challenges related to evidence and jurisdiction. It examines how deepfakes complicate the legal process, especially in terms of evidence authenticity and jurisdictional concerns when crimes span multiple countries. The fourth part of the paper focuses on the legal framework for deepfakes in India and compares it with international approaches. It outlines the current gaps in Indian laws and highlights the limitations in addressing the unique challenges posed by deepfake technology. The fifth part

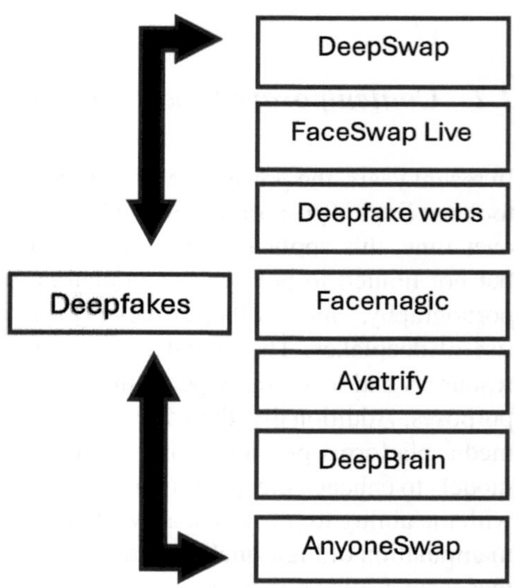

Fig. 1 Websites and apps easily available to create deepfakes

concludes by synthesizing its findings and emphasizing the urgency of regulatory reforms. As deepfakes have advanced and transformed from being a novel technology into tools that hold capabilities to present serious threats by manipulating reality, the question of trust in digital era poses serious challenges. The potential to create hyper realistic content without even the knowledge and consent can result into severe damages to one's privacy and reputation. The current legal framework contains gaps that have resulted into serious exploitation by users and as a result have failed to adequately address the threats posed by deepfakes. In order to ensure individual rights and social values, an urgent need for the development of legal frameworks becoming a priority. However, before delving into these engagements, it is essential to understand how deepfakes are generated.

2 The Process

To understand how deepfakes are created, it is essential to grasp the concepts of machine learning and deep learning as these are the bases of deepfake. Machine learning is the ability of a computer to learn patterns from data and enhance its performance without being explicitly programmed to do so. However, initially, these systems are trained to perform specific tasks and therefore also known as traditional human-assisted model [20]. Deep learning, on the other hand, is a subset of machine learning that involves a neural network, which is basically a computational unit that mimics the function of a biological neuron of the brain, process large amount of high-quality data and automatically learns patterns, without human intervention [21]. For instance, in AlphaGO Zero, the basic rule of the complex board game Go was initially taught, which makes it a classic example of machine learning as it relied on human input. However, through that data, the system began to evolve innovative strategies autonomously and became a grandmaster without human intervention. This shift from human assisted to self-optimization is where deep learning came into play [22]. Similarly, deepfakes function through deep learning without human intervention by utilizing various models which include Autoencoders and General Adversarial Network:

Autoencoders (1980).

An autoencoder is an artificial neural network designed to capture the essential features of the input data for the purpose of reconstructing and recreating an output result. The initial process begins with providing two input images: the first image is the one that is integrated into the second image, which serves as the base image. This process goes through various stages. The first stage being called encoder which extract key characteristics from the input images like the shape, texture, and other essential features. These compressed features of both the input images are then stored in the latent base which is the second stage. In the final stage, the decoder takes the compressed representation from the latent space and attempts to reconstruct the original input. Now after Image A is encoded into the latent space, it is decoded alongside decoded Image B, resulting in a reconstructed image that combines features from both Image A and Image B [23].

Generative Adversarial Network (2014).

Generative Adversarial Network (GAN) works with two neural networks: a generator and a discriminator. Both networks based on knowledge of neural network work together with the ultimate objective of generating data that closely reflects the characteristics of the original data. The generator creates realistic looking data (the illusion), while the discriminator on the other hand evaluates the output produced, keeps giving feedback like an advisor. This loop continues with the generator refining the output based on the evaluation of the discriminator, until the generator achieves the goal of creating exceeding realistic illusion data that becomes indistinguishable from the real one [24].

2.1 Application of Deepfakes: The Benefits and the Risk

The image of a technology is often shaped by its application and its impact on society. Deepfake technology has gained significant popularity due to its capabilities to convincingly manipulate media, which makes it capable of having both negative and positive impacts depending on its interaction with humans. Although deepfake technology shows promising potential for positive outcomes, its dual-edged nature and ease of availability raise significant ethical, legal, and security concerns.

Positive applications of deepfake technology: Deepfakes technology shows promissory potential for creating positive outcomes across various sectors. One of the benefited sectors is the entertainment industry. Filmmakers now can apply this technology to produce, among other things, incredibly convincing visual effects, allowing them to realize creative ideas that were previously not known to them. This includes de-aging actors, reviving performers who have passed away, and producing lifelike computer-generated characters. The potential of this technology to improve storytelling and the cinematic experience for the audience can lead to whole new advancement in the entertainment industry. Furthermore, deepfakes can be useful for simulation and training, especially in fields like the military and aviation. By generating realistic training scenarios that replicate challenging environments, and surroundings, deep training for pilots or soldiers are now possible. Not only this, in medical training also deepfakes can do wonders by making it possible for the learners by visualizing intense medical situations, improving their skills and decision-making abilities without real-world risks [9].

Negative applications of deepfake technology: As mentioned earlier in this paper, the popularity of deepfakes is primarily due to the negative impact. This technology has enabled widespread creation, consumption, and distribution of nonconsensual fabricated pornographic video, resulting in a clear violation of one's privacy. Over 90% of female influencers on Instagram have fallen victim to deepfake pornography in 2024 [9]. Our brains are naturally inclined in a way that we always trust what we see with our own eyes. Deepfakes exploit this natural phenomenon tendency by portraying false content realistically, resulting in the spread of misinformation and reputation damage. Although, the identity of a person is one of the most private possessions, the easy availability of this technology for malicious purposes has threatened that as well which makes it deeply disturbing. The fact that people avail this technology for purposes such as revenge, blackmail, and defamation for one's own advantage or fun, the psychological toll victims face are devastating. Even the judiciary, being the highest authority for delivering justice to innocents, has become vulnerable to the impact of this technology. The possibility of submitting a fabricated evidence—such as fake video and audio—poses serious threat to the integrity of legal proceedings. This not only jeopardizes individual rights but also affects public trust over the judiciary [25].

Beyond individual harm, deepfakes present a serious threat to border security and the political sector by altering audio and video to make it appear as though the subject is saying something authentic, deceiving public to manipulate into believing something that is false. This may result in massive confusion, fear, and political instability [26]. As evident from a recent case during Russia's invasion of Ukraine, a video went viral in which the president of Ukraine-Volodymyr Zelenskyy appeared to urge his forces to surrender before Russia, putting the country in a vulnerable position [27]. Additionally, cybercriminals have taken a beating advantage of deepfake technology to further manipulate people and accomplish their wrongful agendas. Creating fake identities and talking to someone using deepfake audio to mimic someone's voice, creating panic to gain undue advantage has become easier due to this technology. As deepfake technology continues to evolve, the risks associated with its misuse are also increasing, requiring immediate attention. To effectively address and manage this technology, strong strategic legal framework is needed to regulate and ensure the use of this technology for positive applications [6].

Figure 2 represents the growth of deepfake videos available online over two distinct years: 2020 and 2023 [18]. In 2020, the statistic stood at 49,081 deepfake videos. By 2023, this number had significantly increased to 95,820, reflecting a sharp rise in the prevalence of deepfake content. This indicates that the proliferation of such videos has continued to grow at an alarming rate over the years. This statistic raises serious ethical and legal concerns.

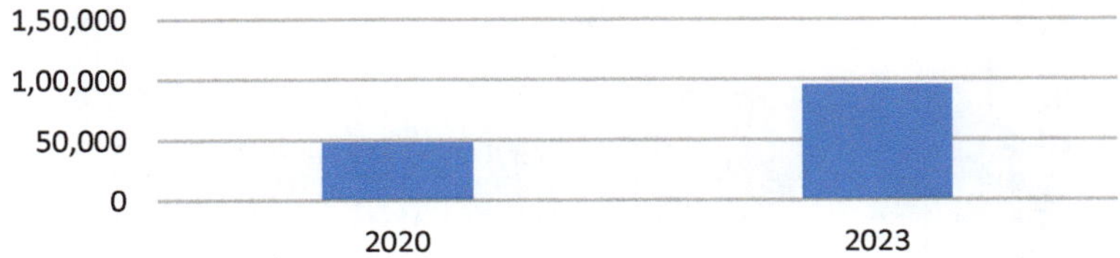

Fig. 2 Deepfake videos available online

3 The Issue of Evidence and Jurisdiction

We have all, at least once in our lifetime, heard the phrase "the judiciary is blind," meaning that judicial decisions are based solely on evidence. Section 3 of the Indian Evidence Act, 1872 (now Sect. 2 of Bharatiya Sakshya Adhiniyam, 2023) categorizes evidence into two types: oral evidence which includes—"*All statement which the court permits is required to be made before it by witnesses, in relation to matters of fact under inquiry; search segment," and documentary evidence which includes—"documents including electronic records produced for the inspection of the courses documents are called documentary evidence.*"[28] In any legal proceeding, the evidence stage holds valuable importance, with the power to move the case in any direction. However, with technological advancements, fraudsters can now use VPNs to conceal their online activities by rerouting internet traffic through remote servers, thereby encrypting the data and making it difficult to trace or interpret. Once deepfake content is uploaded in a social media platform, and people start sharing it, it becomes impossible to find the origin. As a result, the victim really falls in a position of helplessness [29].

Case Study 1 (Political Deepfakes).

Recently, in November 2024, a video surfaced showing Donald trump saying, "*Hello my Pakistani American friend, I promise if I win, I will try my best to get Imran khan out of jail. He is my friend. I love him. I will support him to take over the government again and rock and we will work together to make our trials stronger. Imran Khan zindabad*" [31]. This case spreads all over the world within no time.

Case Study 2 (Financial Deepfake).

In another incident, a UK-based energy company was defrauded of $243,000 when a fraudster used a voice-generating AI software and called the CEO of the UK company

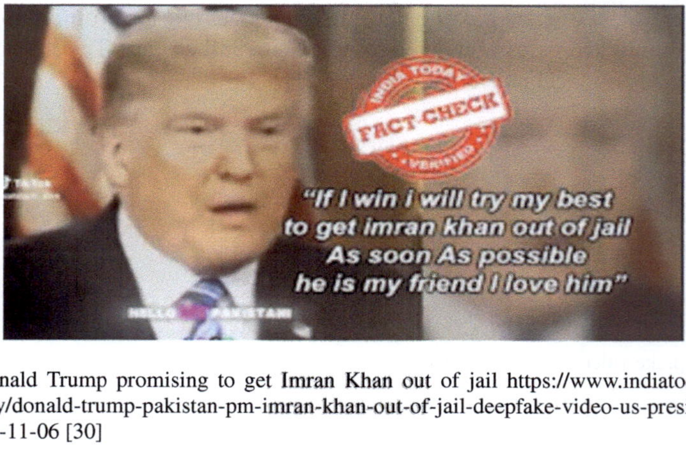

Fig. 3 Donald Trump promising to get Imran Khan out of jail https://www.indiatoday.in/fact-check/story/donald-trump-pakistan-pm-imran-khan-out-of-jail-deepfake-video-us-president-2629371-2024-11-06 [30]

pretending to be the CEO of the Germany-based parent company to facilitate an illegal fund transfer [32].

However, the important question arising from both these cases is: under which jurisdiction would that person be prosecuted? The victims here are from one country whereas the person responsible for this could be from anywhere in the world. Due to world access of this technology, one single offence can span multiple jurisdictions, each with its unique legal framework and regulations. As a result, most of the cases remain unresolved just like the two cases mentioned above.

Now, according to Sect. 197 of the Bhartiya Nagrik Suraksha Sanhita 2023, "every offence shall ordinarily be inquired into and tried by a court within whose local jurisdiction it was committed."[33] Though Sect. 1(4) of the Bhartiya Nyaya Sanhita 2023 provides an exception, stating, "Any person liable, by any law for the time being in force in India, to be tried for an offence committed beyond India shall be dealt with according to the provisions of this Sanhita for any act committed beyond India in the same manner as if such act had been committed within India" [34]. However, this create a complex scenario as countries like India, which do not have specific provision for deepfakes. If the creator of the deepfake content is from a jurisdiction with no laws accountable for such crimes, it may be impossible to pursue legal action.

The principle of universal jurisdiction may give some tranquility as this principle allows a foreign state to prosecute a criminal in their national court for certain offences that affect the community as a whole regardless of where it has been committed such as crimes against humanity, war crimes, genocide, and torture. Deepfake-related crimes must be considered crimes against humanity, given the nature of deepfakes related crimes. It has created complex issues, and to deals with it requires a strong legal framework. Moreover, cooperation and unity globally are required to fill the gaps. Though some countries have implemented laws to deal with deepfake crimes but the issue that they are not globally recognized leaves gaps in its total enforcement [35].

3.1 The Ethical Perspective of Deepfake Technology

Aristotle's ethical framework, as presented in his work "*Nicomachean Ethics*," emphasizes the importance of character and virtues like honesty, justice, and respect to achieve ultimate happiness or well-being [36]. The debate of ethical validity surrounding deepfake technology is due to its nature to appear something look true when it is not. The ethical part becomes more complex when such content is circulated without the consent and knowledge. Moreover, the possibility to use this technology whether for harm, manipulation, or exploitation raises the issue, as it put people's reputations, identities, and mental well-being at significant risk. From both technological advancement and legal perspective standpoint, the technology thought capable to provide positive outcomes, the ethical concerns are much more complex [37].

Our brains are often inclined to trust what we see, even when the content is clearly fabricated. This technology is particularly troubling because it creates a mental resistance to disbelief, making it harder for individuals to accept that what they are viewing is not real. The concern of violating privacy is also a significant issue. According to article 12 of the Universal Declaration of Human Rights, 1948 and article 17 of the International Covenant on Civil and Political Rights (ICCPR), 1966, persons are legally protected against "arbitrary interference" with one's privacy, family, home, correspondence, honor, and reputation. However, deepfake technology has increased the possibility of violating such right [38, 39].

Recently, the Delhi High Court acknowledged the danger associated with deepfakes, passing statement that "*deepfake going to be serious menace in society, antidote of fake AI would be technology only*" [40]. In today's digital world where 63.8% of individuals have social media, the challenge that once a deepfake video goes viral, it becomes incredibly impossible to stop its spread [41]. The saying "lies travel faster than truth" has actually become a reality due to the consequences of deepfakes.

In November 2021, UNESCO adopted the Recommendation on Ethics of Artificial Intelligence, implementing the first ever global standards for AI ethics, binding all 194 member states including India being one of them. This recommendation emphasizes the importance of protecting human rights and dignity while ensuring peaceful living. The core principles include: 1. Proportionality and do no harm; 2. Safety and Security; 3. Right to Privacy and Data Protection: 4. Multi-stakeholder and Adaptive Governance and Collaboration; 5. Responsibility and Accountability; 6. Transparency and Explainability; 7. Human Oversight and Determination; 8. Sustainability; 9. Awareness and Literacy; 10. Fairness and Non-discrimination [42].

However, the misuse of deepfake technology often violates these principles. Deepfakes provide serious security issues by facilitating impersonation, cybercrime, and political instability. They also propagate false information, which goes against the idea of avoiding harm. By utilizing someone's voice or likeness without permission, they violate privacy rights and erode control over personal information. Ethical issues are made worse by the lack of accountability systems to control or track down deepfake abuse. Furthermore, deepfakes violate the transparency principle since they are fundamentally dishonest and due to technology advances it becomes more difficult to reveal that they are synthetic. They avoid the crucial function of human scrutiny by influencing public opinion or judgments.

The ethical obligation is to prevent harm and ensure accountability. Figure 4 shows four parties identified in the responsibility framework being: the user, the developers, the distributors, and the general public.

First and foremost, comes the developers, who can be held responsible for creating deepfake apps and websites without integrating safeguard protocols such as watermarking, which embeds unique identifiers into AI generated content to help trace the origin and ensure authenticity. Secondly, the users who create fraudulent content without giving a thought of the aftereffects. Responsible use of deepfake technology plays a major role in the deepfake ecosystem. The distributors also known as the social media platforms should implement automated AI detection systems and content management tool to able to identify and remove harmful content. Additionally, the

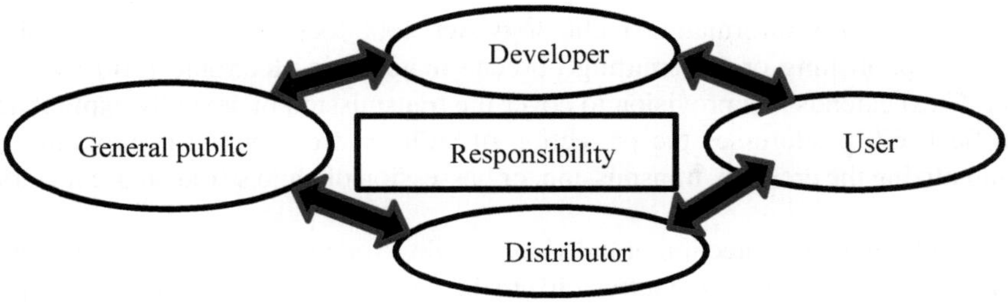

Fig. 4 Parties involved in the responsibility framework

general public, who consume or share deepfake content without critically assessing its authenticity, must adopt a more cautious approach. Addressing the ethical issue posed by deepfakes requires a collaborative effort of all the parties involved to ensure the protection of the dignity of individuals affected by the technology.

4 Legal Framework for Deepfakes in India

At present, unfortunately, Indian regulation does not provide a specific definition of deepfake and neither accommodate any specific law for deepfake related crimes. However, certain existing legal provisions do provide protections that passively offer protection against misuse of deepfake technology. These regulations are not enough to cover the evolving multi-dimensional aspects of deepfake technology. The following discussion explores the major laws that indirectly engages with the malice of deepfakes:

4.(a) Information Technology Act, 2000.

In India, Information Technology Act, 2000 serves as the primary legislation that was formulated to specifically address crime in cybersecurity and electronic commerce. Given that deepfake technology primarily involves computer related offences, this act can serve as a crucial legislation in addressing the misuse of the technology. Section 66(d) provides punishment to those who impersonate individuals through communication device or computer resources with the intent to cheat. It prescribes a penalty of up to three years and fine that can extend to 1 lakh rupees [43]. Since deepfake technology manipulates through impersonating individuals, Sect. 66(d) provides relevant protection in addressing the threat. Crimes related to deepfakes foremost attack the privacy of an individual. Section 66(e) of Information Technology Act, 2000 criminalizes acts that violates the privacy of an individual by publishing and transmitting images of private areas without consent, thereby infringing punishment to an offender for three year or fine up to 2 lakhs rupees [44]. Political manipulation can be covered under Sect. 66(f) which provides punishment for cyber terrorists

[45]. Section 67 of Information Technology Act, 2000 focuses on providing punishment for publishing or transmitting obscene material in electronic form whereas Sect. 67(a) extends this provision to cover the transmission of sexually explicit act, and Sect. 67(b) addresses the protection of children from online exploitation by criminalizing the creation, transmission, or possession of child sexual abuse material [46–48].

Privacy refers to protection of individuals right to control their personal information and prevent unauthorized intrusion into their private lives and while some section of this legislation provides some protection, it remains insufficient to comprehensively address the harm caused by deepfakes, specifically when the victim reputation is at risk. Thus, lack to address deepfake content which may result in incompatibility of the act to fully address the issue evolving deepfake technology. Deepfake crimes are not limited to sexual content or personation thereby holds ability to create content to mislead or harm reputation, credibility, and public image in ways that go beyond obscenity or cheat through impersonating. For instance, deepfakes can be used for political manipulation, cyberbullying, or defamation, none of which are specifically covered by the existing provisions of IP laws, 2000. Additionally, Sect. 77(b) of IT Act, makes offence which are punishable with imprisonment under three years as bailable offences. This leaves the victim in a vulnerable position while failing to reflect the seriousness of the crime [49].

4.(b) Bharatiya Nyaya Sanhita, 2023.

BNS is now the official comprehensive code which covers the substantive aspects of criminal law, replacing the IPC. Section 336 provides punishment to whoever commits forgery through false electronic record which includes digital files, documents, images, emails among others, making it applicable is cases when deepfake offenders uses this technology with deceptive intent to create fake media [50]. Section 356 provides protection against defamation. Deepfake technology creates videos or images that are not real in nature to falsely portray a person and when such videos go viral, it harms the reputation by misleading the public and therefore in such situation, this section might provide some protection [51]. Nowadays, deepfake technology has become a tool to intimidate victims, often used either for revenge purposes or to extort money. Offenders often use fabricated videos or photos, threatens to viral such compromising content to extort money from the victims. In such situations, Sect. 351 which criminalizes criminal intimidation by anonymous communication might be a considerable solution [52]. Additionally, Sect. 79 criminalizes any act that intent to insult the modesty of an women, which are applicable in cases where it is used to particularly harm or degrade female victims [53].

Though BNS does provide provisions to protect one's from deepfake crime, directly it falls short to fully address the broader spectrum of harms posed by deepfake technology. The current framework does not include the scope of cybercrime, the potential of political manipulation, and the gender-neutral impact of deepfake, as it is not limited to any particular gender.

4.(c) Copyright Act, 1957.

Deepfake technology creates audio and videos content from data of others as a result the issue of deepfake when it comes in respect to copyright ownership raises significant issues. Since deepfake technology generates fabricated content by manipulating original data, the question arises whether it leads to copyright infringement? Technically deepfakes require one original content and one base content to morph the characteristics of the original content in the base picture. The scenario is not like it is learning from the data but it is actually using the data of others to create an identical fake content. According to copyright act, the owner of the original data holds exclusive right and therefore one can say that it does lead to copyright infringement [54]. However, another question pops up which is does deepfake generated content even have the right to get copyright ownership. WIPO states that if the deepfake contents are completely contradictory with the victim's life, the deepfake contents should not be rewarded with copyright protection [55]. For instance, platforms like Facebook or Instagram allow users to report copyright infringement if someone posts their original content such as their face or other protected material without consent [56, 57].

4.(d) Digital Personal Data Protection Act, 2023.

Section 2(t) of the act defines personal data as "*any data about an individual who is identifiable by or in relation to such data*" [58]. Section 2(e) defines personal data breach as "*any unauthorized processing of personal data or accidental disclosure, acquisition, sharing, use, alteration, destruction or loss of access to personal data that compromises the confidentiality integrity availability of personal data*"[59]. This act primarily protects the personal data of individuals which works like a fuel for the operation of the deepfake technology. Section 4 of the act emphasizes processing personal data must be fair and lawful, requiring clear, explicit consent from the data principal. Apart from this data can be used for any legitimate use. Since deepfakes often use unauthorized personal data for illegitimate activities, such activities may fall under the violation of this section [60]. Furthermore, Section 9 provides additional protection for the personal data of children or disabled persons by ensuring consent from the parents or legal guardian before processing personal data [61]. Section 14 introduces safeguards for the data of deceased or incapacitated individuals by allowing them to nominate a person to protect their rights [62]. Section 16 provides a mechanism for the Central Government to restrict the transfer of personal data outside India, which could offer protection against cross-border misuse of personal data, including in the context of deepfake technology, which transcends geographic boundaries. This section could potentially serve as a legal safeguard against the global misuse of personal data in the creation and distribution of deepfakes [63].

However, Section 3(c) provides scenarios when the provisions of the act will not be applied to [64]. The first clause 3(c)(i) exempts "*personal data processes by an individual for any personal or domestic purpose.*" Now this raises critical concerns: who are the "individual" here? Can a deepfake creator be qualified as an "individual" who illegitimately acquires the data of the data principal? Furthermore, this act has not defined the terms "personal" and "domestic." This lacks clarity as this may lead to deepfake creator misusing personal data of others for unethical personal reasons

[65]. Section 3(c)(ii) further exempts protection of personal data that has been made publicly available by the data principal. This excludes actors and politicians, who, being public figures, from receiving any protection, thereby giving a free pass to the deepfake content creators [66].

4.1 Gaps in the Existing Laws

The primary issue with current legislation is that it was not created expressly to handle the special difficulties presented by deepfake technology. Due to this, current regulations fall short in adequately dealing with the intricate and dynamic problems brought forth by this technology. Additionally, the onus of proof, which often falls on the victims, is also one of the major gaps. Victims of deepfake-related harm are needed to present proof that the content is fraudulent, which is especially difficult in the light of rapid technological advancement that makes it harder to discern between authentic and manipulated media. Furthermore, it is becoming increasingly difficult to identify the people responsible for producing and disseminating fraudulent content due to the increasing complexity of deepfake technology. Traditional investigative techniques and legal frameworks fall short in keeping up with the increasing realism and difficulty of detecting deepfakes. As a result, the judiciary also fails to take appropriate and effective action due to a lack of concrete evidence and failure to trace and prosecute offenders. This puts victims of deepfake crimes—like identity theft, harassment, or defamation—in a vulnerable position where they can't successfully seek restitution because of existing legal and technological barriers. The absence of adequate legal protection and the challenges in identifying and tracking down the creators of deepfakes worsen the pain experienced by victims. This emphasizes the pressing need for more focused legal changes to address the issues raised by this new technology. It is important to look into the laws available in some countries to deal with this issue for a clearer understanding of the strategies that may be taken to deal with it in India more effectively. Some of the countries which have their sui generis legislation on AI in general and deepfakes in particular include:

4.2 Legal Approach Adopted by Other Countries

China: On January 10, 2023, the Cyberspace Administration of China, the ministry of Industry and Information Technology and the Ministry of Public Security together introduced a five-chapter regulatory framework consisting of 25 articles to govern the misuse of Deep synthesis technology known as "Provision on the Administration of Deep Synthesis of Information-based Information Services or the Deep Synthesis Provisions." China has adopted an approach that encourages healthy growth of technological advancement by mitigating the potential risk associated with deepfake content. Users and service providers of deep synthesis are required to comply with relevant laws and regulations, ensuring they uphold the integrity of digital content and prevent malicious uses. It ensures data security and personal information protection along with transparency, management of content, labeling, and lastly technical security. China's regulatory framework reflects the government's broader vision of controlling the use of advanced technologies, allowing for innovation while preventing the disruptive or malicious effects of synthetic media on social, political, and personal levels [67].

European Union: The European Union (EU) AI Act is the first comprehensive regulatory framework for AI that seeks to govern the development and use of AI in a manner that improves the internal market by prioritizing the fundamental rights to promote and ensure secure and trustworthy ethical AI. Under the EU AI Act, systems that generate and manipulate images, audios, and video content are required to adhere to minimum transparency standards. This includes requirements for clear disclosures regarding the manipulated nature of the content, ensuring that users are fully informed when content has been artificially generated or altered. The Act places particular emphasis on ensuring that these technologies do not mislead or harm individuals or groups, particularly when used to manipulate information in ways that can impact public opinion, elections, or societal trust. Though EU doesn't have a regulatory framework specifically for deepfake content, it seeks to foster responsible use of AI to ensure that the society is also protected from other subsets of AI [68].

UK: The UK has amended their Criminal Code Act 1955 to address the offence relating to deepfake technology. The Criminal Code Amendment (Deepfake Sexual Material) Bill 2024 criminalizes the creation and distribution of deepfake sexual content. The attorney general Mark Dreyfus indicated that this bill is introduced as a commitment to recognize the urgent and collective need to tackle the growing challenges associated with artificial generated sexual material. Additionally, the Online Safety Act criminalizes the sharing of deep fake intimate images. However, both these acts didn't address the issue of gender-neutral impact of deepfakes as well as the fact that deepfake technology is not limited to only sexual orientation [69].

5 Conclusion and Suggestions

Geoffrey Everest Hinton, known as the "*godfather of AI*" regrets the contribution he made in the field of AI. On asking whether he believes if humanity knows what it's doing, he answered "for *the first time in history we have something more intelligent than humans, capable of understanding and having experience of its own in the same sense as humans do*" [70]. AI has evolved into a technology that exceeds human capabilities to fully comprehend, thus, becoming difficult to regulate. One prime example of this is deepfake. The universe of deepfake is vast, yet people have very little understanding of the consequences of this technology. Deepfakes can be described as a dark technology, as the foundation of this technology is based on lie. This technology is capable to play with the brains of the human as it exploits our understanding to differentiate between fake and real. The gap between this technology and its regulatory framework still has miles to cover.

India like most of the countries does not have a focused and specific legal framework for deepfakes. The current legal frameworks that exist were not designed by keeping in mind the consequence of deepfakes and were enacted with different objectives. From the judicial perspective, the main concern regarding deepfake is that it could compromise the public trust in the judiciary, especially when courts struggle to take act due to lack of concrete evidence and jurisdiction challenges. The need for active legal framework to keep pace with technological advancement must be the foremost focus on and for those collaborative efforts among lawmaker to promote responsible AI. For instance, China's approach of providing the creator as well as the users responsibility while using deepfakes can be a potential regulatory model to be adopted for India. Similarly, EU transparency approach, mandating AI generated content to be visibly labelled, offers a practical solution. Additionally, measures like the ones implemented by Netflix—where users cannot take screenshots of content—could be adopted by social media platforms to limit the misuse of deepfake technology.

Though Article 19 of our Indian Constitution gives us the freedom of speech, however, this right must not extend to deepfake generated content that violates one's reputation and integrity. Such harmful content must not be allowed to be posted online [71]. The parties contributing to the negative impact of deepfake such as the developer of deepfakes apps and websites, users who generate negative deepfake content, distributor meaning the social media platforms, and the general public, who often share content without verifying its credibility of the content must take responsibility to mitigate the harmful effect of the deepfakes. Furthermore, the punishment for creating or sharing deepfakes must be harsh enough to frighten potential offenders. As the jurisdiction of deepfakes is not limited, international collaboration is required, with global conventions and agreements being crucial to addressing the problem globally and holistically.

Recommendation for Combating Deepfake Risks

Justice delayed is justice denied. The consequence of deepfake which first emerged in 2017 has long been a concern. To effectively mitigate the threats associated with deepfake technology, a comprehensive legal approach is needed. First and foremost a detailed specific legislation to address the issues of deepfakes, which may include:

1. Enforcement of legal mandates to label AI generated content.
2. Implementation of end-to-end encryption and data protection to ensure strict encryption standards, preventing unauthorized access and manipulation of personal data.
3. Ensuring adoption of IP tagging on AI generated content for easier tracking.
4. Assigning responsibility to creator and distributor of deepfake to ensure possible application of the deepfake regulations.
5. Regulating the freedom of speech to ensure that the freedom doesn't extend to deepfake generated content that violates an individual's reputation and integrity.

6. Adoption of possible AI tools for deepfake detection and removal by social media platforms.
7. Establishment of strict penalties for individuals or entities found guilty of deepfake related offenses to ensure legal accountability.

In addition to these measures, conducting large scale awareness campaigns to educate people on identifying and reporting deepfake content can yield positive results. Furthermore, international collaboration also plays an important role, strengthening the global regulatory framework by establishing conventions and treaties for deepfake control measures will have greater impact. To conclude, while the potential consequences of deepfakes are continuously evolving, the regulatory framework remains severely lagging. Without the implementation of effective action, the risk of deepfakes will only enhance and will keep causing significant harm to individuals and society.

References

1. University of Cambridge The best or worst nothing to happen to humanity Stephen Hawkings launches centre for the future of Intelligence. https://www.cam.ac.uk/research/news/the-best-or-worst-thing-to-happen-to-humanity-stephen-hawking-launches-centre-for-the-future-of
2. Lalla V, et al (2022) Artificial Intelligence: deepfakes in the entertainment industry. WIPO Magazine. https://www.wipo.int/web/wipo-magazine/articles/artificial-intelligence-deepfakes-in-the-entertainment-industry-42620
3. Shekhar S, Ransom A (2023) Ethical & legal implications of deep fake technology: a global overvie. Sci Eng J 11(1):2226–2235
4. U.S. Government Accountability Office (2020) Science & tech spotlight: deepfakes. https://www.gao.gov/assets/gao-20-379sp.pdf
5. BBC News (2015) Bill Gates insists artificial intelligence is a threat to mankind-whatever microsoft research chief says. https://www.bbc.co.uk/news/live/business-33111722?page=4
6. U.S. Department of Homeland Security Increasing threats of deepfake identities. https://www.dhs.gov/sites/default/files/publications/increasing_threats_of_deepfake_identities_0.pdf
7. ReMishra A (2024) A comprehensive review of artificial intelligence and machine learning: concepts, trends, and applications. Int J Sci Res Sci Technol 11(5):2395–6011
8. Wason R (2018) Deep Learning: evolution and expansion. Cogn Syst Res 52:701–708
9. Westerlud M (2019) The emergence of deepfake technology: a review. Technol Innov Manag Rev 9(11):39–52. https://doi.org/10.22215/timreview/1282
10. Ramluckan T (2024) Deepfakes: the legal implications. In: Proceedings of the 19th International conference on cyber warfare and security. https://pdfs.semanticscholar.org/6f93/5d299c7f4f76fad19c5be3c2219e4bef921c.pdf
11. Dixon HB Jr (2019) Deepfakes: more frightening than Photoshop on steroids. Judges J Am Bar Assoc https://www.americanbar.org/groups/judicial/publications/judges_journal/2019/summer/deepfakes-more-frightening-photoshop-steroids/
12. Garg A (2023) Rashmika Mandanna deepfake: 3 years jail, Rs 1 lakh fine, govt sends rule reminder to social media platforms. India Today. https://www.indiatoday.in/technology/news/story/rashmika-mandanna-deepfake-3-years-jail-rs-1-lakh-fine-govt-sends-rule-reminder-to-social-media-platforms-2460104-2023-11-07

13. The Economic Times (2023) Ratan Tata calls out deceptive deepfake video of him recommending investments on social media. https://economictimes.indiatimes.com/news/new-updates/ratan-tata-calls-out-deceptive-deepfake-video-of-him-recommending-investments-on-socialmedia/articleshow/105804628.cms
14. Gambin FA, et al (2024) Deepfakes: current and future trends. Artif Intell Rev 57(3):1–32. https://doi.org/10.1007/s10462-023-10679-x
15. Al-Khazraji SH, et al (2023) Impact of deepfake technology on social media: detection, misinformation, and societal implications. Euras Proc Sci Technol Eng Math (EPSTEM) 23:429–441
16. Vilesov A (2024) Solutions to deepfakes: can camera hardware, cryptography, and deep learning verify real images? vol 1. https://arxiv.org/pdf/2407.04169
17. Computer History Museum. Timeline of computer history. https://www.computerhistory.org/timeline/1950/
18. Ajder H (2019) The state of deepfakes: landscape, threats, and impact. Deeptrace. https://regmedia.co.uk/2019/10/08/deepfake_report.pdf
19. Singh R (2023) Top 10 deepfake websites and apps to create videos online. 91mobiles. https://www.91mobiles.com/hub/how-to-make-deepfake-videos/
20. Kühl N, et al (2022) Artificial intelligence and machine learning. Electron Mark 32(3):1–16. https://doi.org/10.1007/s12525-022-00598-0
21. Nguyen TT, et al (2022) Deep learning for deepfakes creation and detection: a survey. Comp Vis Image Understand 223
22. Chesterman S (2021) We, the robots? Regulating Artificial Intelligence and the limits of the law. Cambridge University Press
23. Gaur L (2022) DeepFakes: creation, detection, and impact. CRC Press. https://doi.org/10.1201/9781003231493
24. Aissa FB (2023) An overview of GAN-DeepFakesdetection: proposal, improvement, and evaluation. Multimedia Tools Appl 83:32343–32365
25. Buo AS (2020) The emerging threats of defect attacks and countermeasures. https://doi.org/10.13140/RG.2.2.23089.81762
26. Gupta S, et al (2024) Deepfake threats: legal challenges in combating AI-generated misinformation. Int J Novel Res Dev 9(2)
27. Wakefield J (2022) Deepfake presidents used in Russia-Ukraine War. BCC. https://www.bbc.com/news/technology-60780142
28. Section 3 of Evidence Act (1872)
29. Cinar B (2023). Deepfakes in cyber welfare: threats, detection, techniques and countermeasures. Asian J Res Comp Sci 16(4): 178–193. https://doi.org/10.9734/ajrcos/2023/v16i4381
30. Saxena S (2024). Fact check: this video of trump promising to grade Imran Khan out of jail is a Deepfake. https://www.indiatoday.in/fact-check/story/donald-trump-pakistan-pm-imran-khan-out-of-jail-deepfake-video-us-president-2629371-2024-11-06
31. Frank BJ (2024) Altered clip shows trump vowing to free Imran Khan from prison fact check. USA Today. https://www.usatoday.com/story/news/factcheck/2024/03/29/donald-trump-pakistan-doctored-clip-nbc-fact-check/73146069007/
32. Damiani J (2019) A voice deepfake was used to scam a CEO out of $243,000. Forbes. https://www.forbes.com/sites/jessedamiani/2019/09/03/a-voice-deepfake-was-used-to-scam-a-ceo-out-of-243000/
33. Section 197 of Bhartiya Nagrik Suraksha Sanhita (2023)
34. Section 1(4) of the Bhartiya Nyaya Sanhita (2023)
35. Philippe X (2006) The principles of universal jurisdiction and complementarity: how do the two principles intermesh? Int Rev Red Cross 88
36. Crisp R (2000) Nicomachean ethics. Cambridge University Press
37. DSCL. Exploratory note on deepfakes and policy considerations. https://www.dsci.in/files/content/knowledge-centre/2023/DSCI%20Exploratory%20Note%20on%20Deepfakes_0.pdf
38. Article 12 of the Universal Declaration of Human Rights (1948)
39. Article 17 of the International Covenant on Civil and Political Rights (ICCPR) (1966)

40. Thapliyal N (2024) Deepfake going to be serious menance in society, Antidote of fake AI would be technology only: Delhi High Court. https://www.livelaw.in/high-court/delhi-high-court/delhi-high-court-deepfake-ai-technology-267918#:~:text=Journalist%20Rajat%20Sharma's%20plea%20states,identity%20theft%20as%20well%20as
41. Statista (2024) Number of internet and social media users worldwide as of October 2024. https://www.statista.com/statistics/617136/digital-population-worldwide/
42. UNESCO Ethics of Artificial Intelligence. https://www.unesco.org/en/artificial-intelligence/recommendation-ethics
43. Section 66(d) of Information Technology Act (2000)
44. Section 66(e) of Information Technology Act (2000)
45. Section 66(f) of Information Technology Act (2000)
46. Section 67 of Information Technology Act (2000)
47. Section 67(a) of Information Technology Act (2000)
48. Section 67(b) of Information Technology Act (2000)
49. Section 77(b) of Information Technology Act (2000)
50. Section 336 of Bharatiya Nyaya Sanhita (2023)
51. Section 356 Bharatiya Nyaya Sanhita (2023)
52. Section 351 Bharatiya Nyaya Sanhita (2023)
53. Section 79 Bharatiya Nyaya Sanhita (2023)
54. Copyright Act (1956)
55. Colak B (2021) Legal Issues of Deepfakes. Institute for Internet & the Just Society. https://www.internetjustsociety.org/legal-issues-of-deepfakes#:~:text=As%20a%20response%2C%20WIPO%20states,to%20the%20inventor%20of%20deepfakes.
56. Help Centre Copyright. https://www.facebook.com/help/1020633957973118/
57. Help Centre How can I make sure the content I post to Instagram doesn't violate copyright law? Instagram https://help.instagram.com/354736791367645
58. Section 2(t) of Digital Personal Data Protection Act, 2023
59. Section 2 (e) of Digital Personal Data Protection Act (2023)
60. Section 4 of Digital Personal Data Protection Act (2023)
61. Section 9 of Digital Personal Data Protection Act (2023)
62. Section 14 of Digital Personal Data Protection Act (2023)
63. Section 16 of Digital Personal Data Protection Act (2023)
64. Section 3(c) of Digital Personal Data Protection Act (2023)
65. Section 3(c) (i) of Digital Personal Data Protection Act (2023)
66. Section 3(c) (ii) of Digital Personal Data Protection Act (2023)
67. China briefing. China to regulate deep synthesis deepfect ecology starting 2023. https://www.china-briefing.com/news/china-to-regulate-deep-synthesis-deep-fake-technology-starting-january-2023/
68. EU AI Act (2023)
69. Ministry of Justice and Laura Farris (2024) Government cracks down on 'deepfakes' creation. GOV.UK. https://www.gov.uk/government/news/government-cracks-down-on-deepfakes-creation#:~:text=Despicable%20people%20who%20create%20sexually,today%20(16%20April%202024).&text=Under%20the%20new%20offence%2C%20those,record%20and%20an%20unlimited%20fine
70. Pelley S (2024) Godfather of Artificial Intelligence. In: Geoffrey Hinton on the promise, risks of advanced AI. CBS News. https://www.cbsnews.com/news/geoffrey-hinton-ai-dangers-60-minutes-transcript/
71. Article 19 of the Constitution of India
72. Ministry of Electronics & IT (2023) Interaction of Minister of Railways, Communications and Electronics and IT Shri Ashwini Vaishnav with stakeholders on issue arising out of deepfakes. https://pib.gov.in/PressReleasePage.aspx?PRID=197904

AI and Law

Machinised Mediation: Apprising the Role of Artificial Intelligence in Existing Dispute Resolution Paradigm in India

Prashna Samaddar and Victor Nayak

Abstract Artificial Intelligence has touched and transformed every aspect of human civilisation, including the dispute resolution scenario, which often is criticised for being inadequate and not justice-oriented due to bias or its cumbersome nature at the domestic level. Mediation against this backdrop, even though not a very new concept for the Indian legal system, its formal introduction through separate legislative enactments or special provisions being made for it has made it an inclusive dispute resolution model. It has brought some respite due to its 'win–win' situation. Still, it has also been criticised for lack of proper skill or effective resolution, especially when an impasse paralyses this 'win–win' model in India. In this context, introducing Artificial Intelligence (AI) creates a positive impact, especially when providing resolutions or helping with the nitty–gritty of the simple yet complicated mediation process. Still, again, the issues of lack of empathy and neutrality prevent experts from relying on AI-generated solutions or interventions. This issue becomes even more acute, especially when the laws on mediation are utterly silent on the scope of AI intervention and the validity of the AI-generated or AI-aided resolutions. The research paper examines this scope of AI intervention in the mediation process by analysing the stages at which it can be integrated to make it more robust. The paper shall also examine the legal challenges in using AI in the mediation process and how it is being done at the institutional level by the various private players offering mediation and ADR services by evaluating the model dispute resolution clauses and guidelines issued by these institutions.

Keywords Artificial intelligence · Dispute resolution system · Amicable settlement of disputes · Mediation · Mediator

P. Samaddar
Xavier Law School, St. Xavier's University, Kolkata, India

V. Nayak (✉)
Department of Law (IIIJS), University of Engineering and Management, Kolkata, India
e-mail: victor.nayak.25@gmail.com

M. Chakraborty et al. (eds.), *Proceedings of Sixth International Ethical Hacking Conference*, Lecture Notes in Networks and Systems 1513,
https://doi.org/10.1007/978-981-96-8632-2_20

1 Introduction

A dispute is a disagreement among two or more individuals or groups. Richard Lempert and Joseph Sanders assert that legal disputes encompass ‘conflicting interests’. Typically, one individual possesses something desired by the other, leading both sides to assert claims of entitlement. If both assertions are irreconcilable, a ‘true conflict’ exists. In most cases, there are three distinct phases to a dispute:

a. Pre-conflict Stage: In the grievance or pre-conflict stage, one feels ‘wronged or injured’. The slight could be imaginary or actual. At this point, the person has to decide whether to continue taking the complaint to a confrontation level or to just walk away.
b. Phase of Conflict: Someone who feels wronged (the aggrieved party) will approach the person or people responsible for the injustice and express how they feel. Both parties are now cognisant of the fact that they are at odds with one another. The party that started the fight may try to defuse it by settling the disagreement, proposing to negotiate, or forcing the other side to accept reality.
c. Conflict Resolution Phase: A third-party step into the argument once the conflict becomes public knowledge. One or both sides or their supporters may ask the third party to step in, or the third party may decide to do so on their own.

Becoming one of the crucial means within the system of the Indian Legal Rules, mediation occupies a special place in the generally accepted principles of ADR. History of mediation in India has laid down requirements and legislation aimed at solving the problem of congestion of the courts and the desire of the opposing parties to voluntarily resolve disputes contributed to the development of mediation in India. The Indian legislation has, over the years, cemented the importance of mediation, although the Commercial Courts Act enacted in the year 2015 clear promoted mediation as the method of first resort for commercial litigations. This change of law aligns with other laws of other countries which consider mediation as better way of solving disputes than confrontation [1].

A comparative analysis of Indian mediation processes with the procedures of other jurisdictions, such as Ireland, reveals important findings concerning the possible directions for the reform and enhancement of the mediation system in India. The paper presented by Manchanda recognises that even though both the countries came under Common law jurisdiction, Ireland has built a more elaborate system of mediation which can act as a model for India. In light of this, the essay seeks to recommend that acquiring certain features from the Irish mediation system would help transform the regulatory structure currently governing mediation practices within India such that more parties embrace mediation as an efficient way of solving their conflicts [1].

Besides, the socio-legal context in the Indian environment presents differing challenges for mediation as well as different opportunities. For instance, in Roychowdhury’s study on domestic abuse law it is clear that other disproportionate extra-legal settlements often emerge in a situation where official legal systems are considered to be inadequate or untrustworthy. This case underscore challenges that surround the introduction of mediation into the legal consciousness of a person especially one who may have little trust in the formal legal systems [2]. These findings highlight the importance of mediation as a process not only as a procedure to replace the existing litigations but also as the one that respect/cultural sensitivity as a mediation process of the people who participate in the judicial processes.

The 'other' bodies and local governance institutions through which and by which mediation is made possible remain paramount. In the seminal work, Banerjee assesses how farm struggles and lawsuits influence the mediation architecture—including kisaan struggles and land fight in the rural areas where cry is likely to be most active. This highlights the role of mediation as a means of social justice for limited input, particularly for marginalised groups in the society, in enabling of government programs interfaced with grassroots campaigning [3]. Further, mediation's integration with personal laws in India, especially for divorce and family related matters makes clear complexity of legal pluralism. Thus, analysing the actual relations between the 'state' law and 'religious personal laws', Saxena proves that 'mediation' appears to be one of the most important facilities which helps to discuss such delicate issues and find a mutual decision without violating cultural and religious beliefs [4].

AI is a rapidly growing field that potentially has the ability to alter many aspects of the social experience with society and specifically the practice of law and legal process including dispute resolution. With the develops in the AI technology, it can be seen that AI might replace human cognition and interaction and raise concern and questions upon the potential application of AI in ADR.

One of the main concerns connected with the integration of AI into ADR is their ability to reduce or even fully remove human presence within the framework of mediation. Mediation mostly relies on the human factor as the latter is often connected with relationships existing during the mediation. Therefore, the use of AI in mediation may be considered as eradicating human touch required to perform the mediation function.

However, there are some potential benefits pinpointed to the use of AI in the process of the Alternative Dispute Resolution (ADR). The era of AI has potential of providing a means to speed up claims processing and provide decision-making in a certain situation which acts as a genuine benefit. Moreover, AI seems to be capable of assessing human feelings and PCs certainly could be used to build and resolve relationships.

Despite these potential benefits, there is concern that parties to a case may be antipathetic to a decision made by a machine, even if the thinking process that underpins it appears to be sound.

One of the key advantages of AI in this context is its ability to replace human thought processes and interactions, enabling faster and more efficient resolution of disputes. However, for those operating in the field of ADR, such forecasts may seem worrisome and even frightening. The question then arises: what could AI likely do in an environment that places a heavy emphasis on concepts such as sense of justice or human interaction, particularly in the context of mediation?

If you peruse most of the literature on mediation and technology, you won't see anything about it in the canonical texts on methods and techniques for mediating disputes. Many of these books were written before the technological revolution of the previous generation really took hold. Interestingly, even books published much later or revised versions of the originals that were published far into the twenty-first century either completely disregard or barely touch upon technology.

2 Conflict Resolution: Understanding the Root of the Problems

Conflict can be defined in various ways. It can be defined as 'opposing ideas and actions of different entities, resulting in an antagonistic and hostile state of affairs'. The clashing of thoughts and ideas is a part of the human experience but if not properly controlled then it can lead to destruction. As outlined in the Mediation Training handbook of the Supreme Court of India, there are certain stages that a conflict goes through and which can even lead to war (Fig. 1).

Thus, it can be seen that *dispute* is something which escalates into *conflict* but starts from some a very trivial matter which can be summarised as *difference.* In this backdrop, it can be said that effective implementation of conflict resolution can lead to positive outcomes. American psychologist Daniel Katz identifies three primary reasons of conflict:

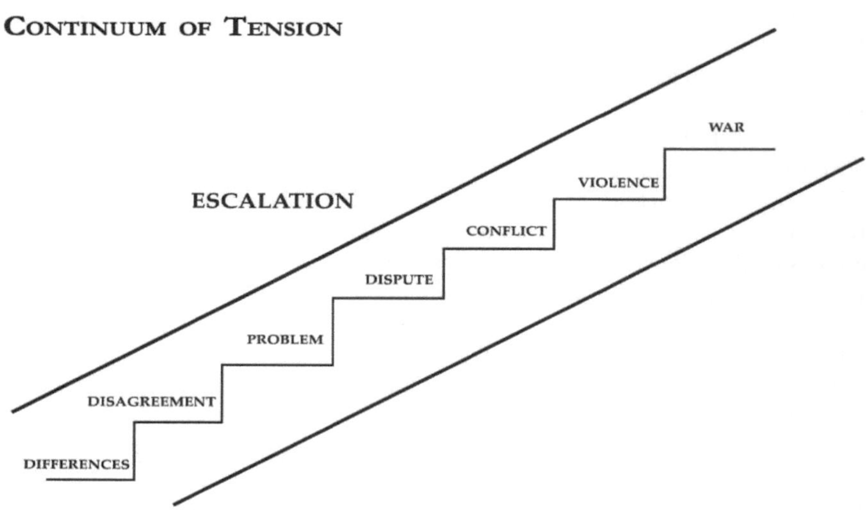

Fig. 1 How 'differences' escalate into 'dispute' and then into "war'

Fig. 2 Categorisation of the types of conflict

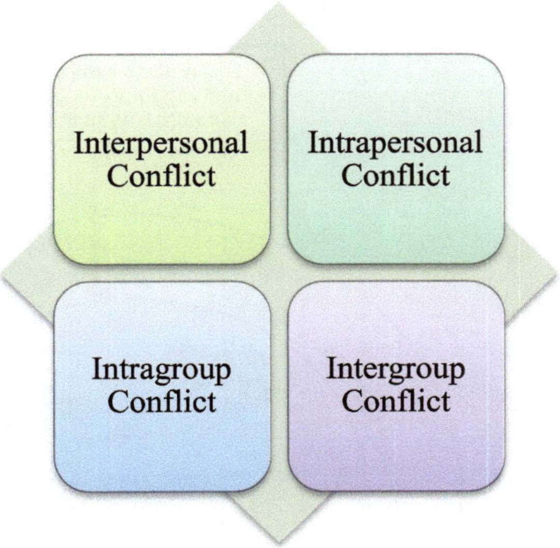

a. Economic Conflict: It arises from resource scarcity, typically including organisations or individuals competing to secure the largest possible share of these resources, resulting in confrontations among the factions involved.
b. Value Conflict: This occurs owing to differing preferences and ideologies that individuals hold as principles, resulting in declarations by distinct groups and expressions of aggressiveness to advocate for their positions.
c. Power Conflict: This scenario arises when the parties concerned seek to optimise their influence within social contexts.

If one tries to categorise the types of conflicts it may be done as shown in Fig. 2.

Thus, it can be said that before delving into mediation it's important to understand the contours of conflict resolution and at the heart of this are certain important dimensions:

a. **Conflict Core and Spiral**: The Conflict Core and Spiral diagram (Fig. 3) depicts the essence of conflicts, which emerge from perceived threats to people, organisations, communities, or nations. This threat can emerge in numerous forms, including dread, suspicion, helplessness, frustration, embarrassment, rage, pain, humiliation, distrust, desperation, and retribution. As a conflict escalates, the original tensions radiate outward, impacting individuals, relationships, tasks, decisions, organisations, and communities. Individual reactions to conflict may result in worry, wrath, animosity, sadness, and retribution, hindering the pursuit of effective solutions. Community reactions and legal counsel may exacerbate tensions and hinder the parties' capacity to manage the situation independently. When a quarrel is publicised, each party adopts inflexible stances and enlists supporters, complicating the resolution of the initial dispute.

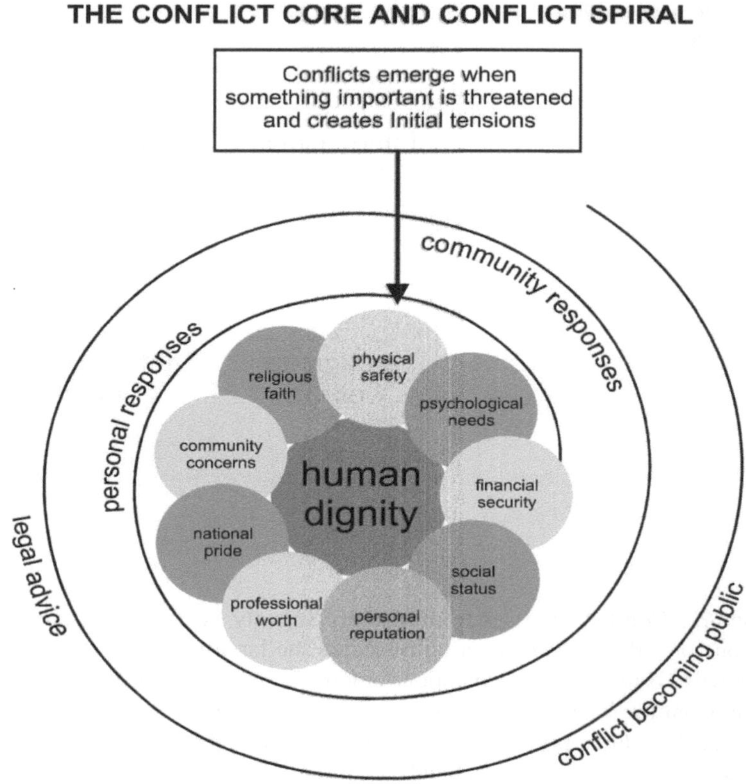

Fig. 3 Conflict core and conflict spiral

b. **The Conflict Triangles**: Illustrated in Fig. 4, delineate a framework for comprehending and resolving disputes, emphasising the dimensions of People, Process, and Problem. The dimensions of the triangle differ among individuals, contexts, and issues. Engaging all parties involved in the triangle can facilitate a more effective resolution of the problem, transforming perspectives and emotions. The initial step in conflict resolution is to ascertain its origin and formulate a strategy to tackle it. This methodology transcends simple problem-solving and conflict management.

c. **Conflict Cycle**: While the mediator's role is to understand the attitudes of individuals towards conflict, focusing on the Conflict Triangle it is important to understand the conflict too as it focuses on the past experiences of the parties including their beliefs as well as attitudes in life. Negative attitudes, such as fear of conflict or avoiding discussing problems, can lead to negative outcomes. The mediator's job is to shift these attitudes to a positive one, ensuring effective conflict resolution. Positive attitudes, on the other hand, encourage growth and learning, leading to peaceful resolutions and improved relationships. Successful

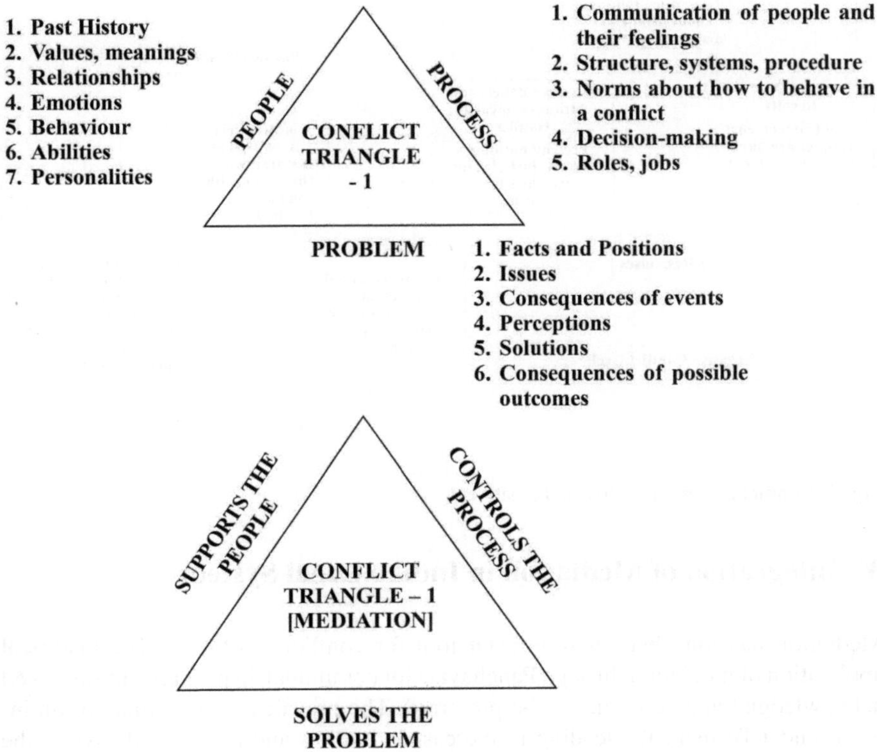

Fig. 4 Conflict triangles

mediation not only solves the current conflict but also helps parties develop positive beliefs and attitudes for future disputes. This transformational approach ensures successful conflict resolution (Fig. 5).

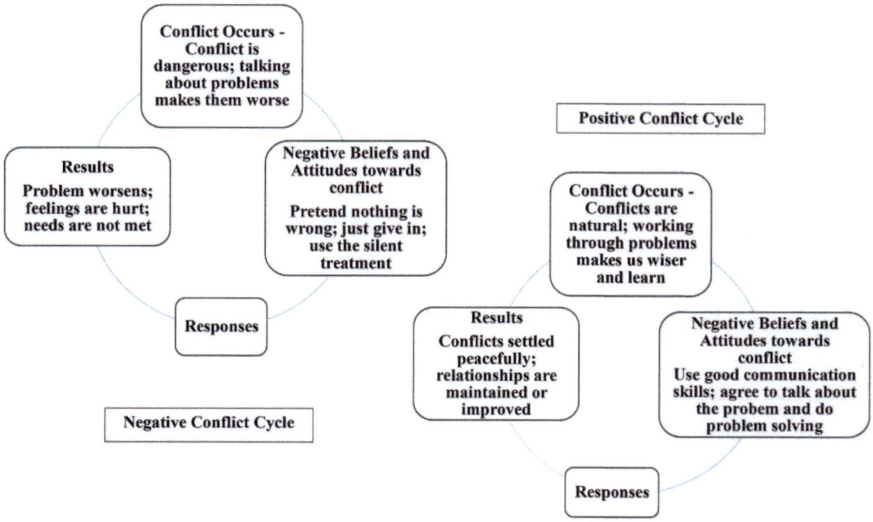

Fig. 5 Conflict cycles—negative and positive

3 Integration of Mediation in Indian Legal System

Mediation has long been a significant tool for conflict resolution. The historical application of mediation through Panchayats for communal dispute resolution is well acknowledged and continues to be preferred. The prevalence of mediation diminished under British rule, leading to increased conflicts and extended delays in the adversarial court system. Mediation has historically been used and is widely utilised in familial conflicts. In January 2020, the Supreme Court established a team chaired by respected mediator Shri Niranjan Bhat to finalise a draft statute that would provide legal validity to conflicts resolved via mediation. It was subsequently submitted to the Union Government as a recommendation by the Supreme Court. Subsequently, the Government published the draft law on mediation on its website for public input on 5 November, 2021.

A summary of mediation case settlements across various centres in India can be seen from the data given in Table 1 (Figs. 6, 7) Also, the scenario of pre institutional mediation is quite dismal as it can be seen from Fig. 8, that only approximately 2% of case (only commercial disputes) admitted to this mechanism reached the settlement stage through the process of mediation and majority of the cases failed at the initial stage itself (i.e. cases where mediation couldn't even begin).

Table 1 Mediation settlement status by major mediation centres

Mediation centre	Cases mediated	Cases settled	Settlement rate (%)
Supreme court mediation centre (SCMC)	1,163	222	20.90
Karnataka mediation centre	61,927	40,854	66
Bengaluru mediation centre	Data not specified	Data not specified	64
Kerala mediation centre	Data not specified	Data not specified	27.7

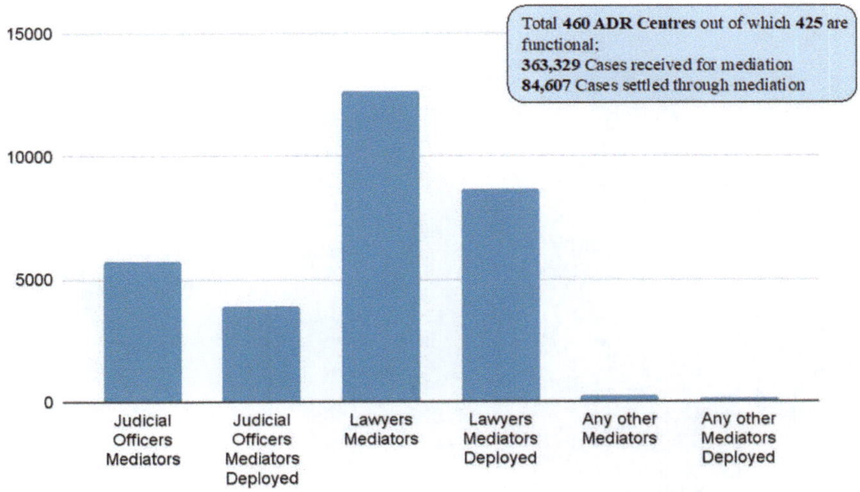

Fig. 6 Total cases received for mediation and its settlement status by the NLSA

The Mediation Bill 2021 was presented in the Rajya Sabha on December 20, 2021, and subsequently submitted to the Standing Committee on Personnel, Public Grievances, Law and Justice for evaluation on December 21, 2021. On July 13, 2022, the Standing Committee published its 117th Report, offering recommendations regarding the provisions of the Mediation Bill. The Rajya Sabha and Lok Sabha enacted the Mediation Bill 2023 on August 2 and 7, 2023, respectively, following the Union Cabinet's approval of certain suggestions. The Mediation Bill, now referred to as the Mediation Act, was enacted by the President on September 15, 2023. On October 9, 2023, the Mediation Act was notified to include provisions related to: (i) non-applicability to Lok Adalat and Permanent Lok Adalat proceedings; (ii) Mediation Council of India; (iii) Mediation Fund; (iv) Short Title, extent, and commencement; (v) Definitions; and (vi) Power to make rules and regulations. The mediation process as provided in Fig. 9 highlights the major aspects of the legislation while the mediation process is undertaken.

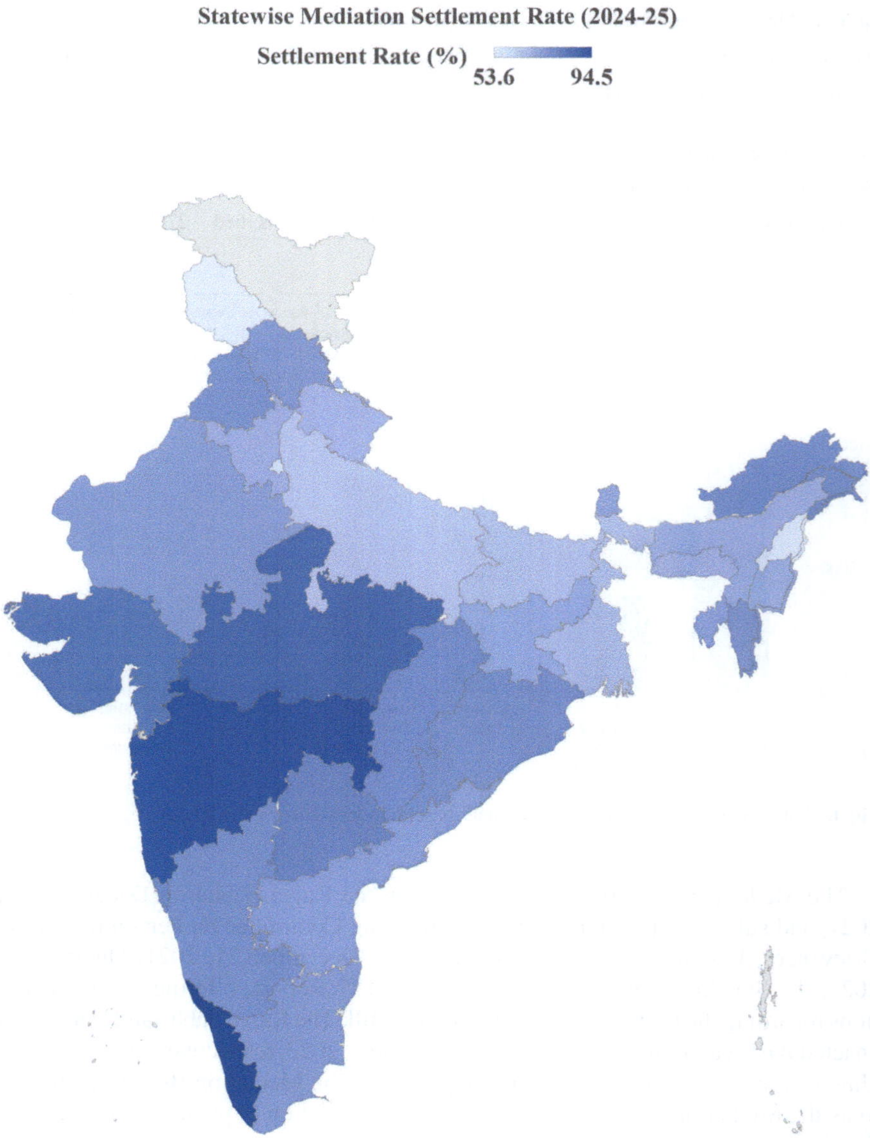

Fig. 7 Settlement ratio of the major ADR centres in India

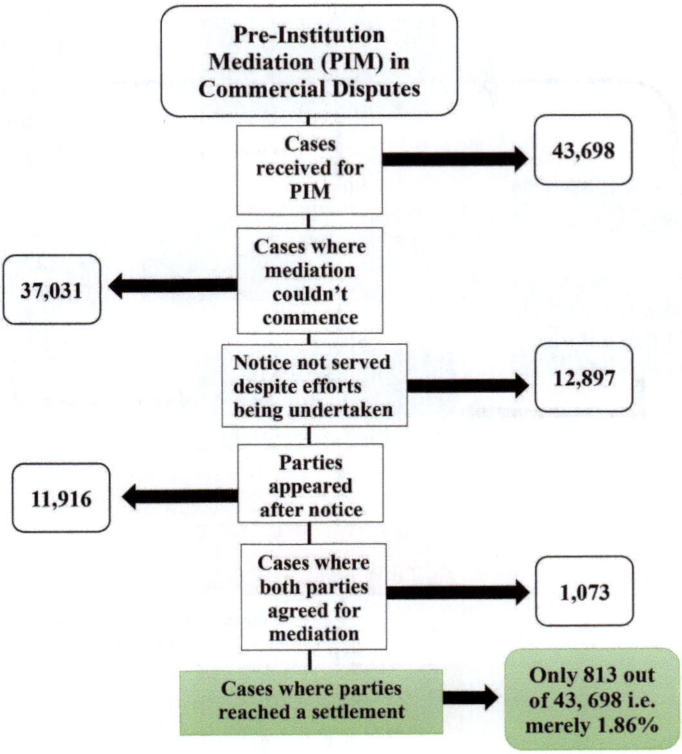

Fig. 8 Status of pre-institutional mediation cases after its recent introduction under Indian legal framework

4 Integrating AI with the Mediation Landscape in India

The incorporation of Artificial Intelligence (AI) into global mediation techniques is revolutionising conventional conflict resolution methods. The interdisciplinary use of capable elements of AI, including probability-based prognosis, data analysis, and stochastic simulation modelling, are being applied in order to enhance different types of mediation processes in various situations, from the scope of international relations up to local disputes. From this paper, it becomes vivid that AI plays an important role in predicting conflict tendencies and facilitating solutions that will work to bring permanent peace and order in the world [5].

Such a possibility is especially valuable in the present period that is marked by diverse international conflicts where new approaches are required. Technologies such as AI can make mediation a data-focused approach since it implies high analysis on information that is useful for decisions. Cognitive science can help mediators when it comes to realising major concepts of conflicts through analysing the patterns of interpersonal communication and conduct between the parties. It is necessary for

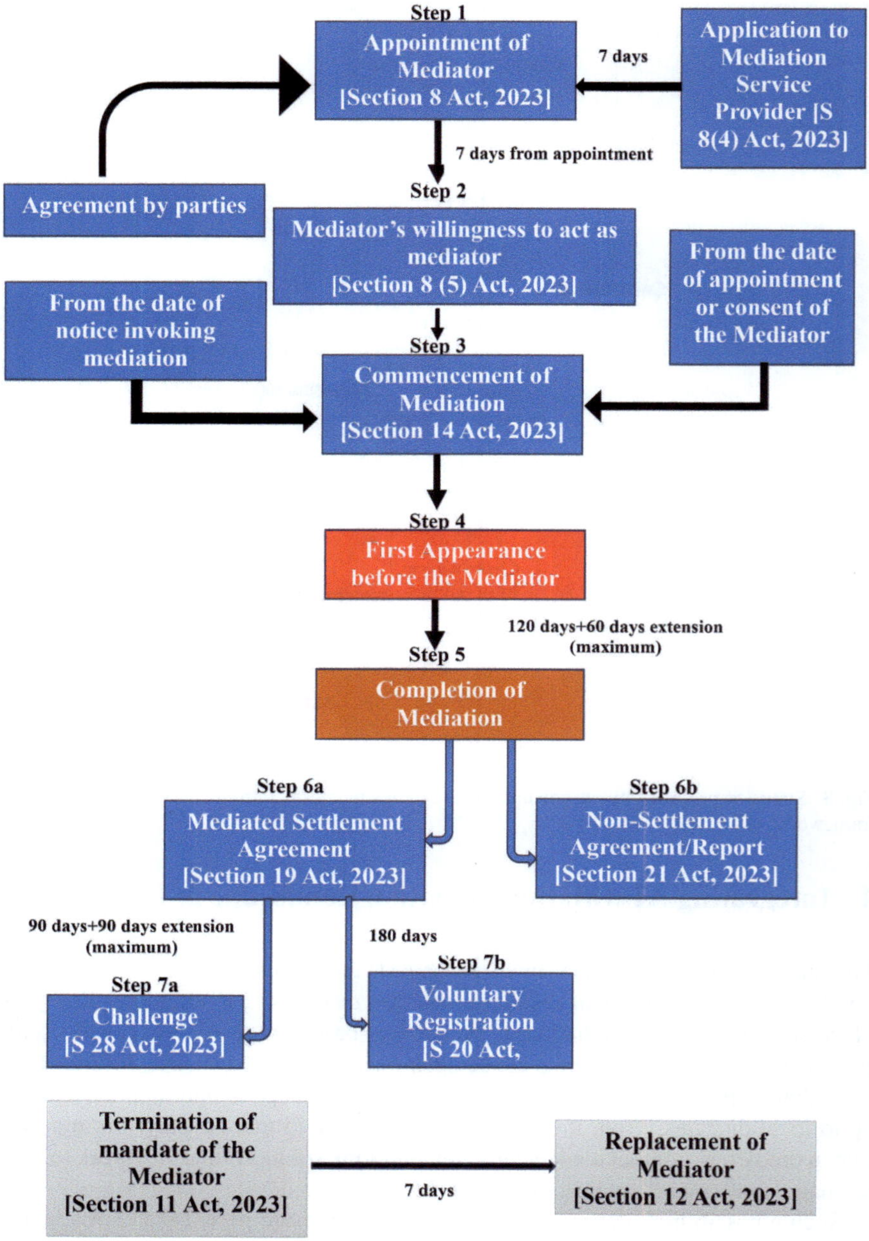

Fig. 9 Mediation process as provided by the Mediation Act 2023

mediators so as to resolve constitutional sources of disputes rather than symptoms only [6].

In addition, it is also noted that it is possible to increase the efficiency of mediators and set up an integrated AI system to help cope with such problems due to the fact that its free mediators from many trivial tasks to address the most important issues of mediation. AI in mediation refers to the design of integrated products by which negotiators are supported in real time. Automated platforms can recreate many negotiating situations that allow participants to view synthetic feelings and make proper judgments. These platforms may contain reinforcement learning elements in order to constantly enhance and adapt its results in accordance to the changing nature of the negotiation. As these advances go alongside the mediation process, it also promotes more constructive conversations between the parties. However, integrating AI in mediation has unique challenges.

The consequences of incorporating AI in conflict-salvage processes are of ethical concern, as AI's biases present the parties concerned to biased results. Furthermore, strong frameworks are need on the part of higher authorities that could check the biased behaviour and indemnify the mediated processes by AI. As AI advances, all the stakeholders have to discuss them in order to build trust among the users.

The primary benefits of employing AI in the mediation process might be listed as follows:

Advantages of AI in Mediation

Efficiency & Scalability | Cost-Effectiveness | Neutral & Impartial | Data Analysis & Decision Support | Automated Communication

NLP | VR & Simulations | Predictive Analytics | Data-Driven Insights | Online Dispute Resolution | Personalized Conflict Resolution Plans

Emotion Recognition | Communication Enhancement | Blockchain for Transparency | Continuous Improvement

AI in mediation is being recognised in India as one of the effective tools to improve the effectiveness of the highly appreciated mediation practice in the country. These technologies are being incorporated in numerous different ways into the mediation process including as a case management tool and a decision aid tool for transforming traditional mediation systems and for handling the problems faced by the legal systems in India. One of the most important benefits of the application of AI in mediation is the capacity of the AI system to analyse massive sets of data and possibly provide mediators with some patterns that would presumably underlie the conflicts. AI means the accumulation of prior case records, etc. and then analyses them to estimate outcomes for current similar cases so that the mediators can get important insights in to it [7]. This predictive ability is especially useful in the Indian legal environment because the dispute may occur at the cultural, social and economic level. Originally, using AI, mediators will be in a position to adopt strategies and processes that best fits the involved stakeholders [8].

AI also has potentials of positively impacting on the communications between the parties in a dispute particularly where translation is required for any party. Since India is a place where multilingual people live, the mediation process and the terms to be negotiated for the given parties can also be explained by using AI translation solutions. This capability improves the quality of information processing while making the negotiation environment less opaque [9]. Additionally, AI can assist in the scheduling as well as management of mediation as a scheduling process, thereby improving on the value and effectiveness of administrative tasks that tend to hinder a mediator and elongate some processes [8].

The insertion of the Artificial Intelligence into the mediation was followed by important ethical and regulatory implications. In that sense, according to Ramachandran, the legal structure of India requires responding to the implications based on the application of AI technology responsibility so that it does not deepen social disparity [7]. Bias in such algorithms also raises the question of control and setting of standards on how the mediated cases should be redressed. This is especially important if in a country like India, the user's position in the socio-economic hierarchy will affect their chances of justice [8]. The integration of AI in mediation is affected by the broad contours of digitalisation in the interior of the Indian judicial system. Although many business entities have realised that AI can bring in lots of advantages, there is rising need for training and educational activities which can help the legal industry gain the necessary knowledge to correctly apply these tools. This is necessary to facilitate innovation amongst the practitioners with a view of using capabilities of skills in AI to enhance mediation [10].

5 Conclusion, Suggestion, and Recommendations

The integration of mediation as part of the Act means that it has become one of the essential components that characterise modern India legal system, more specific, ADR system. Mediation has served to perform its duty in an efficient and culturally sensitive manner from Panchayati system to the organised regulatory system. The passage of the Mediation Act of 2023 is therefore a major development, and brings the legal framework to strengthen mediation practice for various civil society issues ranging from business transactions, family matters, and grassroots community.

This paper carries a comparative study of the Indian mediation system with the Irish system which shows that there are huge potentialities for change. The structured mediation form of complaint mechanism existing in Ireland may help to set up better regulatory measures in India and also enhance public awareness. Recent advances in the deployment of Artificial Intelligence (AI) can change the whole approach of mediation through predictive analysis, real-time decision-making and administrative support. Such innovations hold a lot of potential to mitigate problems afflicting India that include scarcity of resources and number of cases on the docket to optimise the system.

There is a number of challenges that arise from use of AI in mediation. Concerns which are briefly EIPs call for robust regulations, including Algorithmic biases that threaten the core human aspect in mediation. An advance plan that links the technical or scientific approach with the procedural and substantive justice as well as cultural compendium is crucial. When these issues are addressed, mediation becomes a tool for dispute solution in India which improves courts operation as well as social harmony.

For the improvement of mediation as an agreed method of dispute resolution in India, the following should be done: Firstly, legal and policy changes must seek to review the enactment of the Mediation Act, 2023 including, establishing common standards, accreditation of mediators, which are all informed by the best timeless of appropriate international models, for instance, Ireland's mediational system. Thus, there are vast opportunities for the implementation of Artificial Intelligence into mediation procedures, though it raises such challenges as the necessity to adapt conceptual and program solutions to India's socio-cultural and multilingual realities.

This integration of the two has to come hand in hand with another set of ethical norms and oversight mechanisms to avoid bias and or influence. Thus, it is crucial to create training that aims at teaching mediators and legal professionals about AI. It will be important for advocacy work for mediation to increase the understanding for mediation as a cheaper and accessible approach for intolerance for mediation by lower money classes, minorities, etc., and bottom-up mediation schemes need to be supported to create corresponding formal systems.

Culture and diversity should not leave out the loop, mediation programs that reflect legal pluralism, and increased embrace of women and the marginalised. For this purpose, there is a need for a specialised centre or Mediation Council of India to review the performance, receive feedback, and make the appropriate changes in the system so that change is constant, and the mediation process is most effective. These can improve the efficacy, facility, and indigenous modes of mediation in India.

References

1. Manchanda H (2021) Mediating the Irish way: taking an alternative approach to alternative dispute resolution in India. AMDM 87(Issue 1):67–93. https://doi.org/10.54648/amdm2021005
2. Roychowdhury P (2019) Illicit justice: aspirational-strategic subjects and the political economy of domestic violence law in India. Law Soc Inq 44(2):444–467. https://doi.org/10.1017/lsi.2018.13
3. Banerjee A (2017). Agrarian crisis and accumulation in rural India. https://doi.org/10.1093/oso/9780198792444.003.0005
4. Saxena S (2022). Divorce and democracy. https://doi.org/10.1017/9781108653459
5. Mandokhail A (2024) The transformative role of artificial intelligence in conflict resolution and peacekeeping. Nust J Int Peace Stab 104–109. https://doi.org/10.37540/njips.v7i1.166
6. Meleouni C (2023) Artificial intelligence and its impact in international relations. J Polit Ethics New Technol AI 2(1):e35803. https://doi.org/10.12681/jpentai.35803
7. Ramachandran D (2024) Artificial intelligence for legal system: jurisprudence in the digital age. Int J Adv Acad Stud 6(5):03–13. https://doi.org/10.33545/27068919.2024.v6.i5a.1158
8. Madaoui N (2024) The impact of artificial intelligence on legal systems: challenges and opportunities. Probl Leg 1(164):285–303. https://doi.org/10.21564/2414-990x.164.289266
9. Artificial intelligence and language. Edumania-An Int Multidiscip J 02(04):113–123 (2024). https://doi.org/10.59231/edumania/9077
10. Chatterjee S (2023) Evaluating the effects of AI-powered training programs on skill development and career growth. Int J Adv Res Sci Commun Technol 941–945. https://doi.org/10.48175/ijarsct-12743d
11. Dash A, Mohanty S (2023) The mediating effect of a firm's corporate reputation and sustainability practices in translating CSR into competitive performance in Indian ESG companies. Soc Bus Rev 18(4):691–709. https://doi.org/10.1108/sbr-02-2023-0045
12. Kowalski J (2016) Ordering dependence: care, disorder, and kinship ideology in north Indian antiviolence counseling. Am Ethnol 43(1):63–75. https://doi.org/10.1111/amet.12263
13. Pradhan R, Dash S, Jena L (2017) Do hr practices influence job satisfaction? examining the mediating role of employee engagement in Indian public sector undertakings. Glob Bus Rev 20(1):119–132. https://doi.org/10.1177/0972150917713895
14. Sarkar S (2019) How independent is India's labour law framework from the state's changing economic policies? Econ Labour RelatS Rev 30(3):422–440. https://doi.org/10.1177/1035304619863550
15. Singh V (2023) Judicial activism as an essential tool for the protection and expansion of human rights in India. Kutafin Law Rev 10(1):88–109. https://doi.org/10.17803/2713-0533.2023.1.23.088-109
16. Navigating peace: harnessing artificial intelligence for conflict resolution. Int Res J Mod Eng Technol Sci (2024). https://doi.org/10.56726/irjmets49337
17. Alhosani K (2024) Opportunities, challenges, and benefits of AI innovation in government services: a review. Discov Artif Intell 4(1). https://doi.org/10.1007/s44163-024-00111-w
18. Benítez J, Gómez-Reino M, Suárez A (2020) Data-driven democracy? automated decision-making, difference and preferences. https://doi.org/10.33774/apsa-2020-rx8fs
19. Brittain M (2020) A deep multi-agent reinforcement learning approach to autonomous separation assurance. https://doi.org/10.48550/arxiv.2003.08353
20. Morley J, Floridi L, Kinsey L, Elhalal A (2019) From what to how: an initial review of publicly available AI ethics tools, methods and research to translate principles into practices. Sci Eng Ethics 26(4):2141–2168. https://doi.org/10.1007/s11948-019-00165-5
21. Tran P, Pham D, Goh S, Alam S, Dương V (2020) An interactive conflict solver for learning air traffic conflict resolutions. J Aerosp Inf Syst 17(6):271–277. https://doi.org/10.2514/1.i010807

Enhancing India's Anti-Money Laundering Framework: Transforming Integration of Explainable AI for Better Transparency and Compliance

Sourodip Roy, Ankan Bhattacharya, and Victor Nayak

Abstract The case of Anti-money laundering (AML) is still a threat to the Indian financial system, overemphasising the failure of the traditional AML approaches at the legislative, interpretability, and operational levels. Indeed, when financial institutions employ Artificial Intelligence (AI) in AML, its limited interpretability, whereby AI models act like 'black boxes,' creates a problem in comprehending the decisions made in AML. To mitigate this problem, this study proposes a research question: How can Explainable AI (XAI) be applied in the Indian regulatory environment to enhance model interpretability and decision explainability? This paper will focus on how XAI enhances the credibility and effectiveness of compliance and operations and provides valuable suggestions for institutional and policy adoption of accurate, explainable AI-ML systems into the Indian AML framework. XAI is an area of interest in understanding if it can make AI-driven models effective in AML, thus enjoying regulatory compliance in India. The rationale for doing so is that incorporating XAI will further enhance interpretability, decrease the prevalence of false positives, and increase the likelihood of compliance due to the improved observations regarding decision-making gained from integrating the former. This study develops an academic understanding and future prospect for the initiation of an framework, which will assist in examining India's AML regulations derived from the Prevention of Money Laundering Act, 2002 (The PML Act, 2002), the Reserve Bank of India (RBI), and the Financial Intelligence Unit-India (FIU-Ind).

Keywords Anti-money laundering · Explainable AI · Compliance · Regulatory frameworks

S. Roy · A. Bhattacharya · V. Nayak (✉)
Department of Law (IIIJS), University of Engineering and Management, Kolkata, India
e-mail: victor.nayak.25@gmail.com

M. Chakraborty et al. (eds.), *Proceedings of Sixth International Ethical Hacking Conference*, Lecture Notes in Networks and Systems 1513,
https://doi.org/10.1007/978-981-96-8632-2_21

1 Introduction

Due to money laundering schemes, official reports also mention their credibility severely compromised in India, and the stability of its regulatory and institutional systems. However, the country is faced with challenges for which measures under Prevention of Money Laundering Act, 2002 (PML Act) have been put in place with help of the Reserve Bank of India (RBI) and Financial Intelligence Unit -India (FIU-Ind). Current Anti-Money laundering, techniques and tools based purely on rule-based methods have been proven to be inefficient in handling the current styled money laundering processes. Such techniques frequently result in high false-positive outcomes and organisation dysfunction burdening both financial organisations and regulators [1].

Lately, Artificial Intelligence (AI) has been recognized as a possible Anti-Money Laundering (AML) disruptive technology that can improve capabilities for identifying activities of suspicious nature through application of increasingly sophisticated machine learning (ML) algorithms. However, the integration of AI in AML poses a burden of its own which includes the explainability or transparency of the AI outcomes. These models have been described as 'black boxes' because they do not reveal how they arrived at the final output: this is problematic in terms of claiming and achieving traceability, responsibility and compliance. This has caused uneasiness among financial institutions and regulators from the ability to explain how, why, and when the AI decides as it is used in critical decision-making processes [2].

This paper seeks to understand how Explainable AI (XAI) can help solve the above difficulties and improve the credibility of India's AML framework. XAI can help financial institutions enhance AI/ML systems as the end-user will be able to understand why a specific transaction has been flagged for suspicion or not, as the case may be. XAI not only reduces false positives but also improves the agencies' trust among regulators because AI practices are aligned with compliance [3]. This research establishes an interpretability framework based on the AML regimes of India that includes the PML Act, RBI guidelines, and FIU-IND directives. The aim of this paper is to suggest approaches to integrating XAI into institutions and policies and to underline the role of explainable models in enhancing operational effectiveness and compliance with rules [4].

With India's financial structure becoming more layered and larger, there is no better time than now for the country to adopt better AML systems that are reliable, clear, and efficient. The use of Explainable AI coupled with the AML framework in India will ensure that the nation provides leadership to the rest of the globe on how to use superior technologies without compromising on the essence of the law. This paper is designed to present policy implications and specific recommendations for policy makers, banking institutions and scholars who are interested in enhancing the anti-money laundering efforts in India [5].

1.1 What is Money Laundering

Money laundering is the process of 'cleaning' the 'dirty' money, received from the proceeds of any illegal act with an intention of making that money appear as legally earned from legitimate sources. Money laundering is a sophisticated process of disguising the proceeds of crime including drugs and narcotics trade, protection rackets, embezzlement and fraud, human trafficking and money laundering for terrorists; money gained from unlawful business by making it appear as legal. It enables the criminals to incorporate their unlawful incomes to the official economy without a proven possibility of being apprehended. Presently, this crime is legally fought through the "Prevention of Money Laundering Act (hereinafter referred to as the PMLA) 2002" (as amended up to date) in India. Section 3 of the PMLA, 2002, defines the offence of money laundering as "Whosoever directly or indirectly attempts to indulge or knowingly assists or knowingly is a party or is actually involved in any process or activity connected with the proceeds of crime including its concealment, possession, acquisition, or use and projecting or claiming it as untainted property shall be guilty of an offence of money laundering" [6]. The PMLA put into force requirements on reporting suspicious transactions with money laundering, empower authorities to seize and forfeiture properties associated with money laundering while imposing severe penalties on persons and firms involved in the conduct. New enquiries of money laundering persist with more advanced technology such as digital currency and payments thus becoming a challenge for international regulatory authorities [5].

1.2 Understanding XAI (Explainable Artificial Intelligence)

With the evolution of Artificial Intelligence (AI) practices and modelling in the diverse field including the financial system of a country, the emergent difficulties have substantially created complex-opaque actions that are difficult to understand for humans. This lack of transparency, interpretability, and explainability led to the rise of the concept of Explainable AI, which will bridge the gap between the black box AI and human understandability (in relation to the financial sector). XAI is defined as the special class of AI approaches that are intended to reveal explanations of AI decisions to the end-users [3]. By offering a window into how the AI model is thinking, arriving at a conclusion, or making a decision, XAI can allow for understanding, and therefore acceptance of the system's outputs. It concentrates on making raw data understandable to users and hackers, and exposing the inner workings of the whole system. XAI is also useful for the assessment and control of responsibility by tuning AI systems to best practices and laws and is critical in sectors such as finance and healthcare, in which risks are high. Further, it assists in revealing biases, which increases objectivity making it becoming more credible and accurate. Thus, making AI more understandable and trustworthy by reducing the gap between its wonders and what is understandable by human beings that enables the users to make the right decision based on the AI provided results [7].

1.3 Evolution of Money Laundering

The current phenomenon of money laundering has changed geographically and in India in response to additional regulation and the financial landscape. Originally laundering strategies came with gangster groups at the beginning of the 1920s in the United States where they used operations such as intended laundering services like for example laundry. The laundering activities received increasing global attention during the 1970s and 1980s with references to narcotics trade and terrorism financing. Another response that has come in the form of legislation include the USA's Bank Secrecy Act of 1970 and the formation of the Financial Action Task Force (FATF) in 1989 with aim at seeking to create standard policies in the world against money laundering [8].

Money laundering has its roots in India and was once associated with the HAWALA system of transferring funds, which is far older than most modern banking systems. This meant that hawala transactions were anonymous and were popular for avoiding taxes or endorsing cross-frontier smuggling. The Indian government has been following international principles formally starting in the 1990s and officially with the PMLA in 2002. But with criminals that could easily bypass them, over the years, the Indian AML regime evolved, undergoing changes several times to further stiffen and check techniques of washing money [9].

The Hawala system is the system of transferring money without physically shifting cash or documents establishing its ownership:

Hawala is an Arabic term with direct or indirect meaning of "transfer" or "trust", is a system that is not controlled by fixed banking networks. It has been implemented most frequently in countries like India and Pakistan, predominantly those in which the development of the necessary banking framework has remained relatively low. The system uses "hawaladars"–operators who make use of a bond of trust and relationship to transfer money. A customer pays cash to a hawaladar in one country and the hawaladar forwards instructions to another hawaladar in another country to tender an equal amount in local currency to the beneficiary. This system highlights some Key Characteristics, which are discussed as given below:

a. Trust-Based System: The structure of the system relies solely on trust and the business relationships rather than signatures and contracts. If memory is taken, then records are marked in such a way that they are not easily noticeable.
b. No Physical Transfer of Money: Honestly, we do not see actual money crossing the borders. The accounts are cleared by future credit and debit between the hawaladars.
c. Speed and Cost Efficiency: It is faster and cheaper than the traditional banking systems hence widely used by migrant workers [10].

1.4 Uses and Misuses

Legitimate Uses

Migration earners' remittance to relatives in rural regions. In areas where sound banking systems are still a thing of the past, the ability to transfer funds.

Misuses

a. Money Laundering: One of the purposes of money laundering which is the act of making assets stems from criminal activities to look legitimate.
b. Terrorist Financing: Enables highly invisible money transfers for the intended foul-play.
c. Tax Evasion: Skirts the usual channel reporting to reduce paying tax.
d. Smuggling and Corruption: Fosters other vices such as drug and substance trafficking, bribery by not being subordinate to government institutions.

Challenges and Risks

Due to this hawala lacks some regulation as well as availability of information, which have brought its misuse associated with money laundering, terrorism capital, and black market products. They are informal structures, and their operations tend to make it difficult for the police to track funds or even prosecute any offenders.

Counterpart

Governments globally have attempted to regulate or dismantle hawala networks through:

a. Anti-Money Laundering Laws: It initiates the forcing disclosure of financial activities.
b. Increased Financial Monitoring: Identification of large transactions by cash.
c. Cross-Border Collaboration: Cooperating in wishing and providing information to eliminate unlawful organizations.

Hawala systems, though organized and part of hábits of some regions and countries are dangerous if used for unlawful purposes. The two forms of civil control over corporate behavior thus pose the following Questions; It is always hard to achieve the correct level of legal regulation of conduct and cultural acceptance of such regulation.

1.5 Hawala Regulatory Frameworks

The Hawala system is regulated at the national and/or international levels to fight the practice of using it in the unlawful purposes and for money laundering and terrorism financing. Here's how it has been regulated:

Legal and Institutional Frameworks

a. India's Prevention of Money Laundering Act (PMLA), 2002:
b. Makes financial entities responsible for identifying their customers, documenting and reporting suspicious transactions.

The hawala operators, whenever arrested, are charged with PMLA and the related plethora of unlawful activities.

a. Foreign Exchange Management Act (FEMA), 1999:
b. Concerns itself with the smuggling of cash across their borders; it includes hawala business [8].
c. FEMA regards violations as only civil infractions.
d. Enforcement Directorate (ED): Investigate the hawala transactions under the PMLA and FEMA laws and act as e. Reporting Mechanism.
e. Suspicious Transaction Reporting: In this case, financial bureaus must declare big or strange operations towards the Financial Intelligence Unit-India (FIU-IND).
f. Current threshold reporting is cash transactions over ₹10 lakh or FCY.
g. Mandatory Record Maintenance: Customer's details have to be stored for at least 10 years. These records belong to banks and other entities.

Surveillance and Technology

a. International Collaboration: India shares information with international agencies like the FATF to detect hawala networks, and prevent their operation. It uses cross-border intelligence and mutually assisting treaties, legal or else known as MLATs, to track IFFs.
b. Digital Monitoring: Currently, governments are using Big data analytics as well as other artificial intelligence and pattern recognition to identify transactions that are characteristic of hawala business.

Punitive Measures

a. Seizure of Assets: The properties and accounts that are associated with hawala business of any sort are classified as illicit and therefore fall within PMLA.
b. Penalties and Imprisonment: The penalties that one can incur if involved in the hawala are very severe bearing in mind that in cases of offenses, one can be fined or imprisoned depending on the gravity of the offence.
c. A similar system is practiced in India by the Reserve Bank of India RBI. The same work is done under the Consumer Protection Act 2019.

Awareness Campaigns

The measures which are intended to raise public awareness inform people about the threats and illegality of hawala operations. Expanding the utilisation of the ordinary banking procedures to facilitate remittance and other monetary dealings helps in avoiding hawala type activities.

Challenges in Regulation

a. Trust-Based and Informal Nature: Due to poor documentation and personal reliance, hawala transfers cannot be easily traced.

b. Jurisdictional Issues: Hawala networks usually operate across boundaries of sovereign countries and this feature makes it difficult to undertake effective enforcement action since legal standards may vary from country to country.

Though hawala systems are still in place, a mixture of stricter laws and more effective watch, together with theoretical international cooperation will do away with hawala but for now, the elimination of the hawala systems is still in the long-term future [11– 14].

1.6 Types of Money Laundering

Money laundering can be in different aspects, based on the type of crime and the level of the mechanism being used. The primary forms include:

a. *Hawala Transactions*: An old-world, and rather commonplace form of money laundering in India, hawala refers to the transfer of cash without the actual physical movement of the money. Criminals like hawala more because it is outside the standard banking sector and has low accountability [8].
b. *Structuring or Smurfing*: This procedure is done to avoid specified thresholds set by laws within large volumes of cash by making sub-deposit intervals. In India this is widespread in real estate where volumes of money are split and remitted to beat certain ceilings.
c. *Shell Companies and False Invoicing*: Organized crime operates pre-incubator firms or trade-based money laundering, whereby they manufacture invoices detailing incorrect quantities of goods. Through manipulating invoices and/or the amounts of these invoices, they are able to push through several millions of dollars across borders while concealing the sources of those funds.
d. *Investment in Real Estate and Luxury Goods*: Indian money launderers often put their 'tainted' money in purchasing property like 'real estate' or other 'asset' like jewellery or ART which is later sold to convert black money into white money.
e. *Cryptocurrency Transactions*: As more and more people start using cryptocurrencies, it has become much easier to launder money as it is done using blockchain [15].

1.7 Stages of Money Laundering

The entire process of Money Laundering is done in various stages, three of which are given below:

a. Placement: The first step involves the laundering of 'bribe' or 'dirty money' into the legitimate economy. This may be done in the form of bank deposits, purchase of valuable assets or property, through casino and betting services. In

India this is done through Structuring where people deposit small amounts in multiple accounts or buy high value properties, which can be sold in small lots.

b. Layering: This stage is described by making many levels of the transaction with the view of concealing the money transactions. Accounting transfers may be used to move money from one account to another or to encode the funds into other financial instruments that will disrupt the paper trail. Indian money launderers even switch from one or many banks, international transfers, and shell companies which further makes it difficult to track [16].
c. Integration: The last part is to circulate back the money into the economy as 'clean'. This is normally achieved by buying into legitimate businesses in the law firms, acquiring property or establishing phoney firms that have legal appearing incomes.

Further, there is an exploration of the Indian perspective towards the control of money laundering. India has made substantial effort in redressing money laundering particularly through The Prevention of Money Laundering Act (PMLA) that came into operation in 2002. The roles that have been played by PMLA are the duty of maintaining records, reporting on suspicious transactions, and identifying the customer through Know Your Customer (KYC). India is also a member of Financial Action Task Force (FATF) and thereby the country has put into practice all these recommendations and thus categorising its regulations at par with international standards. Nonetheless, implementing these statutes continues to be an uphill task given the high levels of corruption, inadequate resource endowment and the large size of the informal economy respectively [17].

India has put digital surveillance systems for counteraction and has international cooperation for cross-border laundering, but new problems of technologies and decentralised finance exist. Battles against money laundering have been unfolding in India and are based on an ongoing focus region; digital forensics; AI, and cooperation with global law enforcement agencies to ensure effective tracking and prosecution of such individuals. Consequently, money laundering in India is a complicated process, which has its origins in the British colonial period and involves both manual and computer-aided schemes. This is especially true because the Indian government has recently increased its measures regarding certain key business sectors, while at the same time new techniques for money laundering continually emerge [11].

2 Legal Framework: Global and National Prospective

2.1 Global Standpoint

Money laundering is a prominent problem in the world today and needs assistance from other countries. The effort to build international standards against money laundering started at the beginning of arguments surrounding drug trafficking and evolved to integrating all sorts of crimes and terrorism financing.

2.1.1 United Nations Conventions (UN)

i. Vienna Convention (1988): Together with its innovative framework for arresting drugs' monetary effects, a milestone treaty that entails countries to criminalise money laundering and foster mutual legal assistance on asset confiscation [18, 19].
ii. Palermo Convention (2000): Is extended from the drug trafficking activities to the money laundering crimes.
iii. UN Convention Against Corruption (UNCAC 2003): It is dedicated to the fight against corruption and its financial aspects, promotion of assets returns and cooperation with other countries [11].

2.1.2 Financial Action Task Force (FATF)

Originally beginning in 1989, FATF is the international organisation that sets standards and promotes implementation of policies against money laundering and terrorism financing.

i. FATF Recommendations: An anti-money laundering and counter financing of terrorism framework that contains 40 recommendations that are global best practices in the areas of prevention, transparency and legal basis.
ii. Mutual Evaluation Process: Provides an evaluation of member countries' implementation of requirements set by the FATF with the purpose of determining deficiencies [20].

2.1.3 Basel Committee on Banking Supervision

This discusses specific requirements of AML for financial institutions, including due diligence requirements, reporting obligation of suspicious activities, and supervisory requirements.

2.1.4 Regional Bodies

a. European Union (EU) AML Directives: They try to achieve optimality and uniformity in AML measures in member states in the area of customer due diligence, suspicious transactions reporting, and beneficial ownership.
b. Asia–Pacific Group on Money Laundering (APG): Specialises at applying FATF standards in Asia–Pacific states as well as fostering cooperation.
c. Egmont Group of FIUs: Helps in coordination between various Financial Intelligence Units (FIUs) and enables cross-border sharing of information on financial crimes.

2.1.5 Country-Specific Frameworks

a. United States: The Bank Secrecy Act (BSA) relates to secret banking, conducting, record keeping and reporting of large transactions. To some extent, the Patriot Act (2001) upheld the AML endeavours, which focused on terrorism financing.
b. United Kingdom: The Proceeds of Crime Act (2002) made consolidated AML legislation and developed civil forfeiture provisions.
c. Switzerland: Famous for its sound banks, it laid down strict rules for Customer Due Diligence (CDD) & immediate reporting under AML Act [21].

2.2 Lacunas in the Legal Framework Concerning Explainable AI (XAI) in AML Compliance

The AML frameworks at both global and national levels which include United Nations Conventions and FATF Recommendations together with EU AML Directives and country-specific laws of Bank Secrecy Act BSA of the U.S and Proceeds of Crime Act PACO (2002) of the UK face a substantial void when it comes to Explainable AI XAI implementation into legal and regulatory standards [7].

2.2.1 Absence of Legal Mandates for AI Transparency in AML Compliance

AI models fail to meet AML compliance requirements because there exist no mandatory legal requirements for explainability. AML systems that use AI remain beyond existing traditional compliance approaches that focus mainly on customer due diligence (CDD) and enhanced due diligence (EDD) along with suspicious transaction reporting (STR). Therefore these "black box" systems go unaccounted. This absence creates the following problems:

Financial institutions implementing AI-driven AML models face a challenge because they do not have to reveal their decision-making processes to regulators nor affected parties.

AI systems operating without legal oversight in risk assessments result in wrongful account freezes along with incorrect classification of valid activities which subsequently causes discriminatory financial restrictions from which customers cannot appeal [2].

2.2.2 Lack of Standardized Regulatory Oversight for AI-Driven Decision-Making

The Mutual Evaluation Process operated by FATF represents international AML guidance that fails to provide detailed direction about monitoring AI models which power compliance systems. The standardization which exists for financial reporting audits does not apply to AI-driven AML systems which operate without specific regulatory requirements to ensure fairness or accuracy of their operational functions or explain their workings. This results in:

Different countries maintain conflicting AML governance policies through which global AML cooperation becomes hindered.

Financial organizations can avoid taking responsibility for AI biases in their decisions because they claim their proprietary methods and regulatory uncertainty about their duties.

2.2.3 Weak Legal Protections Against AI-Induced Discrimination

The AML laws of Switzerland and USA with their Patriot Act from 2001 emphasize precepts of compliance without implementing strategies against AI bias detection. Artificial intelligence tools access financial data from the past that may introduce hidden discrimination against particular population groups. Key consequences include:

AI-powered AML tools select particular categories for suspicious activity monitoring through their exposure to biased training datasets which disproportionately targets ethnic groups along with low-income people and politically exposed persons.

The lack of due process prevents wrongfully categorized people who are identified as high-risk from obtaining legal ways to challenge AI-based decisions under present AML regulations.

2.2.4 Regulatory Silence on Explainability Standards for AI in AML

Current explainability tools represented by LIME, SHAP and GraphSVX function properly yet major AML frameworks have not yet authorized their use. The lack of predefined XAI benchmarks brings several negative implications for the financial sector [22].

Financial institutions obtain authorization to use uninterpretable AI systems through administrative loopholes.

The regulatory organizations including RBI, FIU-IND and FATF operate without established evaluation criteria for AI-based systems designed to manage Anti-Money Laundering [23].

Ineffective AI Auditing Mechanisms in AML Enforcement.

Decision systems created for Anti-Money Laundering (AML) implement financial audits and compliance checks but the tools based on AI technology do not include detailed audit requirements this results in:

AI models that make incorrect transaction classifications may escape detection because regulatory organizations do not conduct audits on their AI systems.

The current regulatory system lacks proper enforcement tools because financial institutions fail to receive any specific penalties for operating AML systems using unclear or tainted AI models [21].

3 Indian Perspective

3.1 The Prevention of Money Laundering Act, 2002 [24]

a. This act was enacted and brought into force for the purpose of negating money laundering and for the recovery of the proceeds of crime which includes concealment, possession and acquisition (as per Sect. 3 of the Prevention of Money Laundering Act 2002). This Act describes money laundering as any transactions or other handling that consists in or assists in, the proceeds of crime to conceal their source.
b. Empowers enforcement agencies to: Take and seize property which is proceeds of crime.
c. Fine them and imprison them for up to 10 years.
d. Section 48 of the PMLA, 2002, recognises the following classes of authorities to fulfil the objectives of the act, which are Director or Additional Director or Joint Director, Deputy Director and Assistant Director.
e. Such other classes of officers may be appointed for the purpose of the act.

3.1.1 Subsequent Amendments to the Prevention of Money Laundering Act, 2002

a. 2005 Amendment: The Prevention of Money Laundering (Amendment) Act, 2005 and its subsequent amendments to implement the PML Rules, 2005 widened the scope to deal with such offences.
b. 2009 Amendment: Some of the other offences along with the offence of Criminal Conspiracy were incorporated in the schedule of the PMLA as given under Sect. 61(2) of the Bharatiya Nyaya Sanhita 2023. This with the flow of time has lexicalized the permission for the ED to intervene in any case being investigated that involves a conspiracy charge, even if this core offence is listed on the PMLA schedule.
c. 2012 Amendment: Extending the previous one, it introduced the notion of reporting entities for example, banks or other financial institutions.
d. Further bills are brought to extend the meaning of money laundering to include the idea of concealment, possession as well as acquisition of proceeds of crime [7].
e. That the accused property could be confiscated without a conviction but with sufficient evidence.

3.2 *Regulatory Authorities*

3.2.1 The Financial Intelligence Unit-India (FIU-IND)

FIU-IND was set up in 2004 under the Prevention of Money-Laundering Act, 2002 (PMLA) to act as the central internal nodal agency to help detect and prevent money laundering and associated economic offenses. It is under the Department of Revenue, Ministry of Finance.

Key Functions

a. Collection of Financial Information: FIU-IND also receives reports on suspicious transactions, specifically; transactions that involve large amount of cash; counterfeit money; and cross border transfers of funds either through bank, financial institutions or intermediaries.
b. Analysis and Dissemination: Carries out operating and strategic examination of the financial information secured with the purpose of revealing the tendencies of economical providing crime. Provides timely information to enforcement agencies namely the ED and the CBI.
c. Global Cooperation: FIU-IND is a part of the Egmont Group, which is an association of the financial intelligence units. This also makes it possible to share and hence coordinate with counterparts from other parts of the world.
d. Regulatory Role: It provides mechanisms for the implementation of AML (Anti-Money Laundering) and CFT (Combating Financing of Terrorism) requirements of financial institutions.
e. FIU IND plays the role of monitoring money laundering activities and also helps in preventing and recovery of the illicit proceeds. They are in favour of the international AML work and enhance the national compliance with the FATF standards.

3.2.2 Reserve Bank of India (RBI)

The RBI was set up in 1935 and is the central bank of India which controls and supervises the nation's monetary system, banking system and foreign exchange market. It plays an important role in ensuring continuing financial stability and AML measures compliance [25, 26].

Key Functions

I. Regulation of Financial Institutions:
a. Regulates banks and NBFCs and payment system operators that implement the provisions of the PMLA, which address AML.

b. Requires compliance with the KYC norms so that no fraud accounts and money laundering take place [27].

II. Monetary Policy Implementation:

a. Mitigates inflation, liquidity as well as credit by adjusting interest rates and issuing of monetary supplies.

b. Foreign Exchange Management: Controls international transactions of various investments or any value under FEMA that would otherwise be used to facilitate money laundering.

III. Issuance of Guidelines:

a. Provides circulars and directives on AML compliance and risk management and cybersecurity to the financial institutions to increase the Financial Operation Transparency.

b. Role in AML.

c. This ability is facilitated by RBI to compel financial institutions to file dubious transactions to FIU-IND.

d. There are checks and balances which include audits and inspections for compliance with the AML or there are penalties for noncompliance.

3.2.3 Securities and Exchange Board of India or SEBI

SEBI is the modern-day Napoleon of security markets in India and was set up in 1988 and was provided statutory power in 1992 based in Mumbai. Its main purpose is to safeguard investors and promote the proper functioning of securities markets and markets for related financial instruments.

The key functioning of the Securities and Exchange Board of India are as follows:

a. Regulation of Market Intermediaries: Monitors compliance by the stock exchanges, brokers, depositories and mutual funds so as to curb undue manipulation.
b. Investor Protection: Observes tight reporting rules to ensure that all investors receive credible information, and to curb fraud.
c. Prevention of Market Abuse: Prevents and sanctions manipulative and fraudulent conduct in connection with securities and markets for those securities.
d. AML Compliance:
e. Prescribes certain reporting requirements from market intermediaries with respect to suspicious transactions.
f. Helps in compliance to KYC standards to avoid usage of the securities market for money laundering.
g. SEBI holds a working relationship with FIU-IND and enforcement agencies to frustrate illicit money trading in the securities market.
h. It helps to prevent securities markets from being used for placing or incorporating the proceeds of a crime.

3.2.4 Enforcement Directorate (ED)

The ED was set up in 1956 as India's premier investigative and prosecuting agency for covering economic offences. This agency is located within the Department of Revenue which forms part of the Ministry of Finance.

The key functioning of Enforcement Directorate:

a. Enforcement of PMLA: Deals with violations of the PMLA to take action for prosecution and pursuits involving tracking and forfeiture of the proceeds of crime.
b. Foreign Exchange Regulation: Implements FEMA to regulate flows of foreign exchange and transactions involving the export/ import of the foreign exchange.
c. Asset Confiscation: To prevent further laundering it investigates and takes over properties purchased with the proceeds of crime.
d. Prosecution: Registers complaints to special courts under the PMLA for crimes such as money laundering, terror financing among others, tax evasion.
e. Role in AML: ED is central to investigation of major money laundering cases that receive attention in the Australian media.
f. Aided by FIU-IND and cooperating with SEBI and RBI to track the money laundering and punish the criminals.
g. It coordinates itself with international agencies in order to fight money laundering across its borders.

3.3 Associated Regulations

a. The Foreign Exchange Management Act, 1999: Where it controls international movements of funds for the purpose of addressing issues related to capital flight.
b. The Companies Act, 2013: Requires companies to declare beneficial owners.
c. The Black Money (Undisclosed Foreign Income and Assets) Act, 2015: Aims at unreported foreign property and income.

3.4 International Conventions Applicable to India

India is a signatory to key international frameworks, including:

a. FATF recommendations: Two international treaties that are known are the Vienna and Palermo Conventions [28].
b. UNCAC for fighting corruption related money laundering.
c. Sometimes the group membership in regional bodies, for instance APG guarantees compliance to international standards.

4 Integration of XAI in Anti-Money Laundering Practices

4.1 Global Standpoint

The more profound conceptualization of Explainable AI (XAI) within the domain of Anti-Money Laundering (AML) remains limited to the theoretical plane, the same brings a rather radical possibility to the realm of research enthusiasts. One of the biggest issues is the ability to bridge the theoretical gap and transfer proposed methods from practice to practice, as most of the work done in XAI for AML, such as using GNNs with Local Interpretable Model-agnostic Explanations or LIME, SHapley Additive exPlanations or SHAP, and Shapley Value Explanations for Graph Neural Networks for GraphSVX, remains mostly theoretical or in controlled test settings [19].

XAI is involved in explaining the black box of the machine learning model and makes them understandable. So, applied to the Anti-Money Laundering (AML) domain, XAI is mandatory for building trust with customers and regulators and achieving accurate fraud detection. LIME, SHAP, and GraphSVX are listed among the most popular ones. Both present distinct methods of explaining the functioning of AI and, as applied to the structures of AMLs, are outlined below [8, 29].

4.1.1 LIME (Local Interpretable Model-Agnostic Explanations)

a. LIME is also a model-agnostic technique and works independent of the characteristic of the model used for the prediction in the dataset. It produces local explanations by defining a small, linear neighborhood around a prediction and approximating the model's behavior within that context. For example, it slightly alters the input data and then compares the results with initially created predictions to determine features that affect a decision [30].
b. Specifically, LIME stands for Local Interpretable Model-agnostic Explanations and would assist in explaining to the employees of an AML system or the firm why a particular transaction or a certain customer was labelled as suspicious The method would help reveal which features affected the model's prediction in light of feed-forward analysis.
c. The regulators can use LIME explanations to check compliance of the AI systems with legal regulation in areas like the Right to explanation under LRD under certain legal principles like Prevention of Money Laundering Act (PMLA) of India [31].
d. It offers value to investigators in guiding them where their suspicion should be placed, while allowing them to zoom in on specific attributes such as the amount of the transaction, or the country that the transaction originated from [29].

4.1.2 SHapley Additive ExPlanations or SHAP

a. SHAP is inspired from cooperative game theory; particular concern goes to the Shapley Value, a collection point concept that assigns 'fair' contribution credit score to the feature in relation to prediction. It offers the global and local explanations using coefficients that measure how much of an impact a feature has to the prediction or how it is likely to dampen the same. SHAP values were found to be consistent and add up to improve model integrity and facilitate model interpretation.
b. Feature Importance in Fraud Detection: In AML systems, SHAP can sort features (for example, transaction frequency and account activity level) by their impact to explain why such customer or transaction has been identified.
c. The SHAP can reveal threats of bias in the classification and alter it to avoid overemphasizing specific features resulting in an unbiased policy on AML.
d. In schemes with multiple stages of money laundering, the SHAP values can demonstrate ways in which addition and multiplication of some features (like multiple small deposits) lead to recognition of suspicious behavior [31].

4.1.3 GraphSVX

a. GraphSVX is also using graph data and has its explanation method for GNNs, normally applied in AML, to enhance the representation relations between entities (accounts, transactions, businesses etc.). It employs perturbation-based methods to measure relevance of nodes within a graph such as account and edges such as transactions in a network and offers explainable insights into model results. It extends over attribute relevance and presents a way to describe the network between elements in terms of a graph-based framework.
b. Transaction Network: Based on the network of relationships, GraphSVX can always justify why a specific node (for instance, an account) or edge (for instance, a transaction, was labelled.
c. Money laundering also involves carrying out transactions in different legal entities at different levels. The specific connections that GraphSVX accentuates are these: By having access to them, GraphSVX makes it easier to detect and analyze supposedly malignant pattern networks.
d. It allows visual representations of transaction linkages and finding the implied logical connection, providing investigators with the means to view the context of the provided data [23] (Table 1).

This gap can be closed by designing combined systems; integrating rule based systems and XAI improved machine learning systems that adhere to present structures while imposing higher operation effectiveness.

An exciting area for future studies is the specificity of AML systems in specific global settings. Deliveries, transactions, and data differ from one jurisdiction to another, and therefore, dependability also requires designing XAI frameworks

Table 1 Comparison and complementarity [32–34]

Tool	Centre of attention	Strengths	Relevance to AML
LIME	Local explanations for any model	Model-agnostic and easy to implement	Identifies key features driving specific flags
SHAP	Global and local feature importance	Fair and consistent feature attribution	Uncovers biases, ranks key AML features
GraphSVX	Explanations for graph based models	Contextualizes relationships in network data	Highlights transactional links and fund flows

compatible with specific regions' rules, i.e., the European Union's Artificial Intelligence Act, 2024 or, if not more stringent, the FATF guidelines. Moreover, the explanations created by an XAI system should be understandable to individuals in legal, linguistic and financial aspects.

The diversity of data and their high level of security become major barriers to the large-scale application of XAI in AML. Using cases like International Business Machines Corporation's (IBM's) AMLSim is beneficial; however, they can be overly simplified, thus, do not incorporate such features as the multi-level ownership or cross-border transactions. These synthetic frameworks could be extended further by the study of other privacy-preserving paradigms such as federated learning integrated with XAI, to let banking and financial institutions work in unison without crossing data privacy barriers. Further, transition from the approach that provides the significance of features to provide contextual explanations in the form of an interaction graph constructively augments AML investigations. For example, besides the methods described above, GraphSVX can also be used to identify new associated transaction patterns or relationships between accounts to give investigators further information on flagged activities.

Regulation and investigation cooperation is also considered as the crucial working sphere. The actualization of these future ideas could be best done through joint creation of XAI frameworks that harmonize with the operation of industries then the creation of evaluation criteria that best fits the given real-life scenarios. This is because the different stakeholders involved in AML could be made to follow the principles of explainability that could be set to determine what makes one explanation acceptable or not. Additionally, the evaluation of XAI explanations in real AML use cases can reveal their reliability and effectiveness for practitioners and uncover potential effects on operational decisions. It also provides an important foundation upon which to construct elaborate assessment instruments, for example, based on the Co-12 framework to assess clarity, response time and investigator decision consistencies [35].

5 Indian Standpoint

In context to the Indian perspective, the Reserve Bank of India (RBI) specifies strict Know Your Customer (KYC) and Anti-Money Laundering standards, which are still based on traditional rule-based approach, widely known for efficiency and high levels of false positives in many jurisdictions. KYC norms are necessary in verifying the identity of customers and to check the standard of financial systems [9].

Key components include [36]:

a. Customer Acceptance Policy (CAP): Financial institutions cannot open anonymous accounts or fictitious accounts. Risk is basically segregated into lows, mediums, and high and is determined by the background of the customer as well as the nature of business and the transaction of the customers. This type of customer requires additional identification to be carried out on high-risk customers like PEPs and non-resident customers.
b. Customer Identification Procedures (CIP): The COs appearing at the banks have to confirm the identity of customers using the valid documents and have to ensure compliance to the PMLA, 2002. Verification indicates legal status, ownership and those who are allowed to act on behalf of the company [5].
c. Ongoing Monitoring: While making anticipations for any sort of fraud, the banks keep close eyes on accounts for abnormal or big transactions especially on risky categories [26].
d. Periodic Updates: KYC records are updated based on the customer's risk profile [25]:
i. High-risk: Every 2 years.
ii. Medium-risk: Every 8 years.
iii. Low-risk: Every 10 years.
e. Integration with Central KYC Records Registry (CKYCR): Some of the benefits are centralization of customer records thus easing in update and access, elimination of duplication of KYC procedures [36].

6 AML Standards Complement KYC Measures and Include

a. Risk-Based Approach: Customer, geographical, and other transaction risks mean policies to address them in terms of risk assessment and their mitigation should be developed by financial institutions.
b. Transaction Monitoring: High risk transactions or transactions out of the ordinary profile of the customers are monitored.
c. Reporting Obligations: Any doubtful transaction needs to be disclosed to the FIU pursuant to the PMLA i.e., the Financial Intelligence Unit–India.
d. New Unique Customer Identification Code (UCIC) has been implemented to minimize tracking difficulties.

e. Increased control activities for customer segments such as multiple small cash deposits or suspicious, random looking transactions [12–14, 36].

The concept of XAI in AML has a promise with numerous possibilities but with certain distinctive challenges and prospects. India being a country with large transaction volume, growing digitalization, and varying socio-economic environments to deal with, financial crimes are hard to contain in the country. All organisations have basic detection frameworks that lack the sophistication needed to investigate the new age financial crimes, let alone with the occupying payment technologies like UPI. The absence of XAI may result in inadequate detection of anomalous transaction patterns in large datasets created by platforms such UPI, digital wallets, or cross-border remittance systems, which could be filled by the integration of XAI with GNNs.

Therefore, the processes of XAI implementation in India's financial sector need to consider the following general aspects of regulation and socio-economy of the country. The requirement for compliance and reporting under the PMLA integrates well, but incorporating XAI into this system would best be done with consultation with the Indian authorities such as the RBI and the FIU-IND. The explanations that XAI systems are to offer should in the Indian financial space involve the need for the creation of novel approaches for safe data sharing. Federated Learning, combined with XAI, which means that Indian banks and financial institutions can cooperate without sharing the customers' personal information. This approach is particularly favourable for India's government information security policies, such as the Digital Personal Data Protection Act, 2023.

Banks have recently opened their doors to financially excluded customers using Jan Dhan Yojana and other similar FIPs, but such customers are even when performing legal activities treated as high risk in AML. XAI could aid in such a case to prevent such biases through providing clear explanations for flagged transactions, therefore making the model fair [37].

The Fintech industry has grown rapidly in India and presents a real-world environment for XAI in AML. Dissemination can be accomplished through individual and combined efforts of university industry, and financial institutions to encourage the formulation of pilot projects. Also, there is the Indian Stack, which is a set of Application Programming Interfaces that facilitates digital identity, payment, and data sharing, which presents a good environment for implementing XAI-based AML systems.

Last but not least, cost aspects are crucial in India, as financial institutions–especially new generation banks and cooperatives–have rather limited financial resources.

To sum it up, India still has a relatively young and developing AML landscape that has its own peculiarities which also include such aspects as high number of transactions, the problem of personal data protection, as well as systemic prejudice. India's digital growth, continuous changes in its laws, and increasing popularity of Fintech offer the developers such tools to build XAI solutions that are transparent, engaging, and easily scalable to bring maximum value to combating financial crime [35].

7 Money Laundering Case Law in India

a. Yes Bank Case: Rana Kapoor of Yes Bank-Founder was blamed for lending bank funds without proper collateral; in return for personal gains or benefits for himself and families. Yes Bank's investments to the tune of Rs. 1 trillion were channelled into DWHL (now DHFL in a very doubtful manner to loans extended to entities identified with Kapoor. These monies were siphoned offshore. ED chargesheeted and confiscated properties valued at ₹2203 crores of Kapoor, besides arresting many people.
b. National Herald Case: The allegation against Sonia and Rahul Gandhi is that they had taken over the assets of Associated Journals Ltd., the National Herald owners, through Young Indian Ltd. as a company in an unlawful manner. ED probed the scam allegations of breach of trust and siphoning money in procurement.
c. Punjab Sand Mining Case: Bhupinder Singh, nephew to former Punjab CM Charanjit Singh Channi has been arrested under allegations for taking bribes regarding sand mining illegally. Cash and other valuables to the tune of ₹10 crores were recovered by the ED, establishing a connection between him and the illicit activities.
d. Jharkhand Mining Scam: IAS officer Pooja Singhal was found to have been involved in MGNREGA funds embezzlement as well as illegal mining. Cash worth ₹20 crores was recovered during raids on her properties, and charges were brought against her under the PMLA.

These cases demonstrate the systematic exploitation of the financial and regulatory systems where usually investigations will uncover sophisticated levels of financial deception. In these cases, litigation manifests the challenges of curtailing financial crimes [38, 39].

8 Suggestions

In this research paper, the author reveals that despite various problems in the AML system in India, the employment of Explainable Artificial Intelligence can be transformational.

1. Development of XAI-Enabled Frameworks:

Among the recommendations one could suggest the following ones: Apply specific interpretative schemes that fit Indian legislative and financial contexts. These frameworks have to be in compliance with the provisions of the Prevention of Money

Laundering Act, 2002 (as amended to date), the Reserve Bank of India and directions of the Financial Intelligence Unit-India (FIU-IND). For these XAI-enhanced frameworks to be in compliance with the above-mentioned legal frameworks and authorities, the first thing to be done is to enact and/or amend such laws to lay down a strict regulatory framework for the XAI based to work to its full possible capabilities.

2. Adoption of Federated Learning and Privacy Measures

Use federated learning approaches to help several financial institutions share data without compromising on the data's integrity. These techniques must abide by the 'Digital Personal Data Protection Act—2023,' so establishing a more secure and trusted AML.

3. Reduction in False Positives

Develop XAI models which will reduce false positives especially in financial markets and regulators. Through this process of evaluating all the potential threats that may surround an institution, these models eliminate unnecessary noise that may hinder operations more than help and enable institutions to focus on the real threats.

4. Training and Awareness Initiatives

As it fails to develop culturally, linguistically, and technologically accessible training programs for regulators and financial institutions. These measures will foster the appreciation as well as the competent application of the information produced by the XAI systems for efficient decision-making among the stakeholders.

5. Cost-Effective Solutions

Increase usable AI for the Indian market, including XAI-based services, that is economical and easily scalable for the numerous segments in the financial industry. Particular attention should be paid to the development of the products comprehensible for small and middle credit unions, cooperative societies, and regional banks.

6. Pilot Projects and Testing

Engage the cooperation with, and seek collaborations from financial institutions, Fintech solutions, and organisations with focus on AML to deploy XAI-based AML systems. Pilot projects should leverage the Indian Stack environment to understand whether they are scalable or have the desired impact in a real-world scenario.

7. Evaluation Metrics for XAI Systems

All these thus require standard measures to assess the efficiency, impartialness, and legal compliance of XAI systems. These metrics should capture the structure of the financial industry of India for the systems to be as transparent, bias free and legal as required [4].

9 Conclusion

Combating money laundering has reached a significant point in India as new challenges appear, and conventional approaches reach their potential. The potential to restore confidence and enhance the capabilities of the current AML architecture through XAI is tremendous: mistrust towards the opacity of AI that sometimes permeates the financial industry; the high levels of false positives, and improvement of efficiency in detecting financial crimes. Thus, by providing the explanation of the AI models, XAI can increase the confidence of regulators and financially oriented organisations, as well as ensure that the new technologies correspond to the guidelines.

From a global perspective, XAI's potential in AML has been acknowledged but has not been fully exploited in practice especially in jurisdictions with evolving compliance laws. The expectations regarding the role of a flexible framework with respect to socio-economic diversification, language issues and different transactional intensity can be illustrated by learning from the India example. The implementation of privacy preserving techniques like federated learning along with XAI, would make it possible for banks and other financial institutions to share their data with each other while making sure the privacy of their data is preserved. In addition, the growth of fintech in India and the implementation of the Indian Stack offer favourable conditions for the use of novel XAI-based AML solutions.

Issues like data accessibility, funding, and systemized prejudice are social issues which should best be addressed by regulation, tests, and budget friendly approaches. Thus, more future studies engaging multiple disciplines such as ethics, economics and social sciences will be needed to address the challenges related to accomplishment of large-scale, long-lasting AML systems.

Altogether, the integration of the XAI into the Indian AML regime implies a breakthrough. If the elements of technology are harnessed appropriately and if the players in the nation come together to fight money laundry threats, then not only can the nation fortify its defence against money laundering effects but also serves as the epitome of anti-money laundry practices in the world.

References

1. Sugunavel (2025) Explainable AI (XAI): bridging the gap between AI and human understanding. Avit. https://avit.ac.in/explainable-ai-xai-bridging-the-gap-between-ai-and-human-understanding/
2. Rossel L, Unger B, Ferwerda J (2021) Shedding light inside the black box of implementation: tax crimes as a predicate crime for money laundering. Regul & GovAnce 16(3):781–800. https://doi.org/10.1111/rego.12407
3. Dwivedi R, Dave D, Naik H, Singhal S, Omer R, Patel P, Ranjan R (2023) Explainable AI (XAI): core ideas, techniques, and solutions. ACM Comput Surv 55(9):1–33
4. Anuradha G (2010) Money laundering and financing of terrorism-a study on methods to fight money laundering in India and USA. J Inst ChartEd AccountS India 58(10)

5. Kumar V (2010) Money laundering: concept, significance and its impact. Eur J 4(2)
6. Prevention of the money laundering act 2002, offence of money laundering, section 3, 2002. https://www.indiacode.nic.in/bitstream/123456789/15402/1/moneylaunderingact2002.pdf
7. Noronha N (2018) Follow the money: India should become an international leader in financial intelligence. https://www.jstor.org/stable/pdf/resrep20721.pdf
8. Faith DC (2011) The hawala system. Glob Secur Stud 2(1):23–33
9. Abhinandan (2024) Understanding KYC and AML: differences, regulations and importance. AuthBridge. https://authbridge.com/blog/kyc-and-aml-regulations/
10. Eilly ART (1992) The evolution of money laundering: a descriptive analysis of agency response and legislation. Univ Ala Birm ProQuest Diss & Theses. https://www.proquest.com/openview/dbe3a724a795c63059ce82876b54f976/1.pdf?cbl=18750&diss=y&pq-origsite=gscholar
11. Kumar P (2015) Money laundering in India: concepts, effects and legislation. In: Money Laundering in India: Concepts, Effects and Legislation, vol 3, pp 51–63. https://www.raijmr.com/ijrhs/wp-content/uploads/2017/11/IJRHS_2015_vol03_issue_07_11.pdf
12. Abhinandan (2024) RBI Updates KYC norms to align with money laundering laws. AuthBridge. https://authbridge.com/blog/rbi-amends-kyc-norms-align-with-aml/
13. Saran P (2004) Guidelines on 'Know Your Customer' norms and anti-money laundering measures. https://www.rbi.org.in/commonman/Upload/English/Notification/PDFs/58981.pdf
14. Author (n.d.) Reserve Bank of India. https://www.rbi.org.in/
15. Money laundering schemes: 7 common criminal tactics (2025) Linkurious. https://linkurious.com/money-laundering-schemes/
16. Singh VK (2009) Controlling money laundering in India—problems and perspectives. https://papers.ssrn.com/sol3/papers.cfm?abstract_id=1351144
17. Org GA (2021) How dirty money is laundered: three stages of laundering. Global Anti-Scam Org. https://www.globalantiscam.org/post/how-dirty-money-is-laundered-three-stages-of-laundering
18. United Nations (1969) Vienna convention on the law of treaties. In: United Nations, Treaty Series, p 331
19. United Nations, Annan KA (2004a) United Nations convention against corruption. In: United Nations. https://www.unodc.org/documents/brussels/UN_Convention_Against_Corruption.pdf
20. Sultana S (2020) Role of financial intelligence unit (FIU) in anti-money laundering quest. J Money Laund Control 23(4):931–947. https://doi.org/10.1108/jmlc-01-2020-0003
21. Pieth M, Aiolfi G, Oxford Analytica Ltd, Lee M, Capus N, Pini M, Padfield N, Low LA, Tillen JG, Abendschein K, Fisher-Owens DM, Miller & Chevalier Chartered (2004) A comparative guide to anti-money laundering (By British Library, Library of Congress, Cambrian Typesetters, MPG Books Ltd, & Stiftung Finanzplatz Schweiz). Edward Elgar Publishing Limited. https://g-city.sass.org.cn/_upload/article/files/4d/a9/476d63434cc59bfbe12ff60d707a/4ee19ca9-51d7-45bd-8e3c-e6d3daa1b021.pdf. https://doi.org/10.4337/9781845421618
22. Kute DV, Biswajeet P, Nagesh S, Abdullah A (2021) Deep learning and explainable artificial intelligence and techniques applied for detecting money laundering-a critical review. IEEE Access 9:82300–82317
23. Duval, A., & Malliaros, F. D. (2021, April 18). *GraphSVX: Shapley Value Explanations for Graph Neural Networks*. arXiv.org. https://arxiv.org/abs/2104.10482
24. Parliament of India, Rajya Sabha Secretariat, Verma DD (2020) Compendium on parliamentary enactments: the prevention of money laundering (Amendment) act, 2012. In: Compendium on parliamentary enactments. Rajya Sabha Secretariat, New Delhi. https://cms.rajyasabha.nic.in/UploadedFiles/ElectronicPublications/Prevention_Maney_Laundering_Act_2012.pdf
25. Reserve Bank of India. (n.d.). https://www.rbi.org.in/commonman/English/Scripts/Notification.aspx?Id=593
26. Reserve Bank of India - Master Circulars. (n.d.). https://m.rbi.org.in//Scripts/BS_ViewMasterCirculars.aspx?Id=9914

27. Lokesh (2025) AML regulations in India: a complete guide for 2025. KYC Hub. https://www.kychub.com/blog/aml-regulations-in-india/
28. United Nations, Annan KA (2004) United nations convention against transnational organized crime and the protocols thereto. https://www.unodc.org/documents/middleeastandnorthafrica/organised-crime/UNITED_NATIONS_CONVENTION_AGAINST_TRANSNATIONAL_ORGANIZED_CRIME_AND_THE_PROTOCOLS_THERETO.pdf
29. Ribeiro MT, Guestrin SCS (2016) Why should I trust you? explaining the predictions of any classifier. In: Why should I trust you? explaining the predictions of any classifier, pp 1135–1144
30. Local interpretable model-agnostic explanation (LIME)—OECD. AI. (n.d.). https://oecd.ai/en/catalogue/metrics/local-interpretable-model-agnostic-explanation-lime
31. Black money in India: present status and future challenges. Int J Sci, Eng Technol Res 1(5) (2012)
32. Ribeiro MT, Singh S, Guestrin C (2016b) "Why Should I Trust You?" explaining the predictions of any classifier. In: Proceedings of the 22nd ACM SIGKDD international conference on knowledge discovery and data mining, 1135–1144. https://doi.org/10.1145/2939672.2939778
33. Lundberg S, Lee S (2017) A unified approach to interpreting model predictions. arXiv.org. https://arxiv.org/abs/1705.07874
34. Nauta M, Koopman M, Van Keulen M (n.d.) Explaining predictions of graph neural networks with SubgraphX. In: Proceedings of the ACM international conference on information and knowledge management (CIKM)
35. Pousette T, Rosendal A (n.d.) Explainable anti-money laundering application of explainable ai on graph neural networks in anti-money laundering. Department of electrical engineering Chalmers university of technology Gothenburg, Sweden 2024. https://odr.chalmers.se/bitstreams/e8e8359e-3154-46f6-830f-0133bcd38d99/download
36. Shaikh AK, Nazir A (2018) A novel dynamic approach to identifying suspicious customers in money transactions. Int J Bus Intell Data Min 14(4):1. https://doi.org/10.1504/ijbidm.2019.10010869
37. Csr I (2024) Understanding anti-money laundering (AML) and combating the financing of terrorism (CFT): fundamentals, definition, importance, historical context, future trends and innovations. India CSR. https://indiacsr.in/anti-money-laundering-aml/
38. Nishka K (2023) Biggest money laundering cases in India. Biggest Money Laund Cases India. https://blog.ipleaders.in/biggest-money-laundering-cases-in-india/
39. Garg R (2023) Money laundering—iPleaders. iPleaders. https://blog.ipleaders.in/money-laundering/
40. Deep learning and explainable artificial intelligence techniques applied for detecting money laundering–a critical review. IEEE JS & Mag|IEEE Xplore (2021). https://ieeexplore.ieee.org/document/9446887
41. Tiwary D, Vishwanath A (2024) How anti-money laundering law came to have a vast scope, granting police powers to ED. The Indian Express. https://indianexpress-com.cdn.ampproject.org/v/s/indianexpress.com/article/explained/explained-law/anti-money-laundering-law-pmla-evolution-9257298/lite/?amp_gsa=1&_js_v=a9&usqp=mq331AQIUAKwASCAAgM%3D#amp_tf=From%20%251%24s&aoh=17316862930035&referrer%3A%2F%2F%3A%2F%2Findianexpress.com%2Farticle%2Fexplained%2Fexplained-law%2Fanti-money-laundering-law-pmla-evolution-9257298%2F

Use of the Autonomous Weapon System: Exploration of Its Legality in Nexus with International Humanitarian Law

Shrabana Chattopadhyay, Snehatit Sett, Khushi Bhagat, and Sonali Mahakud

Abstract The widespread deployment of Autonomous Weapon Systems (AWS) in contemporary conflict presents numerous urgent legal and ethical dilemmas under International Humanitarian Law (IHL). This research article seeks to examine the legality of AWS under the current frameworks of IHL, particularly concerning the concepts of difference, proportionality, and necessity. Key research enquiries will encompass the compatibility of AWS with the fundamental tenets of IHL, any legal impediments associated with their deployment in armed conflict, and recommended regulatory measures to align such technologies with IHL. This research article aims to elucidate the principal legal ramifications associated with AWS, while considering recent academic discourse and prevailing international legal standards. The existing IHL will be evaluated only on its capacity to address the unique issues posed by the utilisation of AWS, as indicated by relevant scholarly literature. Proposals for legal change and international collaboration aimed at placing AWS under competent governance, ensuring it is not perceived as a threat to the humanitarian standards governing armed conflict, will also be evaluated. This paper contributes to the ethical and legal discourse on autonomous warfare, advocating for a framework that upholds human dignity while adhering to IHL.

Keywords Autonomous weapon system (AWS) · Use in war · Legality · International humanitarian law (IHL) · Human dignity

S. Chattopadhyay · K. Bhagat · S. Mahakud
Department of Law (IIIJS), University of Engineering and Management, Kolkata, India

S. Sett (✉)
Symbiosis Law School, Noida, Symbiosis International Deemed University, Pune, Noida, India
e-mail: snehatit2002sett@gmail.com

M. Chakraborty et al. (eds.), *Proceedings of Sixth International Ethical Hacking Conference*, Lecture Notes in Networks and Systems 1513,
https://doi.org/10.1007/978-981-96-8632-2_22

1 Introduction

For millennia, technical advancements and war have been closely associated. Governments and military commanders have searched for weaponry that will lessen the threat to their troops—the dehumanisation of the combat zone. For this reason, weapon systems are getting more sophisticated and people are relocating farther away from the combat zone. The development of Al in particular has led to the creation of armed systems that require little human intervention [1], such as autonomous weapons systems and killer robots [2].

For ages, there has been a connection between war and the advancement of technology. However, governments and military commanders have been looking for weaponry that will reduce the danger to soldiers. Humans are becoming farther removed from the combat, while weaponry is getting more sophisticated [3]. One of the most important technologies for future combat is autonomous weapons systems, or AWS [4]. Another well-known name for "AWS" is "Lethal Autonomous Weapon Systems" (LAWS) [2].

A precise definition of AWS must be established before moving forward. Although there isn't a widely agreed-upon legal definition, autonomous weapons are machines that can decide how to use weapons without significant human input or control [5]. In a nutshell, AWS are robotic weapons that can choose and strike targets without a human operator's assistance or oversight. Since these weapons differ greatly from conventional weapons of mass destruction, they raise legal concerns under IHL. For example, a human operator performs the deadly action in conventional weapon systems that are manned or, in certain situations, remotely piloted. This operator engages in a process of deliberation that culminates in the weapon's deployment. However, autonomous systems do not require human operators to carry out these vital tasks, which raises serious questions regarding whether AWS adhere to IHL's targeting regulations and the law of military conflict. They can make potentially illegal targeting judgments that raise the possibility of civilian casualties since they rely on software algorithms and similar technologies that lack the cognitive sophistication and deliberate thinking that a human mind possesses [6].

Since the First World War, robotic aspects have been incorporated into military systems, and as the years went by, these features gradually improved. Robotics has therefore been used for years to assist military forces in "identifying prospective targets, tracking them, the timing of when to fire and direction or homing them onto targets" [7]. During World War II, for instance, the US Navy benefited greatly from Carl Norden's bomb sight [8]. The bombsight consisted of a basic PC equipped with a device that was originally used by a mechanic to drop bombs with precise timing and location calculations. The German accuracy-guided automaton "Fritz X," which was controlled by a joystick and transmitter, serves as another illustration [9].

Similar to the contemporary fire and forget missiles, the U.S. military later adopted laser-guided cruise missiles and bombs, often known as "smart bombs," in the 1970s. The Fritz X was replaced by smart bombs, which were built with more sophisticated features. The machinist would mark the target using data or a laser before sinking the bomb. The bomb would automatically remain on the designated target after it was dropped until it struck [10].

There have been numerous battles and conflicts in recent years, such as the battle against ISIS, the war in Iraq and the Arab mechanism. Because of the information and communication technology, everyone is fully informed of these wars and conflicts. Without human intervention, AWS is able to choose targets and fire upon them independently. It can be used to evaluate the battlefield's current context and determine the necessary attack based on the processed information. It is devoid of all human decision-making and intelligence traits that subject and hold people to standards and laws [2].

However, the application of Al in military conflicts raises serious concerns for both human protection and compliance with IHL [11]. AWS is programmed and does not need human oversight, in contrast to remotely controlled weapon systems like drones, which are managed by a human from a distance.

Precursors with varying ranges of autonomy and lethality are currently in use, despite the claims of some legal scholars that weapons with complete toxic autonomy have not yet been used. Numerous nations conduct research and provide funding for initiatives aimed at developing autonomous weapons. India, United Kingdom, the Republic of Korea, China, Russia, Germany and Israel are a few among them. The United Kingdom, United States, Republic of Korea and Israel have already implemented robotic systems with different levels of autonomy and lethality [12]. The employment of autonomous weapons is classified under IHL as tactics and strategies of combat; the type of weapon employed in hostilities is the context of means in this law [4]. This phrase particularly describes the physical tactics used by combatants to harm their adversaries. Accordingly, all weapons and military systems are included in the phrase [13]. Generally speaking, the term describes how parties to a military-conflict employ weapons to conduct hostilities, which is different from the warfare approach [14].

There are no particular regulations regarding AWS under IHL. But according to the IHL, as required by AP-I to the "Geneva Conventions", countries must decide whether using any new AWS or weapons of mass destruction that they develop or obtain would be illegal under IHL in one or more situations [2]. It is impossible to hold an AWS responsible for its activities since it lacks moral or another agency [15]. Furthermore, it would be difficult to determine precisely who was responsible for accidents if AWS were only deployed in specific circumstances under the assumption that they could operate with prejudice. According to some academics, the person who ordered the autonomous weapons to go on a mission would be held accountable [16]. That would not be equitable, though, as it might be the responsibility of the individual who designed the autonomous robot, the person who programmed the mission, or the lawmakers or senior employees who decided to utilise it. It might also be argued that the device was damaged or interfered with [17]. HRW acknowledges the accountability of military commanders, programmers, and manufacturers, as well as some potentially accountable human actors in the use of AWS. HRW notes that cases involving serious violations of IHL or war crimes must be prosecuted under the law of war. For instance, deploying an AWS in a way that kills civilians without cause could be considered a war crime or a grave transgression of IHL.

In November 2012, two very distinct documents released by two very different groups sparked a public debate of high-profile about the ethics and legality of AWS. The first of these is the "DOD Directive: Autonomy in Weapon Systems", a policy paper on AWS released by the Department of Defense (DOD) of United States and signed by Ashton B. Carter, who is now Secretary of Defense and was Deputy Secretary of Defense at the time [18]. The fundamental principle of the policy document—in fact, the fundamental principle of DOD for all US military technology—is that automated advances will and should be incorporated into a large variety of weapon systems, both new and old. The Directive assumes that growing automation, even to the point of "autonomy" in some circumstances, is a reasonable goal in weapons design, despite emphasising dynamic and practical policies to mitigate the risks and contingencies that any given system might offer in any given scenario. Nevertheless, a second well-known document specifically challenges that premise. The well-known international human rights organisation "Human Rights Watch" (HRW) released a report and public call to action titled "Losing Humanity: The Case against Killer Robots in November 2012". According to the paper, any "fully autonomous" AWS should be prohibited from being developed, produced, transferred, sold, or used. In order to implement this comprehensive, preventative prohibition, an international treaty was required [19].

However, autonomous weapons also seem to offer advantages, despite these technical problems and the considerable opposition to their development. Among these is the military edge they can provide, which might greatly boost any military's power. It can successfully lessen human error or flaws on the battlefield, completely preventing war crimes and significantly lowering the danger to innocent populations. According to these reasoning, it is doubtful that autonomous weapons would be outlawed and that states will soon actively develop and deploy them [5]. This study will concentrate on the several kinds of AWS in development, the regulations governing them, and their adherence to IHL in light of the debate above [6].

2 Methodology

A comprehensive legal analysis and a detailed literature evaluation are the two main components of the study process. The first stage comprises a thorough analysis of the body of research on autonomous weapons, with an emphasis on operational considerations and associated technology. In order to guarantee a comprehensive yet accurate analysis of the existing literature, this literature review will employ a narrative method and incorporate scholarly journals, books, papers, and reliable internet sources.

The second section of the research methodology includes a comprehensive legal review of pertinent legislation controlling the development of contemporary weaponry, with an emphasis on autonomous weapons in particular. This requires a careful examination of international treaties, agreements, and customary international laws pertaining to the use of weapons in military conflict as well as national laws. In addition to analysing significant documents such as the "Additional Protocols to the Geneva Conventions" and other relevant international legal agreements, the legal assessment also thoroughly examines national laws and regulations pertaining to autonomous weapons.

The literature review and legal analysis are combined with technical and legal perspectives to form a cohesive research process. The synthesis of results from various areas will lead to a comprehensive understanding of autonomous weapons, which will form the basis for additional analyses and discussions regarding the legal implications of these systems [6].

3 AWS System

When it comes to defining particular weapons, such as autonomous or automatic weapons, IHL does not allow for any such definition or classification and does not contain any provisions pertaining to shortcomings on the general development and use of weapons, such as autonomous or automatic weapons. However, it does forbid specific means and methods during armed conflict [20]. Article 36 of the "Additional Protocol to the GC" of 1949, which addresses the safeguard of those affected by worldwide military conflicts involving new weapons, states: "A high contracting party is under an obligation to determine whether its employment would, in some or all circumstances, be prohibited by this protocol or by any other rule of international law applicable to the high contracting party" [21]. This clause, which is a component of conventional international law, has a vague definition that applies to all nations, including those who are not parties to the contract. It requires states to at least bring their armaments that might not violate IHL. Even if the material is relevant, it provides a negative response to our problem [2].

The Greek words "auto" (meaning "self") and "nomos" (meaning "law") are the origin of the word autonomous. In other words, self-governing [22]. "Having the freedom to govern itself or control its own affairs" is how the Oxford Dictionary describes it [2].

There has never been an attempt to define the term "autonomous" correctly before, but the "US Department of Defense" (DOD) did so for the first time when conducting a research study in this area and defined it as "a system that, once activated, can select and engage targets without further intervention by a human operator" [23]. Due to its broad coverage of weapon systems, DOD's definition caused uncertainty.

Human Rights Watch has classified autonomous weapons systems into many categories according to their level of autonomy in order to group the different types of AWS [24]:

(a) Firstly, there are "**human-in-the-loop weapons**". These arms are defined as: "A weapon system that, once activated, is intended to only engage individual targets or specific target groups that have been selected by a human operator". Therefore, only a human command is required for these weapon systems to deliver force after selecting a target or a group of targets. These munitions fall under the category of semi-AWS.

(b) "**Human-on-the-Loop Weapons**" are robots that can choose targets and apply force while being controlled by a human operator who has the ability to control the robots' movements. Thus, these weapon systems have the freedom to choose and attack particular targets. A human can step in and stop the operation if needed, but no human has the authority to determine whether those particular targets should be engaged.

(c) The third classification, referred to as "**Human-out-of-the-loop**" weaponry, include robots that, without human oversight or contact, can choose targets and exert force. In a pre-programmed area, these weapon systems are configured to autonomously select particular targets and attack them within a predefined time frame [4].

ICRC defines AWS as "autonomous weapon systems (also known as lethal autonomous weapons or killer robots) independently search for, identify, and attack targets without human intervention" [25]. There is a need to differentiate between automated and fully autonomous technologies. The definition of "autonomy" might be challenging. It alludes to extremely intelligent robots with the capacity for autonomous decision-making. The reality appears more like every day robotics than science fiction [26]. AWS will function considerably more like a Roomba than a Terminator [27].

3.1 Defining AWS

According to roboticist Noel Sharkey, an automated machine is one that "moves in a structured environment or performs a pre-programmed sequence of operations" [26]. On the other hand, an autonomous machine functions in an unorganised setting. The environment in which a machine works, not its internal operations, is essentially what gives it autonomy. Predictability is the primary distinction between automated and AWS. While a self-driving bot can only be predicted as a sequence of potential events, an autonomous machine is completely foreseeable (apart from a failure). Making this difference crucial to figuring out whether AWS can follow IHL's rules.

"Human out-of-the-loop weapons systems" are the most common description of AWS. Beyond the more general category of circlet weapon systems, some people classify autonomous weapons systems as human since they execute specific activities according to rules that are defined by humans and react to signals detected by their sensors rather than making decisions on their own [4].

3.2 Perspectives of Various Stakeholders Regarding Regulatory Implementation Challenges of AWS

1. Military Strategists' Perspective:

AWS serves as a revolutionary element for modern warfare according to military strategists because it delivers three primary operational advantages consisting of high-precision strikes and minimal casualties and fast decision capabilities. AWS implementation meets numerous challenges related to regulatory implementation which affect stakeholders in the following ways:

- The regulations must establish precise boundaries regarding combat-related autonomy decisions by combining operational advantages with responsible ethical practices.
- The integration of AWS into military operations requires advanced technological security systems because operational compliance with IHL must remain in force without degrading combat efficiency.
- The command structure needs to function as normal even though autonomous systems exist. The regulations need to establish an accountability procedure for AWS systems when they cause civilian casualties or unwanted escalation during military operations.
- AWS needs to integrate with human forces by implementing training and oversight systems as well as legal structures for strict adherence.

2. Policymakers' Perspective:

As part of their duties policymakers create legal structures for AWS which must handle national security needs along with international peace and moral issues. Their challenges include:

- The existing Geneva Conventions along with UN protocols have legal gaps because they fail to provide direct provisions about automated weapon systems. This leads to problems when trying to achieve regulatory alignment.
- Global Coordination with Nations proves difficult because political conflicts between countries can result in regulatory gaps which nations fail to follow.
- Effective monitoring schemes need to be implemented for verifying and enforcing AWS usage inside legal frameworks thus preventing possible misuse.
- AWS regulations must include provisions which handle public concerns together with diplomatic aspects to protect humanitarian values during conflicts.

3. Civil Society Advocates' Perspective:

AWS draw critical attention from both human rights organizations and advocacy groups and legal scholars due to their human rights implications. Their regulatory challenges include:

- Governmental agencies should create rules that require operators to maintain constant supervision over lethal operations through meaningful human intervention.
- AWS generate sustained focus from human rights associations together with legal scholars and advocacy groups due to their human rights consequences.
- Officials from governmental and military institutions must reveal AWS capabilities alongside their operational policies to preserve accountability.
- Rise in destructive technological competition will occur if regulations fail to establish and this will generate worldwide instability and unintentional warfare.

A proper AWS regulation demands collective work from multiple parties who must balance military performance with precise laws and humanitarian protection protections. The development of responsible rules concerning AWS requires collaboration among international bodies together with national governments along with military institutions and civil society groups.

4 Characteristics of AWS

Autonomous weapons are distinguished from traditional weapon systems by a number of special features [6]. These characteristics stem from their sophisticated technological capabilities, particularly in the areas of robotics and AI [28]. Understanding the technological features of autonomous weapons can help us better understand their potential applications and implications.

4.1 AI and Machine Learning

Autonomous weapons are equipped with sophisticated Al algorithms that enable them to learn from data, adapt to changing conditions, and make decisions independently based on preset objectives [29]. Machine learning techniques enable these weapons to evaluate vast volumes of data, spot trends, and gradually enhance their responses. This feature enables autonomous weaponry to respond swiftly and autonomously to changing battle field conditions.

4.2 Sensor Technologies

Autonomous weapons recognize and communicate with their environment using various sensor technologies. The weapons can gather data about their surroundings in real time, thanks to their sensors. Radar systems, lidar, cameras, and other specialized sensors are a few examples of these sensors. The weapons can successfully mancave and engage targets, thanks to these sensors, which provide the necessary information for obstacle detection, situational awareness, and target recognition.

4.3 Communication and Networking

Many autonomous weapons are designed to operate as part of a networked system, which facilitates coordination and communication between various components. The networked capabilities of autonomous weapons provide improved circumstantial awareness, coordinated operations, and cooperative decision-making [30]. Communication protocols and encryption technologies guarantee reliable and secure data exchange, reducing the likelihood of attackers interfering or obtaining unwanted access.

4.4 Decision-Making Algorithms

Autonomous weapons use sophisticated decision-making algorithms to analyze incoming information, detect potential threats, and determine the best course of action. Russell and Norvig state that these algorithms are designed to prioritize objectives, weigh the advantages and disadvantages, and make decisions based on preset standards [29]. Based on real-time factors including target recognition, engagement rules, and collateral damage potential, the weapons are able to make snap decisions.

4.5 Targeting and Engagement Systems

Automated weapons can engage targets precisely because of their sophisticated targeting and engagement capabilities. These systems use precise guiding mechanisms, such as GPS or inertial navigation systems, to ensure accurate location and trajectory control. Advanced tracking and image recognition algorithms help with target acquisition and tracking, increasing the guns' accuracy when attacking moving targets.

4.6 Redundancy and Fall-Safe Mechanisms

Redundancy and fail-safe features are commonly incorporated into autonomous weapons to ensure operational safety and dependability. These include sensors, communication networks, and backup power supplies to lessen the likelihood of single-point failures. To prevent undesirable or unauthorized acts, fail-safe techniques include emergency shutdown protocols, fail-over systems, and integrated safety measures.

5 Classification of AWS

The following are the current examples of AWS [2]:

PHALANX CIWS

This methodical weapon can identify any type of target and launch an autonomous strike without assistance from a human [31]. It provides a heads-up on any impending danger. The weapon has a self-package that often performs multiple functions, including tracking engagement, threat assessment, search detection, and kill assessment [32].

IRON DOME

Rafael Advanced Defense Systems Ltd., an Israeli defense technology corporation, created the Iron Dome Weapon System in 2004. Incoming rockets, artillery, and mortars can be detected, evaluated, and intercepted by it. The Raytheon Company, a significant American defense contractor and industrial company with a focus on weapons and military manufacturing, collaborates with Rafael to produce weapons like Iron Dome interceptor missiles, which can shoot down any incoming threat launched within a 7–70 km range. Since 2011, these guns have intercepted over 1,500 targets, making them the most widely utilized missile defense system in the world [33].

BAE TARANIS

The UK produced the BAE TARANIS prototype drones, which are autonomous combat drones named after the Celtic British god of thunder. Only when instructed by a human operator do they attack [34]. With Taranis, a stealthy intelligence, surveillance, and reconnaissance (ISR) platform, deep weapon strikes against a peer opponent with a first air defence network are feasible [35].

US NORTHROP GRUMMAN X-478

The US Navy's "Unmanned Combat Air System", or "X-47B", is designed to show that it can land on and take off from a Navy aircraft carrier's flight deck on its own while at sea. The first autonomous system is this prototype [36].

5.1 Compatibility Matrices of Different Types of AWS Under IHL

Classification	Compatibility with IHL
PHALANX CIWS	**Partially Compatible**: It recognises and eliminates danger on its own, but it may not be able to tell the difference between military and civilian things. The Phalanx lacks managerial control in fully autonomous mode, which is necessary for IHL to minimise adverse effects
IRON DOME	**Extensively Compatible**: The Iron Dome, which was built for defence, deflects incoming threats like missiles and projectiles. It complies well with the IHL criteria of necessity and proportionality because it only neutralises projectiles rather than attacking aggressively. Unintentional misfiring or misidentifications, however, can cause issues
BAE TARANIS	**Partially Compatible**: Attacks require human control even if it functions independently. Although this satisfies IHL's criterion for meaningful human oversight, there are questions about its capacity to distinguish between legitimate targets of warfare and civilians, particularly in complicated conflict situations
US NORTHROP GRUMMAN X-478	**Mostly Compatible**: This system prioritises autonomous landing and takeoff over offensive attacks. It does not breach IHL because it does not make its own decisions about engagement or targeting. However, the degree of human control used would determine its compliance if it were later altered for combat purposes

6 Utility of AWS

The advantages of AWS were highlighted by a former Joint Forces Command member, Gordon Johnson at the now-defunct Pentagon: "They don't go hungry. They don't feel scared. Orders are not forgotten by them. The fact that the man next to them was recently shot doesn't matter to them. Will they perform better than people? Indeed" [37]. The emergence of lethal AWS is facilitated by several tactical and operational aspects.

i. Compared to human-operated weapons, AWS are less expensive to run and can function continuously without requiring breaks [38].
ii. With advancements in battery and recharge technologies, AWS's potential mission time will only increase.
iii. In a short while, it will be possible for a single administrator to supervise a collection of partially autonomous drones or for a single human administrator to provide AWS mission instructions while monitoring them from a safe distance.
iv. AWS might be able to process combat data more quickly and effectively than people.

v. It's possible that AWS can uphold IHL more effectively than human soldiers. Since they won't need to defend themselves, they might be able to behave more sensibly. Humans will not be as capable of making observations on the battlefield as robotic sensors. The architecture of AWS will be free of human emotions that could impair judgment. The psychological issue of "scenario fulfilment," which occurs when people use fresh facts to support their preconceived notions, won't affect them [27].

In October 2023, the United Nation (UN) General Assembly, in its first committee (Disarmament and International security) approves the new resolution on LAWS [39]. It confirms that AWS are subject to international law, specifically the United Nations Charter, international humanitarian law, and international human rights law. The draft resolution raises concerns about the potential harm that AWS could do to both local and international stability as well as global security, including the possibility of an escalated arms race and the potential for disagreement and proliferation, notably to non-State actors [40]. Although there are presently no explicit conventions on arms control for AWS, the UN General Assembly has recognised the potential risks these systems pose and has called for additional international discussions to accelerate negotiations on a treaty or agreement to govern them. The adoption was also in consensus of Human Rights Council resolution 51/22 [41] of October, 2022 on human rights implications of new and emerging technologies in the military domain [42]. The general assembly emphasises how critical it is that the international community continue to deepen its understanding of the issues at hand and address the difficulties and worries brought forth by AWS, particularly through the Group of Governmental Experts on Emerging Technologies in the Area of LAWS.

7 International Humanitarian Law (IHL)

7.1 Introduction

IHL emphasizes that all tactics, which includes the use of armaments, must follow the principles of IHL when being used on the battlefield. IHL also outlines guidelines for the creation and use of any weapon that results in needless suffering or injury. State parties are required to evaluate the legality of any new weapons under "Article 36 of the 1949 Geneva Conventions' Additional Protocol I". The creation of new technical weapons based on AI programs has emerged as one of the main issues facing IHL in the field of weaponry. Therefore, there are two issues from a legal standpoint: First, the principle of distinction is not followed by autonomous weapons. Second, these weapon systems' noncompliance with the idea of proportionality [43].

Additionally, there are significant doubts regarding whether completely autonomous weapons could adhere to crucial tenets of IHL, a flaw that jeopardizes civilian legal rights. A fundamental element of IHL is the ability to distinguish between potential targets, such as combatants, and potential non-targets, such as civilians. The ability of fully autonomous weapons to consistently distinguish between these types of distinctions is uncertain. This is particularly true in the increasingly prevalent situations of modern warfare where fighters do not wear uniforms or other distinguishing marks. Assessing a person's intents is crucial to figuring out a possible target's condition and level of threat in close combat situations when there are no obvious indicators.

Being able to relate to someone else's emotional state is a crucial approach to identify their intentions. It would be challenging, if not impossible, to teach a robot to possess the inherent human traits that are essential for determining a person's intents. Programming completely autonomous weapons to pass the proportionality test—which forbids assaults in which the predicted harm to civilians outweighs the projected military benefit—would also be challenging. Because there are an endless number of different scenarios, robots could not be designed to find that balance in every circumstance. Furthermore, human judgement is used by IHL to subjectively determine whether an attack is proportionate or not. Doubts arises about fully autonomous weapons that whether they could exercise comparable judgment to assess proportionality in complex and evolving situations or not [2].

Since 2014, diplomats have discussed AWS problems under the United Nations Convention on Conventional Weapons (CCW) Group of Governmental Experts on LAWS [44]. Although these meetings haven't done much to draft an international convention on AWS, they have helped states clarify their stances, raised awareness of the issue, and improved the way that concerns about AWS are expressed. Arms control activists have asked for restrictions and new treaties, but the extent of these proposals varies. Organisations like the Campaign to Stop Killer Robots contemplate all autonomous weapons must be prohibited. Others hold a more sophisticated perspective, emphasising "unpredictable" weapons, such as the International Committee of the Red Cross [45]. According to Human Rights Watch, 30 states have approved a blanket ban [46]. Large military nations have opposed a new arm monitoring system, claiming that autonomous weapons are covered by current international law [47].

7.2 Applicability of IHL to AWS

The fact that certain individuals and most ordinary spectators seem to think that AWS are not governed by contemporary international law, or at least not by a sufficiently strong body of international law, is an important point of debate surrounding AWS that have been going on since 2012. This misunderstanding gives the argument for a new law to address them—whether it be a new protocol to the CCW or a ban treaty—more force and urgency. However, this is untrue; all weapons, including AWS, are subject to LOAC. A requirement of LOAC is that states conduct legal reviews of weapons to determine if they are lawful weapons based on certain longstanding baseline requirements; if there are any legal restrictions on the battlefield environments for which they are lawful; or if there any legal limitations on how they can be used (Thurnher 2013 for a non-technical exposition of these requirements) [48]. This is significant because, in spite of the attention that the NGO campaign for a ban and calls for a new CCW convention on AWS have received, there is already a strong legal framework in place for the legal evaluation of weapons [19].

Furthermore, AWS, automated weapons, and other armaments are already subject to the law of targeting and other basic LOAC standards. In fact, only a small number of weapon types—like chemical weapons—are subject to a unique set of regulations laid down in international treaties. Specialised regulations like those are the exception rather than the rule. A well-established corpus of legislation governs the great majority of weapon systems and their use, and it extends to newly developed weapons as well.

Three basic principles are included in targeting law, which regulates the use of legal weapons: proportionality, discrimination (or distinction), and attack precautions [49]. These are discussed below:

7.2.1 Principle of Distinction

The principle of distinction, sometimes known as discrimination, consists of two parts. Combatants must first be able to tell the difference between enemy combatants and civilians. Second, soldiers need to be able to tell the difference between military and civilian objects [50]. This principle is codified in Article 48 of the Additional Protocol I to the Geneva Convention: "In order to ensure respect for and protection of the civilian population and civilian objects, the Parties to the conflict shall at all times distinguish between the civilian population and combatants and between civilian objects and military objectives and accordingly shall direct their operations only against military objectives" [51].

The principle is applicable as customary international law for states that have not ratified Additional Protocol I. According to the International Committee of the Red Cross's ("ICRC") Commentary on the Additional Protocols, Article 48 of Additional Protocol I reflects the fundamental tenet of the laws and customs of war, which is that civilians must be protected and thus separated from combatants. Furthermore, the ICJ ruled in the "Nuclear Weapons Advisory Opinion" that the prohibition against indiscriminate assaults is a "cardinal" fundamental of IHL [52]. In particular, doubt is a worry of the ban on indiscriminate assaults. When there is enough uncertainty, a target will be presumed safe from assault [53]. This has been recognized as customary international law [54] and is enshrined in Article 50(1) of Additional Protocol I [55]. The presumption will be established in conditions that make "a reasonable attacker in the same or similar circumstances hesitate before attacking," [53] even though a certain amount of doubt is acceptable. Since the nature of military operations has changed from State against State combat to counterinsurgency operations, it has become more challenging to adhere to the principle of distinction. The fundamental idea of the principle of difference—that parties to a war must make a distinction between civilian and military targets—remains unchanged despite the difficulties in putting it into practice [27].

In contrast, the topic of whether autonomous weapons could adhere to the distinction principle is raised. Since they are entirely dependent on AI and are unable to recognize or interpret differences between individuals and fighters, especially in relation to current armed conflicts, the answer to this issue is unmistakably no [56]. These machines are not as capable as people. While on the battlefield, they are unable to assess each person's intentions, such as determining their emotional state [57]. An AWS might observe, for instance, two individuals running away from the robot while holding unknown objects. This is unquestionably a mother chasing after her two children as they play with toy guns. While a robot might just see people running with guns and potentially attack the targets, a human soldier could judge the situation as a safe demonstration and not a threat [58]. These machine robots might also be incapable of identifying soldiers and hors de combat [59]. Although the AWS can only operate in specific dreams or perspectives due to its programming, a human soldier can experience the entire environment.

7.2.2 Principle of Proportionality

According to the proportionality principle, any unintentional civilian casualties or property damage that could be anticipated from an attack must not be disproportionate [60]. Attacks are prohibited if they do not adhere to the proportionality principle. AWS provide a serious problem for IHL since they might not adhere to the proportionality principle since they lack the ability to predict the outcome of their attack [61].

Furthermore, striking a balance between the anticipated military benefit and potential collateral harm requires careful consideration of each situation and is dependent on intricate, value-based judgments. This raises the question of whether AWS complies with the proportionality principle. According to this study, AWS's lack of human judgment will prevent them from adhering to the proportionality criterion since autonomous systems are unable to securely discriminate between hostile combatants and fighters and protected civilians [2].

7.2.3 Principle of Precaution

Precaution, or preventing or reducing harm or loss to civilians, is the third core concept of IHL. The idea broadens the area of considerations that ought to be considered before an assault and calls for constant caution in order to protect the civilian population [60].

Art. 2(3) of the Hague Convention 1907, which mandates that a method be found to avoid any conditional damage, is where this principle initially originated [62]. According to the explanation in Art. 57 of AP-I to GC, special attention is drawn to the requirement, which is correctly identifying the targets to be attacked as military targets. Identification must be done with the utmost care, particularly if the target is far away. It also states that in order to meet the requirements outlined, it is crucial to obtain additional data and make monitoring easier in circumstances where there is the least amount of doubt [63]. Due to their lack of judgment, these weapons are unable to comprehend and change the situation in such complex situations [64]. Additionally, both customary IHL and article 57 of the "1977 Additional Protocol I" confirm that those conducting military operations must prevent themselves from harming civilian objects and populations.

Many people who have responsibilities related to a weapon system are required to take reasonable precautions. Those who provide, compile, interpret, communicate, or present the information that informs target decisions include campaign planners, mission planners, programmers of navigation equipment, personnel who feed data into the mission control software, load ordnance, monitor and/or control platform operations, program the recognition and decision-making software that determines whether an observed object or person should be attacked autonomously, and others [65]. When it comes to an AWS, even if humans take practical planning safeguards, their plans must still be applicable when the machine decides to attack. In static settings, this might be possible, but in dynamic settings and without human intervention, it is unlikely to be feasible. To put it into practice, sensors would need to be able to accurately identify civilian and military objectives and people, assess military advantage, choose the best weapons and tactics, anticipate incidental harm to civilians and their property, and provide warnings in rapidly changing circumstances. In other worlds, AWS lacks the capabilities necessary to adhere to the precautionary rule [65].

In the end, an AWS is a machine by definition, and regardless of how sophisticated its technology gets or how well it imitates us humans, it will never be able to fully resemble us. This raises the question of whether the necessary common sense and good faith could ever be replicated by a machine [66]. In every instance, a person would have a better understanding of the new situation and whether or not it warrants the use of deadly force. It wouldn't just shoot without thinking because a preprogrammed algorithm told it to [67].

7.2.4 The Principle of Military Necessity

One of the fundamental tenets of IHL is that it encourages contracting parties to defend actions that are not prohibited by it by citing factors like military necessity [68]. Francis Lieber's code of April 24, 1863, which stipulated the requirement of actions that are both necessary for ensuring the war's conclusion and legal under current law and wartime customs, is where the concept of military necessity first appeared [69].

From the military necessity concept, two things can be inferred. The first is the ability to recognize a legal target. The potential evaluation of whether the devastation could yield a clear military advantage comes in second. AWS are entirely dependent on AI and do not possess the same level of judgment as humans. Since they lack the ability to discern potential situations, they will be unable to evaluate their target's outcome. "A content-dependent, value-based judgment of a commander" is another term for this military necessity idea [70]. AWS cannot assess whether eliminating the target will meet a military necessity or a military desired advantage if it cannot identify if the target is a cultural artifact, a medical facility, or a military installation [71]. Similar to the proportionality concept, the military necessity principle also necessitates a subjective analysis of a situation.

However, AWS are unable to determine if an enemy soldier has become hors de battle; in such cases, human judgment is required. It would be difficult for a completely autonomous robot sentry to determine whether a gate crasher it shot was feigning harm or just pounded to the ground by the impact. Accordingly, it might needlessly shoot the person a second time [65].

7.2.5 Martens Clause

The paragraph that now bears Friedrich von Martens' name was introduced by the Russian representative to the Hague Peace Conference in 1899 and 1907 as a result of a disagreement among attendees of the conference over the state of resistance activities in seized territory [72]. The sentence served as a reminder to the parties that, in the dearth of an agreement, the delineation of the resistance groups' status was not free simply because the language did not specify anything. Rather, the public conscience and human values have to be invoked in its resolution [73]. While there is ongoing debate on the clause's precise meaning and scope, it might be argued that it indicates that international law is not always silent on a certain issue just because it is not mentioned [74].

Should the Martens Clause now be considered when determining whether autonomous weapons are legal? Regarding the complementarity of weapon systems to the "Martens Clause" and whether or not AWS should be authorized to decide when to launch an attack without human oversight, legal scholars have voiced strong viewpoints. What this actually means is yet unknown. However, a significant majority of people find the concept of an automatic judgment to be frightening and abhorrent.

Three key principles determine whether weapon systems are legal [19]. A weapon system cannot be indiscriminate by design, to start. According to some experts, the rule applies to the weapon's "normal" uses, or the applications for which it was intended, given its nature. Second, a system of legal weapons must be "of a nature" to inflict "superfluous injury or unnecessary suffering". Third, if a weapon system's destructive consequences cannot be "controlled," it may be considered unlawful in and of itself. The prohibition on weapons that have uncontrollably negative effects is mostly applicable to biological weapons, such as viruses, which are uncontrollable and uncontainable once they are launched. Once more, the threshold for making a weapon illegal in and of itself is high, despite the fact that numerous LOAC regulations forbid the use of weapons in situations that could have uncontrollable consequences.

Ultimately, according to certain LOAC specialists, there is no theoretical reason why an autonomous or a highly automated system couldn't meet targeting legislation criteria [75]. However, in practice, weapon systems with full or semi-autonomy that are unable to adhere to all LOAC regulations may still play a significant role in the future, provided that they are subject to a very stringent set of restrictions on target selection and engagement. An AWS may, for instance, be configured with much more stringent limits than those mandated by law; rather than proportionality, it could be configured to cease firing if it detects the presence of any civilians. In other words, just because it is an AWS does not mean that it cannot be used unless it can fully adhere to the LOAC requirements on its own. As AWS players are slowly realizing, the control of machine-human interactions—rather than AWS being unleashed on a battlefield—is the true subject of discussion.

7.3 IHL Principles: Are They Sufficient?

According to Michael Schmitt, the chairman of the "US Naval War College's International Law Department", AWS would already be illegal if they couldn't adhere to IHL's legal requirements [76]. However, the use of IHL standards and the possibility of criminal penalties for violations of them might not be enough to guarantee that AWS are used legally [77].

It's unclear if AWS will impact civilian populations as much as cluster munitions and anti-personal landmines.

7.4 Challenges Under IHL

Though the present legal structure that governs AWS and other kinds of weapon system is primarily LOAC and its weapons review process (and some other bodies of law, such as human rights legislation, might apply in certain particular contexts), proponents of a complete ban usually offer a number of arguments in favour of a comprehensive, anticipatory ban. This section discusses three of the most well-known [19]:

(a) AWS should be prohibited on the basis of the unadulterated moral principle that humans, not robotic machines, should make the moral decisions to kill;
(b) machine programming and Al will never be able to meet the conditions of LOAC, law, and ethics, and as a result, they should be prohibited now.
(c) Because computer decision-making diminishes or perhaps completely destroys the ability to hold people accountable, AWS should be outlawed. For example, a single human soldier may be held accountable for unlawful or even criminal activities.

The challenges under IHL are as follows:

i. Deference with the Principle of Distinction

The idea of differentiation appears to be an objective need. AWS needs to be able to determine with objectivity whether a possible target is a military or civilian target [78]. However, depending on the situation, a target may be categorized as either a military target or a civilian target, which presents challenges.

Armin Krishnan highlights three primary issues with AWS's capacity to differentiate between legitimate and civilian targets in Killer Robots. First, "weak machine perception" could affect AWS. Second, the "frame problem" could arise from AWS's inability to communicate with their surroundings. Third, "weak software" might be the issue [79].

ii. Weak Machine Perception

Separation necessitates a sensory-based assessment. Present-day technology is just getting close to being able to tell a human from a non-human object, much less a combatant from a civilian [80]. Allowing AWS to fire only after being shot upon would be one way to address this issue. Another would be to restrict AWS use to circumstances in which the claimed hostile force is clearly identifiable [81]. Additionally, restricting the number of possible targets to a predetermined list of legitimate targets could help guarantee that AWS follows the principle of distinction [82]. Although theoretically sound, this strategy might not be feasible given the tendency in warfare toward counterinsurgency operations in or close to civilian areas. In practically every situation, this restriction would make AWS unsuitable for usage.

iii. Frame Problem

In a modern battlespace that is fast-paced and complicated, AWS will find it challenging to evaluate all the data required to accurately comprehend the situation. Going through every option might theoretically take an endless amount of time. As a result, it will be necessary to design AWS to differentiate between pertinent and irrelevant data. Programming this kind of differentiation could result in circumstances where the data is misunderstood, which could lead to an indiscriminate attack in an open setting.

The threshold's acceptance in AWS is complicated by the way it is stated in terms of human rationality. Depending on the situation, this assessment might necessitate varying "doubt thresholds." According to Michael Schmitt, AWS won't go against the principles of differentiation as long as human operators don't set the "doubt thresholds" excessively high, increasing the likelihood of an attack [83].

iv. Weak Software

The complexity of software makes it less predictable. The interactions inside the software become unpredictable as well because no single coder is familiar with or understands the complete piece of software [84]. When combined with an open environment, this could result in instances when an unexpected software malfunction causes AWS to use force arbitrarily.

Deep Blue, a chess-playing computer developed in 1997 by Murray Campbell and Feng-Hsiung Hsu, eventually defeated Garry Kasparov, the world's best chess player. The "weak software" issues that arise when programming AWS are best shown by the challenges that Campbell and Hsu encountered when developing Deep Blue.

v. Compliance with the Principle of Proportionality

Comparative weighting of conflicting interests is necessary for the assessment of proportionality. AWS has a number of obstacles to overcome in order to assess a situation and provide a suitable reaction. While people are known to be able to balance complex interests, AWS is not in the same position.

vi. Collateral Damage

When using AWS in public civilian settings, caution must be used to make sure that enough data is gathered and processed to support the assumptions needed to determine collateral damage.

vii. Military Advantage

There isn't a system that can determine military advantage at the moment. It takes sophisticated processing and sensing skills, as well as an algorithm that can quickly and accurately determine proportionality, to do what a human soldier would see as obvious. It will be necessary to include some shortcuts into the software because AWS will not have endless time to perform these computations.

The term "Reasonable Commander" refers to the ability to balance the two. According to one source, AWS's use may become "almost useless except in the narrowest of circumstances" because to the difficulties in developing them [85]. They will never be able to launch a proportionate attack if they are not programmed to satisfy the "reasonable commander" standard while balancing the military advantage against any collateral harm.

8 Recommendation and Conclusion

With the development of technology, current technologies are becoming more and more autonomous every day, especially weapons, whether they are defensive or offensive. Technology is developing quickly, bringing complete autonomy in modern weaponry closer to reality. The world community finds complete autonomy in armaments concerning since military leaders have no influence over the battlefield during a conflict. IHL faces three significant obstacles as a result of such weapon autonomy. The primary ones are the management and use of such weapons from a great distance and, second, the principle of adhering to IHL, as total weapon autonomy would make it difficult to adhere to the fundamentals of IHL.

Lastly, the most significant problem it presents is the question of who will bear responsibility if these autonomous weapons break the rules of war [2].

After analysis, it is evident that giving weapons total autonomy through the use of AI would render them entirely out of date. In order to prohibit them or at least put them under the purview of Article 36 of the CCW, states must enact a new international law. As was previously said, protecting civilians and others who are not lawful targets during armed conflict is the fundamental and cornerstone premise of IHL. Nonetheless, the study's conclusions support the following research recommendations:

i. Because Al-equipped machines make their own decisions, the autonomy program in AWS needs to be restricted and kept under human supervision. This is due to the fact that it can be difficult to determine if a bad choice was caused by AWS's software or by the independent considerations of the Al-equipped (what we refer as smart) machines. Less autonomy could result in their adherence to the fundamental IHL principle.
ii. The requirement for human supervision or intervention over the crucial tasks of target selection and engagement over individual attacks. Make it clear that individual attacks must be subject to meaningful human control, and that weapon systems that lack this kind of human oversight need to be outlawed. Because there will never be any compensation for the human cost of these weapons.
iii. To specifically forbid such deadly weapons, legal measures must be taken and the Convention on Certain Conventional Weapons may be amended. Because, as autonomy has become more prevalent on the battlefield, it will impose an obligation on state members to manage antecedents that have already been developed or deployed.
iv. In addition to prohibiting lethal weapons, the international community must remember that if these weapons are produced and used in combat, laws and guidelines that explicitly outline who is responsible for any harm caused by these AWS during hostilities must be in place. If a human does not decide to fire on a particular target, the attack's accountability will violate the war autonomous weapons system.
v. In order to reduce the fundamental threat posed by these weapons, states should begin actively engaging in diplomatic discussions with one another over killer robots and clearly define the boundaries of expanding weapon system autonomy.
vi. National laws and regulations should be changed to forbid the production, development, deployment, and use of fully self-sufficient weaponry.
vii. Until it is too late, the international community needs to keep up with technological advancements and negotiate and approve a new, legally enforceable ban on it. They won't have an opportunity to stop the development of a weapon that may potentially transform combat in a dangerous way until then.
viii. Although the UN has called for their prohibition, technology has now been developed to produce autonomous weapons that can choose and murder human targets without human oversight.

ix. Human security and freedom would be severely jeopardized by the production and employment of autonomous weapons like drones, tanks, and automatic machine guns, and the window for stopping their development is rapidly closing.
x. States should consider whether the items they are examining are weapons in the strictest meaning of the word when conducting a legal examination of AWS in accordance with Article 36 of "Additional Protocol I to the Geneva Conventions". According to the Martens Clause, states must also consider human rights standards and universal values.
xi. If AWS is designed in accordance with the laws of warfare, it can adhere to international regulations.

References

1. Geneva Academy of International Humanitarian Law and Human Rights 2014, p 3
2. Khan A (2022) Autonomous weapons systems and the principles of international humanitarian law, Kindle Direct Publisher, ISBN: 9781234567890. https://www.researchgate.net/publication/359471177_Autonomous_Weapons_Systems_and_the_Principles_of_International_Humanitarian_law
3. Geneva Academy of International Humanitarian Law and Human Rights, 2013
4. Gunawan Y, Aulawi MH, Ramadhan AR (2019) Command responsibility of autonomous weapons systems under international humanitarian law. Jurnal Cita Hukum 7(3):351–368
5. Crootof R (2015) The killer robots are here: legal and policy implications. Cardozo Law Rev 36:1837–1840
6. Maqbool A, Anwar A (2023) Warfare and machines: an in-depth study of autonomous weapons in the context of international humanitarian law. Soc, Law Policy Rev 2(1):01–14
7. Peter Scharre's presentation at the CCW (2015), page 2. Or Paul Scharre and Michael C. Horowitz, "An Introduction to Autonomy in Weapon Systems," CNAS Working Paper (February 2015). http://www.cnas.org/sites/default/files/publicationspdf/Ethical%20Autonomy%20Working%20Paper_021015_v02.pdf,7
8. Hillenbrand L (2018) Unbroken: a world war ii story of survival, resilience, and redemption, S.L.: Ballantine
9. Singer P Smithsonian national air and space museum, Bomb, Guided, Ruhrstahl Fritz X (X–I); p 48
10. Singer P The national aviation hall of fame, honoring aerospace legends to inspire future leaders: Carl Norden; p 57
11. Leveringhaus A (2016) Ethics and the autonomous weapons debate. In: Palgrave Macmillan UK eBooks, pp 1–30. https://doi.org/10.1057/978-1-137-52361-7_1
12. Anderson K, Waxman MC Law and ethics for autonomous weapon systems: why a ban won't work and how the laws of war can. Columbia Public Law Research Paper 13–351
13. International Committee of Red Cross (2019)
14. Geneva Academy of International Humanitarian Law and Human Rights, Method of warfare. http://www.weaponslaw.org/glossary/method-of-warfare
15. Noorman M (2018) Computing and moral responsibility, the Stanford encyclopedia of philosophy, Spring 2018 Edition
16. Di Nucci E, Sio FSD (2016) Drones and responsibility: legal, philosophical and socio-technical perspectives on remotely controlled weapons. Ashgate, Farnham, Surrey, UK England

17. Human Rights Watch Losing Humanity, Asaro P (2012) On banning autonomous weapon systems: human rights, automation, and the dehumanization of lethal decision making. Int Rev Red Cross 94(886). https://doi.org/10.1017/s1816383112000768
18. DOD Directive (2012)
19. Anderson K, Waxman MC (2017) Debating autonomous weapon systems, their ethics, and their regulation under international law
20. Cassese A, Acquaviva G, Fan MD and Alex (2011) International criminal law, cases and commentary. Oxford University Press, at 134
21. Article 36 of Protocol Additional to the Geneva Conventions of 12 August 1949, and relating to the Protection of Victims of International Armed Conflicts (Protocol I), 8 June 1977
22. Bhuta N, Kreβ C (2016) Autonomous weapons systems: law, ethics, policy. Cambridge University Press, p 165 (herein after, Nehal Bhuta, C.K. AWS, Law, Ethics, Policy 2016)
23. US DoD Directive 3000.09, Autonomy in weapon systems, 21 November 2012, p 13 (herein after US DoD 3000.09, 2012)
24. Davison N (2017) A legal perspective: autonomous weapons system under international humanitarian Lato, UNODA Occasional Papers, No.30, New York, United Nation Publisher
25. ICRC (2014) Autonomous weapon - Q & A
26. Sharkey N (2011) Automating warfare: lessons learned from the drones, 21 JL Inf & Sci 140 at 141 [Automating Warfare]
27. Foy J (2014) Autonomous weapons systems: taking the human out of international humanitarian law. Dalhousie J Legal Stud 23:47
28. Toscano CP (2015) Friend of humans: an argument for developing autonomous weapons systems. J Nat'l Sec L & Poly 8:189
29. Russell S, Norvig P (2016) Artificial intelligence: a modern approach. s.l.:Pearson
30. Bhosal B (2016) Curvelet interaction with artificial neural networks. In: Shanmuganathan S, Samarasinghe S (eds) Artificial neural network modelling. s.l.: Springer International
31. Geiss R (2015) Dimension of autonomous weapons systems, 08
32. Lin P, Abney K, Bekey GA, Ethics R (2011) The ethical and social implications of robotic. MIT Press, at 381 and https://navaltoday.com/2017/04/12/raytheon-tests-new-electric-gun-for-phalanx-ciws/
33. Kelleher C, Dombrowski PJ (2015) Regional missile defense from a global perspective. Stanford University Press, at 144 and https://www.raytheon.com/capabilities/products/irondome
34. Geiss R (2015) Dimension of autonomous weapons systems, 08
35. Paul J (2012) Springer, 68
36. Michael G (2015) United states naval aviation 1911–2014. Pen and Sword, at 177
37. Tim W (2005) GI robot rolls toward the battlefield. International Herald Tribune. Online: http://www.nytimes.com/2005/02/16/technology/16iht-robot.html?pagewanted=all&_r=0
38. Ethical and doctrinal implications, supra note 33 at 2
39. Select a language for A/C.1/78/L.56 (n.d.-b) Retrieved March 28, 2025, from https://undocs.org/A/C.1/78/L.56
40. First Committee approves new resolution on lethal autonomous weapons, as speaker warns 'An algorithm must not be in full control of decisions involving killing'|Meetings coverage and press releases (2023). https://press.un.org/en/2023/gadis3731.doc.htm
41. Document Viewer (n.d.) Retrieved March 29, 2025, from https://undocs.org/en/A/HRC/RES/51/22
42. See Official Records of the General Assembly, Seventy-seventh Session, Supplement No. 53A (A/77/53/Add.1), chap. III, sect. A
43. Fleck D (2010) The handbook of international humanitarian law. Oxford University Press, Oxford, p 30
44. United Nation, Lethal Autonomous Weapon Systems (LAWS). https://disarmament.unoda.org/the-convention-on-certain-conventional-weapons/background-on-laws-in-the-ccw/
45. ICRC position on autonomous weapon systems (2022) International committee of the Red Cross. https://www.icrc.org/en/document/icrc-position-autonomous-weapon-systems

46. Wareham M (2023) Stopping killer robots. In: Human rights watch. https://www.hrw.org/report/2020/08/10/stopping-killer-robots/country-positions-banning-fully-autonomous-weapons-and
47. Geneva UM (2021) U.S. Statement at the GGE on LAWS during the discussion of agenda item 5(d)—U.S. mission to international organizations in Geneva. U.S. Mission to International Organizations in Geneva. https://geneva.usmission.gov/2021/08/05/u-s-statement-at-the-gge-on-laws-during-the-discussion-of-agenda-item-5d/
48. Thurnher J (2013) The law that applies to autonomous weapon systems. American Society of International Law, 17 Insights www.asil.org/insights/volume/17/issue/4/law-applies-autonomous-weapon-systems
49. Boothby W (2012) The law of targeting. OUP
50. LOAC, supra note 42 at 251
51. Protocol additional to the Geneva convention of 12 August 1949, and relating to the protection of victims of international armed conflicts (Protocol I), 8 June 1977, 1125 UNTS 3, art 48, (entered into force 7 December 1979) [Additional Protocol I]
52. Nuclear weapons advisory opinion, supra note 46 at 78
53. AWS and IHL, supra note 43 at 16
54. See Protocols commentary, supra note 52, at 1410–1439
55. Additional protocol I, supra note 50, art 50(1)
56. Sharkey N (2008) Grounds for discrimination: autonomous robot weapons. USI Defence Systems, 87
57. Guarini M, Bello P Robotic warfare: some challenges in moving from Noncivilian to civilian theaters. In: Lin A, Bekey (eds) Robot Ethics, p 131
58. Guarini M, Bello P Robotic warfare: some challenges in moving from Noncivilian to civilian theaters. In: Lin A, Bekey (eds) Robot Ethics, p 130
59. Protocol additional to the Geneva conventions of 12 August 1949, and relating to the protection of victims of international armed conflicts, adopted in 8 June 1977 and entered in force 7 December 1978. Art 41
60. Crowe J, Weston-Scheuber K (2015) Principles of international humanitarian law. Edward Elgar, Cheltenham, UK
61. Guarini M, Bello P Robotic warfare: some challenges in moving from non-civilian to civilian theaters. In: Lin A, Bekey (eds) Robot Ethics, p 131
62. Hague convention (IX) concerning Bombardment by Naval Forces in Time of War, The Hague 18 October 1907
63. Sandoz Y, Swinarski C, Zimmerman B (eds) Commentary on the additional protocols I and II of 8 June 1977 to the Geneva conventions of 12 August 1949, (ICRC 1987), 2195. (Herein after Sandoz, Yvez, Commentary on the Additional Protocols 1987)
64. Human rights watch, losing humanity (2012)
65. Dinstein Y (2017) Autonomous weapons and international humanitarian law. Dehumanization of Warfare 336. Or Jeffrey ST (2013) Examining autonomous weapon systems from a law of armed conflict perspective. New Technol Law Armed Confl 222
66. Commentary AP I, comment no. 2208
67. Krishnan, Killer Robots 2009 p 132
68. Gary DS The law of armed conflict: international humanitarian law in war (New York: Cambridge University Press, 2010) at 53
69. Instructions for the Government of Armies of the United States in the Field, General Orders No. 100 § 1, art. 14, (Apr. 24, 1863) ("Military necessity, as understood by modern civilized nations, consists in the necessity of those measures which are indispensable for securing the ends of the war, and which are lawful according to the modern law and usages of war."); Burrus M. Carnahan, Lincoln, "Lieber and the Laws of War: The Origins and Limits of the Principle of Military Necessity", 92 AM. J. INT'L L. 213, 213 (1998) at 215
70. Kastan BN (2013) Autonomous weapon systems: a coming legal singularity? J Law, Technol Policy 45:17

71. Gillespie T, West R (2010) Requirements for autonomous unmanned air systems set by legal issues. Int C2 J 4(2):9–10
72. Koutroulis V (2013) Martens Clause, Oxford bibliographies online datasets
73. Fleck D (2008) The Law of Non-international armed conflicts" in the handbook of humanitarian law in armed conflict. Dieter Fleck (ed), pp 619–620
74. Christopher, 2nd ed, 2008, p 34–35
75. Schmitt M, Thunher 3 (2013) Out of the loop: autonomous weapon systems and the law of armed conflict, 4 Harvard National Security Journal 234 http://harvardinsj.org/2013/05/out-of-the-loop-autonomous-weapon-systems-and-the-law-of-armed-conflict/
76. AWS and IHL, supra note 43 at 2
77. Convention on the prohibition of the use, stockpiling, production and transfer of anti-personal landmines and on their destruction, 18 September 1997, 2056 UNTS 211 (entered into force 1 March 2009); Diplomatic Conference for the Adoption of a Convention on Cluster Munitions, Convention on Cluster Munitions, 3 December 2008, CCM/77 (entered into force 1 August 2010)
78. Taking Humans Out of the Loop, supra note 17 at 159
79. Killer Robots, supra note 5 at 98–99
80. Automating Warfare, supra note 6 at 143–144
81. Major Jeffrey S Thurnher (2012) No one at the controls: legal implications of fully autonomous targeting 67:4 Joint Forces Quarterly 77 at 83
82. Robot Soldiers, supra note 15 at 41
83. AWS and IHL, supra note 43 at 17
84. Killer Robots, supra note 5 at 100
85. Taking Humans Out of the Loop, supra note 17 at 163

Predicting the Future: The Scope of Artificial Intelligence Integration in Merger Transactions with Special Reference to Due Diligence Process in India

Purbasha Mukherjee

Abstract It is anticipated that there will noticeable new trend of due diligence in the domain of mergers and acquisitions with the advent of artificial intelligence (AI). AI-powered solutions are continuously getting better in automating a variety of processes, including risk assessment data collection, and analysis. Investigating the public sources serves the initial stage in a due diligence procedure. At this point, AI can look for ongoing tax and due diligence conflicts in public sources including official announcements, financial reports, prospectuses, and media coverage. The problems found in such an AI-generated report might be used as personalised input for the management's due diligence information. By automatically categorising the submitted documents, AI systems can expedite the procedure. Additionally, the software could look for sensitive information in the documents. In Merger transactions, the due diligence process includes numerous documents to be examined which contains employee details, sensitive information, and intellectual property details. AI Integration offers significant time saving and efficiency benefits. After a potential risk has been identified, evaluating its impact of the risk follows. For this, people use like their prior experience with identical situations, conversations with authorities and co-workers, practical training, as well as understanding of human reactions to legal and financial matters. In order to understand the patterns and control the risk assessments, AI can make use on its primary strength by identifying gaps and offering analysis of the submitted documents. This paper will discuss the role AI plays when a company merges and also the stance of AI in the detailed checklist identifying the benefits and challenges associated in due diligence with Mergers and Acquisitions.

Keywords Artificial intelligence · Due diligence · Mergers and acquisitions · Risk · Transactions

P. Mukherjee (✉)
St. Xavier's University, Kolkata, India
e-mail: purbashamukherjee11@gmail.com

M. Chakraborty et al. (eds.), *Proceedings of Sixth International Ethical Hacking Conference*, Lecture Notes in Networks and Systems 1513,
https://doi.org/10.1007/978-981-96-8632-2_23

1 Introduction

The advent of AI in recent years has been an eye opener for the business. In every sphere be it health to finance to manufacturing, the pivotal role of AI can be witnessed [1]. Companies are also are not lagging behind in incorporating AI in their corporate transactions. AI undoubtedly brings accuracy and efficiency and two areas that needs special mention are corporate governance and legal sector [2].

Companies use a strategy called Merger to enhance their overall scale and finance. Companies are increasingly using Mergers and Acquisitions strategy in order to capitalize on growth opportunities, adjust to shifting market conditions, and handle regulatory complications [3].

Growth is primarily of two types; Organic growth and Inorganic growth. When the production increases, profit increases, it is organic growth. It is done through deployment of men, money and material [4]. On the other hand, inorganic growth is the avenue where the growth is accelerated by skipping a few steps in the growth ladder. It is attained through M&A and is seen as a faster way to grow as compared to organic growth [4]. An example to explain organic and inorganic growth may be made:

When two companies are growing at different rates of growth, it is obvious that investors will opt for investment in the company which has a higher growing rate. It is further found that the company is growing at a faster rate because it bought out its competitor. Hence it is seen that though company might be growing, yet lot of risks appear with its growth. While the company which has a lesser rate of growth, is still is in a better position as it is growing without acquisition and therefore there is no necessity to take more debt. Thus, some may go with the company with better growth and will be ready to take the additional risk, while the rest would definitely go for a safer investment.

2 Understanding Mergers and Acquisitions

The consolidation of two companies' assets or shares into one is called Mergers and Acquisitions It may be large or small. Merger is the process where one company erges into an existing company or two or more companies may merge together to form a new company [4]. Broadly speaking the company which extinguishes its existence is usually called the target company and the other remaining is called acquirer company [4]. The company which merges is called the Merging or the Resulting company, the target company, or the seller, while the company with which it merged is called Resultant company often known as the buyer [4].

Different ways in which a merger can take place [5].

(i) Purchasing of assets: company 'A's' assets are sold to company 'B'. Company 'A' ceases to exist and company 'B' survives.

(ii) Purchase of shares: shares of 'X' company can be bought by company 'Y'. When it got hold of all the shares company 'X' is dissolved.
(iii) By exchange of shares for assets: company 'A' may give its shares to shareholders of 'B' for assets. Company 'B' is terminated and holds shares of 'A'.
(iv) By exchange of shares for shares: 'X' gives shares to shareholders of 'Y'.'X' is terminated.

The structure of a merger may look like;
A + B = A
A + B = B
A + B = C
Here, in the above examples we see there are two companies 'A' and 'B'. One of the companies has merged with the other existing company. But in the last scenario we see that two companies have merged into a third company, a new company 'C' [5].

Acquisition is the process of purchase of shares and assests of a company by another company. It may be friendly, where the deal is approved by the board of Directors or hostile applying various strategies to succumb the target company [5]. The reasons behind mergers and acquisitions are to achieve these objectives [6]:

(a) Synergistic Operational Advantages- to produce a new thing together which has got enhanced effect rather than as separate units.
(b) To reduce the cost of production, legal and other professional expenses.
(c) To strengthen its financial situation and carry forward its survival.
(d) To optimize the use of factors of production and allocation of resources and capacities.
(e) To avail the tax advantages of a profit-making amalgamated company and to set off losses of the amalgamating company.

2.1 Stages of Mergers and Acquisitions

The mergers and acquisitions process usually starts from the side of the target company. Before formally starting the selling process, the target company ensures that its legal and corporate affairs are in order [7].

The procedure usually starts with the Target company creating an information memorandum in the first stage. The memorandum contains relevant details to help prospective buyers determine their degree of interest without disclosing sensitive company information. At this point, interested parties typically sign Non-Disclosure Agreements (NDAs) to safeguard sensitive information [8].

Negotiation is one of the most important aspect in a merger. The factors that influence negotiation include: financial strength, employees retention, antitrust issues, licencing and registration. Some supplementary clauses may be added while finalising the proposed deal [8].

The prospective purchaser reviews and then bargains over the details of the agreement, including the purchase price, restrictions, warranties, and indemnities. Last but not least, the post-integration phase contains provisions for post-closing responsibilities such as price modifications, asset transfers, and consent acquisition. Parties also engage in integration exercises during this time [8].

2.2 Identifying the Red Flags: Issues and Challenges in Mergers and Acquisitions

There are some challenges with Mergers and Acquisitions and the first major concern relates to the financing of the scheme. Restructuring entails a large financial investment and therefore, increases accountability [2].

A major issue is the timing of the Mergers and Acquisitions. The acquiring company has to decide as to what is the optimum time of striking a deal which can provide maximum value for the company. A major factor affecting the timing of the deal is the level of competition too [2].

Another significant problem is hasty decision-making. Overemphasizing financial concerns while ignoring operational and cultural integration issues could ultimately lead to an undesirable outcome for the merger [9]. In order to take advantage of the anticipated results, synchronization with the combined or acquired company may require some time and tactful foresight. The pursuit of financial gain should not be the sole goal.

Thus, integration of two different undertakings is though complex in nature, it must be aimed to make it fruitful to function as a single unit.

2.3 Effect of Mergers and Acquisitions

2.3.1 The Shareholders

Generally, the newly merged company experiences a higher stock price than compared to the acquiring and target company. In an acquisition the shareholders of the acquired company are benefitted as they receive a high amount for the acquisition [3]. On the other hand, the shareholders of the acquiring company have a negative effect because of the acquisition as a high amount is paid and the acquiring company may suffer certain losses [3]. There may be a change in the voting power of the shareholders because of the increase of shares that took place in the merger process.

2.3.2 On Capital Structure and Growth of the Company

Mergers and acquisitions have a great impact on the capital structure of the company. After a merger and acquisition deal takes place, there occurs a change in the capital structure of the acquiring company [4]. The news of a merger deal will also receive a market reaction.

2.3.3 On Employees

Mergers and acquisitions also have a great impact on the employees of the company. There always lies a need to adhere to the grievances and concerns of the employees to avoid layoff s and cultural disintegration [10]. It may often happen that the merged company already has sufficient employees and there lies no requirement to the same amount of the employees which it used to have earlier. Thus, in a merger there a lot of loss of job for employees occur [11].

3 3. Transformative Phenomenon-AI Integration in Mergers and Acquisitions

Numerous companies have seen a transformation since the advent of AI, which has made it possible to simplify complex procedures and enhance decision-making [12]. AI has become essential as a result of today's technological advancements, which allow businesses to handle and analyse vast amounts of data and produce precise forecasts.

3.1 Classification of AI

The different types of AI are Narrow, General Intelligence and Artificial Super Intelligence [13]. In Narrow Intelligence the AI focusses on some specific work and operates in limited domain but does not have the ability to apply intelligence in other tasks [14].

In AI General Intelligence, surpasses a wider range of abilities like rational reasoning, solving problems, thought process and also learning from situations. ChatGPT is a classic example of such AI [14].

The last type of AI has the highest potential and capacity to grasp knowledge at a faster rate, at par or even better than human intelligence. This AI is a self-aware AI and extremely intellectual [14].

In Merger transactions, there is the traditional AI, which is a type of Narrow Intelligence that focusses on repetitive tasks performed in a structured manner. This increases the accuracy and inefficiency in due diligence. Further, there is Generative AI, which is a super AI that identifies the risks associated with merger transactions in the complex integration process. This facilitates in the negotiation, strengthening of the merger process.

3.2 Role of AI

In Mergers and Acquisitions there has been usage of AI. Use of AI helps in negotiation and also navigation of other grey areas of the merger process [15].

The strategy for choosing whether or not to pursue a Merger and Acquisition deal is the first area where AI can be useful [15]. To come to a conclusion as to whether to proceed with a merger, AI analyses huge amount of data, earlier transaction history of the company. All these information acts as an aid in coming to a decision regarding the process of the Merger.

To discover suitable buyers for the target company is the second most important concern in a Merger. AI after thorough investigation with large amount of data, filters and finds business in that.

Meet strategic requirement. The data used by AI are Google searches, balance sheet, market trends. These help to determine the successful selection process [14].

AI has a significant impact on valuation as well. By analysing a wide range of data, including market conditions, company performance, and growth expectations, technology allows for more accurate assessment [14]. By incorporating real-time data, AI-driven models have a major advantage over conventional approaches in that they can update valuations continuously in real-time the target business's valuation is a critical phase in the Merger and Acquisition process since any error, such as overvaluing or undervaluing the company, could have a detrimental impact on the deal as a whole [14].

To conclude, the legal practitioners can grasp the potential of AI in Merger and Acquisition. This will help in giving an in-depth idea and making better decisions, while reducing costs and time.

4 Due Diligence—Ex Ante Investigation

Due diligence is basically examining a lot of document thoroughly so as to identify the pros and cons of the deal [1].

Successful Merger and Acquisition transactions require thorough planning and analysis since they are complex processes. In general, due diligence refers to a comprehensive analysis conducted by both purchasers and sellers to closely assess the current condition of the company and its various compliances [16].

There are significant ramifications when AI is incorporated into the due diligence procedure. Due diligence in legal domain has already changed as a result of the AI evolution, becoming quicker and more effective. The need to spend endless hours on labour-intensive chores has been eliminated by AI's superior document analysis capabilities, freeing up to productive analysis. With speedier and deeper searches, the work's quality has also increased. This has made it possible for legal teams to complete diligence evaluations more quickly and more cheaply while also obtaining more accurate results [17].

Due diligence is generally done for a number of reasons. By allowing companies to verify and authenticate information supplied during the transaction process, it first ensures transparency [17]. Secondly, it helps identify potential issues that could jeopardize the transaction's success, allowing parties to avoid bad deals. The various ways of this process include- understanding the financial condition of the company, customer feedback, technological soundness, environmental and sustainable concerns and overall production and marketing scenario [17].

4.1 Due Diligence Checklist

The checklist provides a stepping stone in due diligence process [16].

i. Defining the due diligence task

The first step is to identify what the client wants. A budget needs to be determined, identifying the focus areas and the limitations should be clarified [18].

ii. Assembling the Due Diligence Team

There needs to be a structured team with legal practitioners, accountants and marketing specialists who would focus on the various aspects of the deal [18].

iii. Submitting the Due Diligence Request

Many a times, buyers asks for additional information in the form of structured questionnaires and also requests for document verification to substantiate the claims made [18].

iv. Getting Access

Having access to all the information is very crucial for a good due diligence process. It includes not only documents in public domain but also internal confidential documents [18].

v. Distributing and Organising Materials

Sometimes sellers in order to protect the company may restrict use of sensitive information. Buyers must be vigil in ensuring that core confidential matters of the company are not leaked, thereby leading to bad name of the company in the market [18].

Overall, in Merger and Acquisition deals, due diligence is very important for both buyers and sellers. It paves way to the action as to whether the merger will proceed or not. To grab a successful merger, a well-planned diligent due diligence is very essential.

The various AI models designed to be used in Due Diligence are the- Financial Analysis, where by way of Machine Learning Algorithm, the erstwhile financial data is examined and a future financial health is predicted [19]. The framework used in legal Due Diligence is natural language processing whereby, legal documents are reviewed to point out any non-compliance issue [19].

4.2 Post Merger Integration-Ex Post Analysis

The nomenclature post-merger integration is basically the after effect of a Merger and Acquisition. The word integration means combined effect. Therefore, post-merger integration means the overall effect of the much-talked merger deal in reality and how has the market got along with the deal [20].

Predictive Analysis may be integrated in the post-integration process. This will analyse the market trends and ensure that Merger and Acquisition falls in place with the strategic goals involved in the merger process [19].

Different integration approaches exist for the integration of the acquired companies depending on the necessary level of integration. The main principle stands as much as necessary as little as possible. There are different integration approaches [20].

- Completely independent from each other: This approach basically contradicts the idea of integration. When the two companies are different and have different essential qualities, this approach does not want to destroy the business through integration.
- Integration of some parts of a company: The second approach is the integration of some parts of the company, balances between cost benefit aspects, determining whether integration makes sense or not.
- Full integration of the acquisition of the company: Full integration can be done into lot of different ways and in a more cautious manner. However, if it destroys the most valuable aspects of the acquired company, then an alternative approach of integration must be chosen.
- Integration of both companies: With integration of both companies, new organizational and procedural changes can be implemented such as the name of the company and the new business model.

4.2.1 Phases of Integration

The different phases of post-merger Integration are:

(a) Initial planning

The first phase includes the initial planning and activities, decision to be taken by the managerial staff and integration committee. Every successful integration demands a chief integration officer. The person responsible for the integration coordinates the integration process under the decision making authority and communicates the goals and strategies of the integration [19].

The content part of the integration is managed by an integration committee which consists of the integration project leader and members of both companies mainly responsible for the integration as well as external consultants. The task of the integration committee is to put the action plan that was drafted during the integration planning, into concrete terms and to complete and correct if necessary.

(b) Detailed Communication plan

The buyer should have a clear and detailed communication plan. Common elements of communication: includes a clear statement about the reasons for the acquisition including the benefits that will accrue for both companies and also for the customers. This phase is designed to ensure at the same time that the expected savings and synergies are realized. There must be formation of qualified employees in both companies. In addition, key employees, should be kept motivated to achieve the successful completion of important integration steps [19].

(c) Proper documentation

All information and documents have to be maintained in a suitable form so that they are quickly available during each step of the acquisition. Lack of documentation and poor delivery of documents and information may often lead to a delay in acquisition thereby in the integration process. The documents which must be a part of the project documentation are the working papers which are the written documents that summarize the planning or the result of work stages, presentations, memos and status report which includes short written reports and notes on the progress of the project. The other documents required are the written correspondence and e-mails, approval submissions, data room and the documents provided in the data room must be saved in the event of later legal disputes [19].

(d) Final review

A final review must be made after the completion of integration. This gives the opportunity to learn a lot for future acquisitions. Actual conclusions can usually only be drawn two to three years after the purchase. But in reality, only the initial objectives in an acquisition are achieved or not are described [19].

4.2.2 Problems of Post-Merger Integration

Implementing the goals of Mergers and Acquisitions is a very difficult and complex process. The main issue is implementing all of the plans, policies, and strategies and determining the anticipated synergies. Because mergers and acquisitions are complicated issues, it frequently takes a long time to implement [21]. All departments and divisions must be positively integrated for the merger and acquisition to be effective. While certain components may be successfully integrated, others may not be able to achieve the intended results, which could lead to a disintegrating effect.

Putting together a team for post-merger integration is a challenging task [21]. It presents many difficulties, such as choosing members, giving them responsibilities, keeping track of their performance, offering rewards, and gathering feedback. The team must concentrate on all matters concerning the integration of different facilities, and it is also necessary to give the members sufficient instruction in this respect.

In such operations artificial intelligence can be levered in the following manner: AI has the potential to integrate the different parameters of culture within the company and other aspects be it the operation of the company, combined growth of the company [21]. AI can also monitor the entire deal minutely to merger a successful one.

The companies nowadays with a goal of the Sustainable Development must also keep in mind protecting the environment. AI also considers ESG (environmental, and social governance) and EHS (environmental, health, and safety) very significant in evaluating any reconstruction of the company [22]. EHS considerations are becoming increasingly essential. As non-compliance leads to huge legal and financial risks [22]. Mergers and Acquisitions deploy AI in social and environment compliances which act as a safeguard to mitigate risks but also do not hinder development [22].

5 Conclusion

AI being autonomous makes less error compare to human mind. This helps the companies make deal faster and often full proof. After analysing, there lies no ambiguity that AI integration will have far reaching effects in Mergers and Acquisitions but it is associated with some challenges the fact of AI being autonomous brings more efficiency in the Merger process which the traditional due process is unable to.

The traditional Mergers and Acquisitions due process is plagued with the problems of being time consuming, labour intensive and often human fallacies. This has the effect of lingering the Merger and Acquisition whereby the expected results might not be availed. There lies an extensive study of data in a voluminous amount where the traditional Merger and Acquisition process finds it quite daunting.

It is undoubtedly that the AI due process has a great advantage over the traditional due diligence process. But as with great powers comes great responsibility, with lot of advantages of AI, comes a lot of challenges. The first challenge lies in job losses. AI can work for long hours without a break leading to sacking of many people thereby creating unemployment. Another major threat is to the environment. AI causes huge amount of carbon emission and water consumption. Thus, ethical consideration should also be taken into consideration while incorporating AI in Merger and Acquisition.

Bias in the AI data is such a threat that would manipulate in decision making. If the data that is given to AI is biased and discriminatory, that would affect all companies be it big or small. Again, AI must ensure that it does not data breach and infringe privacy of a person. The data collected and stored by AI often collected in an unethical manner, without taking proper consent. Proper care should be taken that no data gets leaked and confidentiality is maintained. The thin line between innovation and privacy must be maintained in order to get the best results.

As discussed earlier that AI does have a huge beneficial value in Merger and Acquisition, but the accountability poses a challenge. Can AI be blamed and held responsible in case of any ambiguity or error? Implementing AI requires a big investment and trained personnel. Not everyone has the requisite expertise to handle AI. These companies must apply AI after maintaining all regulatory compliance, data privacy, adequate training of the employees, then the anticipated result will be a wholesome one. While integrating AI in due diligence, in order to seek best results, it must be started with pilot projects to minimize risks and optimize anticipated results. AI with blockchain technology whereby data gets stored and linked making due diligence more transparent for monitoring the various compliances in due diligence.

Implementing AI into due diligence has a huge potential specifically in the financial sector. Leveraging Machine Learning and productive analysis will enhance accuracy and strong strategic output. Such framework will be more data driven, navigating various complex areas. AI must be accurately employed in order to reshape traditional due diligence, thereby fostering a much informed decision making merger transaction.

References

1. Abbasli S. Enhancing due diligence while upholding compliance: an analysis of AI integration in the context of mergers and acquisitions and its compatibility with the EU AI act
2. Wangerin D. M&A due diligence, post-acquisition performance, and financial reporting for business combinations. https://onlinelibrary.wiley.com/doi/abs/10.1111/1911-3846.12520
3. Fernández S, Triguero Á, Alfaro-Cortés E. M&A effects on innovation and profitability in large European firms
4. The Institute of Companies Secretaries of India. Handbook on Mergers, Amalgamations and Takeovers
5. Kummer C, Molzer W, Ferdinand Eiffe F. Mergers and acquisitions: a comprehensive step by step approach
6. Gaughan PA. Mergers, acquisitions, and corporate restructurings
7. Miller EL Jr. Mergers and acquisitions: a step-by-step legal and practical guide
8. 'Mergers & Acquisitions—The 5 Stages of an M&A Transaction. https://www.pwc.com/mt/en/publications/tax-legal/mergers-and-acquisitions-5-stages-of-MA-transaction.html
9. Krieg DeVault Travis D Lovett and Charles O. Richert. Using AI in M&A Transactions: Key Advantages, Challenges, and Considerations. https://www.lexology.com/library/detail.aspx?g=b010f0ee-6055-46a2-be05-cc632f16acd2
10. Paranjape NV. The Company Law
11. Ernst and Young. Master guide to mergers and acquisitions in India
12. Stuart Russell and Peter Norvig. Artificial intelligence: a modern approach
13. Dickson B. What is artificial narrow intelligence (Narrow AI)?—TechTalks https://bdtechtalks.com/2020/04/09/what-is-narrow-artificial-intelligence-ani/
14. Cazzaro M. AI and machine learning in M&A A quantitative analysis of their impact on deal outcomes

15. AI and new technologies M&A: is it really possible to automate M&A processes? https://winnerge.com/tecnologia-innovazione/ia-tecnologie-ma-automatizzare-processi/
16. Due Diligence Explained. European commission https://single-market-economy.ec.europa.eu/sectors/raw-materials/due-diligence-ready/due-diligence-explained_en
17. Due diligence: investigation or audit of a potential deal or investment opportunity corporate finance institute. https://corporatefinanceinstitute.com/resources/valuation/due-diligence-overview/
18. O'Leary C, Nakhuda R. M&A due diligence checklist for public and private deals. https://legal.thomsonreuters.com/blog/m-and-a-due-diligence-checklist-for-public-and-private-deals/
19. Vogelsang M (2024) How AI will impact due diligence in M&A transactions. https://www.ey.com/en_ch/strategy-transactions/how-ai-will-impact-due-diligence-in-m-and-a-transactions
20. Sridharan R, Arvind Pandian PH. Guide to takeovers and mergers
21. Noghrehkar N. AI and the future of mergers and acquisitions institute for mergers, acquisitions, and alliances. https://imaa-institute.org/blog/the-future-role-of-artificial-intelligence-in-mergers-and-acquisitions/
22. The Impact of AI on M&A. https://www.ey.com/it_it/newsroom/2023/12/ey-2023-positivo-per-l-italia-e-ottimismo-per-il-2024

Assigning Liability for AI and IoT Decisions: Comparative Analysis Between Strict and Shared Liability as a Regulatory Framework

Abhishake Roy

Abstract Generally considered to be self-learning and autonomous, AI uses IoT devices to collect and analyse data for predictive decisions. Due to its invisibility to the end users, opacity of decision-making process and algorithmic or data driven failures, exacerbated by their interconnectedness, a plethora of legal issues arise when AI fails to act in a way that is expected. This unpredictable nature, further complicates the assessment of potential harm done by AI and IoT devices, and dampens the process of assigning liability by the victims to any one entity to claim compensation for their injuries. In-spite of the global trend towards regulating AI, investments regarding development of new foundational models have grown exponentially in the recent years. This paper, doctrinally analyses if the concept of 'Strict Liability' can continue to be a viable regulatory framework for AI and IoT environment, or if the concept of 'Shared Liability', can be used to create a regulatory system, which fosters innovation and safeguards human protection at the same time, by analysing both Indian and international legal frameworks and judicial pronouncements. In the end, this paper suggests the concept of 'Insurance' as a hedge system to guarantee that victims receive compensation and that the companies can continue innovation in the field of AI technologies without the fear of suffering heavy losses paid as compensation for damages caused by unforeseen AI and IoT decisions.

Keywords Artificial intelligence · Internet of Things · Regulatory framework · Strict liability · Shared liability · Insurance

A. Roy (✉)
IEM's International Institute of Juridical Sciences (IIIJS, University of Engineering and Management, Kolkata, India
e-mail: abhishake.roy@uem.edu.in

M. Chakraborty et al. (eds.), *Proceedings of Sixth International Ethical Hacking Conference*, Lecture Notes in Networks and Systems 1513,
https://doi.org/10.1007/978-981-96-8632-2_24

1 Introduction

The progress of IoT from being a technology that controlled devices remotely, to its present role of collecting real-time data, combined with the AI integration to analyse such data, to provide assistance, has revolutionised the way of solving our day-to-day problems [1], but their rapid integration in our contemporary life has posed significant regulatory challenges, prima-facie of which is the assignment of 'liability' [2]. Critical domains such as healthcare, transportation and personal security have adopted and embedded in themselves the ever-increasing autonomous systems of AI and IoT, which pose a challenge on the legal systems to ensure accountability, in cases where these systems cause harm through unexpected malfunctions, data vulnerabilities or unforeseen algorithmic 'black box' decisions [3].

Traditionally, 'Strict Liability', has been protecting the consumers in product liability cases, holding the manufacturers liable without requiring proof of negligence. However, its application faces challenges when applied to AI and IoT technologies due to their ability to learn, evolve and operate autonomously.

The core complexity lies in the question 'Whom to assign liability for these technologies?' as the fact of assigning responsibility spans across diverse parties consisting of manufacturers, software developers and service providers, each of whom contribute to the design of the product, its data processing and operational capabilities. Since, in an AI driven ecosystem, liability may involve user error, software updates, or data breaches beyond the manufacturers control. This creates a situation too diffused to assign strict liability, to product manufacturers only.

The 'black box', which is the impenetrable process in which an AI makes a decision, makes it difficult to trace causation, a fundamental component to claim liability, specifically when decisions are informed by dynamic data inputs from interconnected IoT devices. These complexities, therefore, advocate for a much more nuanced approach in considering public policy implications. Hybrid liability frameworks and alternative liability concepts, such as shared or tiered liability models, might address the unique risks posed by AI and IoT without stifling innovation. Support and suggestion have been increasingly gaining ground regarding the fact that interconnectivity between AI and IoT devices demand a rethinking of the distribution of liabilities on account of the multi-layered dependencies these technologies entail, while recent studies show the importance of harmonising the standards of liability at an international level, keeping in view the developing legislations by the European Union and other Jurisdictions to ensure both consumer protection and innovation within the AI and IoT industries.

2 Complexities Persisting in Assigning Liability

The AI and IoT environment are not only autonomous, opaque and complex, but also open, vulnerable and distributive, thus creating a situation too diffused to assign liability to any one particular person or a particular group of persons.

No matter what the ethical questions are, the autonomous nature of AI sets it apart from traditional systems of coding. Having close similarity to human 'will', the reaction of the software can often become unpredictable regarding situations that are missing from its learning data. No form of training, be it supervised or unsupervised can guarantee that the AI system will not make a decision, beyond what can be strictly termed as 'predictable'. These deep learning methods make it impossible to understand the reason behind any decision made by these systems, further rendering the scenario of assigning liability to any one person opaque.

The interconnectivity of AI and IoT environment, where multiple autonomous systems influence each other's behaviour, proves that assigning the exact cause of damage to any one particular person or a particular group of persons becomes neigh impossible. These systems being open to software updates often undergo change in their relevant safety features and data sharing protocols which are overlooked by users, further opening gateways for cyber-attacks with an increased risk of cyber fraud.

3 The Strict Liability Environment

Activities involving the use of AI and IoT devices are often considered to contain an element of hazardous risk that might involve inherent and constant danger to people, their personal data or property, as can be seen from the incident that occurred in Tempe, Arizona, where Elaine Herzeberg was killed by a self-driving Uber car, while the driver sat watching Hulu [1]. In another case, a 300-pound robotic security guard ran over and injured a toddler in a California shopping centre [2]. AI chatbots making slanderous comments online [3] and discriminating against minorities during hiring procedures [4] has been a common occurrence. The law can take recourse in these cases through three avenues. Firstly, the law might prohibit the use of these systems altogether. Secondly, their use might be allowed because of their social utility but only under statutory provisions having strict sanctions for their non-compliance. Lastly, the activities may be allowed on the condition of compensation regardless of any fault, on the basis of the principle of strict liability.

Scholars advocating for the concept of strict liability primarily focus on the 'black box' decision of AI and IoT devices, which makes the prediction and establishment of the legal nexus between any accident and the entity responsible for it, an impossible task [5]. Thus, the doctrine of 'respondeat superior' ensures that the principle human entities who are in the best position to monitor and control the AI and IoT systems and take pro-active actions for prevention of potential accidents and damages, remain liable for the actions of their AI agents [6].

The argument supporting strict liability in the AI scenario is further strengthened by Fletcher's non-reciprocal approach, which when applied here, would mean that since AI and IoT systems is capable of inflicting greater damages to their environment through rapid simultaneous communications over a vast number of online platforms, rather than getting damaged themselves, they should not possess rights as similar to those of humans to be safe and protected from harm.

4 Shortcomings of the Strict Liability Regime

The strict liability regime is often criticised because of the fact that it stifles innovation [7]. Development of new and emergent technologies is always bound to cause damage to someone or other [8]. Start-up companies that innovate and develop AI and IoT devices, if forced to pay damages under the rule of strict liability, find their revenue generation compromised since they have to pay compensation no matter their efforts to reduce damage caused by their products. This situation may lead the start-up companies to remove themselves from the market due to the fear of becoming bankrupt, leaving the market open only to giant conglomerates, who have sufficient capital and financial resources to bear the burden of compensation for the damages caused by their AI and IoT devices.

It is therefore not too difficult to assume that legislations when framed under the rule of strict liability, and made public, will invariably push the consumers towards purchasing AI and IoT products and services from those branded companies whose stable market identity will ensure the consumers that they will be able to garner compensation from these companies if their AI and IoT devices cause harm to them. This would invariably lead to stifled innovation and a closed monopolistic market system.

5 The Indian Scenario

The legislative frameworks operating in India have taken a three-pronged approach to provide relief from the threats arising out of the AI and IoT environments. These three different laws, i.e. The Information Technology Act, 2000, The Consumer Protection Act, 2019 and The Digital Personal Data Protection Act, 2023, enacted across three different times, when studied together brings into focus that the legislative framework in India has sought to address the risk arising out of the AI and IoT environments through three different avenues as stated above, i.e. the legislature through The Information Technology Act, 2000 as sought to prohibit some actions using the any computer system along with AI and IoT systems altogether because of their harmful nature. The legislature through The Digital Personal Data Protection

Act, 2023 has allowed the continuation of data focused activities in an AI and IoT environment because of their social utility but only with strict statutory provisions having strict sanctions for their non-compliance. Lastly, the legislature through The Consumer Protection Act, 2019 has allowed the marketing, trade and business of AI and IoT devices on the condition of compensation regardless of any fault by the product manufacturer, on the basis of the principle of product liability.

5.1 The Information Technology Act, 2000

This Act was framed to not only provide legal recognition to 'electronic commerce', but also to bring under the legal framework all those activities that can cause harm using a computer system. Chapter IX, titled 'Penalties, Compensation and Adjudication', under Sect. 43A, states that if a corporate entity fails to protect the data collected by it, it shall be liable to pay compensation to the person suffering because of the negligent security practices of the corporate entity. Furthermore, Chapter XI under the heading 'Offences', gives us a comprehensive list of all the criminal offences that can be done using a computer system. This can broadly be construed to involve the potential use of AI and IoT systems to inflict harm and damages to users by third parties.

5.2 The Consumer Protection Act, 2019

This Act was framed to protect the consumers interest and to establish authorities so that timely and effective administration and adjudication of consumer's disputes can be made. With the rapid growth of online trade and commerce and with the explosion of the use of AI and IoT systems, this Act sought to address the issue of who shall be liable when a product suffers from defects due to the actions of the product manufacturer, service provider or product seller.

Chapter VI of this particular Act, addresses 'Product Liability'. Encased within Sects. 82–87 are concepts like, Application of the chapter (Sec 82), Product liability action (Sec 83), Liability of product manufacturer (Sec 84), Liability of product service provider (Sec 85), Liability of product sellers (Sec 86) and Exceptions to product liability actions (Sec 87).

It is however interesting to see that a product manufacturer under Sect. 84(2), has been made liable, even if he proves that he was neither negligent nor committed any fraud in making express warranty of a product, when a product liability action is brought against him.

5.3 *The Digital Personal Data Protection Act, 2023*

This Act was framed, to recognise the right of individuals and also protect their personal data while their digital personal data is processed. This Act also focuses on keeping a way open for the purpose of lawfully processing such data. The particular sections that need to be highlighted in respect of this paper are those of Sect. 8 which states the 'General obligations of a Data Fiduciary', Sect. 10 which states the 'Additional obligations of a significant Data Fiduciary', Sect. 15 which states the 'Duties of a Data Principal' and Sect. 32 which lays down guidelines for a 'Voluntary Undertaking'. Chapter VIII herein deals with the 'Penalties and Adjudication', where the schedule attached under Sec 33 lays down the amount of compensation to be paid by the data Fiduciaries on the occasion of breach of provisions under this Act.

6 Where Do We Stand?

In the context of India, we can clearly see that to control the AI and IoT environments where the threats span across criminal usage, data breaches and product liabilities, keeping the liability too diffused to assign to any one particular person or group of persons, the strict liability regime alone cannot continue to provide a wholesome remedy against AI and IoT environments. Thus, a multi-layered approach is required to address all the threats appropriately. This has been tried to achieve by the Indian legislative by their three-pronged approach to the AI and IoT revolution. The under-mentioned tables would provide a detailed visualisation of where we stand today (Tables 1 and 2).

Though the above-mentioned legislations do not expressly mention the terms AI and IoT in any of them, but their applicability in the AI and IoT environment cannot be denied as each and every offence as mentioned in the tables above can be enhanced by the use of AI and IoT devices. As AI and IoT devices include a computer or a computer system, the application of the Information Technology Act, 2000 into assigning criminal liability to the person using AI and IoT devices to inflict harm, cannot be denied.

AI and IoT devices are undoubtedly products and they do render services to people. Thus, any manufacturing defects or defects related to its warranty will make the manufacturer, service provider or seller liable under the Consumer Protection Act, 2019, but sadly the Act does not address the issue of whom to assign liability if an AI or IoT system takes a decision that is totally unpredictable. The closest this Act comes to assigning strict liability is that to the manufacturer but that too only in the case of express warranty related to the product.

Lastly, the Digital Personal Data Protection Act, 2023 focuses solely on the protection of personal data and holds the Data Fiduciary liable for any breaches, but it does not address the scenario where data breach can occur through the use of AI and IoT systems.

Table 1 Threats in AI and IoT environment that can be addressed through strict liability [9]

Circumstances under Consumer Protection Act, 2019
Liability of product manufacturer
• When the product has a defect in its design or manufacturing • When the finished product fails to satisfy production specifications • In the event that certain warranty conditions are not met by the product • When warnings or instructions regarding proper product usage are inadequate
Liability of product service provider
• If the service provided has flaws, is flawed, insufficient, or poor in terms of quality, kind, or method of performance that was necessary • When an act of carelessness, omission, action, or wilful withholding of information occurs which results in injury • In situations where adequate cautions or instructions are not given to avoid any injury • When the contract's terms and conditions or the express warranty of service were broken
Liability of product sellers
• When the seller exercises significant control over the design of the product, its testing, manufacturing, packaging, or labeling • If the product has been changed or modified by the seller, which was a significant contributing factor to the harm caused • When the seller makes an express warranty for a product not in accordance with the express warranty of the manufacturer, creating disparity with the express warranty of the seller, resulting in harm • When the product has been sold by him and the identity of product manufacturer of such product is not known, or if known, the service of notice or process or warrant cannot be effected on him or he is not subject to the law which is in force in India or the order, if any, passed or to be passed cannot be enforced against him; or • When the seller fails to exercise reasonable care during the process of assembly, inspection or maintenance of such product or fails to pass on the warnings or instructions of the product manufacturer regarding the dangers involved or proper usage of the product while selling such product and such failure becomes the proximate cause of the harm

7 Conclusion

Unprecedented difficulties in the area of legal liability arise from the growing integration of AI and IoT technology into many aspects of our daily lives. Because these technologies are autonomous, opaque, networked, and constantly changing, the old strict liability framework is insufficient as a stand-alone regulatory mechanism. The challenges of determining causality, allocating accountability among various stakeholders, and managing risks associated with outside variables like data breaches or cyberattacks are where this deficiency is most noticeable. Although strict liability has long been a pillar of consumer protection, its rigorous application to the ever-changing AI and IoT environment runs the risk of inhibiting innovation, deterring smaller businesses, and unintentionally strengthening the position of giant firms in the market.

Table 2 Threats in AI and IoT environment that cannot be solely addressed through strict liability [10]

Circumstances under Information Technology Act, 2000
• Tampering with computer source documents
• Computer related offences
• Sending offensive messages through communication service, etc
• Dishonestly receiving stolen computer resource or communication device
• Identity theft
• Cheating by personation by using computer resource
• Violation of privacy
• Cyber terrorism
• Publishing or transmitting obscene material in electronic form
• Publishing or transmitting of material containing sexually explicit act, etc. in electronic form
• Publishing or transmitting of material depicting children in sexually explicit act, etc. in electronic form
• Misrepresentation
• Breach of confidentiality and privacy
• Disclosure of information in breach of lawful contract
• Publishing [electronic signature] certificate false in certain particulars
• Publication for fraudulent purpose
Circumstances under Digital Personal Data Protection Act, 2023
• Breach in observing the obligation of Data Fiduciary to take reasonable security safeguards to prevent personal data breach
• Breach in observing the obligation to give the Board or affected Data Principal notice of a personal data breach
• Breach in observance of additional obligations in relation to children
• Breach in observance of additional obligations of Significant Data Fiduciary
• Breach in observance of the duties
• Breach of any term of voluntary undertaking accepted by the Board
• Breach of any other provision of Digital Personal Data Protection Act or the rules made thereunder

The need for a multi-layered regulatory approach is demonstrated by the Indian legal system, which has three main components: The Digital Personal Data Protection Act of 2023, the Consumer Protection Act of 2019, and the Information Technology Act of 2000, but no specific concerns created by AI and IoT technologies are addressed by the regulatory frameworks. Thus, even when the regulatory frameworks try to control an environment overlapped by the concept of computer system, product liability and data breaches by outlawing harmful behaviour, enforcing compliance through stringent statutory restrictions, and providing compensation under the strict liability concept for consumer protection, the legislative effort is insufficient to fully address liability issues.

This study highlights the necessity of developing liability frameworks that take into consideration the distributed obligations in AI and IoT ecosystems, such as shared or tiered liability models. To maintain legal clarity and promote international innovation, policymakers should think about harmonising international liability norms. By developing flexible measures that reduce hazards without placing an excessive burden on inventors, public policy must strike a balance between innovation and consumer safety.

In summary, alternative liability models and supportive legislative frameworks are necessary to supplement strict responsibility, even if it remains a crucial consumer protection mechanism. In the era of the fourth industrial revolution, only a comprehensive strategy can effectively handle the complex risks presented by AI and IoT technologies while promoting a safe and innovative atmosphere.

8 The Way Forward

Notwithstanding the legal regime operating in the country, technological innovation should not be limited by legislative control. In the context of AI and IoT revolution, insurance may act as a tool to safeguard the existence of start-up companies and giant conglomerates to operate effectively in an environment largely governed by the concepts of strict, shared or tiered liability. Having insurance safeguard will allow the start-up companies to continue innovating and give them a level playing field with the giant conglomerates in the AI and IoT business, while keeping them on their toes regarding the potential sanctions for compensation on the event of their AI and IoT devices take any unforeseen decision leading to unforeseen circumstances.

References

1. Shu L (2025) The evolution of AI: progression from the Internet of Things. https://www.forbes.com/councils/forbestechcouncil/2025/02/10/the-evolution-of-ai-progression-from-the-internet-of-things/
2. AI & ML Tech Trends (2024) Integrating AI with IoT: opportunities and challenges. https://www.rapidcanvas.ai/blogs/integrating-ai-with-iot-opportunities-and-challenges
3. Hassija V, Chamola V, Mahapatra A et al (2024) Interpreting black-box models: a review on explainable artificial intelligence. Cogn Comput 16:45–74. https://doi.org/10.1007/s12559-023-10179-8
4. Griggs T, Wakabayashi D (2018) How a self-driving Uber killed a pedestrian in Arizona, THE N.Y. TIMES. www.nytimes.com/interactive/2018/03/20/us/self-driving-uber-pedestriankilled.html
5. Hoffer S (2016) 300-pound security robot runs over toddler at California shopping center. www.huffpost.com/entry/security-robot-toddler_n_57863670e4b03fc3ee4e8f3a
6. Dastin J (2018) Amazon scraps secret AI recruiting tool that showed bias against women,. www.reuters.com/article/us-amazon-com-jobs-automation-insight/amazonscraps-secret-ai-recruiting-tool-that-showed-bias-against-womenidUSKCN1MK08G

7. Reed J (2018) Is AI an asset to hiring, or will it bring us down a sinkhole of algorithmic bias? DIGINOMICA. www.diginomica.com/2018/12/12/is-ai-an-asset-to-hiring-or-will-it-bring-us-down-asinkhole-of-algorithmic-bias
8. Bloomberg J (2018) Don't trust artificial intelligence? time to open the AI' Black-Box. www.forbes.com/sites/jasonbloomberg/2018/09/16/donttrust-artificial-intelligence-time-to-open-the-ai-black-box/#69410d83b4a7;
9. Government of India (2019) The consumer protection act, 2019 (No. 35 of 2019). Ministry of Law and Justice, New Delhi. https://egazette.nic.in
10. Government of India (2000) The information technology act, 2000 (No. 21 of 2000). Ministry of Law and Justice, New Delhi. https://www.indiacode.nic.in
11. Lior A (2020) AI entitiesas AI agents: artificial intelligence liability and the AI respondeat superior analogy 46 MITCHEL HAMLINE L. REV. 1043 (2020)
12. Gregory Mandel (2009) Regulating emerging technologies, 1 L., INNOVATION & TECH. 75
13. Wickre K, What Startups Can Learn from Facebook's Reckless' Move Fast and Break Things' Mantra, THE MARKER, marker.medium.com/whatstartups-can-learn-from-facebooks-reckless-move-fast-and-break-things-mantra8dcc93ee3437

Automated AI Systems for Legal Compliance Monitoring in Regulated Industries

Rahul Vadisetty, Anand Polamarasetti, and Madhava Rao Kunchala

Abstract The new AI-powered compliance monitoring systems use automated methods to review high-volume data to detect anomalies in operations and assess the risks of non-compliance to the lowest levels while meeting the high standards outlined in the legal environment. GE climates, such as finance, healthcare, and data privacy, are increasingly encumbered with complex and dynamically changing regulatory requirements, raising the cost of their compliance management. The findings reported herein investigate the architecture of AI-based compliance systems: data collection, natural language processing to identify keywords, and machine learning models to spot patterns and potential breaches. It also examines some of the challenges these face, such as data privacy, algorithmic bias, and lack of transparency, which is essential for standards of ethics and integrity within regulations. It also examined real-life applications and various case studies of AI in industries like finance and healthcare, as well as the efficacy and barriers that stand in the way of completely handling compliance. The study concludes the discussion on future directions by calling for advancements in Explainable AI-XAI, guidelines for ethics, and policy updates to rule on AI in compliance monitoring. Mapping out the potentials and limits of AI in regulatory compliance, this study communicates a roadmap for organizations and regulators on navigating responsible adoption, whereby AI would drive efficiency but maintain compliance across fast-shifting regulatory landscapes.

Keywords Automated AI · Legal compliance · Regulated industries · Anomaly detection · Natural language processing · Machine learning · Data privacy · Explainable AI · Regulatory compliance · Financial compliance

---

R. Vadisetty (✉)
Wayne State University, Detroit, MI, USA
e-mail: rahulvy91@gmail.com

A. Polamarasetti
Andhra University, Visakhapatnam, AP, India

M. R. Kunchala
SAP Finance Solution Architect, Mygo Consulting Inc., Aurora, IL, USA

M. Chakraborty et al. (eds.), *Proceedings of Sixth International Ethical Hacking Conference*, Lecture Notes in Networks and Systems 1513,
https://doi.org/10.1007/978-981-96-8632-2_25

1 Introduction

1.1 Overview of Legal Compliance Requirements in Regulated Industries

Compliance has become an integral and essential part of the industries under regulation, such as finance, health, pharmaceuticals, and even data privacy. Each of the sectors mentioned above has unique, strict rules that are generally aimed at ethical practices, consumer protection, and data security [1].

For instance, companies have to conform to AML laws, KYC requirements, and extensive reporting obligations in the financial sector. Because patient confidentiality and data protection laws, like HIPAA, strictly regulate health industries concerning patient information, the pharmaceutical industry has to follow set regulations, not unlike the guidelines of the FDA, in calling for transparency and safety in drug development and distribution. Similarly, laws about data privacy, like the General Data Protection Regulation in Europe, have strictly controlled personal information collection, storage, and sharing. All these non-compliances may result in heavy penalties, legal consequences, and reputation loss; therefore, robust compliance monitoring is vital in those sectors [2–4].

1.2 Role of Technology in Compliance Activities

As regulations widened in breadth and changed unendingly, their compliance methods became increasingly less adequate for most organizations. Due to their manual nature, they are prone to errors and very laborious, unable to keep pace with the speed at which regulatory changes move and the exponential volume of data growing within the current digital ecosystem [3, 4].

On the one hand, it's technology at the core of automating routine managerial processes in compliance activities. This enables real-time monitoring and the ability to make changes needed because regulations change very fast. Advanced technologies such as cloud computing, big data analytics, and machine learning will ensure that organizations can efficiently deal with their compliance requirements at scale. It also allows automated systems to continuously monitor organizations' activities, flag those that may have potential issues, and hence help organizations speed up responses in case of changes to regulations, reducing the risk of non-compliance [3, 4].

1.3 Introduction to AI and Its Capabilities

Simplistically stated, artificial intelligence implants greater sophistication into compliance monitoring. With its data analysis, pattern recognition, and anomaly detection abilities, AI can process volumes at unmatched speeds and accuracy not previously possible for a human [5, 6].

The ability of machine learning algorithms enables AI to identify patterns and correlations within large datasets and highlight possible compliance risks more accurately. NLP further enhances this capability by making AI capable of reading and understanding complicated legal texts, extracting information from them, and keeping a tab on regulatory changes. This becomes extremely important for regulated industries where anomaly detection algorithms can pick up unusual activities that are not normal, such as fraudulent transactions or money laundering attempts, for security aspects. The flexibility of AI makes the technology continuously learn from new information, perfect its analysis over time, and stay tuned to changing regulatory needs [5, 6].

1.4 Rationale for Adopting AI-Driven Automated Systems for Compliance Monitoring

AI-driven compliance monitoring is an automated way out of the inadequacies created by traditional systems. These are operationally cost-effective systems since organizations have to use fewer personnel resources, considering that such systems automate processes that are usually repetitive and data-intensive. Besides, AI dramatically enhances the accuracy of compliance monitoring since the technology eliminates human error and can review continuous data streams to find real-time anomalies [6–9].

Real-time monitoring capabilities mean organizations can quickly respond to regulatory changes, reducing risks associated with protracted periods of non-compliance. AI will be able to adapt more efficiently to the ever-changing nature of regulatory demands, thus letting an organization maintain compliance without much need for human intervention. Also, AI supports organizations with actionable insights extracted through data analysis for better decision-making on risk management and compliance strategy [7, 8].

AI systems help organizations stay compliant in highly regulated industries, such as finance and healthcare, where the rate of change in compliance requirements is very high. The systems are designed to be scalable, proactive solutions to compliance challenges, with machine learning and natural language processing embedded within them. They will also help position organizations for effective navigation in the complex, ever-changing regulatory landscape [1, 10, 11].

One of the best examples of AI compliance monitoring is in the banking industry, where various institutions have implemented it for AML purposes. For instance,

JPMorgan Chase once utilized an AI-powered solution that could track millions of financial transactions in real time and spot suspicious ones.

Traditional compliance—AML relies on manual reviews, often powered with rule-based systems, which, because of the associated propensity for misinterpretation of transactions, end in wastes of time and result in errors and inefficiencies. However, at this juncture, to detect anomalies based on unusual or exceptional changes in transaction traffic, cross-boundary transfers, and quick change-out within accounts, AML AI brings relevant detection capacity via machine-learning algorithms and natural language processes. It primarily reduces the amount of false positivity by about a fifth while advancing recognition ratios above 95% [7, 8, 11].

Real-time monitoring capability empowers the bank to react, with only a little time lag, in the case of suspicious activities, without the help of large teams of compliance officers who must keep abreast with the dynamic AML regulations. By leveraging AI, the platform can also provide actionable insights from transaction data to enhance the bank's overall risk management further and keep steps ahead of emerging threats in financial crime. The following example illustrates how AI-driven systems represent a scalable, cost-effective, adaptive solution for dynamic regulatory environments typical in highly regulated industries like finance [7, 8, 11].

1.5 Scope and Objectives of the Research

This study reviews how AI has changed the face of compliance monitoring across regulated industries. The paper outlines the architecture of the AI-driven compliance system, which contains core components like data collection, machine learning, and natural language processing. The paper then proceeds to outline AI's ability to bring greater efficiency, precision, and economy besides surmounting well-acknowledged challenges such as data privacy, algorithmic bias, and the need for interpretability.

This study aims to analyze successful applications of AI in compliance monitoring, identify the limitations in existing systems, and give practical recommendations about how an organization should adopt AI responsibly. This research will review best practices applicable when using AI technologies in compliance processes and ensure that ethical considerations are taken into account, along with adherence to regulatory standards.

This research maps both the potential and limitations of AI to provide a roadmap for organizations to leverage its capabilities while aligning with industry regulations. The findings are meant to aid organizations and regulators in fostering a well-balanced approach toward AI adoption that nurtures innovation without compromising compliance or ethical integrity.

2 Key Components of Compliance Monitoring Systems Driven by Artificial Intelligence

2.1 Data Collection and Processing

Data acquisition and processing form the backbone of compliance monitoring in AI. Data sources can be many and varied and emanate from structured datasets-customer transaction histories, employee records, operation logs-to unstructured data-communications, contracts, and regulatory filings across regulated industries. Giving all this diversified information some sense, AI systems offer several automated ways to capture and digitize the data with APIs that connect databases of organizational details, document scanning technologies, and advanced Optical Character Recognition (OCR) methods for conversion of physical records into digital format. Such functionality is handy when dealing with unstructured data, which contains information between words and texts of enclosed documents [1, 12].

The data collected then undergo an intensive processing phase, where data is cleaned, categorized, and structured. Instead, it prepares data for analysis and makes them conform to regulatory requirements. Natural Language Processing can legally parse and interpret complex terms, helping an AI system isolate and organize that information. Also, data integration tools compile multiple data sources into a cohesive dataset. This step is essential for a holistic view of compliance, where different data sources lead to gaps or inconsistencies, resulting in non-compliance [13].

2.2 Machine Learning and Natural Language Processing (NLP)

ML and NLP are vital technologies driving AI's predictive and analytics capabilities for compliance. The machine learning algorithms, in turn, get trained over historical data to identify patterns that define compliant and non-compliant behaviors. Labeled datasets in supervised learning help AI understand what does and does not match particular compliance standards so it can then apply that same knowledge to new data. In unsupervised learning, AI identifies clusters or anomalies without predefined labels. This is advantageous when monitoring minor or unexpected compliance non-conformances [4, 12].

While AI is about performance, NLP provides AI systems with the capability of understanding human language, which is indispensable in the legal or regulatory context, where compliance requirements are often couched in intrigue and subtlety. NLP models parse legal documents, regulatory guidelines, and internal policies for crucial information relevant to compliance needs. This technology also allows AI systems to understand the intent behind communications, including performing sentiment analysis to spot language in emails, chat logs, and recorded conversations that may show compliance risks. For example, NLP can identify statements that reflect an intent to circumvent regulatory procedures and provide early warning signals that help prevent violations before they happen [4, 13].

2.3 Pattern Recognition and Anomaly Detection

Equipped with advanced capabilities for pattern recognition and anomaly detection, an AI-driven compliance system can trace irregularities in real time. Algorithms studying historical data for pattern recognition build norms of behavior: typical transaction volumes, communication patterns, or even access levels. This would be a good benchmark for checking ongoing activities against and identifying deviations pointing to compliance risks. For example, within the financial industry, an AI-driven system could establish regular patterns of transactions from a particular client and flag suspiciousness regarding more significant amounts transferred that may be indicative of money laundering or fraud [1, 4, 9].

Anomaly detection is not just about flagging a deviation but involves assessing the risk associated with different anomalies. This capability enables algorithms to analyze the context and severity of detected patterns, classify alerts according to priority, and assist compliance teams with their work on the most high-priority risks. Predictive analytics, as another essential subcategory of anomaly detection, further extends the AI capability to forecast future compliance risks from emerging trends and, in turn, enables organizations to undertake a proactive role in addressing potential problems. The insight from such predictive analytics would be even more valuable regarding those sectors for which the regulatory updates and enforcement actions are unpredictable and help them adapt to changes in compliance requirements much in advance [14].

2.4 Compliance Reporting and Audit Trails

Compliance reporting and audit trails are the most salient features in regulated industries to demonstrate legal standards and facilitate responses in case of any regulatory inquiry. AI-driven systems make reporting more effective by automatically generating compliance reports, reducing human error and documentation consistency. In such reports, information related to the status of compliance, together with detected risks and mitigation strategies, is summarized, making the compliance practices of an organization crystal clear to organizational management and regulators. Automated compliance reporting holds high value in dynamic regulatory environments where manual reporting may be fraught with delays and inaccuracy [10, 11].

Audit trails further ensure transparency in maintaining a record of compliance-related activity, such as data processing, decision-making, and user interaction. Every action done internally in the AI system is captured, making every compliance decision traceable. This audit trail will help internal and external reviews by providing a prominent and chronological record of compliance management processes. Audit trails give an organization rudimentary accountability, enabling the organization to explain to regulators that compliance activities are performed consistently and follow broadly accepted norms of ethical behavior. The recording also aids in improving compliance processes by allowing organizations to study past compliance incidents and adjust their monitoring techniques accordingly in light of their experience [9, 10, 14].

2.5 *Cybersecurity Protocols and Data Privacy Measures*

Due to the sensitive nature of data arising from compliance monitoring, robust cybersecurity protocols and data privacy have become an integral part of AI-driven compliance systems. Compliance monitoring often includes processing sensitive information, such as customer financial data, patient health records, or employee details. Therefore, the systems must maintain different data protection regulations, such as the General Data Protection Regulation, Health Insurance Portability and Accountability Act, and California Consumer Privacy Act, among others, by implementing strict security with personal and confidential information [14–16].

Cybersecurity policies include encryption, access control, and multi-factor authentication to protect against classes of data integrity compromise by ensuring that only authorized personnel can access sensitive information. Regular security assessments and vulnerability testing identify and mitigate the associated high-level risk. Compliance systems use AI to improve cybersecurity through pattern detection that suggests cyber threats, phishing, unauthorized access attempts, or data breaches. These AI-driven systems prevent security incidents that could compromise compliance data through early identification of such threats [15].

Further, data privacy protocols ensure that no data can be used in any manner inconsistent with applicable regulations. To illustrate, differential privacy methods can anonymize personal information during their analysis, thus nowadays reducing the risks of unauthorized disclosure. In addition, AI systems are engineered to comply with data minimization principles by collecting only relevant data to perform the compliance burdens. This approach supports regulatory compliance and makes a business more trustworthy for clients and regulators by showing its responsible attitude toward data management [14].

Data gathering, machine learning, pattern recognition, reporting, and strict security protocols integrated into one solution create a comprehensive environment for sustaining compliance across complex regulatory regimes. These components, each solved together at scale, provide a high degree of accuracy, responsiveness, and accountability for compliance. The robust framework creates proactive monitoring and fast adaptation to regulation changes, making AI-driven compliance systems a game-changing tool for regulated industries [16].

3 Challenges and Limitations of AI for Legal Compliance Monitoring

3.1 Data Privacy and Security Risks

One of the significant challenges with AI in legal compliance monitoring is data privacy and security. The AI-driven compliance systems require access to large datasets, which usually include sensitive and personal information, like financial records, patient health data, or entities' personal identification information. The more this dependency, the more concern it raises regarding data privacy regulations, given laws like the General Data Protection Regulation in Europe, the Health Insurance Portability and Accountability Act in the U.S., and the California Consumer Privacy Act. Inability to comply with such regulations may lead to severe penalties, legal consequences, and loss of reputation.

Data privacy security in AI systems involves careful thinking over data collection, processing, and storage. Most sensitive data security approaches include data encryption, access controls, and anonymization, such as differential privacy methods. However, these measures may add complexity and costs when balancing the approaches to data utility for AI analysis while protecting privacy. The AI systems themselves have to be designed in such a way that they would prevent the data from leaking or unauthorized access on their part. For instance, techniques like differential privacy can anonymize sensitive information to retain its analytical utility, whereby the AI can derive insights without exposing identifiable information. Machine learning models with the preservation of confidentiality are also developing as solutions, though the application is less widespread commercially in compliance applications.

3.2 Algorithmic Bias and Fairness

The most exonerating reason seen to be severely problematic in the operation of AI for compliance monitoring is algorithmic bias. For any AI model, learning arises out of historical data. If such data contain biases, the AI risks replicating or even magnifying those biases, thereby generating biased outputs that can result in discrimination. In this respect, algorithmic bias could appear in compliance monitoring, where unfair treatment or scrutiny of certain groups of persons could be based on such variables as race, gender, and socioeconomic status. For instance, biased algorithms in financial compliance would flag all kinds of transactions from people originating from a particular region as high-risk, using historical biases without any evidence that such people are indeed non-compliant.

To this effect, organizations must be very critical of the datasets on which their AI models are being trained, using various techniques that ensure fairness in machine learning. Biases could also be spotted through constant auditing of the AI systems, popularly known as algorithmic auditing. This auditing involves assessing AI decisions and their outputs against the criteria of fairness; basically, it is used to ensure that the behavior of the AI is not at variance with ethical considerations. Additionally, diverse data is needed to reduce bias in training data, which is not easy to acquire since one would need access to unbiased and representative samples from different demographics and geographies.

3.3 Explainability and Transparency

AI systems, particularly complex machine learning models, are often black boxes, making their decision-making processes hard to interpret. A lack of transparency can erode trust and accountability in regulated industries, where compliance decisions must be justified to regulators and stakeholders.

XAI techniques, such as LIME and Shapley values, are explainability methods for AI decisions. These techniques help compliance teams and regulators make sense of AI outputs that assure accuracy, transparency, and traceability of decisions.

3.4 Regulatory Ambiguities

The dynamic nature of regulatory environments, exacerbated by fragmentation, is a real challenge for AI systems. Variations in legal requirements across jurisdictions make applying uniform compliance strategies difficult. For instance, the privacy requirements under GDPR differ from those under the Dodd-Frank Act or MiFID II. Hence, organizations need to tune their systems for particular regional regulations.

AI-driven compliance systems require the regular updating of emerging legal standards. This is a very resource-consuming process. Legal expertise must be accessed; system development cannot be a one-off event. Upcoming RegTech solutions will automate updating AI systems; however, this is still nascent and has a narrow scope.

3.5 Scalability and Integration

Growing organizations and heightened regulatory demands involve scaling upward, similar to AI, with more significant data volume and a whole complexity of requirements for compliance. Integrating AI systems sometimes faces difficulties and challenges in infrastructure with older database systems and dated software. This requires organizations to invest in scalable architectures like cloud-based platforms and microservices while maintaining compatibility with legacy systems. Overcoming such integration challenges will be a significant step toward deploying AI systems for large-scale compliance operations.

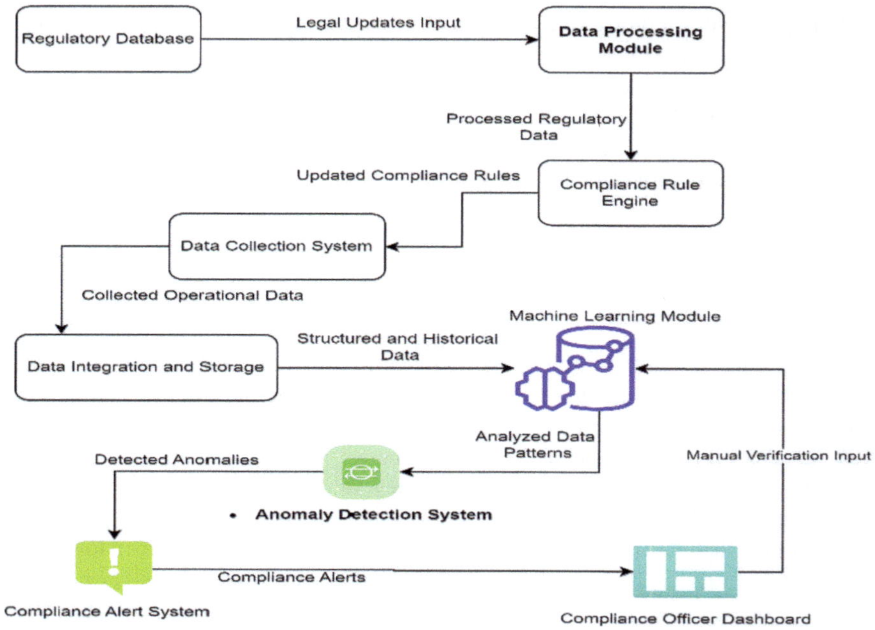

Fig. 1 Conceptual architecture for an AI-driven compliance monitoring system

Table 1 compares several scalability requirements. It describes the data volume and integration complexity expected in each industry. Financial services and healthcare, for instance, represent very high data requirements; scalability challenges often lie in data diversity and regulatory sensitivity for many other industries.

These issues are varied, involving technical, ethical, and even regulatory impediments to the effectiveness and reliability of AI-driven compliance systems. While these challenges most noticeably present themselves as requirements for technology solutions, they also require careful planning and consideration of transparency and ethics issues, coupled with organizational commitments to staying current with

Table 1 Scalability requirements and integration complexity across regulated industries

Industry	Data volume	Integration complexity	Key scalability challenges
Financial services	Very high	High	Transactional volume, real-time anomaly detection
Healthcare	High	Moderate	Patient privacy, cross-institutional data sharing
Pharmaceuticals	Moderate	Moderate to high	Regulatory updates, legacy system integration
Data privacy	Varies	Moderate to high	Legal compliance across jurisdictions

evolving regulations. Each limitation offers opportunities for further innovation in AI compliance as researchers and developers try to overcome the current obstacles to scalable, secure, and transparent compliance monitoring.

4 Case Studies and Applications of AI in Compliance Monitoring

4.1 Finance Sector

AI has become indispensable in finance, especially for compliance monitoring in anti-money laundering, fraud detection, and trade monitoring. Financial institutions fall under stringent laws: the USA PATRIOT Act, the Dodd-Frank Act, and the European Markets in Financial Instruments Directive, popularly known as MiFID II, all require great care in monitoring financial transactions and generating reports. AI has played an essential part in such a requirement through the out-of-capability real-time monitoring of transactions, risk-based assessment, and pattern recognition that may lead to detecting possible illicit activities [15–18].

AI-driven AML systems use machine-learning techniques to identify suspicious activities that could indicate money laundering. These range from strange patterns in transactions and cross-border fund transfers to large, sudden deposits that could not be anticipated based on trends in past transactions. These systems analyze historical transaction data to develop customer behavior profiles and pinpoint deviations indicative of fraudulent or money-laundering activities. Automation taking over these processes reduces the time compliance teams spend identifying and investigating perceived breaches, thus streamlining operations and enabling much faster response times [17, 18].

In addition, the AI-powered systems monitor trading activities regarding the game's rules on market manipulation and insider trading while transferring money or securities. Algorithms analyze trading patterns across large data datasets in almost real time, enabling pattern recognition to determine anomalies that may indicate rule violations. In this respect, institutions can comply with financial regulations and avoid the related fines and penalties that might also include reputational damage. Case studies from large banks prove that AI does help enhance AML compliance, though some limitations also exist in the form of requiring constant adjustments in models of AI to adapt to fresh tactics used in money laundering [17–19].

4.2 Healthcare Industry

The healthcare industry works in a regulated environment, and specific laws, such as the U.S., bind the sector. Among many pharmaceutical regulations, the Health

Insurance Portability and Accountability Act calls for hi-tech data protection and compliance standards. This is where AI-driven compliance-monitoring systems have become indispensable tools for healthcare organizations to maintain HIPAA compliance, patient data security, and adherence to pharmaceutical standards for drug safety and patient treatment protocols [17–19].

Following HIPAA, AI systems will try to ensure that sensitive patient data is accessed and handled as required by law. For example, AI-powered systems can maintain access logs to identify attempts of unauthorized access and usage patterns that differ from ordinary usage. Once the system determines a suspicious trend, such as repeated access to patient files from a location that is not familiar, this is flagged for further review. This capability also helps healthcare providers meet HIPAA standards by preventing possible data breaches [15–18].

These AI-driven systems keep a tab on everything from clinical trial data and the manufacturing process to pharmaceutical distribution, ensuring that the safety standards are complied with. Using machine learning algorithms on real-time clinical data, these would find abnormalities suggesting drug trials or manufacturing issues to address potential compliance issues proactively. In other words, a pharmaceutical company's compliance with the standards of drug safety reduces the risk of fines by the regulatory framework, legal liabilities, and delays in the time-to-market of its products. In the future, AI-driven compliance systems will be crucial to solving the challenges of maintaining patients' trust and ensuring operational integrity within the healthcare industry.

4.3 Data Privacy Compliance

Suppose one considers the recent upsurge in data privacy around the world. In that case, the General Data Protection Regulation in the European Union, right down to the California Consumer Privacy Act in the United States, now stands at a high level of compliance obligation regarding collecting, processing, and storing personal data. AI has emerged as a powerful tool that organizations can employ in monitoring and enforcing compliance with data privacy legislation since it provides them with the automation they need: first, to trace data usage; second, to detect risks to the interested party's privacy; and lastly, to keep them from failing to comply with regulations [17–19].

AI systems facilitate GDPR and CCPA compliance more efficiently by finding and managing personal data within information systems. They can trace data from their collection to storage and apply compliance-related rules according to regulatory requirements for appropriate handling practices in data management. For instance, under GDPR, individuals have a right to request data deletion, also termed the "right to be forgotten," in which AI-driven systems can automatically locate and delete relevant data records upon request. This automation dramatically minimizes the risk of human error in managing personal data and ensures timely responses to requests on privacy-related matters [4, 13, 15, 17, 18].

AI-driven compliance systems can also apply NLP to scan documents and communications for references to the use of personal data, highlighting where applications and processes might not be compliant with regulations related to the privacy of information. Monitoring data flows and maintaining audit trails, AI-driven systems thus offer holistic solutions to data privacy compliance.

4.4 Comparative Real-World Applications Analysis

Looking into real-world applications of AI in compliance monitoring in various industries, examples are mixed; some point to success, while others point to limitations. Within the finance sector, one large bank reportedly achieved a dramatic improvement in fraud detection accuracy and efficiency in compliance, having implemented its AI-powered AML system and showcasing a reduction in manual workload related to a 40% lessening of the work pressures on compliance officers. On the other hand, these also pointed out some challenges in algorithmic bias, as the high-risk groups were disproportionately flagged; hence, discussions were evoked concerning enhancing equity and transparency in AI-based systems [17, 18].

AI was implemented to monitor compliance related to HIPAA in the networking of hospitals in healthcare; it succeeded in detecting some data breaches and unauthorized accesses much quicker. It also underlined how AI can be gainfully utilized to manage privacy risks proactively. It also highlighted the limitations of AI interpretability, with healthcare compliance officers needing help to make sense of AI-driven alerts, which demands more explainability from these tools [17, 18].

However, this comparison analysis indicates that AI systems contribute significantly to ensuring compliance while refining their tuning regularly to pace up with the dynamic nature of regulatory demands and minimize unwanted bias. Success in this domain across industries demonstrates the potential of AI in cost reduction, efficiency enhancement, and improved compliance outcomes. However, interpretability, algorithm fairness, and adaptation remain critical challenges for optimizing the effectiveness of AI in compliance.

4.5 Trends in AI for Compliance

Several emerging trends within the AI technologies sphere promise to transform compliance monitoring further. One of the outstanding trends is the development or establishment of self-regulatory frameworks within the AI compliance systems. These, ideally, will enable AI to make autonomous adjustments to its compliance threshold and parameters with the changing regulations, making somewhat dynamic monitoring that limits human touch on updates. Examples include the finance sector, where this self-regulatory mechanism could mean banks automatically adjust transaction thresholds in real time to match changing anti-money laundering requirements.

Another leading-edge involves cross-jurisdictional compliance capability, which is especially pertinent for multinational organizations. These systems are supposed to handle multiple regulatory standards across various regions, make automatic parameter changes based on regulations in a particular area, and more. With improving NLP, AI modules can interpret and apply multiple legal texts relevant to jurisdictions to keep their organizations compliant across these diverse regulatory environments. This feature is, in turn, indispensable for companies dealing with finance and data privacy, where regulations vary significantly from region to region [4, 15, 16].

Finally, Explainable AI is an emerging domain of research in which transparency for AI-driven compliance monitoring would be increased. Techniques such as LIME and Shapley values in XAI make AI decisions more understandable to compliance officers, auditors, and regulatory bodies. Since the regulators started paying close attention to AI, the demand for explainability has grown increasingly critical in industries where accountability is paramount. This is about compliance with regulations and creating trust among stakeholders with XAI solutions through transparency in the decision-making process [4, 15, 16].

These up-and-coming trends illustrate how AI technology has evolved to meet some of these principal shortcomings of compliance monitoring since times present and will continue to drive solutions that are not only far more adaptable and scalable but also transparent. Organizations will be much more open toward states with more sound systems of compliance, real-time adaptability, and cross-jurisdiction capabilities through their particular leveraging of ever-advancing AI, which will accordingly meet the needs of present and future regulatory landscapes.

4.6 Quantitative Performance Analysis of AI in Compliance Monitoring

The quantitative analysis makes it somewhat more concrete how much better the state of AI-driven approaches performs than the traditional one, followed by increased efficiency, accuracy, and scalability for those respective systems. Based on empirical data, this section will discuss the main performance metrics, compare AI systems with manual approaches, and discuss improvements within industries.

The subsequent analysis describes the tangible contribution of AI in compliance management after analyzing detection accuracy, operational efficiency, cost savings, and reduction in manual efforts. Further, graphs and tables have been used to zero in on creating order and clarity to present the findings systematically. Together, these findings provide a data-driven validation of AI as a game-changer in meeting the demands of regulated industries.

Table 2 Performance metrics—traditional versus AI-driven methods

Metric	Traditional methods	AI-driven methods
Detection accuracy	80%	95%
False positive rate	20%	10%
Processing speed	Baseline	5 × Faster
Cost savings	Baseline	30–50%

Table 2 highlights some critical performance metrics of traditional approaches in compliance monitoring versus AI-driven methods. Some of these quantifying merits include accuracy of detection, false positives, processing speed, and costs saved.

Data shows that AI-powered compliance systems increase detection accuracy from 80 to 95%, with a 50% reduction in false positives, which is crucial to reducing unnecessary investigations. Additionally, AI systems enhance speed, processing up to five times faster to offer real-time compliance monitoring. Translated into dollars and cents, this means up to 30–50% cost savings, reduced need for extensive manual reviews, and the ability to respond to regulatory changes faster.

From Table 3, the finance industry leads in efficiency gains of 40% and a reduction in manual effort of 50% due to the volume of transactions and strict regulatory requirements, which AI addresses most effectively. The healthcare and data privacy industries have seen huge improvements, particularly in processing data and detecting anomalies. Pharmaceuticals lag somewhat, given the unique complexities of drug compliance and challenges with integrating legacy systems.

Figure 2 shows anomaly detection accuracy across various industries, hence, the performance of AI in real-world settings.

Anomaly detection has AI systems with pretty high accuracy throughout industries: Finance at 92 percent and data privacy at 90 percent since both these sectors depend on highly structured datasets. Results from healthcare and pharmaceuticals were close but lower in degree, indicating unstructured data with challenge boundaries-CTRs or clinical trial reports and patient-level records. It is not so much the sheer potential to drive AI in several regulatory environments and maintain its highest standards as it is all about adaptability.

Figure 3 presents the average time for compliance reporting in various industries using traditional and AI-driven methods. AI significantly reduces compliance reporting time: Finance and healthcare had the fastest at 60 and 70 min, respectively. Data privacy demonstrates a moderate reduction, with a benchmark of 65 min, since AI has smoothened out processes like data mapping and the creation of audit trails.

Table 3 Efficiency gains and manual effort reduction by industry

Industry	Efficiency gains (%)	Reduction in manual effort (%)
Finance	40	50
Healthcare	30	40
Pharmaceuticals	25	30

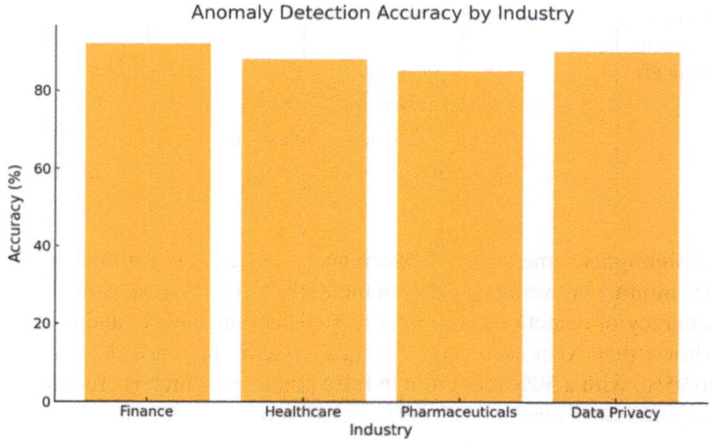

Fig. 2 Anomaly detection accuracy by industry

Pharmaceuticals take the longest 80 min because their quality assurance is quite exhaustive. These findings again demonstrate AI's ability to improve operational efficiency and reduce bottlenecks within the compliance workflow.

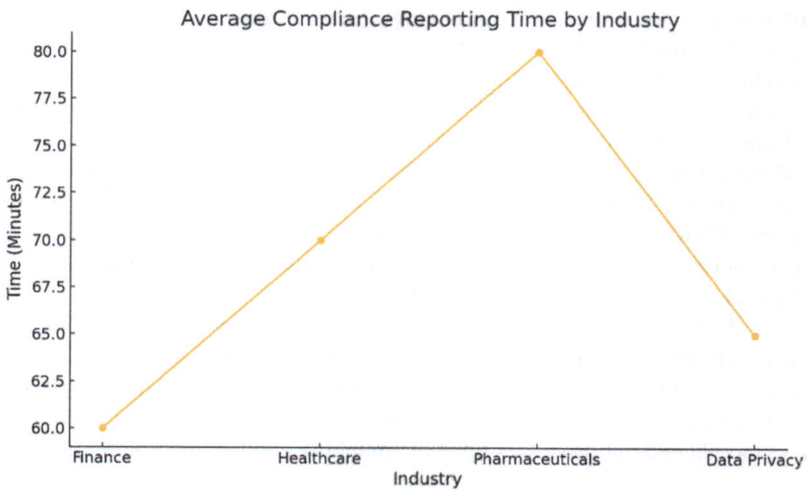

Fig. 3 Average compliance reporting time by industry

5 Future Directions and Recommendations

5.1 AI Ethics and Regulatory Frameworks

With AI further permeating compliance monitoring, the need to develop ethical guiding principles and regulatory frameworks has risen. Ethical AI practices pertain to fairness, transparency, accountability, and the responsible handling of sensitive data. Compliance monitoring, where AI decisions significantly affect individuals and organizations, requires ethical standards that prevent harm, discrimination, and privacy breaches. These should be the priorities of regulatory bodies and organizations in building frameworks that enforce ethics within AI development and application concerning algorithmic bias, data privacy, or the necessity of human intervention in decision-making [4, 15, 16, 18].

Ethical AI could be implemented by establishing an AI ethics board in organizations that will monitor the application of AI in compliance. It would develop policies to ensure that the application is fair and practices the best use of data, transparency, and accountability. Auditing AI systems will also be a critical raiser framework that ensures periodic testing for bias and that AI models are vetted against standards of dynamic ethics. Organizations can also create strict reporting structures through which employees or stakeholders could raise ethical concerns regarding AI-driven compliance monitoring. From that perspective, a setting of ethics in regulatory frameworks may create trust where guides enable AI-driven compliance with alignment to greater organizational and social values.

5.2 Explainable AI (XAI) Advancements

XAI plays a leading role in compliance monitoring, as the learnings and interpretation of AI decisions are necessary to comply with regulations and fully regain stakeholder confidence. Techniques for XAI aspire to make AI decision-making processes more transparent and thus permit explanations that the human user, for example, shall understand a compliance officer, auditor, or regulator. With interpretability, AI decisions come off as arbitrary or obscure; it is easier to justify them in audits or reviews. XAI also targeted one of the critical limitations of traditional AI—its "black box" nature [20, 21].

Effective XAI solutions for compliance could be developed from techniques such as LIME, which produces locally faithful explanations of individual predictions, to Shapley values, which assign a value to each feature for a given AI decision. An organization may enable rule-based explanations, too, translating how AI has reached a particular outcome into human language to the extent that a compliance officer can easily verify the same. XAI investments also allow organizations to meet regulatory requirements for transparency, as many jurisdictions have started developing requirements using explainability in AI applications [18–20].

Beyond just individual tools, the developments in XAI should go toward holistic interpretability frameworks integrated into compliance systems. This would enable compliance teams to understand AI decisions down to a granular level, analyze decision trends over time, and understand why particular alerts were generated due to an anomaly. This will go a long way in helping to instill confidence among the stakeholders in the working of AI systems, besides meeting the demands of accountability and transparency sought by regulators.

5.3 Policy Recommendations

To create this type of environment, the regulatory bodies of compliance need to renew and extend existing policies currently governing AI. These policies must handle the unique challenges of AI-driven compliance: data privacy, algorithmic accountability, and the differentials in cross-jurisdictional compliance requirements. More concretely, this will mean establishing standards for handling and storing data regarding personal information and algorithmic audits to identify and fix biases in these AI systems [17, 18].

Other policy considerations include a harmonized or, at least, standard set of regulatory standards that would cross jurisdictions, making international business compliance easier. Standardization reduces the pain wrought by dealing with different regulations in different regions, such as differences in GDPR for Europe and CCPA for California. Regulatory bodies could also institute safety certifications for compliance AI systems cleared on specific ethical and operational grounds. That gives organizations clear benchmarks around responsible behavior with AI [17–19].

Second, AI oversight frameworks that strike a proper balance between innovation and compliance requirements from policymakers ensure more ethical development in a structured way while enabling organizations to realize the full benefits of AI. Regular consultation of policies with industry stakeholders and AI experts helps them be relevant and adaptive to technological change.

5.4 Continuous Improvement and Training

All this means that with the ever-changing regulation environment, continuous improvement and training are significant for effective compliance systems driven by AI. Also, systems need periodic updates and model retraining to update regulatory issues, changes to industry-specific demands, and new risks. AI models will periodically be required to recalculate to account for new compliance data or modifications in regulatory expectations. It will be essential to have a routine schedule when assessing an AI model for thorough retraining to ensure models remain current and relevant to the latest compliance standards [20, 21].

Here, organizations may use data augmentation techniques to train AI systems in synthetically prepared new regulatory scenarios; the idea is to prepare the AI system for newer regulatory requirements that might still exist. Furthermore, feedback loops should be provided wherein compliance officers can give their inputs about the AI decisions, which could be used to refine the models and improve the accuracy of the entire system.

Of course, no less critical training for compliance teams is needed, as they must understand how to use insights provided by AI systems. Regular training sessions could make the compliance officers adept at the various tools and techniques used in AI-driven monitoring. Compliance officers can use AI insights most effectively this way. Therefore, continuous improvement and training are the other two necessary building blocks of any AI compliance system.

5.5 Conclusion

AI can transform compliance monitoring in regulated industries by equipping organizations with ways to improve regulated demands' efficiency, accuracy, and responsiveness. By automating data analysis, pattern recognition, and detecting anomalies, AI-powered systems can do real-time compliance monitoring that reduces human workload significantly and reinforces regulatory compliance. The profound realization of AI's potential in compliance requires the resolution of some key challenges related to ethics, interpretability, regulatory ambiguities, and continuous model updates.

These challenges require the cooperative development of ethical frameworks, the advancement of explainable AI, updating regulatory policies, and continuous improvement. Such actions will enhance the effectiveness of AI-driven compliance systems and build trust among stakeholders and regulators by assuring them that AI adoption will align with organizational goals and societal expectations.

Although AI will become increasingly crucial in monitoring compliance, new developments can be driven by continued machine learning, data protection, and mechanisms across borders. As AI technologies evolve, organizations that actively address ethical, operational, and regulatory considerations will be better positioned to deploy AI responsibly by the law while fostering innovation. The trajectory of AI in regulated industries is toward a more efficient, transparent, and compliant future, provided that ethical and regulatory frameworks keep pace with improvements in the technological landscape.

References

1. Khinvasara T, Shankar A, Wong C (2024) Survey of artificial intelligence for automated regulatory compliance tracking. J Eng Res Rep 26(7):390–406

2. Mokhtarian E (2018) The bot legal code: developing a legally compliant artificial intelligence. Vand J Ent & Tech L 21:145
3. Lin H (2024) Ethical and scalable automation: a governance and compliance framework for business applications. arXiv:2409.16872
4. Al-Shabandar R, Lightbody G, Browne F, Liu J, Wang H, Zheng H (2019) The application of artificial intelligence in financial compliance management. In: Proceedings of the 2019 international conference on artificial intelligence and advanced manufacturing, pp 1–6
5. Kulkarni V, Sunkle S, Kholkar D, Roychoudhury S, Kumar R, Raghunandan M (2021) Toward automated regulatory compliance. CSI Transactions on ICT 9:95–104
6. Robert A (2024) Automating financial compliance with intelligent process automation: designing AI-powered control systems for regulatory compliance and fraud prevention
7. Amaral Cejas O (2023) Artificial intelligence-enabled automation for compliance checking against GDPR. Doctoral dissertation, University of Luxembourg, Luxembourg, Luxembourg
8. Adeyelu OO, Ugochukwu CE, Shonibare MA (2024) Automating financial regulatory compliance with AI: a review and application scenarios. Financ & Account Res J 6(4):580–601
9. Beach TH, Rezgui Y, Li H, Kasim T (2015) A rule-based semantic approach for automated regulatory compliance in the construction sector. Expert Syst Appl 42(12):5219–5231
10. Magnusson Sjöberg C (2019) Legal automation: AI and law revisited. Legal tech, smart contracts and blockchain, 173–187
11. Zhang J, El-Gohary NM (2015) Automated information transformation for automated regulatory compliance checking in construction. J Comput Civ Eng 29(4):B4015001
12. Wall AM (2021) Guidelines for artificial intelligence-driven enterprise compliance management systems (Doctoral dissertation). https://napier-repository.worktribe.com/output/2850798
13. Thompson S (2023) Artificial Intelligence and project governance: enhancing compliance and accountability through automated monitoring. Aust J Mach Learn Res & Appl 3(2):518–523
14. Hassan M, Aziz LAR, Andriansyah Y (2023) Artificial intelligence's role in modern banking: exploring AI-driven approaches for enhanced fraud prevention, risk management, and regulatory compliance. Rev Contemp Bus Anal 6(1):110–132
15. Wylde V, Rawindaran N, Lawrence J, Balasubramanian R, Prakash E, Jayal A, Platts J (2022) Cybersecurity, data privacy, and blockchain: a review. SN Comput Sci 3(2):127
16. Raul AC (Ed) (2021) The privacy, data protection, and cybersecurity law review. Law Business Research Limited. https://datamatters.sidley.com/wp-content/uploads/sites/2/2019/11/The-Privacy-Data-Protection-and-Cybersecurity-Law-Review-Edition-6.pdf
17. Ashraf M, Haile A (2023) Data protection and AI: navigating regulatory compliance in AI-driven systems
18. Yanamala AKY, Suryadevara S (2024) Navigating data protection challenges in the era of artificial intelligence: a comprehensive review. Revista de Inteligencia Artificial en Medicina 15(1):113–146
19. Muhammad S, Meerjat F, Meerjat A, Dalal A (2024) Safeguarding data privacy: enhancing cybersecurity measures for protecting personal data in the United States. Int J Mach Learn Res Cybersecur Artif Intell 15(1):141–176
20. Chandak T, Jayashree J, Vijayashree J (2024) Trends and advancements of AI and XAI in drug discovery. In: Explainable AI (XAI) for sustainable development. Chapman and Hall/CRC, pp 233–255
21. Hosain MT, Jim JR, Mridha MF, Kabir MM (2024) Explainable AI approaches in deep learning: advancements, applications, and challenges. Comput Electr Eng 117:109246

Navigating 'Accountability Asymmetry' by Re-assessing Individual Criminal Responsibility: Gauging Usage of Artificial Intelligence-Driven Automated Weapon System

Saheli Chakraborty and Nirbindu Banerjee

Abstract Artificial intelligence-driven modern warfare has sparked significant debate regarding their use and the potential violations of international law. The growing reliance on artificial intelligence-powered Automated Weapon Systems by states, coupled with their direct violations of various aspects of international criminal law, presents numerous challenges. Despite their 'automation' and 'artificial' drive, the one who triggers or operates these systems play an undeniable role in the outcomes. This creates a unique legal situation where the connection between key criminal law elements, mens rea (mental state), and actus reus (physical act), becomes difficult to establish. This leads to the major issue of Accountability Asymmetry. Even though individuals are actively involved in the various stages of its manufacturing to deployment, AWS has been considered 'beyond human control', resulting in a lack of accountability. As the use of AWS continues to grow and its complexities increase, it becomes challenging to assign responsibility to any individual due to the various stages of operation and decision-making involved. Article 25(4) of the Rome Statute clarifies that the International Criminal Court can only exercise jurisdiction over natural persons. This paper aims to explore this 'Accountability Asymmetry' from the perspectives of Individual Criminal Liability.

Keywords AWS · AI · International law · IHL · ICL · CCW convention

S. Chakraborty (✉) · N. Banerjee
The West Bengal National University of Juridical Sciences, Kolkata, India
e-mail: sahelichakraborty@nujs.edu

N. Banerjee
e-mail: nirbinduphd2023@nujs.edu

M. Chakraborty et al. (eds.), *Proceedings of Sixth International Ethical Hacking Conference*, Lecture Notes in Networks and Systems 1513,
https://doi.org/10.1007/978-981-96-8632-2_26

1 Introduction

Technological warfare has continued to advance at an alarming rate throughout the twenty-first Century, and the fundamental international laws that govern armed conflict have not parallelly kept up, with possible justification. The focal point of international law has primarily been to stop humanitarian distress and hold individuals accountable for violating standing legal instruments. The advancement of modern weaponry, which has evolved into artificial intelligence, can be seen in the development of sophisticated drone frameworks, satellite cyber-attacks and more subtle and ferocious weaponry systems that are often automated. These developments have strained the existing limitations of international frameworks, pushing the existing boundaries and understandings [1]. Historically, it has always been a catastrophic event of magnanimous proportions like the bombings of Hiroshima and Nagasaki [2] that constitutes a rearrangement, amendment, or overhaul of related international laws. Drawing a historical parallel, the emergence of neoteric technological advancements like AWS, drones, air strikes, etc., have introduced complex discussions around accountability [3], liability and duties of state and non-state origin stakeholders. International criminal law can be fairly ambiguous in clarifying modern problems dealing with robotics and the sentiency of technology. Automated Weapons Systems, or AWS, are the latest incarnation of such technological integration. These weapon systems can operate with minimum or no human involvement, ranging from border enforcement [4] for states to drone and air strikes [5], with human review or supervision being almost 'unnecessary'. Some nations, such as India [6], have already begun to explore the potential of these technologies, while the global discourse on the accountability gap presented by AWS is only beginning to take shape [7]. The idea behind the automatisation of weapons is not novel in recent decades and has existed since the WWI era, although at a much more rudimentary level. An early example of this is the 'Kettering Bug' [8], which was a prototype aircraft designed for bombing missions. This, however, does not compare to the modern design, innovation, and efficiency of AWS, which can prominently increase the efficiency of the authorities using these systems in conflicts. Moreover, these systems of varying degrees are often invented and owned by private entities, creating another complexity for the international legal framework. The United States of America led the AWS invention as a pioneer through technology and governing policy [9], outlining the system's core functionality that reduces the necessity for human intervention once deployed in any scenario. Many scholars and Human Rights Watch have reviewed the 2012 policy, as we discuss in this article. However, the existence of AWS creates a responsibility gap within the international legal framework due to a lack of clarity and the unknown capability and limitation of automated systems. Often, academicians and international law jurists have pointed to this gap, while some scholars argue that the accountability gap can only be addressed by ensuring the AWS safety and protocols accurately and understanding the capabilities and often limiting them [10]. To maintain respect for International Humanitarian Law (IHL) in the creation and use of AWS, notably by avoiding an 'accountability gap' for IHL violations involving

AWS, the regulations controlling state responsibility for globally unlawful conduct and individual criminal culpability for war crimes are crucial [11]. These regulations serve various but complementary purposes. A basis for collective responsibility is provided by the laws governing state accountability.

2 Defining Automated Weapon Systems (AWS)

'Automated', the phrasing, itself originates from the words 'auto' and 'nomos'; often descriptive of any process, system or mechanism that is designed to reduce human labour and can 'run' by itself. Haynes further describes this system in the context of international law as robotic weapons that have the capacity to select and destroy targets [12]. Ironically, the process of selection may be arguably more significant than the process of destruction, as self-sufficiency or non-necessity of human input, in this case, caused the initial conundrum regarding AWS. AWS categorisation has been, therefore, based on the degree of 'independence' of these mechanisms or the necessity of human control, to be exact. In scholarly and legal discussions, terms such as 'autonomy', 'target selection' [13], and 'human intervention' [14] are commonly used to describe AWS. Dr Markus Wagner, in his article [15], attaches two specific fundamental characteristics to AWS, which are the ability to operate independently and the ability to make discretionary decisions on its own. Prof. Seixas-Nunes concludes [16] that a general definition of AWS has to have three characteristics that are specifically and transparently apparent in the text. First, it should be a weapon system and not an amalgamation of various agents or new agents. It also must be designated as a human creation or creation of such an agent. Thirdly, algorithms operating in AWS have to be specifically designed as a military plan or mission in its core specified objective. UNIDR report of 2017 [17] stresses that definitional discussions will not be 'value-neutral' discussions and have factors of political and other motives to be considered. Taddeo and Blanchard [18], in their comparative discussions of the definitions of AWS, provide valuable information regarding the different actors and their definitional approach towards a definition. For example, China, as a state actor in 2018, defined the AWS through a list of characteristics, which are lethality, autonomy, impossibility of termination, adaptability from the environment, and indiscriminate effects. NATO, as an international organisation in 2020, defined AWS as 'a system that decides and acts to accomplish desired goals, within defined parameters, based on acquired knowledge and an evolving situational awareness, following an optimal but potentially unpredictable course of action' [19]. On the other hand, the US Department of Defence provides a variation on the definition [20], which conversely discusses the degree to which autonomy should be counted as part of AWS. According to them, AWS is 'A weapon system that, once activated, can select and engage targets without further intervention by a human operator'. US Defence Primer on Autonomous Weapons Systems of lethal capabilities emphasises that no US policy prohibits the application, usage or deployment of AWS and further posits that human judgement or supervision of 'appropriate degree' can be applied to every Lethal Automated Weapons Systems (LAWS). However, such a subjective measure of 'appropriate degree' is hardly quantifiable, especially when the further explanation in the defence primer, last updated on February 2025, reads, 'human judgment over the use of force does not require manual human "control" of the weapon system, as is often reported, but rather broader human involvement in decisions about how, when, where, and why the weapon will be employed'. This shows that appropriateness, in the case of human control over AWS, might not be a defining trait that can be objectively and uniformly perceived. The definition further expands on that, stating that it would include autonomous weapon systems under human supervision, which enable human operators to take control of the system's functioning while enabling the system to choose and engage targets without additional human input once it has been activated.

The precise definition of AWS is, therefore, fluid in the sense of unanimity and temporal developments [21]. International organisations, such as the International Committee of the Red Cross (ICRC) [22] and Human Rights Watch, differ in their views [23]. The ICRC recognises [24] some existing technologies as forms of autonomous weapons, while Human Rights Watch predicts that fully autonomous 'killer robots [25]' could become a reality within a few decades [26]. While the definitions may vary due to numerous factors that include technological ambiguity and constant 'progress', geopolitical tension and the hierarchical nature of international relationships that play a huge role in the establishment of legal systems, it does provide a space for scholarly discussions of implications in international humanitarian and criminal law. While definitions may vary, categorisation is hard to achieve; the concept is well-known to every international and regional stakeholder and actor. Prof. Livoja [27] considers that while definitional consistency and uniform regulatory frameworks are not pre-established, the positions regarding AWS by different state actors have enough similarities to concur with the overall understanding of issues, challenges and legal implications.

3 The Legal and Ethical Complications of AWS

Under the aegis of the 1980 Convention on Certain Conventional Weapons (CCW Convention), international negotiations have been centred on the security, ethical, and legal concerns brought up by autonomous weapon systems since 2013 [28]. The open-ended Group of Governmental Experts (GGE) on Emerging Technologies in the Area of Lethal Autonomous Weapons Systems has been spearheading the discussion since 2017. AWS and AI have the potential to have a profound, enduring impact on humanity that extends to the different facets of international law due to their intrinsic capabilities. Talking about autonomous weapons only in terms of 'autonomous attack' is insufficient. Certain weapons are prohibited from being used in combat, according to the Geneva Convention's Prohibition on the Use of Certain Conventional Weapons (International Committee of the Red Cross (ICRC), 2004). In 2019, the GGE also adopted a principle that would clearly avoid mentioning AWS specifically but would run similarly to Article 36 of Additional Protocol I, focusing on the legal review of any 'new weapon, means or method of warfare' [29]. Article 36 of Additional Protocol I to the Geneva Conventions, for example, requires countries to examine new weapons to make sure they are not indiscriminate or capable of causing needless harm. Additionally, according to the Martens clause, people will continue to be 'under the protection and rule of law of nations ... from the laws of humanity and the dictates of public conscience'. The purpose of this clause was to guarantee that victims of conflicts, whether they be civilians or combatants, would always be protected under the 'laws of humanity' and that they would never be denied protection during a conflict by taking into consideration situations that were not covered by the current international humanitarian law treaties [30]. Legal reviews have been seen as a practical way of addressing the viability of these weapons, and several states have updated their domestic guidelines, adopted and integrated legal review processes into them, as well as moved towards transparent information sharing (like England and Australia). As expected, discussions surrounding AWS have brought up significant issues regarding applying fundamental principles of the law of armed conflict, such as distinction, proportionality, and the duty to take preventative action, as well as the function of human agency in guaranteeing adherence to these regulations. As opposed to military ammunition, the accuracy with which an AWS operates can effectively reduce suffering, incidental injury, and collateral damage due to the specificity of target identification and self-automated attack. It is built with faster data collection, risk reduction, rapid communication, lethality, persistence, and survivability. Nonetheless, there are dystopian arguments against its limitations, arguing that its application violates the fundamental principles of international humanitarian law, namely the principles of proportionality and distinction. New issues for international law are brought about by the growth of AWS, which is powered by AI technology. AWS raises serious ethical and legal issues even though it might increase strike precision and lessen collateral harm. These issues revolve around the fundamental tenets of humanitarian law: proportionality and distinction. While proportionality requires that the damage inflicted by an attack not exceed the military advantage obtained, distinction refers to the capacity to differentiate between combatants and non-combatants. Automating the decision to use lethal force raises ethical questions about the delegation of life-and-death decisions to machines [31]. Critics of AWS, however, specifically argue that the delegation of the decision-making process in itself will always be unethical, and it would remove the moral sanctity and human element of decision-making in the process. While some countries, maybe driven by different factors, argue or assert that current frameworks do not need a specific overhaul in this scenario, many would rather have a re-evaluation of the legal framework. Since its inception in the early days of the twentieth century, till now, it is forbidden to utilise incendiary weapons, laser weapons, and conventional munitions that contain explosives during combat, as well as weapons such as landmines, boobytraps, and other devices. Since LAWS and other AI-powered weapons systems are new, there isn't a comprehensive international convention or protocol that thoroughly covers this new technology.

The AWS inventions have diluted all existing concepts of human warfare and raised critical questions about human decision-making and oversight, which might soon be a minute part of armed conflicts. The distinction between combatants and non-combatants, participating humans and non-participating humans, only grows towards an opaque phase of legal, administrative, and practical turmoil. Even if an automated AI-driven program with arms can be programmed to follow a preset existing ethical and moral set of principles, it will still lack the human understanding and subjectivity of decisions, not to mention that preset artificial ethical guidelines can be changed, tampered with and bent to suit the interest of the controlling authority. With the accountability gap in international law, it would be hard to prove the control and intention of said authorities under any circumstance. Consequently, the increasing use of AWS in warfare presents serious problems for international criminal law, especially regarding enforcement and accountability. In order to address the legal, ethical, and technical issues that AWS systems raise, academics, decision-makers, and international organisations must participate in interdisciplinary discussions as their use grows. The accountability gap brought about by AI-driven warfare must be considered by international criminal law to maintain the importance of human oversight in military operations.

4 AWS and 'Accountability Asymmetry'

Concepts of 'agency' and 'accountability' are no longer in perfect alignment with the technological press forward, especially so in the case of weapons systems. With the evolution of AWS technologies, the involvement of advanced AI in these mechanisms accentuates the 'accountability gap' [10] in the interactions with ICL and IHL, where the assignment of responsibility or accountability for violations of international law is concerned. As discussed previously, guidelines have been established, although sometimes the exact use of AWS terminology has been avoided. International and regional legal frameworks must also be developed in parallel to frame proper guidelines and limitations for AWS and determine accountability norms as we address the increasing participation of automotive machinery and the degree of human control significance in warfare of various kinds. The 1-year SIPRI project [32] mentions a critical legal framework that can or should be used to address AWS and their application. Few who are aware of the global geopolitical landscape have been surprised by the driven and rapid development of military technology. The emergence of automation can be linked to the sixteenth century war, which saw significant advancements in gunpowder and artillery. During the revolutionary times, mechanised combat came right behind it. During World War II, it was further highlighted by the popularity of remote-controlled drones or their more basic forms. Although heavily dependent on human interaction, these systems had become the seed of imagination and eventually developed into the semi-autonomous weapon category [33]. The dynamic shift towards crewless aerial vehicles and more advanced remotely piloted drones that required less human guidance as technology made even more astounding advances.

Because of the technology, strikes could now be carried out from thousands of miles away with unprecedented precision. This made people question whether or not it was morally right for humans to be so detached from battlefield judgments. Soon, the contestation and arguments began on accountability, as the increasing usage also stretched the gaps in international law, and many academic stalwarts supported the development [34]. To this day, many academicians, state stakeholders, and private organisations argue about the implication of a broad ban on AWS technology, citing the impossibility of international law compliance with this machinery. The legal review system is vaguely attached to AWS through Article 36 of Additional Protocol I and has been relied on by many domestic juridical and legislative frameworks in recent years.

AWS is distinguished by its autonomous operation and capacity to recognise targets and marks without requiring continual human intervention. Since this capacity differs from the widely utilised UAV and does not require human input, it offers a fundamental challenge to the conceptual underpinnings of combat. To understand the Human interaction in the usage of AWS, one has to rely on scholars Amoroso and Tamburini's [35] compartmentalising human engagement into six well-distinct circumstances, which are as follows:

(1) a human engages with, selects, targets and initiates any attack with no substantial input from an AI system;
(2) an AI system suggests alternative targets and a human chooses which to attack;
(3) an AI system selects targets, and a human must approve before the attack;
(4) an AI system selects and engages targets but is supervised by a human who retains the power to override its choices and abort the attack;
(5) an AI system selects targets and initiates attacks on the basis of a pre-existing and discrete mission goal without further human involvement; and
(6) an AI system selects targets and initiates attacks based on pre-existing parameters that are not mission-based but rely entirely on the system's ability to elaborate, act, and improvise like a human being [36].

As we can analyse, human interaction with AWS ceases to exist after four stages. However, AI mediates each stage of its integral functioning irrespective of human inspection and intervention [37]. The reliance on AI is important, from data gathering and processing utilising various algorithms to the entire four stages, where human contact is critical. This is done to determine the target based on the filtered data for human consumption and to allow AI to recommend other targets. Human determination receives little consideration in the purchase, transmission, and data analysis processes. Moreover, the functioning of AWS and its components have unpredictable interactions [38] that can result in consequences where human control ceases to exist. To analyse human control [39] in light of a pragmatic approach to the principles of International Humanitarian Law, one has to study the entire operation of deployment, the Blueprint of manufacturing technology, Scientific Novelty and its embedded capabilities, the distinct components, an algorithm that is fed, and necessarily the governing mechanism that shall be fed for achieving the machine output, which is expected to supersede [40] the human military operations. An accountability gap that occurs in the face of criminal culpability is addressed in this study.

5 Individual Criminal Responsibility in AWS: The Heart of the Machine

International criminal law may be referred to as the judicial intervention to protect the fundamental concerns of the society and global community at large and protect people and certain protected groups from harrowing violations of internationally accepted norms. In this endeavour, the regime has established specific barbaric actions as criminally liable and set specific actions that harm human dignity under the ambit of criminal prosecution [36]. ICL has emerged as the foremost scholarship of international law that attributes penal consequences for certain specific crimes and few IHL Rules. The aftermath of World War II saw the rapid transgression of Ad Hoc Tribunals, which was the edifice of the development of the subject, where individual responsibility was the fulcrum of it being identified as a separate regime of international law. In the landmark case of Einsatzgruppen [41], the criminal liability against the state was cited. At the same time, the ICTY [42] in the Tadic Case [43] extended individual criminal responsibility to international or internal armed conflict. Article 25 of the Rome Statute refers to 'individual criminal responsibility' as 'A person who commits a crime within the jurisdiction of the Court shall be individually responsible and liable for punishment per this Statute. In accordance with this Statute, a person shall be criminally responsible and liable for punishment for a crime within the jurisdiction of the Court if that person: (a) Commits such a crime, whether as an individual, jointly with another or through another person, regardless of whether that other person is criminally responsible'; [44].

This catapults diverse kinds of responsibility, including aiding, inciting, attempting, giving directions or orders, and committing that holds the individual liable for such actions. The convergence of human intervention, while parallel orchestration of autonomy of the weapon system, poses challenges in determining criminal responsibility [45], which is rooted in provisions of ICL. The regime exclusively holds humans accountable, which can be challenging to prove or punish given that humans have little to no control over the effects of AWS. Public international law and criminal law interact to form international criminal law, which has evolved to prosecute international crimes using doctrines of liability interpreted from domestic jurisdictions. The foundation of international criminal law is the edifice of prerequisites, which include (I) provisions for attribution of the action, which determine whether criminal liability or responsibility can be attributed based on the facts and aspects of the law, and (II) sentencing for those found guilty of the crimes. The first criminal law premise is based on identifying the agency. The autonomous responsibility theory is not predicated on the idea that responsibility has human characteristics. The latter philosophy on punishment draws its inspiration from responsibility [46].

Because of its intricate structure and operations, AWS lacks agency and does not adhere to the long-held criminal law principle that, in armed conflicts, the actors would be combatants and commanders under whose direction the crime is committed. However, when it comes to AWS, the actor is as complex as its less straightforward layers. The role of human control over AWS and AI, as well as the necessity of state control to adhere to IHL's provisions, specifically, International Customary, Martens Clause, Article 3 of Additional Protocol I, are at the centre of the entwined discussion surrounding AWS. As previously discussed, among the six stages of Human Interaction and AI-fuelled AWS, there are two categories (Stages 5 and 6) [47] of AWS that function and propel without any human intervention, highlighting the legality of such weapons. Can a machine of such capabilities be allowed to function without human supervision as a prerequisite? In order to understand the obligation of duty and its omissions in such stages, human liability falls to zero. As scholars like Dinsteien mentioned, the weapons that are AI Robots created by humans may put humans on the centre stage of accountability [48]. In the quest to ascertain liability, certain scholars have also suggested the prohibition of AWS, which may function at any level without any human intervention and may even be extended as a violation of IHL. As part of the 11 Principles cited by the GGE in 2019, Principle 2 [49] makes this stance unambiguous. It ascertains human accountability as opposed to machine accountability. Human accountability is, therefore, crucial at every level of the decision-making process. The choice of 'target' and 'attack' would not be arbitrary, as one might anticipate from a killing machine, given human attributes like empathy, compassion, and reason. The general prudence of humans behind the ammunition cannot indeed be superseded by convenient and precise AWS [36].

That does not, however, resolve the issue of 'criminal accountability' for AWS. Moreover, the debate of pointing fingers to blame for AWS's acts originates from the idea that only humans are criminally accountable. According to Principle 4 of the GGE on LAWS, 'any emerging weapons system in the framework of the CCW must be developed, deployed, and used with accountability assured by applicable international law, including through the operation of such systems within a responsible chain of human command and control'.' This principle assumes that AWS has transitioned as a quantitate shift in the military operations, where general existing principles of IHL and thereon extension to the ICL shall be an ordinary adaptation to the new weapon system [50]. In this context, the perceived laws that guide 'individual criminal responsibility' and 'human-controlled international crimes' shall be applicable even in cases of AWS. Such assumptions are not receptive to the qualitative shift that weapons systems are capable of lex lata and de lege ferenda. The definite philosophies of criminal law and criminal liability are outweighed by the criminal mind (mens rea) and their criminal actions (actus reus). For criminal conduct to exist, especially in International Crimes, actions supersede any political or military objective to attain the mens rea [51]. As for human-controlled war, in situations of armed conflict or warfare, the actions that a soldier takes and the weapons he uses are a confluence of his mind and his actions, where conscious awareness of such actions is violative of international law [52]. In various jurisdictions, the state practice has provided divergent procedural connotations; however, what has remained justified is the presence of both the tenets of Criminal Law, which cannot cease to exist in establishing criminal responsibility. The malicious mental state is a pre-requisite in drawing the unambiguous nexus with the act [53]. The situation demands a better conceptual understanding of complexities drawn by AWS and its uncertainty, with or without human intervention. In cases of behavioural unpredictability, it is arduous to attribute any pre-existing mens rea to the capacity of negligence, intent, or knowledge [54]. In AWS, which has human control, the data on which decisions are made are a convergence of human and non-human components; therefore, the mental elements of the engineering system pose a unique challenge in establishing men's rea in exceptional cases. Thus, the accountability structure, traditionally viewed from Article 25 [55], may not be the pragmatic approach to holding an individual criminally responsible for violation of the tenants of ICL. Secondly, in the stance where the AWS performs unpredictably, shall it have outweighed the mens rea of the one who may have deployed the machine? In cases like war crimes, crimes against humanity, and genocide, mens rea of the alleged forms a crucial step in labelling [56] and attributing certain criminal liability. In cases of unpredictable AWS, it is undeniable that the mens rea shall be one such challenge in holding criminal liability and deploying AWS and foreseeing a possibility of harm in an unforecastable circumstance cannot be referred to as men's rea to be held accountable [57].

Thirdly, AWS is a product of diverse components, where various individuals are required to make decisions in its design, manufacturing, programming, and deployment. Therefore, there is a vast difference between mens rea and the one who has designed and the one who has deployed, where the decision at every level may be misinterpreted. On the basis of certain directions, a machine may be designed. In contrast, for an alternate reason, such a machine may be deployed without a rudimentary nexus with the actus reus and the consequence of such actions posed. Some individuals may just be referred to as 'in the background' [37] on whom men's rea and actus confluence is cumbersome to establish. In light of AWS, special consideration should be given to the war crime of indiscriminate assaults. This is caused by a number of factors, including the possible elevated risk of target identification errors linked to AWS use [58]. However, when AWS is involved, a number of problems make it challenging to assign blame for war crimes. No international conventions are discussed here, nor can the Rome Statute clearly define 'indiscriminate attacks', and understandably so. Although criminal courts have functionally treated it as an attack that has no specific military objective behind it, the non-existence of specific provisions is an immensely crippling hurdle [59]. Lastly, due to the unpredictability of the AWS, where the autonomy allows the machine to choose its path, where assumptions and predictions of humans fail, in such stances can 'Act of Machine' be delegated as criminal liability on individuals, as per the provision of ICL. The way that national governments apply the idea of criminal responsibility for war crimes is influenced by the different ways that war crime laws are implemented in national legislation, especially the definitions that each state chooses to include [60].

6 Meaningful Human Control or MHC: The Degree of Autonomy in AWS

To counter the gap or asymmetry, the concept of 'Meaningful Human Control' (MHC) emerged as a possible factor to be assessed and evaluated in the distribution of accountability. The concept of itself appeared through academic analysis. It can be traced back to a 2013 paper [61] arguing that it is necessary to retain human control in a significant way that is in compliance with the International Criminal and Humanitarian Law. Consequently, MHC has progressively become a focal point in the discussions of weapon autonomy, specifically for governance and legal framework application. However, despite its growing importance, the concept remains difficult to define and implement accurately, especially as AWS becomes more sophisticated and autonomous. Since 2014, the United Nations Convention on Certain Conventional Weapons (CCW) has included AWS on its agenda. Even if the Group of Governmental Experts (GGE) initiated political conversations around the potential restriction of AWS in 2017, developments in the technology of automated and autonomous aspects of weapons are still going unaddressed [62].

Three main pillars that constitute the basis of Meaningful Human control (MHC) are accountability, action, and information. These components are doctrinally meant to have a certain hierarchical construct which guarantees that human operators uphold adequate supervision to stop senseless violence and guarantee adherence to the law. However, given how dynamic and unpredictable AWS is, each element has particular difficulties. The first part of MHC highlights how important it is that human operators have access to contextual information regarding targets, mission goals, and the possible consequences of their actions. This requirement mainly reflects current legal duties under the Law of Armed Conflict (LOAC), even though it guarantees that operators make informed decisions. It does not introduce new concepts or sufficiently address the complexities of AWS in combat situations where information can be fragmented or rapidly changing. It fails to adequately handle the intricacies of AWS in combat scenarios, when information may be fragmented or changing quickly, or to propose novel ideas [63]. Ensuring that operators or 'human agents' take precautions or be proactive before engaging in any action with a 'lethal' output of damage accentuates the human involvement part even further. However, that would still be consistent with weapon systems with supervision or constant control and not something that can practically be fully automated without such involvement. Current drone warfare may fall within that semi-supervised area, with the use of force having 'meaningful human control'. However, that probably negates a bit of the efficiency of AWS. These frameworks urgently need to be improved as AWS develops further to maximise the benefits of autonomous systems while maintaining human oversight. One of the most important aspects of guaranteeing human responsibility is gaining a better grasp of how states plan to guarantee their practical capacity to investigate occurrences involving AWS. To uphold human responsibility in the creation and use of AWS, states must ensure that they have the ability to practically connect potential wrongdoers, including individuals and state agencies, to the unlawful activity of IHL violations.

7 A Practical Deconstruction of MHC

MHC, or Meaningful Human Control in the contemporary international law regime, is more than a theoretical concept or an intellectual exercise in discussion, with the tangible ramifications from AWS system usage in military operations. The best way to describe the current positioning of the discussion can be explained as a doctrinal necessity, which establishes that any use of force is not capricious and has ample supervision. However, in the case of AWS, this procedure is made more difficult by the weapon systems' growing autonomy. Decisions made by AWS may be influenced by machine learning [64] algorithms, interactions with other systems, or environmental factors, making it difficult to determine where responsibility lies when things go wrong [65]. Important concerns concerning how to guarantee that these systems function within the parameters of IHL are brought up by the integration of AWS into military operations. In contrast to traditional weapons, where human choices directly cause harm, AWS adds complexity that could make accountability more difficult to understand. It gets more difficult to keep an eye on and manage AWS's activities in real time as they operate in more unpredictable circumstances, which could make it more difficult to assign legal accountability. The spectrum of complexity in the case of AWS goes beyond the extremes of machine autonomy and human control. It can be better adjusted and explained as a continuum of human–machine interactions, and thus, understanding the non-binary nature of any action of AWS is necessary for legal and administrative response by any authority. The absence of a clear delineation of human involvement at each stage necessitates an expanded conceptualisation of MHC that accounts for both ex-ante regulatory controls and ex-post facto liability frameworks.

Determining its boundaries and creating precise rules that work in many situations are two of the biggest obstacles to creating real human control. MHC needs to be customised to the unique responsibilities of the various parties engaged in the deployment and use of AWS to function as a trustworthy legal norm. The U.S. delegation expressed concerns about MHC's ambiguity during the 2014 Convention on Certain Conventional Weapons (CCW) negotiations, claiming that it failed to sufficiently consider the full range of human engagement in military systems. Various stakeholders, like developers and authorised state and non-state users in international conflicts, have a part to play, and any international guideline or discussion around MHC has to recognise these complex dynamic stakeholders. This necessity for a nuanced approach while discussing obligations leads to the fact that combatants are the primary 'users' of AWS during the conflict. However, when an operator lacks the time to supervise a certain action that AWS has decided to execute actively, a problem occurs [66]. The issue of criminal liability is particularly pressing in these situations. Likewise, there is a problem when AWS's actions result in what would otherwise be regarded as a war crime. This does not imply that AWS will 'go rogue' in the sense of defying orders. Under IHL regulations, the use of a weapon system with autonomous capability would be prohibited if it violated certain instructions. Combatants, being the focal point of MHC, have a few supporting pillars, like the direct operational capability of soldiers, ultimately in close proximity to the battlefield. Because of the unreliable nature of AWS and the narrative of distrust in using machines to possibly make decisions affecting human lives and the international status quo between various actors, it is also a lot more difficult to assign mens rea or criminal intent. Although AWS developers and manufacturers might be held partially accountable for their products, they are usually not physically engaged in battle. Therefore, the combatants using AWS in this situation are responsible for guaranteeing adherence to IHL. Determining the roles and responsibilities of combatants in using AWS is made simpler by concentrating on them. This can be the starting point for extending the idea of MHC to additional players, such as designers and manufacturers, guaranteeing that everyone engaged in the development and implementation of AWS is held responsible for their deeds. The idea of MHC will need to change as AWS continues to play a bigger part in contemporary combat to handle the new difficulties these systems provide. Although the existing knowledge of MHC offers a foundation, it can still not handle the complexity of autonomous combat. The emphasis must move to create a more thorough and adaptable architecture to guarantee significant human control over AWS, especially in relation to soldiers' duties. Addressing accountability in the use of AWS will be made easier with a more precise definition of MHC that explains the obligations of combatants. In addition to ensuring adherence to IHL, this strategy will make the system for using autonomous weapons in armed conflict more open and responsible. As technology develops, so must our comprehension and control of its application in combat.

8 Conclusion

Implementing machine warfare has presented significant hurdles to the ICL and IHL. When defining control, numerous inputs and outputs from AWS need to be considered. When the relationship between responsibility, liability, and AWS deployment is viewed as a continuous, complex process with MHC at its intersection, the shortcomings of the current paradigm become evident. The paper has attempted to explore the complex relationship between autonomous weapons systems (AWS) and their potential impact on international humanitarian law (IHL) and international criminal law (ICL). Primarily, these arguments and uncertainties around AWS can be boiled down to the 'accountability asymmetry', where human involvement in the 'decision-making' becomes increasingly diluted and attenuated. It is so much so that it often raises doubts over ever-existing principles like mens rea and actus rea and their applicability in certain situations stemming from the unknown potential of artificial intelligence. In order to assign criminal liability, international tribunals' jurisprudence, especially that of the ad hoc tribunals for the former Yugoslavia

(ICTY) and Rwanda (ICTR), emphasises the necessity of a demonstrable mental state. To say the least, this condition is difficult to meet in AWS situations. Fragmentation of AWS usage, involvement of stakeholders, and lack of expertise in both operations and responses clearly exacerbate this issue. The liability concerns raised by the 'qualitative paradigm shift' anticipated with the launch of AWS cannot be adequately addressed by the current legal framework, which is founded on human-to-human interactions. Today's definitions of international crimes are based on criminal law ideas that lack underlying philosophical underpinnings when it comes to AWS deployment. This is particularly noticeable when AWS functions without human oversight or control, resulting in a lack of accountability on the part of superiors, corporations, and individuals. The 'autonomous accountability gap' results from giving autonomous robots the ability to eliminate the human population since the nexus of liability for international crimes depends on human control. However, by guaranteeing significant human control over AWS during deployment, this 'liability lacuna' might be remedied. Some researchers [60] also see the potential opportunities arising out of the AWS dilemma, like international investigations into assessing responsibility for actions being a beneficiary of action logs of these automated weapons, which can shed direct light on the responsible parties for some specific decisions that might lead to a violation of humanitarian laws. AWS might be incorporated as an extension of military operations in reaction to this paradigm shift, which would reduce their independence. The process of establishing culpability is faulty on several levels when taking superior responsibility into account, which renders it an untrustworthy foundation for accountability. Current changes and modifications to existing doctrines and principles have established the concept of MHC or meaningful human control to better adjust to modern autonomous weapons and application thereof. Even then, MHC remains an 'amorphous' legal standard, where precise criteria can be debated, faltering internal and external investigation and legislative or judicial procedure in international scenarios. While the prohibitionist approach has gained traction among academicians, it is highly unlikely that such a ban would be properly maintainable internationally in the face of rapid technological development, not to mention the diplomatic and geopolitical factors that would render such theories unimplementable. Rather, the most practical course of action seems to be a hybrid strategy that incorporates both ex-ante regulatory protections and ex-post accountability systems. In the end, this article makes the case for changing the law to protect against possible risks to autonomy by preventing it from becoming a 'kill machine'.

References

1. Weigend T (2023) Convicting autonomous weapons? criminal responsibility of and for AWS under international law. J Int Crim.Justice 21(5):1137–XX. https://doi.org/10.1093/jicj/mqad037. Accessed 25 January 2025

2. District Court of Tokyo–Japan (1963) Shimoda et al. v. The State, 7 December 1963, 32 ILR 626. For an English version of the judgment, see Cassese A et al (2011) International criminal law: cases and commentary. Oxford University Press, pp 132–137
3. Heller KJ (2013) One hell of a killing machine: signature strikes and international law. J Int Crim Justice 11
4. GNIUS. Guardium UGV. http://g-nius.co.il/pdf/brochures/GuardiumUGV.pdf
5. Macfarlane K, Christie L (2022) Automation in military operations. Research Briefing, 7 October 2022. Accessed 25 Jan 2025
6. Early Steps in India's Use of AI for Defence. International Institute for Strategic Studies (IISS). https://www.iiss.org/online-analysis/online-analysis/2024/01/early-steps-in-indias-use-of-ai-for-defence/
7. Acquaviva G (2022) Autonomous weapons systems controlled by artificial intelligence: a conceptual roadmap for international criminal responsibility. Mil Law & Law War Rev 60:89–121
8. Fahlstrom PG, Gleason TJ (2012) Introduction to UAV systems, 4th edn. John Wiley & Sons, Ltd, pp 3–15. https://doi.org/10.1002/9781118396780
9. Human Rights Watch (2013) Review of the 2012 US policy on autonomy in weapons systems: US policy on autonomy in weapons systems is first in the World. Background Briefing. Accessed 25 Jan 2025
10. Weigend, supra note 1
11. Bo M, Bruun L, Boulanin V (2022) Retaining human responsibility in the development and use of autonomous weapon systems: on accountability for violations of international humanitarian law involving AWS. SIPRI, Stockholm. https://doi.org/10.55163/AHBC1664. Accessed 25 Jan 2025
12. Letendre LA (2020) Lethal autonomous weapon systems: translating geek speak for lawyers. Int Law Stud 96:278–282
13. Ekelhof MAC (2017) Complications of a common language: why it is so hard to talk about autonomous weapons. J Conflict Secur Law 22(2):311–319
14. Merriam-Webster Dictionary. Autonomous 39 (2016)
15. Wagner M (2014) The dehumanization of international humanitarian law: legal, ethical, and political implications of autonomous weapon systems. Vanderbilt J Transnat'l Law 47. University of Miami Legal Studies Research Paper No. 15–1
16. Seixas-Nunes A (2021) Autonomous weapons systems and the procedural accountability gap. Brook. J. Int'l L. 46
17. UNIDIR (United Nations Institute for Disarmament Research) (2017) The weaponization of increasingly autonomous technologies: concerns, characteristics, and definitional approaches. UNIDIR Resources
18. Taddeo M, Blanchard A (2022) A comparative analysis of the definitions of autonomous weapons systems. Sci Eng Ethics 28(5):37. https://doi.org/10.1007/s11948-022-00392-3
19. NATO (2020) AAP-06 Edition 2020: NATO glossary of terms and definitions. NATO Standardization Office
20. U.S. Department of Defense (2012) DoD Directive 3000.09 on autonomy in weapon systems. https://www.esd.whs.mil/portals/54/documents/dd/issuances/dodd/300009p.pdf
21. Jain AG (2023) Autonomous weapon systems, errors and breaches of international humanitarian law. J Int Crim Justice 21(5):1005–1032. https://doi.org/10.1093/jicj/mqad043
22. ICRC. Views of the international committee of the red cross on autonomous weapon systems' working paper www.unog.ch/80256EDD006B8954/(httpAssets)/B3834B2C62344053C1257F9400491826/$file/2016_LAWSþMX_CounyPaper_ICRC.pdf4
23. For more details on the ICRC during the UN CCW Meeting of Experts General Exchange on 13 April 2015 and 11 April 2016, Geneva, Switzerland. www.unog.ch/80256EDD006B8954/(httpAssets)/4CE346B40DDBF000C1257E2600616A59/$file/ICRCþgeneralþstatementþCCWþLAWSþexpertþmeetingþ13þ04þ2015þFINAL.pdf4and5www.unog.ch/80256EDD006B8954/(httpAssets)/9324B81015529E3DC1257F930057AF12/$file/2016_LAWSþMX_GeneralExchange_Statements_ICRC.pdf4

24. ICRC Expert Meeting on Autonomous Weapon Systems, 15–16 Versoix, Switzerland. For the full report see: www.icrc.org/en/publication/4283-autonomousweapons-systems#4accessed ek
25. See details pertaining to the Campaign to Stop Killer Robots available at www.stopkillerrobots.org/learn/4. Accessed 18 Sept 2024
26. Human Rights Watch and International Human Rights Clinic
27. Liivoja R (2024) AWS legal review series–protracted debate. Incremental Progress, Unexpected Outcomes
28. Convention on Prohibitions or Restrictions on the Use of Certain Conventional Weapons which may be Deemed to be Excessively Injurious or to have Indiscriminate Effects (CCW Convention, or 'Inhumane Weapons' Convention), entered into force on 2 (1983)
29. United Nations (2019) Convention on prohibitions or restrictions on the use of certain conventional weapons which may be deemed to be excessively injurious or to have indiscriminate effects: group of governmental experts on emerging technologies in the area of lethal autonomous weapons systems, Geneva, 25–29 March 2019 and 20–21. Item 6 of the Provisional Agenda: Adoption of the Report (Second Session)
30. For more details, see considerations for the report of the group of governmental experts of the high contracting parties to the convention on certain conventional weapons on emerging technologies in the area of lethal autonomous weapons systems on the outcomes of the work undertaken in 2017–2021
31. Lewis J (2015) The case for regulating fully autonomous weapons. Yale Law J 124(4):1309–1325. http://www.jstor.org/stable/43617052
32. Boulanin V, Goussac N, Bruun L (2021) Autonomous weapon systems and international humanitarian law: identifying limits and the required type and degree of human–machine interaction. SIPRI, Stockholm. Accessed 25 January 2025
33. Crootof R (2015) The killer robots are here: legal and policy implications. Cardozo Law Rev 36:1837. https://ssrn.com/abstract=2534567
34. Schmitt MN (2013) Autonomous weapon systems and international humanitarian law: a reply to the critics. Harv Natl Secur J Featur
35. Amoroso D, Tamburini G (2020) Autonomous weapons systems and meaningful human control: ethical and legal issues. Curr Robot Rep
36. Ibid
37. Acquaviva, supra note 7
38. Mathias A (2004) The responsibility gap: ascribing responsibility for the actions of learning automata. Ethics Inf Technol 6:175
39. Boulanin V, Davison N, Goussac N, Peldán Carlsson M (2020) Limits on autonomy in weapon systems–identifying practical elements of human control. ICRC, SIPRI
40. Verdiesen I, Santoni de Sio F, Dignum V (2020) Accountability and control over autonomous weapon systems: a framework for comprehensive human oversight
41. Trials of War Criminals before the Nurnberg Military Tribunal under Control Council Law No. 10 (1950). United States v. Otto Ohlendorf, vol 4. Springer, p 462
42. United Nations Security Council (1993) Statute of the international criminal tribunal for the former Yugoslavia, adopted 25 May 1993 by UNSC Res 827, amended 27 May 1999, UN Doc S/RES/827
43. Prosecutor v. Tadić, Case No. IT-94-1-AR72 (1995) Decision on the defence motion for interlocutory appeal on jurisdiction, 129 (Int'l Crim. Trib. for the Former Yugoslavia, Oct. 2, 1995)
44. Rome Statute of the International Criminal Court, adopted 17 July 1998, entered into force 1 July 2002, UN Doc A/CONF.183/9, art. 25
45. Bo M (2018) Who is criminally responsible for the commission of war crimes when lethal autonomous weapon systems are deployed in armed conflicts? Geneva Graduate Institute, Geneva
46. Lubell N, Pejic J, Simmons C (2019) Guidelines on investigating violations of international humanitarian law: law, policy and good practice. Geneva Academy of International Humanitarian Law and Human Rights and ICRC, Geneva

47. Amoroso, supra note 35
48. Dinstein Y (2016) The conduct of hostilities under the law of international armed conflict, 3rd edn. Cambridge University Press, p 99
49. United Nations (2019) Guiding principles affirmed by the group of governmental experts on emerging technologies in the area of lethal autonomous weapons systems. UN Doc. CCW/MSP/2019/9
50. See US commentary on the Guiding Principles of 1 September 2020. https://documents.unoda.org/wp-content/uploads/2020/09/20200901-United-States.pdf
51. Prosecutor versus Tadic, supra note 44
52. Badar M (2013) The concept of mens Rea in international criminal law: the case for a unified approach, vol 52. pp 234
53. Schabas WA (2017) An introduction to the international criminal court, 5th edn
54. Amoroso D (2017) Jus ad bellum and jus in bello arguments against autonomy in weapon systems. Quest Int Law 43:19–20
55. Rome Statute of the International Criminal Court, 2187 UNTS 90. Adopted 17 July 1998, entered into force 1 July 2002
56. Rome Statute of the International Criminal Court, 2187 UNTS 90. Adopted 17 July 1998, entered into force 1 July 2002), art 8(2)(b)(xx)
57. Acheson R, Pytlak A (2022) Autonomous weapons and questions of ethics. Control and accountability. CCW Report 10(4)
58. Ohlin JD (2016) The combatant's stance: autonomous weapons on the battlefield. Int Law Stud 92(1). Accessed 25 Jan 2025
59. Dörmann K (2003) Elements of war crimes under the rome statute of the international criminal court: sources and commentary. Cambridge University Press, pp 131–132
60. Bo and Brunn, supra note 11
61. Article 36 (2024) Killer robots: UK government policy on fully autonomous weapons. https://article36.org/wp-content/uploads/2013/04/Policy_Paper1.pdf. Accessed 1 March 2024
62. Ferl AK (2023) Imagining meaningful human control: autonomous weapons and the (De-) legitimisation of future warfare. Global Society 38(1):139. https://doi.org/10.1080/13600826.2023.2233004. Accessed 25 January 2025
63. Zerilli J (2019) Algorithmic decision-making and the control problem. Minds Mach 29:555–578. https://doi.org/10.1007/s11023-019-09513-7. Accessed 1 March 2024. at 559 (and cited literature since the 1980s)
64. Ibid; Johnson J (2022) The AI commander problem: ethical, political, and psychological dilemmas of human-machine interactions in AI-enabled warfare. J MilY Ethics 21:246–271. https://doi.org/10.1080/15027570.2024.2175887. Accessed 1 March 2024
65. Santoni de Sio F, van den Hoven J (2018) Meaningful human control over autonomous systems: a philosophical account. Front Robot AI 5(1) at 10
66. Liu H-Y (2012) Categorisation and legality of autonomous and remote weapons systems. Int Rev Red Cross 94:627–629

Artificial Intelligence and Law: Navigating the Convergence of Subsets

Prem Kumar Agarwal and Namrata Chakraborty

Abstract Recently, while OpenAI is facing several lawsuits alleging potential infringements of privacy and intellectual property rights in multiple jurisdictions across the world, it simultaneously spotlights the ethical conundrums surrounding the intersection of artificial intelligence and law and underlying the shadow of technological advancement in legal frontiers. It is hence indispensable to appreciate the contours of such convergence at this juncture in order to assess whether the growth of artificial intelligence in legal sector satisfies the test of time and effectiveness. The present research article thus focuses upon analyzing the bright sides, the challenges and the far-reaching impact associated with the intersection of artificial intelligence and law.

Keywords Artificial intelligence · Legal practice · Judicial system · Law enforcement · Intellectual property rights · Accountability

1 Introduction

Over the past few decades, the regime of modernization and technological invention has not only transformed the living standards across the world, but also made a revolutionary impact upon every industry or field of discipline. One such latest breakthrough is artificial intelligence or AI—a concept coined by Professor John McCarthy in 1956 [1]. There are multiple ways to define it, but in simple sense—AI is an umbrella term referring to the technology used in enabling the machines to automate specific tasks by way of simulating human intelligence, problem solving, spatial recognition, cognitive abilities, comprehension, and decision-making skills.

P. K. Agarwal (✉)
WBES, Government Centre of Legal Education (Hooghly Mohsin College), Chinsurah, Hooghly, India
e-mail: premkagarwal1968@gmail.com

N. Chakraborty
Judicial Officer, West Bengal Judicial Service, Kolkata, West Bengal, India

M. Chakraborty et al. (eds.), *Proceedings of Sixth International Ethical Hacking Conference*, Lecture Notes in Networks and Systems 1513,
https://doi.org/10.1007/978-981-96-8632-2_27

Even though AI is largely presumed as a subset of computer science, the scope of artificial intelligence in reality is extended to a wide network of logical rules, machine languages, automation techniques, and research ideologies extracted from multiple disciplines which include but do not limit to statistics, neuroscience, mathematics, logic, robotics, linguistics, and engineering. In short, AI represents an interdisciplinary system primarily focused on automation of a real-world process in two approaches, namely—(1) machine learning, and (2) the representation of underlying knowledge and logical rules concerning the target activity [2]. Presently, with the steadfast pace at which AI has been continuously evolving and permeating the nooks and crannies of daily human lives, it results into the emergence of various kinds of legal and regulatory implications that mandate due and expeditious supervision. Therefore, it is pertinent to delve deeper into the intersection of AI and law while simultaneously assessing the ethical concerns revolving around the convergence of both spheres.

2 Intersection of AI and Law: The Convergence

It is of no surprise at this juncture that the application of AI into every industry has now become a natural phenomenon and the field of law is no exception. Even though the growth in the usage of artificial intelligence among the legal researchers and lawyers has been axiomatic since the mid-twentieth century, the origin dates back to the 1600 s when Gottfriend Leibniz, a mathematician and trained lawyer, experimented for the first time on the application of mathematical formulas and techniques to make the law more accessible and layman-friendly [3]. The early researches on the intersection of AI and law involved the experimentation on the second approach, i.e., the representation of knowledge and logical rules-based legal formalism. But, fast forward to the present scenario, the decades since 2000 witnessed a shift of focus toward the machine-learning approach in the legal field. As a result, the legal-tech companies came forward to launch the AI-enabled computer systems in order to reshape the efficiency, accessibility and accuracy in a myriad of tasks ranging from the automation of routine legal documentation to the application of predictive analytics so as to challenge the existing legal frameworks. The convergence of AI and law can further be simplified through the following categories.

2.1 AI in Legal Practice

Integration of AI with legal practice showcases wonders. The buzz that artificial intelligence is taking over the legal service is derived from the fact that the lawyers have finally started embracing this drastic shift despite portraying initial reluctance. A recent legal survey conducted by LexisNexis International in 2023 reveals that out of 7950 survey participants comprising of law students, lawyers, and consumers across four countries (USA, Canada, United Kingdom and France), a stark majority of 65% believe in higher potential of generative AI-powered systems in legal research, whereas 56% participants believe the capacity of generative AI to transform the documents drafting, 44% for the documents analysis, and 35% for email writing [4].

Some of the core areas of legal practice where AI is being applied for the purpose of reducing the burden of cost-expensive services, mounts of workload, the need for efficiency, expeditiousness and effective human resources, are as follows.

2.1.1 Document Review and Analysis

One of the core aspects of the field of legal practice is the document review and analysis, which involves the detailed examination and investigation of the voluminous documents for the purpose of identification and extraction of relevant information and material facts. By leveraging AI-powered tools such as Everlaw, Relativity Trace, Kira Systems through machine-learning algorithms and advanced automation, the law professionals can swiftly minimize the risk of human-made errors in tasks like contract review, due diligence, document classification and analysis, thus leading to more emphasis on the high-yield tasks that demand expert insights and judgments [5].

2.1.2 Legal Research and Knowledge Management

The effectiveness of legal research primarily relies on the relevant data input wherein artificial intelligence comes into play by offering approaches like AI-enabled search engines, predictive analytics, legal research database, generative AI assistants, machine-learning algorithms, and language processing tools. AI-powered tools like Manupatra, LexisNexis, Westlaw, Casetext, CaseMine, etc., are few of such examples which aid the researchers and legal professionals in analyzing the large-scale up-to-date legal data, identifying the relevant sources, and predicting the litigation outcome with the help of persuasive precedents. For example, "Manupatra," an India-based AI-powered legal database uses features like Machine Learning (ML), Natural Language Processing tool (NLP), analytics, and AI-enabled search engines to accelerate the process of managing and analyzing legislative, procedural and regulatory database [6]. These tools enable the legal researchers and lawyers to deliver a fruitful research outcome by offering accuracy in identifying the relationship between variables and in turn, enhance their capability to facilitate effective legal strategies and argumentation.

2.1.3 Litigation Discovery and Dispute Resolution Prediction

Apart from the above applications, AI is also harnessed in facilitating the litigation discovery process by way of applying predictive coding and analysis. For example,

"Relativity Trace," an AI-powered tool, uses Technology Assisted Review (TAR) in combination with the predictive coding approach for the purpose of e-discovery that enables the lawyers to identify the key focus areas after analyzing large volume of data and helps to formulate effective case strategies [7]. Predictive coding is an AI technology which creates a relevance score based on the algorithms for the documents being reviewed. Further, in the context of alternative dispute resolution, the use of AI also enhances the probability to drive accurate and effective outcomes. Although in India, the efficacy of artificial intelligence in dispute resolution sector is often under-utilized, NITI Aayog's Report on Online Dispute Resolution Plan for India (2021) can be foreseen as a ray of hope as the role and potential of AI was duly recognized first time ever in the revitalization of an effective ODR system in India [8].

2.1.4 Client Management and Communication

The convergence of artificial intelligence with client management services produces personalize results and high client satisfaction. Law firms and chambers therefore use AI automation tools and chatbots as a medium to gain more clients and deal with their queries effectively at the same time. AI-powered chatbots such as Lawpal, Automio, DoNotPay provide 24/7 client support and assist the law firms to help their clients in routine documentation as well as answering legal queries. Harvey AI is also another example of AI legal chatbot currently partnering with Allen and Overy law firm and other legal giants [9]. These legal AI chatbots are trained with offering the support to clients on the templates used by law firms, resulting in reducing the time and effort required to ensure client management and communication.

2.2 AI in Judiciary

While the usage of artificial intelligence in the legal service sector is becoming prevalent across the countries, the judiciary is no exception to embarking on this transformative journey. Application of AI in judiciary brings with it a myriad of benefits including the predictive analytics, online dispute resolution, translation of judgments in various languages, streamlining judicial processes, and documentation involved, offering virtual assistance to litigants, guiding for research, helping the courts to monitor the case records and pendency and so on. The courts all over the world have presently stepped forward to focus on unleashing the full-fledged potential of artificial intelligence in judicial systems and started using AI-powered tools for the purposes of document analysis, case predictions, and offering accessibility as well as flexibility to the judicial processes. Following are some of the real-world examples of AI applications in the judiciary:

1. The Supreme Court of Singapore has implemented AI-powered system known as the Court Audio Services System (CAS) to transcribe the audio recording of the court hearings for the convenience of the judges and the litigants [10].
2. The courts and tribunals in United Kingdom are using AI in a variety of tasks. For example, the judges in UK can now monitor the risk of recidivism through the AI-powered systems like The Offender Assessment System (OASys) and The Offender Group Reconviction Scale (OGRS) and accordingly, pass sentencing orders. The UK courts also take assistance of AI in writing judgments and summarizing the material facts from the voluminous documents. Even the administrative sides of the courts in UK are the direct beneficiaries to the usage of AI [11].
3. Like UK, even the US courts also use generative AI-powered systems like "Correctional Offender Management Profiling for Alternative Sanctions (COMPAS)" software to assess the track record of the convicts and decide the extent of sentencing based on the potential risk of recidivism [12].
4. "Prometea," an AI-powered system performing swift identification of urgent cases, is currently used by the courts in Argentina [13].
5. In India, the judiciary has recently been proactive in recognizing the need for cutting-edge technologies especially the artificial intelligence in order to bring efficiency and speed in justice-delivery process. The "eCourts Project" and the constitution of Artificial Intelligence Committee by the Supreme Court of India are the prime examples of the efforts to transform such need into reality. While the eCourts Project and its phases are an initiative of the Ministry of Law and Justice to make the District Courts in India Information and Communication Technology (ICT)-enabled, the AI committee of the Supreme Court of India has been actively involving in adopting the AI-powered technologies for translation of the Apex Court judgments into regional languages through "Supreme Court Vidhik Anuvaad Software (SUVAS)," automation of judicial processes and offering assistance to court through "Supreme Court Portal for Assistance in Court Efficiency (SUPACE)" so as to minimize the pendency [14–16]. "The National Judicial Data Grid (NJDG)" is another AI breakthrough, launched by the Supreme Court of India to facilitate the citizens with quick access to the judgments, pendency tracking and case status. The adoption of AI-powered virtual court hearings during the outbreak of COVID-19 was a significant turning point, and since then, the virtual hearings as well as the use of AI technologies in e-filing or case management have witnessed a drastic rise.

Artificial intelligence has been turning into becoming a part and parcel of the judicial system globally in the twenty-first century. Despite the ethical concerns and challenges associated with data privacy and misuse of AI posing questions on its far-reaching impact, the efficiency and accessibility it offers to the judiciary suffices the much-needed dynamics in judicial system in the era of digitization.

2.3 AI in Law Enforcement

With the rapid evolution of technological innovation and the advent of AI, a need has been felt to revolutionize the functioning of law enforcement agencies to adapt the cutting-edge technologies in order to ensure public safety and showcase efficiency in crime prevention. Likewise, the law enforcement agencies across the countries have been increasingly adopting the integration of AI-powered technologies in their operational policies. AI in the law enforcement sector functions in various forms, ranging from facial recognition and data analytics to the predictive policing through predictive analytics software. AI technologies thus unlock new ways to reduce crime rate, ensure operational efficacy and re-establish public trust and safety time and again. Some of the examples of AI application traced in the law enforcement sector are:

1. Since 2012, the UK police forces have been using various types of AI tools to make the custodial decisions through algorithms. These tools include facial recognition technology, predictive analytics, risk assessment technology, etc. For example, "the Harm Assessment Risk Tool (HART)" is an AI-powered system being used by the Durham Police for the purpose of predictive policing [17]. The technology classifies the risk profile of a person by keeping a track record on the 34 categories including offending history, age, address, and other background details of the person through machine learning algorithms and predictive analytics. The HART algorithm has been in continuous usage since 2017 and updated time-to-time in order to avoid any scope for bias or prejudice. Besides, another example of AI application by the UK Police forces is PredPol—a predictive policing tool that alerts the police forces using the past crime records the areas where crimes are likely to occur and additional police protection is needed [18].
2. Similarly, a growth in usage of AI is observed in the US law enforcement agencies with a vision to accelerate operational efficacy and decision-making process by way of identifying the crime patterns and also in maintenance of traffic safety. For instance, "The FACE Program, i.e., the Facial Analysis, Comparison, and Evolution Program" conducted by the FBI is used in a series of high-profile cases involving terrorism, human trafficking, organized crimes, etc., and also to build the Next Generation Identification database for the purpose of crime investigation and identification of suspects through AI-powered facial recognition and identification technology [19]. The predictive policing AI technology, "PredPol," is also used by Los Angeles Police Department to identify crime patterns and the areas vulnerable to likelihood of crimes [20]. Likewise, "Heat List" by the Chicago police or CompStat Program by New York Police are also the instances of predictive policing AI system developed with an objective to link the crime patterns with the potential victims or offenders and the areas at risk of being involved [20]. Besides, several US police departments have already launched AI-powered chatbots with natural language processing technology to offer emergency support and assist the citizens in filing of crime reports. "Chip"—the LAPD Chatbot

developed by the Los Angeles Police Department is a renowned AI system in this context [21].

3. In India, although the full-fledged reformation of police forces at par with the era of digitization is still a far-fetched dream, the ongoing decade has observed a significant progress made by the Indian police force in terms of using AI to crime prevention and enhancing public safety. For instance, the Telangana Police has launched an AI chatbot called "Cop Connect" to provide speedy assistance and information sharing service to the citizens and received an overwhelming response exceeding 10,000 queries in the first month [22]. Similarly, "Public Eye" is an AI-powered chat assistant launched by the Bengaluru Police to provide assistance on traffic violation reports and road safety sensitization to the citizens [23]. Further, in the context of facial recognition technology, one notable example is the "Trinetra" mobile application of the Uttar Pradesh Police which is primarily used to identify the suspects and monitor their movements followed by arrest of several criminals and proclaimed offenders [24]. Besides being a tool for crime prevention, the AI technologies are also applied by Indian police in traffic management system. Recently, Kolkata Traffic Police has run a trial project of a traffic management software and geared up to develop an AI-powered system that can provide congestion alerts or predictions and any road safety violation reports immediately through automatic camera facility [25]. The similar type of software is already in limited use by the Mumbai and Bengaluru police.

Apart from the above, the police forces in other countries have also stepped forward to adopt initiatives to integrate AI technologies with the maintenance of law and order. The predictive policing system, facial recognition technology, CCTV camera, and AI-powered chatbot are some of the globally prevalent applications of artificial intelligence in the context of law enforcement. Therefore, by no stretch of imagination, it is axiomatic that the artificial intelligence has taken the law enforcement field by storm and continues to play a profound role in prevention of law and order across the world.

2.4 AI and Intellectual Property Laws

Lately, the rise of artificial intelligence has sparked profound debates with respect to its intersection with various fields. Apart from actively being involved in legal and judicial systems, artificial intelligence has a widespread application in the fields of science and technology, thereby substantially contributing to the development of IP Laws. As the artificial intelligence gradually converges with the subsets of intellectual property rights, it hosts plethora of new challenges and complexities germane to the IP law and practice. Questions therefore may be posed on the patentability of AI-driven inventions, the legality of copyright protection in reference to the AI-generated works, the impact of AI in trademark protection, etc. Therefore, it is pertinent to delve deeper into the conundrums revolving around the convergence of both the domains.

The convergence of AI and intellectual property laws is further elaborated through three subsets: (1) Patentability of AI-driven inventions, (2) Copyrightability of AI-generated creative contents, (3) Protection of AI-generated trademarks and industrial designs.

2.4.1 Patentability of AI-Driven Inventions

AI-driven invention generally refers to the novel inventions such as products, processes which are generated and developed through the application of AI algorithms. The purpose behind granting patents serves as a safeguard to the novel, useful inventions and confers exclusive rights to the inventor to use and commercialize the invention. The application of AI in the context of patents thus directs toward the questions as to whether AI-driven inventions are patentable inventions and to whom the rights of an inventor can be conferred exclusively. A classic case study in this domain is the "DABAUS AI system" invented by Dr. Stephen Thaler. Device for the Autonomous Bootstrapping of Unified Sentience, shortly DABUS, is an AI-generated invention that highlights two novel categories, namely, a flash-light system for emergency alerts and a novel food container design [26]. This case brought into spotlight the lack of uniformity of patent laws across the world. Following the application of patents by Dr. Thaler in multiple countries in the name of DABUS as an inventor, several jurisdictions such as USA, UK, and European Union rejected the patent application on the ground that the human involvement is pre-requisite to the application for inventorship. However, in countries like South Africa where the patent laws are less rigid, the patents were granted listing AI as the recognized inventor. In Australia, although initially rejected, the Australian Court ruled in favor of granting patents to DABUS, citing that the patent laws must undergo changes to adapt the technological advancements in public interest. This case highlighted the sheer divergence and the need of a uniform global standard in terms of patentability of AI-generated inventions. The reasons behind such legal uncertainty are the strict requirements under the patents laws that leave little scope for flexibility at par with meeting the needs of technological evolution. The domestic patent law and rules in majority countries mandate the inventor to be a natural person and hence, rules out the inventorship of AI-generated systems. Besides, there are certain domestic legal frameworks where the nature of person to be the inventor is not explicitly mentioned. For example, in India, Sect. 6 of the Patents Act 1970 states that true and first inventor to be any person who can file the patent application [27]. Thus, the scope of such patent regimes to cover the AI systems remains up to the interpretation of the authorities and courts.

Secondly, to determine whether the inventions qualify the novelty and inventive step criteria, a thorough examination of the other associated factors as to the extent of transparency and potential infringements must be conducted. Last but not the least, an array of disputes may arise with respect to clarifying the stance on ownership of an invention. This is because the claimants to the ownership rights may be either of the three parties, namely, the organization where AI-driven invention is developed, the programmer developing the invention and the end user who applied the AI to create the invention. Hence, navigating the intersection of AI-driven inventions and patent reveals a myriad of challenges and grey areas which need immediate attention and resolution so as to move forward.

2.4.2 Copyrightability of AI-Generated Creative Contents

A glance at the contents posted in the social media networks today portrays the increasing usage of AI to generate creative musical, literary, and art works. It brings with it a series of questions on the status and legal stance of copyright protection and ownership as to those AI-generated creative works. The research reveals that the AI-powered creative works have always experienced resistance in terms of recognizing their originality, the reason being the ability of AI-powered algorithms to easily imitate and reproduce the copyrighted original content making it an arduous task to differentiate between the original and unauthorized reproduced works. Further, in almost all the western jurisdictions and European Union, the copyright protection laws stipulate human authorship and creation of works only by a human author as the prerequisites to claim for protection. China, on the other hand, has evolved their approach toward dealing with AI-generated creative works. In "Li v. Liu" (2023), the Beijing Internet Court pronounced for the first time in favor of the copyrightability of an AI-generated image using the Stable Diffusion Software, although ruled out the scope of authorship of an AI system itself [28].

In India, although the Copyright Act 1957 does not explicitly cover the AI-generated works, an extended interpretation of "computer-generated works" mentioned under Sect. 2(d)(vi) of the Act may be used to govern the sphere providing that the person causing the creation of the works is deemed to be the author [29]. This results into creating the myriad complexities revolving around the ownership and liability of AI-generated content. The "author" therefore can either be the person who created or operated the AI system, or the person who employed the AI system to generate creative content—in both cases being the human authorship as the criteria. An Indian case study in this context is an AI-generated artwork "Suryast" for which, two copyright applications were filed [30]. The first application which listed the AI painting app "Raghav" as the sole owner was rejected on the ground of lack of human authorship. The second application where along with Raghav, Ankit Sahani (the owner of Raghav) listed himself as the co-author, was registered initially. Subsequently, the objections were raised by the copyright office leading to the cancellation of the registration. Hence, it is evident that the existing legislative framework in India for copyright protection does not explicitly cover the ownership of AI-generated contents due to the ambiguities involved in the lack of "personhood." However, on the application of the fair-use doctrine codified under Sect. 52 of the Copyright Act 1957, it is the AI developer and the end-users on whom the onus should be placed to ensure that the contents generated by AI do not infringe the copyrighted materials at any cost.

2.4.3 Protection of AI-Generated Trademarks and Industrial Designs

The application of artificial intelligence in the sphere of trademark protection is two-fold. On one hand, the artificial intelligence is being leveraged to use machine learning algorithms for analysis of the trademark database, assistance in trademark search and clearance process, and prediction of potential trademark infringement issues. For instance, "TrademarkVision," an AI software for trademark search, has assisted the IP offices and revolutionized the process of assessing the search data on trademark and industrial design as well as the classification and prediction analysis as to the cases of trademark infringements [31]. On the other hand, AI-generated trademarks are steadfastly becoming common and at the same time, pose questions on authorship, ownership, novelty, and descriptiveness likewise other intellectual property subsets. Like patents, the trademark laws across the world also highlight a rigidity toward the legal recognition of AI-generated trademarks. The trademark applications related to AI-generated logos are often dismissed citing the concerns on novelty and distinctiveness of the logo, the absence of a human authorship and the ambiguity in determining the ownership of the logo. Further, in "M/S Kibow Biotech v. M/S The Registrar of Trademarks," the Hon'ble Delhi High Court held that only a person and not an AI system can be considered as the applicant and registered proprietor of the trademark in accordance with the Trade Marks Act, 1999 [32, 33].

3 Ethical Concerns Surrounding the Intersection of AI and Law

The evolution of AI technologies has not only transformed the work format in every industry but also outpaced the development of legislative and regulatory safeguards for their effective governance. Consequently, the countries are now grappling with the issues as to data privacy, accountability, transparency, intellectual property infringements, and potential biases in AI algorithms. Breaking down each of such concerns associated with the application of AI in law.

3.1 Accountability

One of the significant challenges in the periphery is the unclear position on who bears the liability in case of any erroneous information or analytics shared by AI system in the legal context because the AI systems operate mostly as black boxes where the process to arrive at a specific decision without the presence of human rationale is often peculiar to understand. For instance, in circumstances like predictive policing where AI system analyzes datasets on potential suspects and predicts the involvement of a person in a crime, there underscores a need to establish a clear stance on accountability if a person is falsely accused due to a mistake made by AI system. The lack of a proactive regulatory framework on the governance of AI makes it more difficult to hold any individual or organization accountable for misleading through erroneous AI-driven decisions. The only comprehensive AI-centered tiered framework existing as of now is the Artificial Intelligence Act (AIA) implemented by the European Union in 2024 [34]. Despite sincere efforts to introduce a robust regulatory regime, the AIA and AI Liability Draft Directive lack practical approaches to mitigate the risks associated with AI governance. The provisions therein lay more emphasis on the tangible harms involving monetary losses, disregarding other intangible damages such as fabrication of facts, privacy invasion, copyright infringement, algorithmic biases, etc. [35, 36]. Therefore, the EU's AI Regulation creates plethora of avenues to facilitate the evasion of accountability for AI systems and developers through half-baked enforcement process.

3.2 Bias in AI Algorithms

Bias in AI algorithms may likely lead to discriminatory results. Such bias if found in AI technologies used in the legal or judicial or law enforcement system can prove to be fatal. Additionally, when AI algorithms are trained on such biased material without any initiative to remove the bias, it results into formulating a faulty judgment or decision time and again, exasperating racial-gender disparities and targeting marginalized communities even to the extent of costing lives. For instance, the COMPAS, an AI-powered risk assessment tool used by US Police, has been receiving criticisms due to the influence of racial biases in its results. Even in "State v. Skaff," the Wisconsin Court expressed concerns as to increasing usage of COMPAS and its biased datasets in formulating sentencing decisions [37]. Further, it appears from a report published by the National Institute of Standards and Technology that the facial recognition technology often used in law enforcement to identify the criminals suffers from biased algorithms which are likely to result in illegal arrests or convictions and perpetuate discriminatory outcomes in form of colorism, stereotyping, and racism [38].

3.3 Transparency

Transparency is one of the pressing concerns in the context of convergence between AI and law. Often it remains in dark how AI algorithms operate and interpret the rules of logical representation when deriving into a conclusion. Thus, a reliance on AI-driven legal decisions can be prove to be dangerous and lead to unfairness in due process and miscarriage of justice in cases where the lawyers, judges, and law enforcement agencies find it difficult to comprehend the logic and reasoning behind such opaque decisions. The sheer lack of human oversight and autonomous character in the black-box decision-making process of AI systems is one of the key factors that aggravate questions on the lack of transparency and due process [39].

3.4 Data Privacy

AI systems are data-hungry in nature. To elaborate, the training and operation of AI algorithms and making of AI-driven decisions rely on the collection and critical analysis of massive amount of datasets available on internet. Thus, the use of AI raises ethical concerns regarding the compromise of individual privacy in the process of personal data collection. These data are collected from multiple sources, such as social media networks, ecommerce shopping, online transactions and hence, the collection and storage of such data raises significant ethical concerns on data breach, potential misuse of such data, invasion of privacy, hike of deepfakes and unauthorized access granted to the third party. Even though the newly-implemented Digital Personal Data Protection Act, 2023 in India stipulates restrictions on the automated collection, processing, usage and storage of personal data, a nuanced analysis of the Act transpires the explicit exemption of publicly accessible personal data from its application [40]. This undeniably bespeaks that the AI systems' autonomous processing and training on the personal data available in public domain require no significant compliance under the Act. Using this defense, the AI software, more specifically, the deepfake creators continue to operate scot-free without due supervision.

3.5 Infringement of IP Rights

The recent lawsuits filed against OpenAI in several jurisdictions are based on varied grounds which mostly are related to the infringement of copyrighted materials [41]. As AI continues to copy, generate, and reproduce the copyrighted materials, trademarks and patents available in the internet, ethical concerns as to IP infringements are on rise. AI-powered algorithms, under the garb of "fair use" can easily generate creative works and in that process, the systems can unknowingly and likely reproduce and train on any copyrighted artworks or musical or literary works which tantamount to the potential infringement of copyrights. Likewise, the process to generate AI-created logos or advertising designs by AI-software may cause infringement of existing trademarks, leading to losses of businesses and trademark infringement lawsuits. An instance can be taken from the recent concerns expressed by the Walt Disney Company regarding the potential dilution of "Disney-Pixar" logo by Microsoft Bing's AI imaging software through which the social media users were engaging into a trend of generating "Pixar"-styled pet images. The threat was not indicated to implicate the pet images but the probability of AI-software to reproduce the trademarked logo inadvertently within the generated results, likely resulting in trademark infringement [42]. Further, AI algorithms can also violate the patent rights by way of generating inventions which may be closely similar to the existing patented inventions on the basis of methods, processes and utility. These concerns as to IP infringements by AI technologies not only cause harm and suffering to the end-users but also the existing right-holders.

4 Conclusion

It is quite evident that existing legal frameworks have been outperformed by the rapid evolution and upgradation of AI technology. As artificial intelligence gradually permeates the frontiers of law, it is of utmost importance for the legal systems to address the concerns like bias, privacy, accountability, and transparency on one hand and take necessary arrangements to adapt and embrace the opportunities created by the efficiency of AI systems on the other hand. Especially in law enforcement and judicial service, the potential of AI as a pivotal pillar to revolutionize the pace of access to justice can be harnessed with proactive and timely human-intervention. Besides, it is also a need of the hour for the countries to develop a robust legal framework to give due emphasis on the complexities underneath the intersection of artificial intelligence, law and intellectual property rights that demand careful navigation and a balanced integration. In domestic level, for instance, the existing Indian IP legislations such as the Patents Act 1970, the Copyright Act 1957, the Trade Marks Act 1999 underscore a necessity of deliberate changes to clarify the role of AI in fostering innovation and ensure essential safeguards in order to differentiate the infringement of IP rights by AI systems from the flexible application of fair-use doctrine. Simultaneously, as the algorithmic colonization transcends transnational boundaries and continues to exacerbate ethical conundrums, the countries must act in harmony to develop a standardized global practice and approach to regulate the evolution of AI in legal landscape. Furthermore, the onus also lies on the AI developers, programmers, and end-users to administer and ensure that the functioning of AI models and the sampling used to train those models are unbiased and inclusive in nature. The post-automated decisions must also be monitored to maintain fairness and adherence to the existing privacy regulations and standards in contrary to which strict penalty must be imposed. At this juncture, it is hence a settled position that the more AI is considered as a supplement and not a replacement of human oversight, the higher the efficiency will be witnessed in harnessing its full-fledged potential in law and justice.

References

1. Artificial Intelligence (AI) Coined at Dartmouth. Accessed 12 Dec 2024. https://home.dartmouth.edu/about/artificial-intelligence-ai-coined-dartmouth

2. Sudden H (2019) Artificial intelligence and law: an overview. GA ST U L Rev 35(4):1305–1337. Accessed 12 Dec 2024. https://readingroom.law.gsu.edu/gsulr/vol35/iss4/8
3. Sartor G (2005) Legal reasoning: a cognitive approach to the law. Springer, Berlin
4. LexisNexis (2023) International legal generative AI report. Accessed 12 Dec 2024. https://www.lexisnexis.com/pdf/lexisplus/international-legal-generative-ai-report.pdf
5. Machuskyy V (2024) The role of AI in legal document review: pros, cons, and safeguarding client confidentiality. LinkedIN. Accessed 13 Dec 2024. https://www.linkedin.com/pulse/role-ai-legal-document-review-pros-cons-safeguarding-client-ciquf/
6. Manupatra (2021) Powering your legal practice using technology. Manupatra Brochure. Accessed 13 Dec 2024. https://www.manupatrafast.in/Resources/pdf/Manupatra-Brochure-2021.pdf
7. Copperpod IP (2023) Document review: popular TAR platforms and algorithmic insights. Medium. Accessed 14 Dec 2024. https://copperpod.medium.com/document-review-popular-tar-platforms-and-algorithmic-insights-52ecd2aad332
8. NITI Aayog (2021) Designing the future of dispute resolution: the ODR policy plan for India. Accessed 14 Dec 2024. https://www.niti.gov.in/sites/default/files/2023-03/Designing-The-Future-of-Dispute-Resolution-The-ODR-Policy-Plan-for-India.pdf
9. Wakeling D (2023) A&O announces exclusive launch partnership with Harvey. A&O Shearman. Accessed 15 Dec 2024. https://www.aoshearman.com/en/news/ao-announces-exclusive-launch-partnership-with-harvey
10. SG Courts (2024) Use of technology at the SICC. Accessed 27 Mar 2024. https://www.judiciary.gov.sg/singapore-international-commercial-court/forms-and-services/use-of-technology-at-the-sicc
11. Grindle H (2023) The techno-judiciary: sentencing in the age of artificial intelligence. Sentencing Academy UK. https://www.sentencingacademy.org.uk/sentencing-in-the-age-of-artificial-intelligence/
12. Prabhu A (2023) Artificial intelligence in the context of the Indian legal profession and judicial system. Bar and Bench. Accessed 16 Dec 2024. https://www.barandbench.com/columns/artificial-intelligence-in-context-of-legal-profession-and-indian-judicial-system
13. OECD/CAF (2022) The strategic and responsible use of artificial intelligence in the public sector of Latin America and the Caribbean. OECD Public Governance Reviews. OECD Publishing, Paris. https://doi.org/10.1787/1f334543-en
14. Ministry of Law & Justice (2023) Answer to Lok Sabha Unstarred Question No. 2248: findings on national judicial data grid. Sansad. Accessed 17 Dec 2024. https://sansad.in/getFile/loksabhaquestions/annex/1714/AU2248.pdf
15. PIB India (2024) Artificial intelligence in judiciary. Accessed 17 Dec 2024. https://pib.gov.in/PressReleaseIframePage.aspx?PRID=2043476
16. Rana V, Nagpal N, Gopal A (2023) Harnessing the power of artificial intelligence in arbitration: a comprehensive analysis of Indian jurisprudence. Bar Bench. Accessed 17 Dec 2024. https://www.barandbench.com/law-firms/view-point/artificial-intelligence-arbitration-analysis-indian-jurisprudence
17. Burgess M (2018) UK police are using AI to make custodial decisions–but it could be discriminating against the poor. Wired UK, 1. Accessed 21 Dec 2024. https://www.wired.com/story/police-ai-uk-durham-hart-checkpoint-algorithm-edit
18. Kelion L (2019) Crime prediction software 'adopted by 14 UK police forces. BBC News. Accessed 21 Dec 2024. https://www.bbc.com/news/technology-47118229
19. Greco KD (2019) Facial recognition technology: ensuring transparency in government use. FBI. Accessed 21 Dec 2024. https://www.fbi.gov/news/testimony/facial-recognition-technology-ensuring-transparency-in-government-use
20. Asaro PM (2019) AI ethics in predictive policing: from models of threat to an ethics of care. IEEE Technol Soc Mag 38(2):40–53. https://doi.org/10.1109/MTS.2019.2915154
21. Microsoft (2017) How the city of LA serves citizens with Chip the chatbot. Microsoft in Business Blogs. Accessed 21 Dec 2024. https://www.microsoft.com/en-us/industry/microsoft-in-business/customer-experience/2017/07/14/how-the-city-of-laserves-citizens-with-chip-the-chatbot/

22. Staff Reporter (2018) TS police launch cop connect. The Hindu. Accessed 21 Dec 2024. https://www.thehindu.com/news/cities/Hyderabad/ts-police-launch-cop-connect/article24195433.ece
23. Karthikeya A et al (2022) An ML-based citizen-driven traffic violation reporting android app. In: IEEE fourth international conference on advances in electronics, computers and communications, ICAECC 2022, Bengaluru, India, January 10–11. pp 1–6. https://doi.org/10.1109/ICAECC54045.2022.9716674
24. Correspondent HT (2023) Op 'Trinetra' helped solve 295 incidents of crime in 45 days in U.P. Hindustan Times. Accessed 21 Dec 2024. https://www.hindustantimes.com/cities/lucknow-news/operation-trinetra-u-p-dgp-vijaya-kumar-s-cctv-initiative-helps-solve-295-crimes-in-45-days-101692896031504.html
25. Ghosh D (2024) KP runs 1st AI trial on bypass to manage traffic flow. The Times of India. Accessed 21 Dec 2024. https://timesofindia.indiatimes.com/city/kolkata/kolkata-launches-groundbreaking-ai-traffic-management-trial-on-em-bypass/articleshow/114440887.cms
26. AP (2021) Meet DABUS: the world's first AI system to be awarded a patent. Economic Times Brand Equity.com. Accessed 23 Dec 2024. https://brandequity.economictimes.indiatimes.com/news/digital/meet-dabus-the-worlds-first-ai-system-to-be-awarded-a-patent/85149000
27. The Patents Act, No. 39 of 1970
28. Li V. Liu, Beijing Internet Court Civil Judgment (2023) Jing 0491 Min Chu No. 11279
29. The Copyright Act, No. 14 of 1957
30. Abraham K (2024) AI's right to copy. India Bus Law J. Accessed 23 Dec 2024. https://law.asia/generative-ai-copyright-law/
31. Coldewey D (2016) TrademarkVision uses machine learning to make finding logos as easy as a reverse image search. TechCrunch. Accessed 30 Dec 2024. https://techcrunch.com/2016/12/12/trademarkvision-uses-machine-learning-to-make-finding-logos-as-easy-as-a-reverse-image-search/
32. Jain A, Sharma V (2023) Navigating the legal challenges posed by AI on intellectual property'. Legal Service India. Legal Service India. Accessed 30 Dec 2024. https://www.legalserviceindia.com/legal/article-12029-navigating-the-legal-challenges-posed-by-ai-on-intellectual-property.html
33. The Trade Marks Act, No. 47 of 1999
34. Regulation (EU) 2024/1689 of the European Parliament and of the Council of 13 Jun 2024 laying down harmonised rules on artificial intelligence and amending Regulations (EC) No 300/2008, (EU) No 167/2013, (EU) No 168/2013, (EU) 2018/858, (EU) 2018/1139 and (EU) 2019/2144 and Directives 2014/90/EU, (EU) 2016/797 and (EU) 2020/1828 (Artificial Intelligence Act)
35. Proposal for a Directive of the European Parliament and of the Council on adapting non-contractual civil liability rules to artificial intelligence (AI Liability Directive), COM(2022) 496 final, 2022/0303(COD)
36. Wachter S (2024) Limitations and loopholes in the EU AI act and AI liability directives: what this means for the European union, the United States, and beyond. Yale J L & TECH 26(3):671–718
37. State v. Skaff. Court of Appeals of Wisconsin (1989) 152 Wis. 2d 48
38. Schwartz et al (2022) Towards a standard for identifying and managing bias in artificial intelligence. NIST Spec Publ 1270. https://doi.org/10.6028/NIST.SP.1270
39. Cheong BC (2024) Transparency and accountability in AI systems: safeguarding wellbeing in the age of algorithmic decision making. Front Hum Dyn 6:1421273. https://doi.org/10.3389/fhumd.2024.1421273
40. The Digital Personal Data Protection Act, No. 22 of 2023, s. 3(c)
41. Kalra A, Chaturvedi A, Vengattil M (2025) OpenAI faces new copyright case, from global book publishers in India. Reuters, 24 Jan 2025. Accessed 2 Feb 2025. https://www.reuters.com/technology/artificial-intelligence/openai-faces-new-copyright-case-global-publishers-india-2025-01-24/

42. McElwaine JC (2024) AI-generated content and trademarks. Natl Law Rev XV(59). Accessed 2 Feb 2025. https://natlawreview.com/article/ai-generated-content-and-trademarks

Ethical Challenges and Opportunities in Artificial Intelligence: A Comprehensive Exploration

Raghavan Krishnasamy Lakshmana Perumal and Mohuya Chakraborty

Abstract This study examines the ethical promise and challenges of Artificial Intelligence (AI). The analysis highlights key ethical ideas like responsibility, transparency, justice, and privacy, all of which are supported by significant findings from the corpus of recent literature. In order to address concerns regarding the use of ethical AI and its limitations, the study systematically synthesises information using the PRISMA technique. The study offers recommendations and practical strategies for incorporating ethics into AI development while highlighting pressing problems like bias, insufficient rules, and a lack of collaboration among stakeholders.The analysis highlights the need for interdisciplinary approaches, continuous monitoring, and robust governance to guarantee that AI serves social goals responsibly.

Keywords Ethics in AI · PRISMA methodology · Fairness in AI · Transparency in AI · AI in sectors · Responsible AI

1 Introduction

The unparalleled advantages that Artificial Intelligence (AI) brings to the fields of healthcare, education, law, finance, manufacturing, and more are revolutionising entire industries. However, the quick adoption of AI technologies has brought up ethical issues that require urgent consideration, such as issues of bias, privacy violations, accountability, and transparency. This study's central focus is the tension between ethical reflections and technical advancement. Businesses have seen significant changes since the advent of AI, but there have also been challenges. Applications

R. K. L. Perumal
Atlassian, San Francisco, USA
e-mail: rlakshmana@atlassian.com

M. Chakraborty (✉)
Cybersecurity Centre of Excellence, Institute of Engineering and Management, University of Engineering and Management, Kolkata, India
e-mail: mohuyacb@iem.edu.in

M. Chakraborty et al. (eds.), *Proceedings of Sixth International Ethical Hacking Conference*, Lecture Notes in Networks and Systems 1513,
https://doi.org/10.1007/978-981-96-8632-2_28

of AI, such as predictive policing and facial recognition technologies, usually entail the risk of violating user privacy or reinforcing institutional biases. Presenting a comprehensive review of the ethical dilemmas and gaps in the development and use of AI is the aim of this work. It also seeks to provide practical ways to properly address these issues and ensure that AI systems adhere to social standards and human rights.

Although AI is transforming many industrial business sectors globally, there are serious ethical questions raised by its quick adoption. According to one case study, the Amazon recruitment system failed to follow ethical guidelines when processing job applications. In this case, it was discovered that the system discriminated against female candidates [1]. Technically sound, ethically sound, and compliant with all relevant laws and regulations are requirements for AI systems. The ethical principles of AI are as follows.

- **Transparency**—Confidence and trust building in AI systems requires transparency. The 'black box' nature of machine learning models has led to the development of explainable AI (XAI) techniques. For instance: In healthcare diagnostics, transparent decision-making guarantees that patients are aware of their available treatment alternatives.
- **Accountability**—Clear accountability mechanisms are essential for addressing AI failures. Ethical frameworks like IEEE 7000–2021 emphasise the need for defined roles and responsibilities across the AI lifecycle.
- **Fairness**—Bias mitigation strategies, such as balanced datasets and fairness-aware algorithms, have been proposed. Case Study: Recruitment AI systems have faced criticism for perpetuating gender and racial biases.
- **Privacy**—Data protection frameworks, such as GDPR, play a pivotal role in safeguarding user privacy. Example: AI applications in biometric surveillance raise significant concerns about consent and misuse.

Google, SAP Association of Computing Machinery (ACM), Access Now, Amnesty International, AI HLEG, IEEE ISO/IEC have taken initiative to set up policies to develop ethically aligned AI systems [2–5].

There is a dearth of comprehensive analysis of AI ethics in literature. Using the PRISMA (Preferred Reporting Items for Systematic Reviews and Meta-Analyses) [6] methodology, this research presents a Systematic Literature Review (SLR) [7] investigating the ethics of AI application across numerous sectors. The review analyses the consequences of major ethical issues in industries like healthcare, education, finance, manufacturing, and public governance. These issues include fairness, transparency, accountability, privacy, and bias. This study attempts to identify gaps, provide solutions, and provide practical advice for guaranteeing ethical AI practices by synthesising the body of existing work. The results help academics, practitioners, and policymakers adopt more responsible AI systems by providing a thorough grasp of the ethical environment in AI.

1.1 Contributions

In addition to discussing how the findings can impact industrial practices and policymaking, this subsection emphasises the paper's academic and practical significance by highlighting the significance of connecting theoretical ideas with workable frameworks. The key contributions to the paper are as follows.

- A methodical identification of Research Gaps (RGs) in literature.
- Development and formulation of Research Questions (RQs) from gap analysis.
- A thorough RG-RQ analysis emphasising on focus research areas.
- A comprehensive literature survey and critical analyses of selected related works using PRISMA.
- Recommendations for interdisciplinary partnership and stakeholder engagement.

1.2 Organization of the Paper

After the introduction in Section I that provides an overview of AI's impact and ethical challenges. It further contextualises the current AI landscape and ethical concerns, and explains the rationale behind exploring these issues comprehensively. Section 2 highlights the methodology of systematic literature review using PRISMA model, which describes the process of identifying and selecting relevant literature and explains the criteria used to ensure rigor and reliability. Research flow including Research Questions (RQs), Research Gaps (RGs), and their analyses are provided in Sect. 3. It further expands on the key RQs guiding the study and discusses the implications of the identified gaps for future research. Section 4 highlights related works synthesising findings from key studies reviewing prominent ethical frameworks and case studies, and examining their effectiveness in addressing real-world challenges. The paper concludes in Sect. 5 with inference, call to action, and conclusion that summarises insights and suggests future directions, offers actionable recommendations for stakeholders, and proposes pathways for ongoing research and collaboration.

2 Methodology—Systematic Review Process

The study employs the Preferred Reporting Items for Systematic Reviews and Meta-Analyses (PRISMA) framework to systematically review and analyse relevant literature.

1. **Identification**: Comprehensive searches were conducted across research databases such as IEEE Xplore, Springer, and ACM Digital Library. Keywords included 'AI ethics', 'transparency', 'accountability', 'bias', and 'privacy'. 104

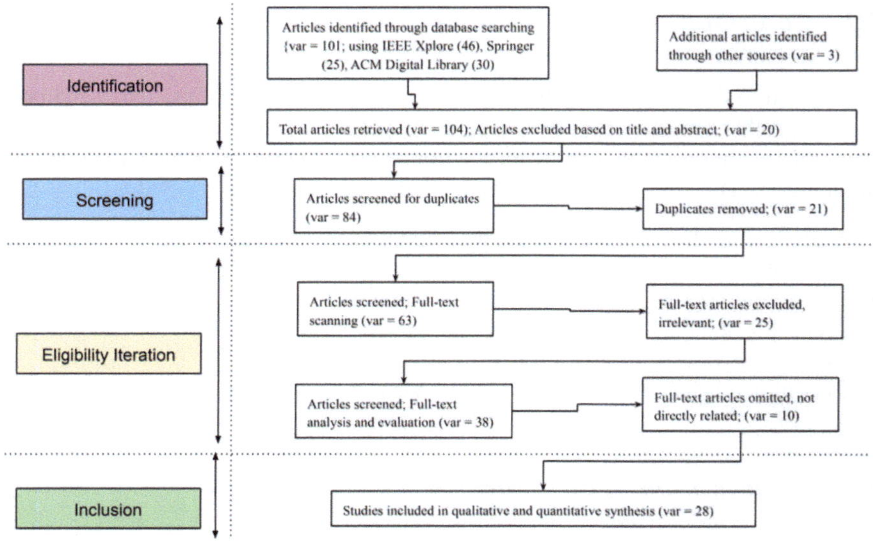

Fig. 1 PRISMA flowchart

papers were initially identified, spanning various subdomains of AI ethics. Based on title and abstract 20 articles were excluded. 84 articles remained at this stage.

2. **Screening**: Inclusion criteria focused on peer-reviewed articles addressing AI ethics from 2015–2024. Exclusion criteria removed duplicate studies, non-English papers, and those lacking empirical evidence. Approximately 63 papers remained after the initial screening.
3. **Eligibility Iteration**: Further refinement was carried out by assessing each study's relevance and methodological rigor. Studies focusing on actionable insights and practical implementations were prioritised. 38 papers were selected for the final review.
4. **Inclusion**: Finally, 28 papers formed the foundation for identifying key trends, challenges, and solutions in ethical AI.

A detailed PRISMA flowchart visualising the identification, screening, and inclusion process was created to ensure transparency and replicability as depicted in Fig. 1.

3 Research Flow

The research flow comprises identifying RGs, development and formulation of RQs and making a thorough RG-RQ analysis based on the studies from the PRISMA flowchart.

3.1 RGs and Analyses

1. **Implementation Gap in AI Ethics**: Many studies (e.g. Huang et al. [8]; Kazim et al. [9]; Zhou et al. [10] highlight key ethical principles (e.g. transparency, fairness, accountability) but struggle to provide concrete methodologies for their implementation. There is a lack of practical, enforceable frameworks that bridge the gap between abstract ethical guidelines and real-world AI development.
2. **Structural versus Technical Approaches to AI Ethics**: Critical works (e.g. Saxena et al. [11–13] emphasise that ethical AI requires structural change, while mainstream AI ethics research focuses on technical solutions like bias audits. Limited studies provide a hybrid approach that integrates both technical solutions (algorithmic fairness, explainability) and structural interventions (corporate governance, legal frameworks).
3. **Sector-Specific AI Ethics Challenges**: Different sectors (healthcare, law, business) have unique ethical challenges, yet many AI ethics frameworks are generalised (e.g. Huang et al. [8]; Zhou et al. [14]. A lack of sector-specific AI ethics models tailored to different industries (e.g. AI in healthcare vs. AI in finance).
4. **Policy-Practice Misalignment**: Studies like Khan et al. [15] highlight the disconnect between policymakers and AI practitioners, making it difficult to align AI governance with real-world development. There is insufficient research on how AI practitioners and regulators can collaborate to create workable ethical policies.
5. **Ethical Decision-Making and Accountability in Organizations**: Ali et al. [13] discuss how ethics in AI organizations is often individualised, with 'ethics entrepreneurs' bearing responsibility rather than institutional structures enforcing compliance. There is a lack of research on organizational accountability frameworks that ensure AI ethics is institutionally embedded rather than reliant on individuals.

3.2 Formulating RQs and RG-RQ Analysis

In this section, RQs are developed and formulated based on the RGs followed by an RG-RQ analyses as depicted in Table 1, which highlights the need for practical, sector-specific, and governance-oriented AI ethics frameworks. Moving beyond theoretical discussions, future research should focus on bridging implementation gaps, integrating structural and technical solutions, improving policy alignment, and strengthening accountability in AI ethics.

4 Related Works

Based on the RG-RQ analysis and the PRISMA structure, this section highlights and critically analyses a few of the related works.

Table 1 RG-RQ analysis

Sl. No	RG	RQ	RG-RQ analyses (potential research focus)
1	Implementation gap in AI ethics	RQ1: how can ethical AI principles be operationalised into practical decision-making processes for AI system development? RQ2: What methodologies can be developed to assess the effectiveness of AI ethics frameworks in real-world applications?	Developing a systematic approach or tool to integrate ethical principles into AI system design and decision-making
2	Structural versus technical approaches to AI ethics	RQ3: How can AI fairness algorithms be combined with structural governance mechanisms to create a comprehensive ethical framework? RQ4: What are the limitations of current AI fairness audits, and how can they be improved through organizational interventions?	Investigating how structural governance models can be combined with AI-driven fairness mechanisms to create a robust ethical framework
3	Sector-specific AI ethics challenges	RQ5: What are the unique ethical challenges faced by AI adoption in healthcare, law, and finance, and how can sector-specific ethical models be developed? RQ6: How do industry-specific AI risks (e.g. bias in medical diagnostics vs. bias in hiring algorithms) require different regulatory approaches?	Developing contextual AI ethics frameworks that address industry-specific risks and ethical dilemmas
4	Policy-practice misalignment	RQ7: How can collaboration between AI developers and policymakers be improved to create effective and enforceable AI regulations? RQ8: What governance models can be used to ensure that AI regulations remain adaptive to technological advancements?	Investigating participatory governance models where AI developers, ethicists, and policymakers co-create regulatory guidelines
5	Ethical decision-making and accountability in organizations	RQ9: How can organizations institutionalise AI ethics to prevent reliance on individual ethics champions? RQ10: What role do internal governance structures (e.g. ethics committees, compliance audits) play in ensuring ethical AI deployment?	Proposing corporate AI governance models that distribute ethical responsibility across teams rather than placing it on individuals

4.1 Overview of Related Works

Huang [8, 9] provide comprehensive examination of the ethical considerations surrounding AI. It delves into the historical context of AI ethics, identifies key ethical principles such as transparency, accountability, and fairness, and discusses challenges in implementing these principles in AI systems. The authors also explore various ethical frameworks and propose strategies for integrating ethical guidelines into AI development and deployment. Saxena [11] introduces a critical realist emancipatory perspective to AI ethics, addressing existing gaps in ethical AI discourse. By leveraging the stratified ontology of critical realism, the authors argue for modifying structural conditions to activate mechanisms that promote desired outcomes, such as bias-free AI. In order to change the fundamental dynamics impacting AI development and use, the study highlights the necessity of an ethical framework that goes beyond superficial concerns. Deepak [12] investigates the incorporation of ethical machine learning techniques to promote moral AI in commercial and industrial settings. It addresses the significance of openness, responsibility, and equity in machine learning models and offers suggestions for managers and engineers to successfully apply moral AI solutions. The authors also point out issues and suggest future lines of inquiry and application for ethical AI implementation.

Through the lens of algorithmic governmentality, this conceptual study investigates the relationship between algorithmic profiling and legal practice [16]. It makes the case that because these technologies allow for indirect control through economic calculations, they are consistent with neoliberal government. With a focus on the creative aspect of law as a tool to contest and change algorithmic influences in legal contexts, the study also addresses the possibility of critique and reform. In order to foster public trust, [17] investigates the shift from abstract ethical concepts to real-world AI system applications. It discusses the challenges of operationalising ethics in the creation and application of AI, emphasising the importance of stakeholder accountability, transparency, and involvement. The authors offer methods for incorporating ethical considerations into AI systems in an attempt to bridge the gap between moral values and real-world applications. The authors of [14] evaluate a range of ethical principles proposed for AI systems and investigate how they are implemented in many applications by grouping ethical standards into basic subjects including fairness, accountability, openness, and privacy. The report evaluates the effectiveness of existing frameworks, identifies gaps between ethical ideals and practical applications, and makes recommendations for additional research and policy development.

This position paper [10] addresses the disconnect between high-level ethical notions and their real-world implementation in AI systems. It criticises the overemphasis on abstract concepts in the lack of clear application pathways or fruitful stakeholder participation. The authors highlight fundamental obstacles to operationalising AI ethics and provide preliminary examples to begin real-world implementations in order to increase the applicability and significance of AI ethics in practical contexts. Similarities, differences, and priorities between these groups are highlighted by the authors of another study [15]. The findings highlight the challenges of balancing technological innovation with moral principles and legal requirements, and they offer recommendations for combining the efforts of legislators and engineers. Vakkuri [18] investigates the interpretation and incorporation of values like beneficence, fairness, and autonomy into AI systems. The authors highlight the necessity of interdisciplinary cooperation to successfully handle ethical quandaries and offer a conceptual framework to direct the moral development and application of AI technology. The ethical aspects of AI have been studied in [19], with an emphasis on the difficulties and ramifications of incorporating AI into different industries. It talks on the possible drawbacks of AI, including prejudice, privacy issues, and the loss of human labour. The authors support the creation of thorough ethical standards and legal frameworks to guarantee that AI technologies are created and used in a way that respects human rights and social values.

This article employs the Delphi technique to gather expert opinions on the connection between AI, ethics, and human rights [19]. It draws attention to points of agreement and debate among experts about the ethical application of AI. The results highlight the importance of including human rights concerns into AI development and provide policy recommendations to support ethical AI practices, ensuring that technical advancements do not infringe upon fundamental human rights. This scoping review looks at the ethical implications of applying generative AI in healthcare settings [20]. It draws attention to potential benefits, such as improved diagnostic tools and customised treatment plans, as well as drawbacks, such as algorithmic biases and invasions of data privacy. By providing an ethics checklist to assist AI developers and healthcare professionals in identifying and resolving ethical concerns, the authors promote the appropriate integration of generative AI into medical practice. Clio [21] discusses the moral dilemmas raised by the application of AI in the legal domain. It addresses issues including bias in AI systems, the transparency of AI decision-making processes, and possible impacts on client confidentiality. The paper emphasises how crucial it is for attorneys to thoroughly assess AI tools to ensure that they uphold ethical standards and protect the integrity of the legal profession.

Sanna et al. [13] examines the practical challenges that companies face with implementing AI ethics. Qualitative research reveals that while many companies articulate ethical AI principles, there is often a mismatch between stated guidelines and practical applications. The phrase 'ethics entrepreneurs' was created to describe those who advocate for decent business practices. In order to bridge the gap between moral aspirations and actualisation, the article describes the difficulties these individuals encounter, including organisational resistance and the personalising of moral obligation.

4.2 Critical Analysis of the Related Works

4.2.1 Broad Ethical Frameworks and Principles

Huang et al. [8] and Kazim et al. [9] provide a comprehensive overview of AI ethics, covering the core concepts, background, and challenges of implementation. Both pieces offer useful information, but they largely avoid getting into the specifics of how difficult it is to put abstract ideas like accountability, transparency, and fairness into practice. Zhou et al. [10, 14] and Chen et al. [17] also look at the gap between ethical ideals and execution, pointing out that operationalising ethics in AI research presents recurrent difficulties. These articles highlight the continuous challenge of translating AI ethics from abstract ideas to practical frameworks. Vakkuri and Abrahamsson [18] present a conceptual framework for integrating values such as autonomy and beneficence into AI design. However, the framework's applicability is still limited in the absence of precise methodological standards. A significant critique of these studies is their focus on theoretical frameworks rather than empirical evidence for moral AI.

4.2.2 Structural and Critical Approaches to AI Ethics

Saxena et al. [11] offer a critical realist emancipatory approach to AI ethics, arguing that ethical AI must address structural conditions rather than just surface-level fairness or transparency. This perspective is different from common tactics that prioritise technological solutions, such as explainability tools or bias audits. Harkens [16] also criticises the connection of AI with neoliberal governance, emphasising the need for legal reform to counteract algorithmic biases in economic systems. Despite their significant systemic critique, these works may face resistance in industry contexts where structural change is usually deprioritised in favour of incremental improvements. Ali et al. [13] further on this critique by looking at how ethics are personalised inside businesses and show that 'ethics entrepreneurs' are typically in charge of AI ethics. This study highlights a noteworthy vulnerability in AI governance in spite of having ethical standards.

4.2.3 Empirical and Sector-Specific Studies

A variety of research studies have been conducted into AI ethics in different domains. Although Saxena et al. [11] mention that responsible machine learning serves as a strategy to achieve ethical AI, they focus more on managerial issues than on management technology. Issues such as decision-making, biased algorithms, and other dangers are highlighted in the reports of Harkens [16] and Clio [21] regarding AI's impact in law. Yet, for the most part, their responses remain more of an analysis rather than a suggestion. Ning et al. [20] present a scoping review of generative AI in medicine and its ethical considerations along with providing an ethics checklist to work with AI in healthcare. Although the checklist method has its merits, it may not adequately resolve fundamental systemic issues such as medical data management or regulatory compliance enforcement. However, in a comparative study of AI ethics between policymakers and practitioners' perspectives by Khan et al. [15, p 1] highlighted significant awards between normative principles and practical realization. While this study emphasises how important alignment of policy is for AI governance, it doesn't offer any practical solutions for bridging that gap.

4.2.4 Delphi Method and Expert Consensus

Stahl et al. [22] use the Delphi method to elicit expert opinions on human rights and AI ethics, identifying areas of agreement and disagreement. The Delphi method is based on subjective expert judgements that don't always align with broader societal issues, though again the study provides really valuable expert-driven insight.

Even with an abundance of studies investigating the ethics of AI, very little work appears to have been done on the implementation of ethical AI at scale. Saxena et al. [11] and Ali et al. [15] are two examples of structural critiques. Difference between structural critiques and principle driven critiques like Huang et al. [8] and Kazim et al. [9] shows that the ethics of AI is more intricate as it is divided into different levels: industry level details, high ethical standards, and governance systems that can actually be executed.

4.3 Key Findings

Transparency and accountability are prioritised in the IEEE 7000–2021 model, which incorporates ethical considerations into AI system design [23]. By examining the 'Ethically Aligned Design' document, which outlines values including human rights, well-being, and accountability, IEEE demonstrates its efforts to create worldwide standards for ethical AI research [24]. Research indicates notable biases in AI applications, ranging from predictive policing to recruitment tools [25]. Few of the many frameworks have been successfully operationalised [26].

5 Inference and Conclusion

5.1 Key Inferences

This highlights the need for closely collaborative, interdisciplinary organizations and solutions in efforts to develop ethical Artificial Intelligence solutions (both morally and technically), these theoretical positions cannot be bridged without ethicists, developers, and policymakers at the table together.

5.2 Recommendations

Ethical AI principles need to be thoroughly incorporated from step one, capturing iterative tracking and contained learning. Politicians should work on creating Regulatory Harmonization to have cohesive global principles around AI ethics and allow for transnational cooperation. There must be strong mechanisms in place to mitigate algorithmic bias and ensure diversity in the training datasets for AI.

5.3 Future Directions

The most important focus of further studies should be the development of scalable instruments and indicators of ethical compliance. Empirical research on interdisciplinary cooperation's impact on ethical AI outcomes is pivotal in this stage. Filling these gaps is essential if AI technologies are to be developed in a responsible manner while upholding human rights and societal values.

5.4 Call to Action

The goal of this paper is to encourage stakeholders across domains to devote the utmost attention to these ethical challenges during the development of AI, which could produce technology that enhances human welfare while reducing risks. To ensure that AI is developed and used in an ethical manner that benefits society as a whole, it is imperative that all involved parties—academics, business executives, legislators, and the general public—work together and take proactive measures. Organizations should enforce robust accountability frameworks; legislators should craft global, harmonised regulations; and developers should build in ethics at the design phase. New challenges warrant new approaches—it is necessary to foster interdisciplinary research and involve continuous learning of AI ethics. By placing a high priority on justice, openness, and inclusiveness, we can design AI systems that can not only facilitate innovation but also protect for social value and fundamental rights.

References

1. Dastin J (2022) Worldwide devops software forecast update 2017–2021. Accessed 5 Apr 2022. [Online]. https://www.reuters.com/article/-amazon-com-jobs-automation-insight-idUSKCN1MK08G
2. (2022) High-level expert group on artificial intelligence, ethics guidelines for trustworthy AI. Accessed 6 Apr 2022. [Online]. https://www.aepd.es/sites/default/files/2019-12/ai-ethics-guidelines.pdf
3. Jobin A, Lenca M, Vayena E (2019) The global landscape of AI ethics guidelines. Nature Mach Intell 1(9):389–399
4. (2017) IEEE global initiative on ethics of autonomous and intelligent systems, ethically aligned design: a vision for prioritizing human well-being with autonomous and intelligent systems. Accessed 6 Apr 2022. [Online]. http://standards.ieee.org/develop/indconn/ec/autonomous_systems.html
5. ISO (2017) ISO/IEC JTC 1/SC 42 artificial intelligence. Accessed 26 Feb 2022. [Online]. https://www.iso.org/committee/6794475.html
6. Selcuk A (2019) A guide for systematic reviews: PRISMA. Turkish Arch Otorhinolaryngol 57:57–58. https://doi.org/10.5152/tao.2019.4058
7. Kumar A (2023) Systematic literature review (SLR) BT-meta-analysis in clinical research: principles and procedures. In: Kumar A (ed) Springer Nature Singapore, Singapore, pp 7–14. https://doi.org/10.1007/978-981-99-2370-0_2
8. Huang C, Zhang Z, Mao B, Yao X (2023) An overview of artificial intelligence ethics. IEEE Trans Artif Intell 4(4):799–819. https://doi.org/10.1109/TAI.2022.3194503
9. Kazim E et al (2021) A high-level overview of AI ethics. Patterns 2(9):100314. https://doi.org/10.1016/j.patter.2021.100314
10. Zhou J, Chen F (2023) AI ethics: from principles to practice. AI & Soc 38:2693−2703. https://doi.org/10.1007/s00146-022-01602-z
11. Saxena D, Wall PJ, Lewis D (2023) Artificial intelligence (AI) ethics: a critical realist emancipatory approach. In: 2023 IEEE international symposium on technology and society (ISTAS). Swansea, United Kingdom, pp 1−5. https://doi.org/10.1109/ISTAS57930.2023.10305995
12. Saxena D, Lamest M, Bansal V (2021) Responsible machine learning for ethical artificial intelligence in business and industry. Handb Res Appl Data Sci Artif Intell Bus Ind. https://doi.org/10.4018/978-1-7998-6985-6.ch030
13. Ali SJ et al (2023) Walking the walk of AI ethics: organizational challenges and the individualization of risk among ethics entrepreneurs. In: FAccT '23: proceedings of the 2023 ACM conference on fairness, accountability, and transparency. pp 217–226. https://doi.org/10.1145/3593013.3593990
14. Zhou J, Chen F, Berry A, Reed M, Zhang S, Savage S (2020) A survey on ethical principles of AI and implementations. In: 2020 IEEE symposium series on computational intelligence (SSCI). Canberra, ACT, Australia, pp 3010−3017. https://doi.org/10.1109/SSCI47803.2020.9308437
15. Khan AA et al (2023) AI ethics: an empirical study on the views of practitioners and lawmakers. IEEE Trans Comput Soc Syst 10(6):2971–2984. https://doi.org/10.1109/TCSS.2023.3251729
16. Harkens A (2018) The ghost in the legal machine: algorithmic governmentality, economy, and the practice of law. J Inf Commun Ethics Soc 16(1):16–31. https://doi.org/10.1108/JICES-09-2016-0038
17. Chen F, Zhou J, Holzinger A, Fleischmann KR, Stumpf S (2023) Artificial intelligence ethics and trust: from principles to practice. IEEE Intell Syst 38(6):5–8. https://doi.org/10.1109/MIS.2023.3324470

18. Vakkuri V, Abrahamsson P (2018) The key concepts of ethics of artificial intelligence. In: 2018 IEEE international conference on engineering, technology and innovation (ICE/ITMC). Stuttgart, Germany, pp 1–6, https://doi.org/10.1109/ICE.2018.8436265
19. Inglada Galiana L et al (2024) Ethics and artificial intelligenceÉtica e inteligencia artificial. Revista Clínica Española (English Edition) 224(3):178–186. https://doi.org/10.1016/j.rceng.2024.02.003
20. Ning Y, Teixayavong S, Shang Y, Savulescu J, Nagaraj V, Miao D, Mertens M, Ting DSW, Ong JCL, Liu M, Cao J, Dunn M, Vaughan R, Ong MEH, Sung JJ-Y, Topol EJ, Liu N (2024) Generative artificial intelligence and ethical considerations in health care: a scoping review and ethics checklist. Lancet Digit Health 6(11):e848−e856, https://doi.org/10.1016/S2589-7500(24)00143-2
21. https://www.clio.com/resources/ai-for-lawyers/ethics-ai-law/
22. Stahl BC, Brooks L, Hatzakis T, Santiago N, Wright D (2023) Exploring ethics and human rights in artificial intelligence–a delphi study. Technol Forecast Soc Chang 191(2023):122502. https://doi.org/10.1016/j.techfore.2023.122502
23. (2021) IEEE standard model process for addressing ethical concerns during system design. IEEE Std 7000–2021. pp 1–82. https://doi.org/10.1109/IEEESTD.2021.9536679
24. Morandín-Ahuerma F (2023) IEEE: a global standard as an ethical AI initiative, In: Normative principles for an ethics of artificial intelligence. Council of Science and Technology of the State of Puebla
25. Ferrara E (2024) Fairness and bias in artificial intelligence: a brief survey of sources. Impacts, Mitig Strat Sci 6(1):3. https://doi.org/10.3390/sci6010003
26. Corrêa NK, Santos JW, Galvão C et al (2024) Crossing the principle–practice gap in AI ethics with ethical problem-solving. AI Ethics. https://doi.org/10.1007/s43681-024-00469-8

AI and Industry 6.0

Critical Analysis of Legal Framework and Market Structure for the Future of Virtual Power Plants Using Artificial Intelligence in India

Avishikta Chatterjee, Rajiv Ganguly, and Abhisikta Basu

Abstract The energy industry is the one area of our life where environmental degradation is currently pushing us for paying closer attention to the adverse impacts of fossil fuels. Our centralized, aging power grid lacks the resilience needed to resist the catastrophic weather-related occurrences that have become more frequent in recent years, as seen by the frequent scheduled and unplanned outages. Virtual power plants (VPPs) are leading creator of renewable energy initiatives. During moments of high demand and extreme weather, which are occurring more frequently, the conventional electricity grid is prone to failure. It is becoming more and more obvious that major infrastructure changes, including substation renovations, are needed for our electrical grid. As an alternative to such upgrades, VPPs have been effectively implemented, allowing for the deferral of some modifications while addressing others. Internationally, German business Sonnen produces a battery meant to meet 75% of a home's annual energy demand with energy from renewable sources while sharing the rest with the community at large in a VPP context. In the USA, Tesla debuted Powerwalls in 2015. With the capacity to scale up to 1,000 assets, Central schweizerische Kraftwerke AG (CKW) in Switzerland developed a sizable VPP with 110 assets. This paper proposes use of artificial intelligence for prediction and virtual power plants can offer users more economical power and a more affordable energy alternative and also proposes the legal and policy reform to build the infrastructure to promote competition in the market.

A. Basu
IEM's International Institute of Juridical Sciences (IIIJS), University of Engineering and Management, Kolkata, India
e-mail: abhisikta.basu@uem.edu.in

R. Ganguly
Centre of Excellence of Renewable Energy, Institute of Engineering and Management, University of Engineering and Management, Kolkata, India
e-mail: rajiv.ganguly@uem.edu.in

A. Chatterjee (✉)
Department of Law, University of Stockholm, Stockholm, Sweden
e-mail: avishiktachatterjee.law@gmail.com

M. Chakraborty et al. (eds.), *Proceedings of Sixth International Ethical Hacking Conference*, Lecture Notes in Networks and Systems 1513,
https://doi.org/10.1007/978-981-96-8632-2_29

Keywords Virtual power plant · Electricity policy · Artificial intelligence · Conventional electricity

1 Introduction

The concept of AI-driven virtual power plant (VPP) emerged as a response to the challenges posed by the integration of renewable energy. VPP is a revolutionary approach which addresses challenges by providing modern energy management and reflects a transformative shift in energy management. It began to originate in the early 2000s and was conceived to address the challenges that were associated with the traditional, centralized power generation system, which was struggling to cope up with the intermittent nature of renewable resources like solar and wind. The virtual power plants, unlike the conventional, aggregate different decentralized resources like the wind turbines, solar panels, and batteries. It demands the response system to be a single coordinated network with the help of digital and IoT technologies, which will enable real time monitoring along with enhanced stability and reliability across the grid [1]. VPPs, with the support of the advancements of smart grid technology, provide an agile and scalable alternative to the traditional system along with low-cost facilities, improved resilience which in totality promotes decarbonization [2]. AI adds on to these benefits as it enhances the VPP's capabilities. It provides facilities like analyzing vast data sets which predict demand, balance supply with unprecedented precision. This predictive ability of the AI minimizes the wastage of energy and gives more scope to maximize the use of renewable resources along with effective reduction of carbon emissions and lower operational costs [3]. AI powered VPPs have the ability to offer real time responses to demand and helps in maintaining a supply balance with the needs of the user. Thus, it fosters sustainability, and as more and more regions will adapt this decentralized solution, it will create a scalable pathway towards a resilient and low carbon energy system in future.

1.1 Integration of AI into Modern Energy System

There are several ways that AI technology is being applied to increase energy efficiency. Machine learning algorithms analyze vast amounts of data from sensors and smart devices to identify patterns and immediately reduce energy consumption. By modifying lighting, heating, and cooling in accordance with occupancy and usage patterns, these systems significantly decrease energy waste. AI-powered building management systems, for instance, may predict and respond to variations in occupancy, weather, and energy prices to guarantee optimal energy use. In many industries, artificial intelligence plays a key role in optimizing energy use. In industrial contexts, AI systems monitor machinery and equipment to ensure optimal performance. Predictive maintenance is one important AI-driven innovation that reduces energy loss and downtime by employing data analytics to foresee equipment failures before they occur. In residential settings, AI-powered smart home appliances can identify user patterns and adjust energy usage to conserve energy and improve convenience. A vital aspect of contemporary energy management is artificial intelligence, and smart grids are no different. These networks use AI algorithms to balance energy supply and demand, incorporate renewable energy sources, and better manage energy distribution. By anticipating energy production from renewable sources like solar and wind, AI-powered energy forecasting models assist grid operators in better planning and energy distribution. This reduces the likelihood of blackouts and improves grid resilience overall by making the electricity supply more consistent and reliable.

AI-driven innovations have been revolutionizing sectors such as predictive maintenance, smart grids, energy forecasting, autonomous energy systems, demand response optimization, renewable energy integration, and energy efficiency optimization. Predictive maintenance enables tracking of equipment performance to predict potential problems, thereby reducing the time lost and energy consumed. Smart grids improve the energy distribution network by controlling renewable energy sources, enhancing demand response capabilities, and optimizing electricity flow. AI in energy forecasting predicts energy output and consumption, enhancing resource allocation and planning. Thus, autonomous energy systems use AI to monitor energy use; balance supply and demand; and react instantaneously to conditions in the grid. Demand response optimization is applied using AI to predict energy-use patterns and automate responses to changes in demand, making the grid more stable and less likely to experience blackouts. The integration of renewable energy employs artificial intelligence to project the output of renewable resources and optimize their integration into conventional energy systems. It involves the optimization of efficiency through machine learning algorithms, hence always evaluating and improving its energy consumption in buildings and industries as well as transportation nets. This way, a constantly learning and adapting system will automatically conserve energy and reduce emissions to ensure sustainable energy utilization with minimal environmental impacts.

The VPP system is an advanced framework designed for optimized energy generation, use, and distribution within a decentralized energy context. As shown above in Fig. 1, the artificial intelligence integration model in virtual power plants commences first with data collection as the initial process stage; the distributed energy resources would be solar panels, wind turbines, and batteries that feed data into this system at real-time. It also comprises other data, including market electricity pricing, weather forecasts from APIs, and energy use from smart meters. Preprocessing of data further resolves issues regarding noise and missing values while getting the data cleaned and normalized so that such data is accurate. After refinement, this data is fed into AI models using the application of neural networks or machine learning (ML). These models are mainly charged with three key responsibilities: energy forecasting, which involves both forecasting energy generation and consumption; optimization of demand response, which calls for improving responses at points of peak demand; and supporting trading decisions based on the market. The second stage is decision-making whereby artificial intelligence insights are utilized to decide whether surplus

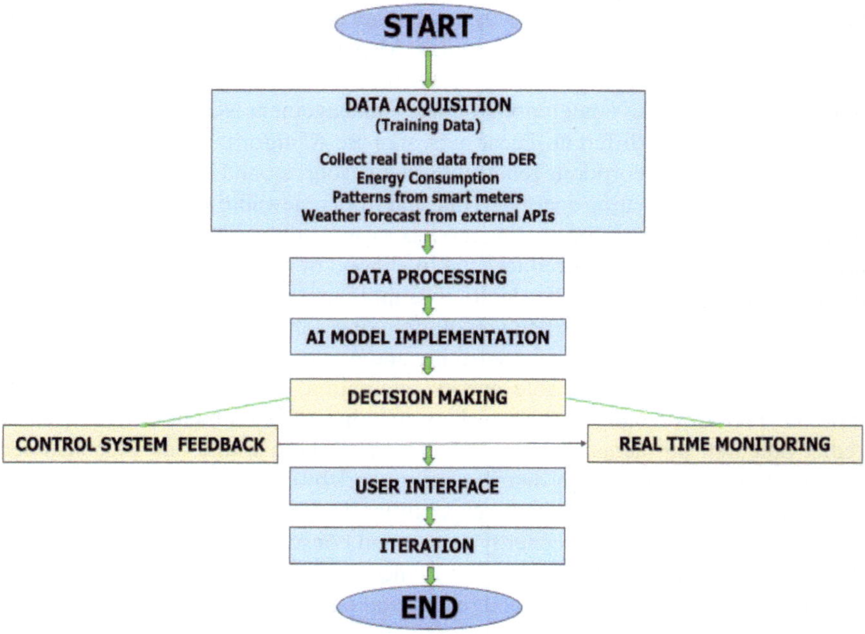

Fig. 1 Working mechanism of integration of AI in virtual power plant

energy should be stored, used, or marketed, and how energy should be allocated among the available resources. This further optimizes through reinforcement learning for adaptive decision-making and evolutionary algorithms to effectively distribute resources. This stage again is a loop as the feedback mechanism of the control system automatically regulates energy by changing the DER output and other loads or storage systems when required. Currently, predictive maintenance serves the purpose of detecting anomalies before they escalate, while real-time monitoring ensures continuous oversight of system functionality. Stakeholders interact with the system via a dashboard or user interface, which provides insights, alerts, and tools for manual intervention. The system is very strong in its iteration focus, with the idea of feeding AI models updated data to constantly improve accuracy and performance. The VPP model manages energy effectively, reduces costs, and promotes sustainable energy practices in real time with AI, optimization, and automation.

2 Implementation of AI-Driven VPPs

AI's role in the functioning of the VPP is fundamental as it enables sophisticated management of energy resources by enhancing real time decision making, dynamic load balancing and predictive maintenance. With the advancing data analytics and machine learning systems AI can optimize energy flow across multiple sources, it can enhance the grid stability by anticipating fluctuations through its predictive ability and can manage energy distribution according to the changing needs [4].

2.1 Benefits of Using AI-Driven VPPs

AI-driven VPPs are seen to transform the entire energy landscape with the integration of a variety of distributed energy resources like solar panels, electric vehicles, batteries, and smart devices. The traditional rule-based algorithms follow the preexisting instructions which ultimately limits their ability to offer new solutions with the changing needs. Whereas VPPs powered by AI have the ability to offer solutions based on real-time problems [5].

With the sophistication of AI, VPPs are able to analyze data from various sources including weather patterns, market conditions, and past trends, which subsequently allows it to forecast energy generation and usages along with a much higher accuracy rate. This capability specifically is beneficial for addressing fluctuations in supply of energy and its demand along with the management of consumption costs.

VPPs have been seen to be critical in handling climatic changes however its true potential will only be unlocked when AI-driven software are deployed. The integration of EVs which are equipped with the vehicle to grid technology is one of the most promising developments in VPPs. EVs offer a significant, variable energy storage resource with the ability to draw power as well as supply power back to the grid. As unlike the stationary batteries the EVs, mobile storage is dependent on its primary function that is transportation. The ability of AI to manage all the associated variables through its probabilistic or predictive analysis and advanced machine learning ensures that the system can make the most of the available resources [6].

VPPs driven by AI are also critical for the management of electricity markets where the prices keep on fluctuating depending on a variety of factors like supply, demand, weather etc. Its predictive ability enables VPPs to give real time responses and help in decision making depending on the current situation.

Along with the evolution of the energy sector the need for scalability and an adaptable AI system has become more evident. The growing complexity of managing a varied range of energy resources coupled with increasing demand for flexibility and resilience, puts focus on the advantages of AI in VPPs and various initiatives are already being undertaken to implement such.

2.2 Initiatives Undertaken in India Aligning the Renewable Energy Goals

India in order to modernize its power sector has undertaken many initiatives and programs which collectively aims to establish an advanced as well as resilient energy infrastructure which supports AI-driven VPP and DERs promoting a clean and efficient grid. The Smart Grid Forum initiatives serve as a collaborative platform which involves various stakeholders like the industry experts and the policy makers. It promotes and ensures innovations to enhance the reliability of the power grid as well as its efficiency. It puts focus on the smart grid technologies and plays a pivotal role by supporting AI driven initiatives like VPPs that are dependent on a responsive data grid infrastructure [7].

The Smart Grid Task Force has been established by the Ministry of Power, to facilitate and monitor the proper implementation of the smart grid projects nationwide. This force has guided various pilot projects that integrate DERs like solar panels, electric vehicles and battery storages. These projects implement AI-based technologies to create a balance between the demand and that supply by focusing on the interoperability, data privacy and cyber security aligned with the global standards [8].

The National Smart Grid Mission or the NSGM is another initiative launched in 2015 which promotes grid modernization across the country. It provides a strategic framework in order to develop smart grids focused on automation and advanced metering infrastructure and real time data management. This program supports projects that implement AI usage and machine learning to optimize the energy use, and as a result enabling effective integration of renewable resources into the grid [9].

Together these efforts reflect the efforts that have been made under India's commitment towards a smart, adaptable as well as environmentally sustainable future. By providing both policy and technical support these initiatives are setting a groundwork for the large-scale deployment of AI-driven VPPs in accordance with the legislative frameworks.

3 Existing Legal Frameworks and Policy on VPPs

Ensuring proper integration and operation of the virtual power plants within the energy sector requires effective legislative frameworks and regulations. The frameworks are known to regulate different aspects like market participation, consumer rights, grid integration, and to ensure the VPPs contribute to a more resilient energy future. Countries across the globe are seen adopting tailored policies that tend to address the challenges posed by the technological shift to VPPs like cyber security, data protection, etc. The European Union, Australia, and the USA have taken significant steps in order to progress; India is still progressing to adapt proper regulatory measures to fully harness the potential.

3.1 *European Union's Unified Framework for Energy Aggregation*

The EU has adopted robust frameworks for proper integration of the virtual power plants that focuses on enhancing the grid stability and empowering the energy consumers. One of its key initiatives under this is the *"Clean Energy for All Europeans Package"*, it focuses on encouraging businesses and households to act as "energy prosumers" and "aggregators" driving the decentralized management of energy [10]. The said approach shows a paradigm shift from a centralized to more of a decentralized generation within the energy system, setting stage for innovations like VPPs. These formative changes were laid down through the "Clean Energy Package" that was adopted in the 2019, which consisted of several directives that aimed to achieve the EU's energy and climate targets. Among them the EU Electricity Directive (2019/944) plays a crucial role in the energy ecosystem, which requires the member states to take forward the establishment of the said aggregators and the energy communities. Innovative models like the VPPs are supported by the directive, enabling them to enhance grid stability and ensure utilization of the renewable energy [11]. Many nations including Germany and the Netherlands have started adopting strategic measures that are in accordance with the directives. Germany's Renewable Energy Sources Act introduced in the 2000 plays a huge role in the country's shift to the sustainable energy sector and thereby served for the development of the virtual power plants. The act does not explicitly deal with the VPPs but its concept is in accordance with the laid down objective of promoting renewable energy integration and to enhance the grid stability under the act [12]. Among the several amendments the 2012 amendment introduced the concept of 'direct marketing' which is an alternative to the 'fixed feed in tariffs'. It allows the renewable generators to become a part of their wholesale market for selling electricity. This particular change facilitates the development of the virtual power plants with the enablement of the aggregation and the optimized dispatch of decentralized renewable energy sources [13]. In addition to these, the 2023 amendments further aims to raise the share of renewable energy in gross electricity consumption to at least 80% by 2023.

This collaborative approach of the EU highlights its commitments towards fostering a unified energy market that focuses on creating a balance between innovation, efficiency and sustainability, setting a benchmark for others.

3.2 *United States' Advanced VPP Policies and Incentives*

The USA stands as a global leader in integrating the virtual power plant within the energy systems, along with advanced policies and regulatory frameworks that fosters its growth. In the development of the VPP the Federal Energy Regulatory Commission (FERC) Order No. 2222, issued in 2020 is a pivotal regulation which enables the DERs or the distributed energy resources including the VPP to take part in the wholesale energy markets. The Order No. 2222 removes the barriers to enter for the DERs (Distributed Energy Resources) aggregation which ensures that the VPP contributes to the grid reliability, cost effectiveness, and market efficiency. This marks a significant step towards a more decentralized as well as a flexible energy system [14]. Similarly, California at the state level has also taken proactive approaches that establish VPP friendly policies through the California Public Utilities Commission (CPUC) [15]. The commission established in 1911 ensures safe, affordable services. Initially it regulated transport and water but its scope further expanded to include both telecommunication and energy. During the 2000's the California Public Utilities Commission is known for implementing reforms to stabilize the markets and to promote renewables. In recent times it has been leading initiatives like the Self Generation incentive program that fosters energy storage, VPPs [16].

Under the Self-Generation Incentive Program (SGIP) is an initiative that provides for financial incentives for the energy storage systems which is one of the critical components of the VPPs. Along with supporting the adoption of the distributed energy resources it also encourages their aggregation, which enhances the grid's resilience and optimization [17]. The state additionally focuses on integrating the VPP objectives with the renewable energy sources that provides for a sustainable energy future. Further the New York's REV or the Reforming the Energy Vision focuses on the modernization of the grid with the promotion of the integration of Distributed Energy Resources through the distributed system implementation plans [18]. This mandates for outlining the strategies to incorporate Distributed Energy Resources like the energy storage, solar panels as well as a demand response systems within the grid systems. Distributed System Implementation Plans (DSIPs) also emphasizes the enhancement of the grid's reliability, customer engagement, and operational efficiency.

It has also launched the NY- Sun Program which is flagship under New York's energy reform initiative. It is designed for increasing the adoption of solar energy across the states. It provides for educational resources, financial incentives, and for streamlining the process of making solar installation much more accessible [19]. Hawaii has also embraced VPPs to be a part of their Hawaii Clean Energy Initiative (HCEI) that aims to achieve 100% of the renewable energy by 2045 [20]. Hawaiian Electric's VPP projects enhance the stability of the grid and support the renewable energy integration by aggregating the residential solar panels and the battery storage systems.

All these measures reflect the United States' commitments toward fostering advanced regulatory measures for the adoption of the VPP along with innovation and sustainability.

3.3 Australia's Pioneering VPP Pilots

Australia has also emerged in this sector with the deployment of large scale VPP projects along with innovative regulatory measures and strategic initiatives.

The Australian Energy Market Commission has played a crucial role by enabling distributed energy resources like the VPPs to be an active participant in the wholesale energy market. The 2020 amendments to the National Electricity Rules have provided a playing field for the VPP operators by systemizing the market rules along with transparency and reliability [21]. The provisions try to standardize the way Distributed Energy Resources are aggregated by ensuring that similar to the traditional generators the Distributed Energy Resources can also bid into the wholesale energy market.

The South Australian VPP project is a standout example of one of the largest operating VPPs globally supported by the Australian Renewable Energy Agency. This particular initiative aggregates thousands of residential solar panels along with battery systems that form a virtual power plant. The project not only reduces the grid stress, it also lowers the cost of energy for participants which showcases the dual benefits of the grid stability [22].

This highlights Australia's commitments in integrating renewable energy resources alongside enhancing resilience.

3.4 Legal Gaps and Challenges

With the transition of the global energy system the virtual power plants have emerged as a crucial and effective solution that integrates the distributed energy resources. However there exist significant gaps in the legislation and regulations which are as follows:-

The lack of proper definitions and uniform standards which stands to be one of the primary challenges in the implementation of the VPPs. The absence of a universally accepted definition leads to ambiguity in their understanding which leads to several interpretations in different jurisdictions which ultimately causes inconsistencies in the regulatory measures and approaches. This ambiguity further leads to not having uniform standards of regulations across the jurisdictions. It is seen that the United States and Australia put more emphasis on the participation in the wholesale market, whereas on the other hand the European regulations put emphasis on the empowerment of the consumer thereby creating a fragmented regulatory framework that in the end hinders global cooperation and standards [23].

The existing frameworks are seen to fail in ensuring an equitable access to the market for the operators in many countries. It has also been highlighted under various studies that the current structure of the electricity market is seen to put much more focus and prioritization to the traditional utility over the advanced innovations like the VPPs. If we consider the U.S FERC Order No. 2222 although it aims to bridge the said gap but at the event of implementation it has been seen to face various hurdles like the different interpretations at the state level. This ultimately limits their ability to effectively foster and provide for innovative solutions that can be adapted globally [24].

The existing frameworks at times are seen to adequately fail in addressing the aggregation of the DERs which is one of the core functions of the VPPs. Australia's amendment to the National Electricity rules from its outlook seems commendable but in reality, its scope has still remained limited. It primarily does benefit the large-scale projects but it tends to neglect the smaller aggregators [25]. Similarly if we see in India, the Electricity Act, 2003 and its ancillary regulations do not provide for any provisions relating to the aggregation ultimately stalling the growth of the VPPs.

Renewable Energy Policies.

The concept of the VPPs is in accordance with the global targets relating to the renewable energy goals. But the misalignment that exists in the regulatory frameworks hampers the process of the integration with the Renewable energy policies. The Germany's Renewable Energy Sources Act (EEG) is seen to support renewables but it does not provide for any proper mechanisms for the same. This lacks the optimized participation in the VPPs, creating an inefficient resource that lacks proper utilization of the resources [26].

The feasibility of VPPs on the economic aspect still remains as a concern due to the unclear tariff structures and the mechanism of the cost sharing. And the present consumer protection laws do not address or deal with the concept of the VPPs and thus it does not provide for the potential associated risks. One such risk would be the financial losses during the failure of the system [27].

The VPPs are digital in character which makes them highly vulnerable towards the risks of data security and privacy concerns. Research in recent times shows the need for stringent legal protections that have the ability to safeguard the consumer's data thereby preventing the cyber-attacks on the aggregated system. The GDPR or the General Data Protection Regulation of the EU safeguards data and ensures protection of privacy but the act has no provisions under it that deals with the new innovation of the VPPs. This shows the existing global regulatory lack of laws and policies that deals with the functioning and the implementation of the VPPs [28].

4 Case Studies on Role of AI in Energy Ecosystem

Tesla's initiative in South Australia and TEPCO or the Tokyo Electric Power Company's VPP in Japan demonstrates the pivotal role of artificial intelligence in the enhancement of energy storage and in the optimization of the VPPs functionality. The said project shows how AI-driven virtual power plants can efficiently manage the decentralized systems of energy and simultaneously reduces the dependency on the grid and promotes sustainability. Tesla's project of VPP in South Australia stands to be a prime example that shows how AI can transform the storage solutions of energy. It integrated powerwall batteries and rooftop solar panels across thousands of homes which creates large decentralized networks of energy. AI's algorithm of predictive analysis predicts the patterns of the demand and supply which helps in ensuring an efficient energy distribution with a stabilized grid. The project has resulted in the reduction of the power outages, improvement in the security of the energy, along with cost reduction for the participants. In addition to these it also provides for a scalability model that demonstrates how the broader grid operations can be supported by the localized energy systems [29].

Similar to Tesla's project TEPCO's VPP in Japan also uses artificial intelligence for managing the sources of renewable energy like the wind and solar. It analyzes the real time data and optimizes the storage of energy and its distribution and keeps it in accordance with the fluctuating demand and supply. The system provides for reduction in the wastage of energy and also enhances the resilience of the grid system. It also uses AI's predictive analysis that helps in forecasting the needs of energy thereby contributing towards a sustainable development in the urban areas and also lowering the carbon emission [30]. These case studies paves way for India for the implementation of virtual power plant.

5 India's Emerging Virtual Power Plant Frameworks

The energy sector in India has been going through significant transformations, specifically shifting toward cleaner and smarter grid systems. As a part of such transformations VPPs are seen to emerge as a crucial innovation although the regulatory frameworks relating to it are still in their nascent stage. Existing laws like The Electricity Act of 2003 acts as the cornerstone facilitating power generation, trading and distribution [31]. It emphasizes on the open access and competition though the act does not specifically deal or mentions about the VPPs but its focus of decentralizing the energy generation and integration of the renewable energy fosters an environment that enables their deployment. This shows a regulatory gap that hinders the full-scale deployment as well as the optimization of the virtual power plants in India. The said gap can be bridged with the introduction of specific policies for the VPPs that focuses on their incentives for the Distributed Energy Resources aggregation and their integration into the energy market.

In 2015, the National Smart Grid Mission was another significant step towards the modernization of India's electricity grid, which fosters integration of the smart technologies along with the deployment of the virtual power plants. The mission provides guidelines for enhancing the efficiency of the grid, its reliability as well as sustainability. It highlights the adoption of the demand side management and advanced metering infrastructure [32]. Although the mission supports these advancements, there is a lack of explicit frameworks that directly addresses the VPPs. Therefore, establishment of stringent regulations can enable the development and deployment of the VPPs for fully realizing its potential. Along with this the Central Electricity Regulatory Commission (CERC) has also established certain regulations for the demand response as well as ancillary services aligning the VPP's operational requirements [33]. However there can be seen lacks in a stringent regulatory framework specifically tailored to VPPs, this necessitates proper policy reforms for addressing critical issues like data management, aggregation, and cybersecurity.

5.1 Role of Virtual Power Plant in Upholding India's Commitment to SDGs

Virtual power plants enable sustainable and decentralized energy systems and thus are known to be instrumental in the advancement of the United Nations Sustainable Goals. The goals drive innovations, integrate the renewable energy sources, and provide for the optimized utilization of resources that contributes to a cleaner, resilient energy source.

The SDG 7 explicitly deals with "Affordable and Clean Energy" which focuses on enhancing the integration of the renewable energy sources and provides for improving the efficiency of the grid system. One of the studies highlights that the VPPs accelerates the transition to a cleaner energy system by aggregating the distributed energy resources, increasing the accessibility of renewable energy [34]. The goal aligns the broader push for the adoption of clean energy as it is seen that the nations in the recent times are seeking transition from the fossil fuels for reducing the greenhouse effects, combatting the climatic changes and ensuring sustainability.

Additionally, the SDG goal 9 provides for "Industry, Innovation, and Infrastructure". It provides for modernizing the energy systems and enhances the stability of the grid through technological advancements like blockchain, IoT and artificial intelligence. The advancements provide for facilities like optimization, real time monitoring and integrating the renewable energy resources. Research suggests that VPPs are the key stakeholders in the transformation to the grid systems efficiency and its reliability [35].

Further SDG 11 ensures "Sustainable Cities and Communities" which aligns the VPPs transformative role of creating a resilient and a sustainable urban environment. It enables a decentralized system of energy, and allows the shift from the reliance on centralized and large power plants by reducing the losses of transmission and improving the accessibility of energy for the local communities. The process of decentralization ensures greater control to the urban areas over their energy resources, enabling quicker adaptation to the fluctuations in the supply and demands. Recent studies show that the virtual power plant improves the efficiency of the energy with the aggregation of the distributed energy resources like wind turbines, battery storage, solar panels to a unified virtual network [36]. This highlights an integrated approach that not only supports the energy needs of the urban sectors but is also in accordance with the set renewable energy targets, reducing the carbon footprints fostering and ensuring the sustainable growth.

The SDG goal 12 deals with "Responsible Consumption and Production". The optimized energy management is in accordance with the VPPs that plays a pivotal role by fostering the practice of sustainable energy. It is done with the aggregation of the DERs like wind turbines, solar panels, and battery storage, which creates a unified and efficient system that ensures minimum waste along with the generation, storage and distribution of energy. The VPPs integrate seamlessly within the supply chain along with a sustainable consumption of energy sources. The VPPs enhance the sustainability of the environment with the improvement of the efficiency of the energy which lowers the carbon emission that reduces the dependency on fossil fuels. The VPPs also provide for the demand response system or programs which adjusts the consumption of energy in the real time with optimized resource utilization [37].

The VPPs are also in accordance with SDG goal 13 that deals with "Climate Action" as it accelerates the global shift toward a lower carbon energy system. The VPPs enable the efficient integration of the renewable energy within the power grid by aggregating the distributed energy resources like the wind turbines, energy storage, and solar panels. The said integration focuses on reducing the reliance on the power generation by fossil fuels, which lowers the emission of the greenhouse gases ultimately mitigating the climate changes. Studies show that the VPPs facilitate the adoption of clean energy, optimizing the use and the minimization of the wastes [38]. This said global transition to the low carbon technology that are driven by the VPPs are expected to be crucial in the achievement of the climate targets like the net zero emissions which includes the integration of the advanced technologies along with fostering the adoption of the renewables ultimately ensuring a sustainable energy future.

As part of its remit, the MNRE encourages a variety of sources for renewable energy, which covers small hydro power, solar, wind, and bioenergy. The largest percentage growth in installed capacity occurred within the Solar Power sector out of all installed capacity, from 3.99 GW in 2014–15 to 81.81 GW in 2023–24, representing 1950.38% growth throughout. The solar power installations are ground-mounted and represent 81.88%. There was a 95.78% increase in the installation of capacity under wind power.

6 Current Performance of Energy in the Indian Sector

According to the Statistic Report 2023–2024 published by Ministry of New and Renewable Energy, Government of India, the country's total installed capacity has been 441.97 GW as of March 31, 2024 against 275.90 GW in 2014–15, showing an incremental growth of 60.19% over the preceding nine years. The solar, wind, bio-power, and small hydro installed capacities have grown by as much as 259.55%, while the respective installed capacities under renewable energies, including large hydro power, have increased by around 134.63%. In 2023–2024, the non-renewable sector installed 7.35 GW while the renewable sector peaked at 18.56 GW. For the past six years from 2017–18, the annual growth in RE capacity installation has also been more than the growth of 3.01% for non-RE. At the end of 2023–2024, the installed capacity shares were 44.97% for the non-fossil and 43.12% for renewable energy sectors (Table 1).

Between 2014–15 and 2023–24, installed capacity of renewable energy grew by a staggering 134.63%, surpassing the overall installed capacity growth of 60.19%. Moreover, installed capacity of renewable energy including solar, wind, biopower, and small hydropower has grown by a more spectacular 259.55%, thereby underlining the immense strides taken in these areas.

Table 1 Cumulative installed capacity since 2014–2015 (Data has been retrieved from MNRE publicly available). *Source* Ministry of New and Renewable Energy, Government of India

Year	Mode-wise breakup (GW)					Grand Total	Growth (%)	Share of RE (%)
	Non-RE		Renewables (RE)					
	Thermal	Nuclear	Hydro	RES*	Total RE			
2014–15	188.90	5.78	41.27	39.95	81.22	275.90	10.62	29.44
2015–16	210.68	5.78	42.78	47.09	89.87	306.33	11.03	29.34
2016–17	218.33	6.78	44.48	58.56	103.04	328.15	712	31.40
2017–18	222.91	6.78	45.29	70.65	115.94	345.63	5.33	33.54
2018–2019	226.28	6.78	45.40	79.41	124.81	357.87	3.54	34.88
2019–2020	230.60	6.78	45.70	88.26	133.96	371.34	3.76	36.07
2020–21	234.73	6.78	46.21	95.80	142.01	383.52	3.28	37.03
2021–22	236.11	6.78	46.72	109.89	156.61	399.5	4.17	39.20
2022–23	237.27	6.78	46.85	125.16	172.01	416.06	4.15	41.34
2023–24	243.22	8.18	46.93	143.64	190.57	441.97	6.23	43.12
Gr (2014–15 to 2023–24)	28.76%	41.52%	13.71%	259.55%	134.63%	60.19%		
CAGR (2014–15 to 2023–24)	2.85%	3.93%	1.44%	15.28%	9.94%	5.38%		

Over the past nine years, the renewable energy sector has made a noteworthy contribution. Figure 2 depicts the total installed capacity, rising from 29.44% in 2014–15 to 43.12% in 2023–24.

As part of its mandate, the Ministry of New and Renewable Energy (MNRE) promotes a range of renewable energy sources, including small hydro power, solar, wind, and bioenergy. The capacity installation under the Solar Power sector saw the most significant increase out of all installed capacity under this sector, rising from 3.99 GW in 2014–15 to 81.81 GW in 2023–24, or a growth of 1950.38% throughout that time. Ground-mounted installations account for 81.88 percent of the installations of solar power. There was an increase of 95.78% in the installation of capacity under wind power (Table 2).

The total installed capacity of renewable energy from solar, wind, biopower, and small hydropower as of March 31, 2024, was 143.64 GW, which was much more than the 40.04 GW installed capacity as of March 31, 2015. In recent years, there has been a noticeable increase in the capacity for renewable energy, with solar power becoming the primary driver of this growth.

Figure 3 has been made after collecting the data from MNRE which portrays that by the end of March 2024, solar power made up 56.95% of the total installed capacity of renewable energy from solar, wind, bio-power, and small hydro power,

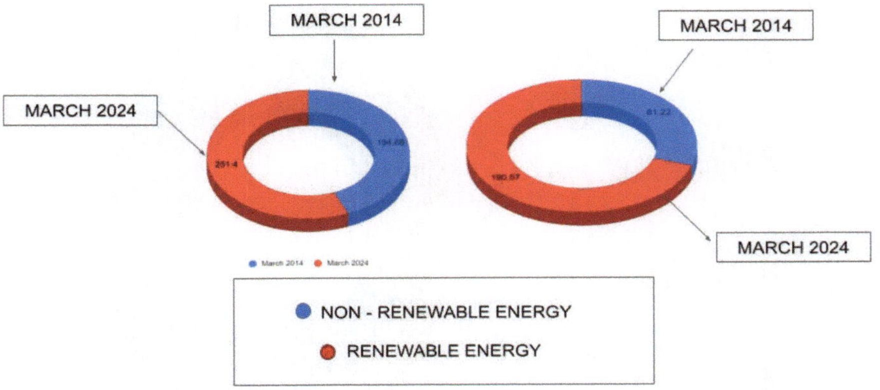

Fig. 2 Share of renewable energy as on March 2014 and March 2024. *Source* Ministry of New and Renewable Energy, Government of India

Table 2 Cumulative installed capacity under RES since 2014–2015 (Data has been retrieved from MNRE publicly available). *Source* Ministry of New and Renewable Energy, Government of India

Year	Mode-wise breakup (GW)					RES capacity total
			Bio-power			
	Small hydro power	Wind power	BM power/ cogeneration	Waste to energy	Solar power	
2014–15	4.06	23.44	8.31	0.24	3.99	40.04
2015–16	4.27	26.78	8.67	0.25	7.12	47.09
2016–17	4.38	32.28	8.84	0.28	12.78	58.56
2017–18	4.49	34.15	9.36	0.31	22.35	70.65
2018–2019	4.59	35.63	9.78	0.32	29.10	79.41
2019–2020	4.68	37.74	9.88	0.35	35.60	88.26
2020–21	4.79	39.25	10.15	0.39	41.24	95.80
2021–22	4.85	40.36	10.21	0.48	54.00	109.89
2022–23	4.94	42.63	10.25	0.55	66.78	125.16
2023–24	5.00	45.89	10.36	0.59	81.81	143.64
Gr (2014–15 to 2023–24)	23.15%	95.78%	24.67%	145.83%	1950.38%	258.74%
CAGR (2014–15 to 2023–24)	2.34%	7.75%	2.48%	10.51%	39.88%	15.25%

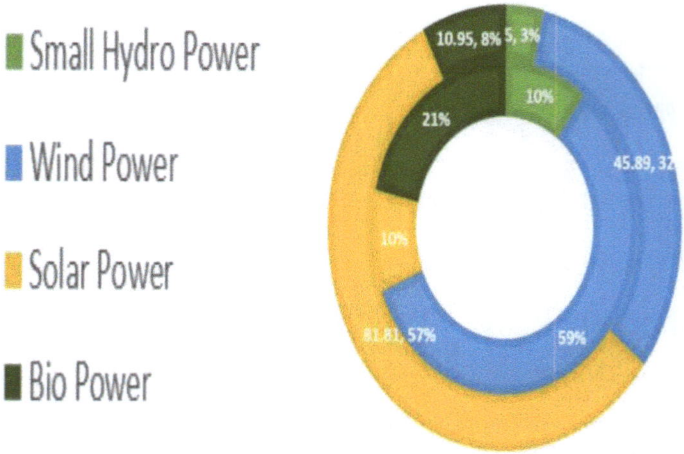

Fig. 3 Share of various sources in RES cumulative installed capacity. *Source* Ministry of New and Renewable Energy, Government of India

reaching a cumulative capacity of 81.81 GW. The second-highest installed capacity was seen in wind power, which reached 45.89 GW and accounted for 31.95% of the total installed capacity. With 10.95 GW of installed capacity, biopower constituted 7.62%, while small hydropower constituted 3.48% with 5 GW of installed capacity. In comparison with 2014–15, when the total installed renewable energy capacity was 40.04 GW, this capacity installation development signifies a huge change. At that time, with 23.44 GW installed and a 58.54% share, wind power dominated the market. Bio-Power came in second with 8.55 GW installed and a 21.35% share. Small Hydropower had a 10.14% share having installed 4.06 GW and solar power. The least share was held by 9.97% with a cumulative installation of 3.99 GW.

By 2030, India has set an ambitious goal of 450 GW of renewable energy capacity, of which 280GW are anticipated to come from solar PV. The most recent report is that there is currently 89.4 GW of installed capacity. With a compound annual growth rate (CAGR) of about 50% from 2015 to 2022, the solar PV market in India has expanded quickly in recent years. It is anticipated that the 2030 target will be readily accomplished given the lofty goals, strong policies, and desire to achieve them. India has recently surpassed Japan to be among the top 5 nations in the world for solar PV installation. Solar PV currently generates around 12% of India's electricity. By 2047, it is estimated to be around 856 GW, which would account for almost 46% of India's power capacity. It is presented with an analytical tool developed by NITI Aayog as "India Energy Security Scenarios 2047". Since it would have to reach the level of net zero carbon emission by 2070, India has to take the solar PV generation up to 5630 GW. It is expected that during the next 25 years, the energy transition of India will center on solar energy, which will have a very deep impact on decarbonization and energy security in a positive way. Ecosystem development around solar PV has shown that it is primed for continued growth through the increased demand of electricity from clean sources, good and ambitious government policies, and the declining costs of the system. All these propitious conditions notwithstanding, however, a few challenges remain which need to be taken care of. Sometimes, land acquisition and infrastructural constraints can pose challenges to big projects in a country like India. Finance and policy issues often could be blocking the way to progress. Additionally, grid stability along with transmission and distribution issues will also create a big problem at times at remote locations.

6.1 Impact of Incorporation of VPP in Indian Electricity Market

Virtual power plants in India will transform the electric market into an efficient, sustainable, and resilient sector. VPPs aggregate solar panels, wind turbines, batteries, and electric vehicles to one central network for optimized energy generation, storage, and distribution. This decentralization eliminates inefficiencies, decreases the wastage of energy, and enhances grid stability. The use of AI-based real-time forecasting helps the VPPs to reduce dependency on expensive peaking plants while offering a stabilized electricity price; it is thus critical to the management of increased countrywide energy demands and to minimizing any form of grid imbalances. VPPs speed up the process of renewable energy integration into India's grid by including clean energy sources and thereby reducing the use of fossil fuels, thus decreasing carbon emissions. VPPs also help consumers and small energy producers to compete and innovate with each other as well as in terms of price. Consumers benefit through demand response programs wherein the consumers alter their usage according to price signals received by them, which would ultimately result in cost savings and active participation in grid management. VPPs also unlock new revenue streams for DER owners by selling surplus back to the grid or trading their energy in markets. Thus, this will positively impact the adoption of renewable systems in alignment with India's renewable energy targets. However, this development poses challenges for traditional utilities, which should be integrated into regulatory reforms and necessary grid modernization in order to safeguard fair competition and secure operation. Decentralization of energy production and improvement in the resilience of the grid will reduce the likelihood of widespread blackouts, especially in those regions that are most vulnerable to natural disasters. The potential of VPPs in India can be unlocked by filling regulatory gaps, defining clear policies for the integration of DERs, and scalable pilot projects. In an effective implementation, VPPs could redefine India's energy landscape, with cleaner, more reliable, and economically viable electricity.

7 Conclusion and Recommendations

The global energy ecosystem is being revolutionized by virtual power plants (VPPs), which are driven by advanced technologies such as artificial intelligence (AI), machine learning (ML), and the Internet of Things (IoT). VPPs provide a decentralized method of optimizing energy production, consumption, and distribution as the globe shifts to a cleaner, more sustainable energy paradigm. The deployment of renewable energy sources, including solar panels, wind turbines, and energy storage systems, requires the integration of these technologies in the grid. In addition, the capacity to collect and control DERs is vital in responding to changing market requirements and climate-related issues in order to effectively manage energy. The definition of the new configuration of VPPs by the researchers gives a more complex framework for attaining energy optimization. During the initial data acquisition process of the system, DERs offer real-time knowledge of energy production and usage. All this information is taken into the decision-making process coupled with weather forecasts, wholesale price quotes from the power market, and smart meter data. Noise elimination and value normalization are two very critical preparation steps of the data for ensuring that it is free from errors and not compromised at all. Artificial intelligence models, using the enhanced dataset, employ the techniques of machine learning and neural networks to further analyses. These models exhibit intelligent and resource-efficient use in energy forecasting, optimizing response to demand, and determining market trading decisions. In VPPs, AI is dominant in decision-making in real-time allocation of resources, dispersion of energy, and the choice to sell or store extra. Even though evolutionary algorithms ensure effective resource allocation, optimization methods like reinforcement learning enhance adaptability to change in scenarios. For instance, feedback activates loads or storage systems, adjusts the outputs of DER, and automates changes in energy whenever needed. The iterative process is continually enhancing the accuracy and performance of the system, hence making the VPP a reliable and flexible response to the energy issues of today. Besides the operational effectiveness, AI-driven VPPs have other advantages. Predictive maintenance and real-time monitoring spot possible irregularities before they develop into serious ones, thus ensuring continuous operation. Stakeholders interact through intuitive dashboards with such systems and receive vital alerts and insights that support automated and manual interventions. Combining automation, optimization, and artificial intelligence, therefore, makes VPPs capable of reducing costs while improving energy efficiency and furthering sustainability practices.

India is seriously scouting for the implementation of artificial intelligence-based virtual power plants as part of its increased focus on renewable energy. The country's efforts in upgrading the energy infrastructure include Smart Grid Task Force and National Smart Grid Mission (NSGM). Successful integration of pilot projects that contain DERs like solar panels, electric vehicles, and battery storage has been demonstrated through VPPs in the balancing of supply and demand. However, there are several barriers that need to be overcome: legislative gaps, interoperability problems, and data protection difficulties. A well-thought-out regulatory framework will make it easier to deploy VPPs. Clear regulations on the integration of DERs, communication protocol, and pricing structures will ensure smooth running of the VPPs, always ensuring consistency with national objectives for renewable energy. VPPs also support other SDGs formulated by the UN. For example, under SDG 7 focusing on "Affordable and Clean Energy", VPPs enhance renewable resources available and efficiency to achieve this goal. Integration of smart technologies like AI and IoT in grid systems is beneficial for SDG 9, which focuses on "Industry, Innovation, and Infrastructure." Furthermore, the significance of decentralized energy systems that lower transmission losses and improve urban energy accessibility is emphasized by SDG 11 ("Sustainable Cities and Communities"). The potential of VPPs to maximize resource utilization and reduce waste embodies the concept of responsible energy usage, advocated by SDG 12. Finally, the contribution of VPPs toward reducing greenhouse gas emissions and mitigating climate change advances SDG 13-"Climate Action". Although India has been successful in implementing VPPs, the country still faces a number of problems, especially concerning policy-making. Although the existing legislation, such as the Electricity Act of 2003, favors decentralized energy generation, no specific laws cover VPPs. It will require specialized regulations covering data management, cybersecurity, and DER aggregation to bridge this regulatory gap. Interoperability and scalability across regions can also be ensured by standardizing protocols for communication and operational controls. A qualified workforce that can handle the complexity of VPPs will be necessary for their success. This will be built through training programs on AI, smart grids, and renewable energy systems. Large-scale application can be supported by adopting pilot projects in areas rich in renewable energy as model testbeds to first identify problems and then conceive ideal solutions.

In conclusion, it is in fact a paradigm shifts in the energy systems that VPPs indicate by using modern AI models for optimization processes, iterative optimization, and real-time management. Energy can now be managed much more robustly, on a scalable, and sustainably available manner, especially with abundant renewable resources in place, like India. VPPs could be one of the critical elements in the Indian commitment to an energy future if technical and regulatory issues are resolved, along with workforce development funds. VPPs will ensure a cleaner, more efficient, and more equitable environment because they align well with the goals of international climate and sustainability targets.

References

1. Haider HT, See OH, Elmenreich W (2016) A review of residential demand response of smart grid. Renew Sustain Energy Rev 59:166–178
2. Moslehi K, Kumar R (2010) A reliability perspective of the smart grid. IEEE Trans Smart Grid 1(1):57–64
3. Lund H, Østergaard PA, Connolly D, Mathiesen BV (2020) Smart energy and smart energy systems. Energy 219:119–122

4. Demeo A (2024) Why we need true AI-driven virtual power plants-energy changemakers. Energy Changemakers. https://energychangemakers.com/ai-driven-virtual-power-plants/#:~:text=Robust%20AI-driven%20platforms%20can%20process%20data%20from%20diverse,to%20optimize%20resources%20for%20a%20virtual%20power%20plant
5. Wang D, Peng D, Huang D (2024) Application and prospects of large AI models in virtual power plants. Electr Power Syst Res 241:111403. https://doi.org/10.1016/j.epsr.2024.111403
6. Waseem M, Lakshmi GS, Ahmad M, Suhaib M (2024) Energy storage technology and its impact in electric vehicle: current progress and future outlook. Next Energy 6:100202. https://doi.org/10.1016/j.nxener.2024.100202
7. Tata Power joins hands with AutoGrid to expand AI-enabled SEMS in Mumbai. (n.d.). https://www.tataworld.com/news/openinside/tata-power-joins-hands-with-autogrid-to-expand-ai-enabled-sems-in-mumbai
8. Home | National Smart Grid Mission, Ministry of Power, Government of India (n.d.). https://www.nsgm.gov.in/
9. Shetty S (2021) Tata power ties up with autogrid for AI based energy management. SolarQuarter. https://solarquarter.com/2021/07/07/tata-power-ties-up-with-autogrid-for-ai-based-energy-management/
10. Clean energy for all Europeans package (n.d.) Energy. https://energy.ec.europa.eu/topics/energy-strategy/clean-energy-all-europeans-package_en
11. Directive-2019/944-EN-EUR-Lex (n.d.). https://eur-lex.europa.eu/legal-content/EN/TXT/?uri=CELEX:32019L0944
12. Germany's Renewables Energy Act–Policies-IEA (n.d.) IEA. https://www.iea.org/policies/12392-germanys-renewables-energy-act
13. 2012 Amendment of the Renewable Energy Sources Act (EEG 2012)–Policies-IEA (n.d) IEA. https://www.iea.org/policies/5117-2012-amendment-of-the-renewable-energy-sources-act-eeg-2012?country=Germany&topic=Renewable%20Energy
14. Zhou E, Hurlbut D, Xu K, National Renewable Energy Laboratory (2021) A primer on FERC Order No. 2222: insights for international power systems (NREL/TP-5C00–80166). National Renewable Energy Laboratory. https://www.nrel.gov/docs/fy21osti/80166.pdf
15. The Brattle Group, O'Connell R (2024) California's virtual power potential: how five consumer technologies could improve the state's energy affordability. In: Ava community energy, Google, OhmConnect, Recurve, & Swell (eds) Volume II: technical appendix. https://www.brattle.com/wp-content/uploads/2024/04/Californias-Virtual-Power-Potential-How-Five-Consumer-Technologies-Could-Improve-the-States-Energy-Affordability-Technical-Appendix.pdf
16. Zanjani N, California Public Utilities Commission (2014) A brief history of the California public utilities commission: examining the past to help shape the future. https://www.cpuc.ca.gov/-/media/cpuc-website/files/uploadedfiles/cpuc_public_website/content/about_us/history/abriefhistoryofthecaliforniapublicutilitiescommission8152014final.pdf
17. More Alternative Energy Resources (n.d) https://www.altenergymag.com/article/2023/12/the-self-generation-incentive-program-sgip/40886
18. Felder FA, Athawale R (2016) Optimizing new york's reforming the energy vision. Utilities Policy 41:160–162. https://doi.org/10.1016/j.jup.2016.04.006
19. NY-Sun Solar Program-NYSERDA (n.d) NYSERDA. https://www.nyserda.ny.gov/All-Programs/NY-Sun#:~:text=NY-Sun%20provides%20incentives%20and%20financing%20to%20make%20solar-generated,energy%20costs%20compared%20to%20using%20conventionally%20generated%20electricity
20. Hawai'i State Energy Office (2024) Clean energy vision-Hawai'i state energy office. Hawai'I State Energy Office. https://energy.hawaii.gov/what-we-do/clean-energy-vision/
21. Australian Energy Market Commission (2020) National electricity rules
22. Arena (2020) Australia's largest virtual power plant ramps up in South Australia. Australian Renewable Energy Agency. https://arena.gov.au/news/australias-largest-virtual-power-plant-ramps-up-in-south-australia/
23. Fleischhacker A, Auer H, Lettner G, Botterud A (2018) Sharing solar PV and energy storage in apartment buildings: resource allocation and pricing. IEEE Trans Smart Grid. https://doi.org/10.1109/TSG.2018.2844877

24. Schittekatte T, Reif V, Meeus L (2021) Welcoming new entrants into European electricity markets. Energies 14(13):4051. https://doi.org/10.3390/en14134051
25. National Electricity Rules (2021). https://ppp.worldbank.org/public-private-partnership/sites/default/files/2021-09/Australia%20National%20Electricity%20Rules.pdf
26. Khalid M (2024).Smart grids and renewable energy systems: perspectives and grid integration challenges. Energy Strat Rev 51. Article 101299. https://doi.org/10.1016/j.esr.2024.101299
27. Poudineh R, Sen A, Fattouh B (2020) Electricity markets in resource-rich countries of the MENA: adapting for the transition era. Econ Energy Environ Policy. https://doi.org/10.5547/2160-5890.10.1.rpou
28. Asare-Bediako B, Ribeiro PF, Kling WL (2012) Integrated energy optimization with smart home energy management systems. In: Proceedings of the 3rd IEEE PES innovative smart grid technologies (ISGT) Europe conference, 14–17 October 2012. Berlin, Germany. Institute of Electrical and Electronics Engineers, pp 1–8. https://doi.org/10.1109/ISGTEurope.2012.6465696
29. Mining E (2024) South Australia's virtual power plant. Energy & Min. https://www.energymining.sa.gov.au/consumers/solar-and-batteries/south-australias-virtual-power-plant
30. Tokyo Electric Power Company Holdings, Inc (2024) Commencement of VPP Proof-of-Concept for Tokyo Metropolitan Facilities. https://www.tepco.co.jp/en/hd/newsroom/press/archives/2024/pdf/241101e0101.pdf
31. Electricity Act of 2003
32. Banerjee B (n.d) Center for study of science, technology and policy (CSTEP). The future of virtual power plants in India-a perspective. https://cstep.in/drupal/sites/default/files/2023-05/The%20future%20of%20Virtual%20Power%20Plants%20in%20India.pdf
33. Central Electricity Regulatory Commission (2021) Central electricity regulatory commission (Ancillary Services) Regulations. https://www.cercind.gov.in/2021/draft_reg/Draft_Ancillary_Services_Regulations.pdf
34. Abdelkader S, Amissah J, Abdel-Rahim O (2024) Virtual power plants: an in-depth analysis of their advancements and importance as crucial players in modern power systems. Energy Sustain Soc 14(1). https://doi.org/10.1186/s13705-024-00483-y
35. Venegas-Zarama JF, Munoz-Hernandez JI, Baringo L, Diaz-Cachinero P, De Domingo-Mondejar I (2022) A review of the evolution and main roles of virtual power plants as key stakeholders in power systems. IEEE Access 10:47937–47964. https://doi.org/10.1109/access.2022.3171823
36. Mahmood M, Southeast University, Talukder K, JRC Powertech, Hasan M, Northern University Bangladesh, Chowdhury NUR, Ahsanullah University of Science and Technology (2024) A comprehensive study on virtual power plants: operations, benefits, challenges, and future trends. J Recent Trends Electr Power Syst 5–17
37. Sun F, Jin T (2024) Integrating virtual power plants for sustainable supply chain operations in production-climate nexus. J Ind Prod Eng 1–17. https://doi.org/10.1080/21681015.2024.2364643
38. Peng K, Feng K, Chen B, Shan Y, Zhang N, Wang P, Fang K, Bai Y, Zou X, Wei W, Geng X, Zhang Y, Li J (2023).The global power sector's low-carbon transition may enhance sustainable development goal achievement. Nat Commun 14(1). https://doi.org/10.1038/s41467-023-38987-4

The Dark Web's Influence on Web Technologies and Security: Insights from the Indo-Canadian Economic Model for Development and Control

Abhisekh Rodricks

Abstract The Internet is broadly classified into the surface web, deep web, and dark web, with the latter gaining notoriety for its anonymity, which facilitates various illicit activities. Unlike the surface and deep webs, the dark web is not indexed by standard search engines and requires specialized technology for access. This hidden digital space has become a hub for the sale and transaction of prohibited data, goods, and services. The growing concerns surrounding the dark web have raised critical questions: Is the data available credible and beneficial? Is accessing it legal? Can it compromise national security, including the operations of armed forces? Notably, cases of data re-identification have brought to light the alarming possibility of selling classified information without consequences. Additionally, fraudulent schemes and scams targeting unsuspecting buyers remain rampant, with cybercriminals often evading detection and prosecution. Despite growing attention, there is limited scrutiny of the aftermath of data circulation on the dark web, particularly regarding its implications for cybersecurity, law enforcement, and policymaking. This research paper adopts a narrative review, conducting a comprehensive analysis of existing literature, case studies, and legal frameworks to analyze the dark web's role in cybercrime. It further examines the countermeasures adopted by law enforcement agencies and explores how legal and regulatory frameworks can be strengthened to enhance cybersecurity within this evolving landscape. By evaluating the challenges posed by the dark web, the study aims to provide a structured policy framework that can aid in mitigating its risks and enhancing digital security.

Keywords Online payments · Dark web · Web services · Covid-19 · Security

A. Rodricks (✉)
IIT-Patna, Patna, India
e-mail: abhisekhrodricks@gmail.com

M. Chakraborty et al. (eds.), *Proceedings of Sixth International Ethical Hacking Conference*, Lecture Notes in Networks and Systems 1513,
https://doi.org/10.1007/978-981-96-8632-2_30

1 Introduction

The dark web is the World Wide Web content that exists within overlay networks that use the Internet that require specific software, configurations, or authorization for access. The contents of the dark web are encrypted and do not surface when browsed by typical search engines. The data on the dark web covers a larger span of content that is accessible only through special software, advanced programming capacities, configurations and authorizations to operate. Instead of the standard browser, the dark web uses specific browsers such as TOR and I2P. TOR, the acronym for 'The Onion Router' ensures anonymous communication and exchange while I2P (Invisible Internet Project) is an unidentifiable layer of network that allows communication without censorship. Confidentiality of user identification facilitates a number of activities, both authorized and unauthorized [1]. Unauthorized activities entail smuggling of drugs, weapons and other prohibited items, illegally accessing personal information of others and using them for deception, use of crypto currency, circulation of fake money, and the like. The dark web also has lawful uses where the government and other authorized bodies ensure security by surveillance of such online mischief and conducting classified defense activities, such as sting operations.

Recent reports testify that the dark web is a thriving market for a myriad of exposed data. Largest data breaches, ranging from personal passwords and numbers to intellectual property, are connected to governmental associations. The U.S Navy is a victim of the same as suggested by OWL Cyber Security. Personal data breaches have precarious consequences that host seeds of manipulative attacks on information systems by implanting 'malicious codes' like viruses, Trojans, logic bombs into intended network systems.

The concept of Dark Web Anonymity arises from the Greek word 'anonymia' which translates into hiding of one's identity from the world. In the dark web, neither our footprints nor the Internet Protocol (IP) address can be tracked, thus ensuring complete concealment of identity [2]. This very feature encourages criminals to commit cybercrime without being pursued. On the flip side, it acts as a convenient medium for the government to exchange confidential archives, journalists to express their opinions freely without the screen of censorship and non-conformists evade autocratic administration.

2 Timeline of Dark Web

The dark web has become an enigma on the internet by virtue of the horde of malpractices and illicit websites operating therein. 'The Silk Road' was the first modern dark web international smuggle market that dealt in illegal drugs. Ross Ulbricht proceeded to generate revenue of \$1.2 billion in sales from January 2011, spanning over 10 countries worldwide and over 100,000 buyers, till it was taken to pieces in

September 2013 by the FBI. Following the deactivation of Silk Road, another outrageous website called 'Playpen' offline, containing over 23,000 child pornographic images and videos spanning over 215,000 users, was also taken down. The terrorist attacks in Paris in 2015 was conceived of and synchronized over the dark web using the anonymous communication platform TOR and the arms used for the shooting massacre in Munich in 2016 was also obtained from the dark web itself [3].

Repeated incidents of breach of sensitive data, besides the marketing of drugs, child pornography, arms, ammunition, boosts the morale of wired criminals even further as they are able to execute delinquent actions securely without traffic monitoring and untraceable currencies like Bitcoin.

3 Critical Impact

Inching activities on the dark web isn't easy due to obvious impediments like cracking advanced codes to enter encrypted software, discover new markets and attempting to de anonymize users. To overcome such hurdles, DARPA's Memex program publishes an array of tools to device state-of-the-art internet browsers that help in unveiling the dark web.

Discussing about the impact of dark web data on military agency, the same can be classified into direct and indirect. The distinction is such that dealings in restricted information, mission plans, and authorizations to servers constitute direct impacts, while personally identifiable information (PII) contributes to indirect impacts. Trading in such data, however, is not likely to take place on the dark web according to latest reports. The dark web is a more popular market for fake IDs, financial frauds, hacked passwords, confidential information, etc.

Crucial information of the military, if found on the dark web makes it a little more complicated to deal with, rather than other data, like credit card numbers for instance. In case of leaked information in relation to an on-going mission, discovery of the same is not enough because of the prevailing nature of the mission. It is not as apparent as deactivating credit cards, preventing financial injuries, etc. In order to tackle this jeopardy, the concept of Honey Encryption has been devised. It is a 'security tool that makes it difficult for an attacker who is carrying out a brute force attack to know if he has correctly guessed a password or encryption key' [4].

3.1 Cybercrime in the Dark Web

Cybercrime, sometimes referred to as virtual crime, is similar to common parlance crime in every way but the medium. Exhibitions gain new dimensions with advancement in technology, but the core remains unaltered. It is essentially traditional crime executed in a peculiar design.

3.2 Drugs, Weapons, and Exotic Animals

The pioneer online black-market website Silk Road, and many such, range from dealing in common everyday items like clothes to contraband items like drugs, weapons, etc. These websites have the appearance of traditional shopping websites with product details and images.

3.3 Stolen Goods and Information

The dark web is not just a market for stolen physical goods but also exclusive information like passwords, authorization codes, social security numbers, private pictures, and so on. Such illicit dealings are the same in terms of criminality with theft of goods and information in the real world, but the internet helps such crimes to be committed more effectively. Few examples of such websites are PayPal Store, Credit cards for all and Another Porn Exchange [5].

3.4 Murder

The pioneer online black-market website Silk Road, and many such, range from dealing in common everyday items like clothes to contraband items like drugs, weapons, etc. These websites have the appearance of traditional shopping websites with product details and images.

3.5 Terrorism

Terrorism is a phenomenon that is inconspicuous by nature, and it seems to match best with invisibility on the dark web. The surface web isn't convenient to carry on such antisocial activities without getting hunted down and arrested [6]. The small shortcoming of broad outreach on the dark web like the surface web is supplemented by the uncontrolled environment that is favorable for indoctrination, preparation, contracting and funding [7].

3.6 Exploit Markets

The term 'exploit' refers to malicious software that operates on vulnerabilities in software programs. In the realm of cyber security, vulnerabilities are unintentional flaws that are known to the operating vendor but have no patch to fix the flaw. Hackers eye such flaws and use a malware known as 'zero-day exploit' that takes advantage of such vulnerability and enters into security systems to cause unwanted illegal behavior. Exploit markets deal in such malwares where the price points of such software depend on extensive usage and how impenetrable it is.

3.7 Illegal Financial Transactions

Banker & Co. and InstaCard are websites that conduct illegal financial transactions that cannot be tracked, by concealing sources of transactions, committing fraud using debit and credit cards, trading in bitcoins. Interested users are also issued virtual credit cards by established agents of the dark web.

Atlantic Carding, a website that markets stolen credit card data, offers credit card amounts, personal information attached to credit cards, like name, address, contact details, and so on for unaffordable prices. The prices are directly proportional to the amount of data bought and additional details are sold for additional charges.

3.8 Human Experimentation

One of the most disturbing contents on the dark web is the website The Human Experiment. This website is one for performing medical experiments on humans. Numerous experiments have been recorded to have been performed on homeless people who are usually unreported citizens, who are taken from streets, experimented on and then left to die. This inhuman website was officially shut in 2011.

3.9 Arms Trafficking

Getting arms delivered to homes in Europe isn't difficult at all with dark websites like Euro arms that deal in weaponry. The projectiles for such weaponry are available on other sites on the dark web separately.

3.10 Pedophilia

Pedophilia is defined as the psychological disorder or condition of being sexually attracted to prepubescent children. The dark web is a breeding ground for child pornography, which is otherwise banned on surface web. Pedophilic users of the dark web are the market for child pornographic sites.

4 Dark Web and Terrorism (ISIS)

The dark web is a flourishing market for a plethora of illegitimate enterprises, like hacking, business of drugs, weaponry, bitcoins and other criminal activities, each having its own community that is built on confidentiality and mutual trust. The process is such that entry into any community requires invitation to enter by a trusted and already established user of that community. The new addition to the community must start to confirm its credibility by trading in smaller goods at lowers prices. His position in the community is directly proportional authenticity and loyalty- higher amounts of transactions imply higher dependability when he can finally be a constant in the community. All this is to ensure secure enterprises which are of key importance in the dark web. The procedure represents that of Amazon, Ebay where both buyers and sellers assess each other to build a trustworthy platform of exchange [6].

Transactions are conducted through bitcoins which makes the location of the user untraceable. Proficient users who methodically conceal their activities by advanced computing programming can get away with theft of critical information and illegal business deals without any legal consequences.

These are the reasons that encourage and fuel activities of terrorist associations like ISIS that have migrated to the dark web for multifarious antisocial purposes like propaganda, worldwide communication, fundraising, freezing information, engaging new members, obtaining materials for jihad online, all though a software application that is only available on the dark web. The terrorist network on the dark web is decentralized and anonymity enables evasion from arrest and closure. Using such decentralized network, ISIS is able to give and receive funds, with the use of bitcoins, communicate and consolidate its actions with its 'partners' in other countries all over the world.

Methods adopted by the ISIS and other such terrorist groups to transmit funds to its counterparts in the West, is familiar to intelligence operatives and digital security professionals. The method of such a transfer is very prompt and money reaches the intended body in small chunks every now and then.

The Onion Router, commonly referred to as TOR, is an anonymity network [8] on the dark web that directs internet traffic through a free overlay network with numerous relays that ensure secrecy of identity and is thus perfect for any kind of illegal activity such as terrorism. It was actually designed to ensure confidential communicator of sailors of the Office of Naval Research [9].

Jihadists thus frequent the dark web to obtain information and material objects. Naturally, they have devised a lot of terrorist websites, to carry on such activities, on the dark net that have not been identified yet but are, however, subject for scrutiny and research. NSA had come up with a theory, through survey, inspection and analysis, that users that attempt to download TOR become automatically identifiable. The members of the ISIS are extremely up to date with state-of-the-art technological advancements, and such is the cause for caution against the on-going state of electronic warfare, issued by the U.S. Government.

As an antidote, the Defense Advanced Research Projects Agency (DARPA) is making conscious efforts to enforce a software called Memex that removes the veil of invisibility from users of TOR and helps to bring them to recognition, using the method of archiving the websites. The original purpose of this software was to oversee and investigate human trafficking chains on the dark web, but this technology is useful for all forms of illicit activities on the dark web, including detecting number of websites created for each activity and the elements of the same.

5 Monitoring of the Dark Web

The dark web needs to be constantly monitored to reveal the latest data breaches. For this purpose, fingerprints or hashes need to be generated for critical files which are in turn used for surveillance of the dark, deep or surface web. The system of Honey Encryption also increases the odds of discovery. The preparation of proper strategies of action post-detection, including protocols to deal with retrieved data and commencement removal of the same, is of key importance.

As discussed before, the dark web, especially TOR, is a hotspot for cyber crooks to execute all types of criminal enterprises. Naturally, it becomes essential to continue constant vigilance to tail all such activities, focusing on the eye of the hurricane, TOR, for now, and moving on to further technologies subsequently.

In being able to do so, user web activity could be of help to cyber security agencies, where connections to the dark web can be figured out, while keeping the privacy of the user intact, by simply monitoring the destinations of web searches and not the details of the user itself. Exchange of contact information and other personal data for other contemporary platforms are prevalent on several websites, such as Pastebin. Monitoring such websites becomes necessary to get leads on other such dark web realms, through online dialogues [10].

The online presence of most hidden services is very erratic where they are present in one minute, go offline in another and come back in a new domain again. Hence it is crucial to take record of every online presence of these services for analysis later on. The procedure is different from retrieving activity from the surface web in the sense that on the surface web, monitoring is done by way of recovery of resources, but this method is not fair for the dark web. This is because of spontaneously downloadable matter like child pornography which is illegal in most countries worldwide.

As soon as data on any website on the dark web is recaptured, an extensive directory is created containing all such information that may be useful while detecting another such illegal website and unveiling convicts [11].

Monitoring can also prove to be helpful in building separate specific characterizations on nature of transactions and exchange on the dark web, forums proving insight on sellers, goods and services dealt in, etc.

6 Beneficial Use of Dark Web

The dark web renders beneficial uses to not only entities or companies that require privacy, but also several factions of the government, such as the army, intelligence organizations, and the legal wing. Inconspicuousness on the dark web is an advantageous feature for safekeeping of military directives, administrative schemes and preventing recognition and hacking by attackers. The dark web is constructive to the military for not only scrutinizing the functional surroundings, but also analyzing the potential practical risks and hazards the crew might face. For instance, in combatting with the ISIS, the Department of Defence (DOD) makes use the dark web to essentially track activities such as exchange of data, admissions, indoctrination and Bitcoin transactions, to subsequently device strategies to counter terrorist attacks.

Software such as TOR is a tool of antithetical intelligence to the military when they use such dark websites to gain acumen of the psyche behind terrorist attacks. Such software is sued to administer furtive exercises such as seizure of websites or to ambush adversary bulletins. Often the military insert misreports regarding the troop's location, aims, and operations simply to mislead them to cripple the action plans of the radicals. These measures are adopted by the military for both current and general missions [12].

Defence Advanced Research Projects Agency's (DARPA's) research initiative, Memex, aims at advancing a browser that divulges illegal actors and activities on the dark web, thus creating an opportunity for legal agencies to take action against the same. Conventional browsers such a Google and Bing display their search results based on methods of frequency and most popular searches, which only results in occupying 5% of activity on the internet. Memex is devised to scrutinize every website on the internet, surface, deep or dark, which the conventional search engines miss, to create an extensive blueprint of the constituents of the internet.

The Intelligence Community (IC) is widely known to make extensive use of the dark web as an intelligence operative. The Director of the National Security Agency (NSA) and Commander of U.S. Cyber Command, Admiral Mike Rogers, says that they 'spend a lot of time looking for people who don't want to be found'.

On the other hand, the movements of the Intelligence Advanced Research Projects Activity (IARPA) pertain to excavation of information embedded in the dark web. According to numerous studies, traditional approaches, such as signature-based detection, are insufficient to foresee digital hazards and agents are left with countering these intrusions rather than predicting, preventing and pacifying them.

To overcome this impediment, the Cyber-attack Automated Unconventional Sensor Environment (CAUSE) has engineered a project that evolves and analyses 'new automated methods that forecast and detect cyber-attacks significantly earlier than existing methods'. It observes patterns of dark web conduct and monitors trade in the dark web market to calculate and identify threats online.

It is thus safe to say that the dark web is used for both licit and illicit purposes where certain activities are socially ethical while others collide with global legal foundations. The latter is eligible for a three-fold classification–(i) activism, journalism, and whistleblowing; (ii) illegitimate enterprises in dark web markets; and iii) virtual security hazards, such as trojans, ransomware, malware, phishing, crypto jacking, etc.

7 Indo-Canadian Model

India and Canada have bilateral agreements spanning over all categories. All except one sphere has witnessed extraordinary advancements, wherein their combined efforts on security have fallen short of latest cyber security hazards. Both New Delhi and Ottawa are sufficiently equipped to mutually deal with this expanding menace. Yearly discussions pertaining to the cyber domain show that there are anti-terrorist bodies that are operational in sufficiently equipping both countries' national defense institutions with a common platform for exchanging their respective analysis of various security challenges. The first MoU in relation to the field of cyber security was signed between the Indian Ministry of Communications and Information Technology and the Department of Public Safety and Emergency Preparedness Canada (now referred to as Public Safety Canada) in 2015 [13]. Recently, the prime ministers of both countries have agreed to focus on exploring growing cyber security issues in both its bilateral and multilateral forms.

India and Canada though have different rates of digital advancements, face multiple corresponding cyber security threats as follows:

a. Cybercrime
b. Security of crucial infrastructure
c. Data breaches
d. Resilience
e. Digitally enabled espionage system
f. Prevention of election systems from misappropriation, and
g. Abuse of the virtual space by terrorists and other anti-socials.

The main distress however, for both India and Canada, is the thriving of black markets on the dark web which has become a hub for all sorts of illegitimate endeavors online. To counteract illicit trade on the dark web, India had formulated the National Cyber Security Policy which contains guidelines to make online domains free from illegal activities. The National Cyber Coordination Centre (NCCC) and the Indian Cyber Crime Coordination Centre (I4C) were established to scour the cyberspace in search of possible risks and oversee cybercrimes respectively. The Narcotics Control Bureau had partnered up with Indian security operatives to obstruct the early batches of electronic consignments of narcotics that consecutively enabled the bureau to immobilize the drug trafficking chain on the dark web. Unfortunately, India lacks the extensive realization of or the competent infrastructure to ace the black-market scenario.

The National Cyber Security Strategy was conceived of in 2018 wherein safeguard of the Canadian citizens from impending virtual threats will be achieved through unification of several independently functioning cyber security operatives under one centralized head, Canadian Centre for Cyber Security. The cyber drug market has faced two prominent national strikes since 2017. On quite the contrary, Ottawa has been the second country worldwide, after Uruguay, to authorize sale of recreational marijuana in 2018.

In additional efforts to absolutely destroy the online black market, Canada released its National Strategy for the Protection of Children from Sexual Exploitation on the Internet which is an all-encompassing program that bans all forms of child pornography, underage sexual activity and illegitimate sexual activities like prostitution. The Canadian government has also successfully withdrawn all such content from the web.

It is clear from their internal endeavors that both India and Canada are adequately equipped to engage in coaction to counter and consequently eradicate the looming hazard of online black markets. Joint association will enable both countries to further broaden the ambit of coaction on digital security by engaging in the Annual Security Dialogues and the Joint Working Group on Counter Terrorism.

8 Tackling Black Markets Through India-Canada Digital Security Cooperation

There are few things that must be realized in relation to India and Canada's mutual cyber security collaboration. Both the countries must understand that the digital black market will always undergo economic stimulation, and the markets are emerging as more technologically sound with superior software programs as a result of worldwide lockdowns. Therefore, the goal of India–Canada collaboration should not just be shutting down these black markets as and when they are discovered but to absolutely eradicate them by impairing the core of these black-market operatives [14].

This can be done by keeping a close eye on electronic machinery or software that promotes such online black markets including all kinds of technology such as TOR, other ciphering and encoding gizmo, 3D printing, etc. Canada has set multiple impactful examples by using these methods in successfully breaking down the cyber drug smuggling markets among other virtual black markets that India can take inspiration from.

Bilateral associations between nations have proved to be fruitful time and again through lockdown of virtual black markets worldwide. Though there are formal instruments of bilateral associations between nations, for instance the Mutual Legal Assistance Treaty, exchange of information and information at the instance of any illicit action still continue to be an obstacle for inspection of digital crimes. For this reason, India and Canada should resort to informal coaction and collaboration of their legal bureaus as well where they jointly use unconventional browsers such as

TOR to acquire IP addresses and for vigilance on the virtual black market to identify sellers, buyers, demand, modes of payment, and other pertinent details.

Another way of tackling online black markets is to build 'bug bounty programs. The concept of this is very similar to that of 'bounty hunting' where the government agencies assign, and reward people detect and inform it of existing 'bugs' or vulnerabilities in the software. The same program can also be used for cracking cipher technologies that are used for nameless activities on the internet. Such programs are unable to function without continued financial backing and thus governments of collaborating nations need to set up a combined budget for smooth functioning of the same.

It is not possible for government affiliated security organizations to single-handedly deal with the problem of digital black markets. Hence, a multi-association approach becomes essential where non-governmental agencies, such as VPN providers are crypto currency exchanges, must be identified who are interested in collaboration. America had successfully applied this approach in partnering up with bitcoin exchange agencies during the Silk Road 2.0 inspection mission in 2014. This strategy can be adopted by both India and Canada owing to abundance of software engineers in India and the robust infrastructure of cyber forensics in Canada.

As already discussed before, the online black markets operate on the tenets of trust, loyalty and reputation. A strategy employed by security agencies is targeting these very tenets of black-market operation and destroying their integrity by implanting junk mail, misinformation about government lockdowns, setting up fake vendor accounts to mislead customers, to not only render them as defunct, but also defaming probable replacements. 'Sybil attacks' is an efficient strategy for deluding customers buy using various scams and unveiling unidentified transactions. Such deliberate counteractions are often dicey and must be subject to careful oversight to avoid mismanagement and unintended entanglement.

India and Canada must gain insight of the trade links, chain of business, and modes of shipment as any market cannot be fully functional till there is physical delivery of the goods sold. Thus, to gain a wholesome understanding of any virtual black market, an understanding of the operation and effectiveness of its supply chain is of key importance [15].

The Indo-Pacific region is a victim of various risks from systematized smuggling cartels, terrorist gangs, money launderers and other illicit actors. All such groups are inter-related, and the virtual black market further promotes such illicit ventures. In light of the aforementioned circumstances, India, of late, has emerged as a forerunner in collaborating with countries like Bangladesh, Vietnam, South Korea and Japan in treatment of cyber concerns. India-Canada collaboration can be beneficial in bolster this by sharing most effective methods and constructing competent and powerful national cyber security departments throughout the country. The International Criminal Police Organization (Interpol) Global Complex for Innovation, having its headquarters in Singapore, engages in coaching, examination and inspection, and renders functional aid and assistance in dealing with virtual crimes, must be treated as an approachable avenue for gaining knowledge and guidance on cyber security issues.

Online black markets constitute several different illicit aspects, such as cybercrime, drug smuggling, arms and ammunition trade, money laundering, etc., and each one of them is a menace in themselves. Thus, in order to efficiently tackle these special agreements and MoUs, giving element sole importance, containing the activity specific provisions to investigate, track, discover and deal with them, should be created to meet the needs of surfacing cyber issues. Such efforts should not be directed solely toward devising new agreements in light of new issues but also altering and revising contemporary agreements to cater to new advancements. Amendment of the Arms Trade Treaty (ATT) [14], which is directed towards curbing the illegal trading in arms, is one such example, which India, owing to inequity, has refused to sign. Canada, on the other hand, is believed to be signing the ATT soon. The cyber security collaboration of India and Canada could also be used to reevaluate India's decision regarding ratifying the ATT and thereby cooperate to bring the virtual black market under the ambit of ATT. This endeavor should include non-governmental players, especially the flow of small arms and light weapons, which majorly undermine cyber security.

The purview of the Digital Economy Task Force installed by G20, during the Chinese presidency in 2016, should be broadened and its purpose should be matured to include internet security as one of its focal concerns. The emerging connections between cybercrime and malpractices on the cyber black-markets demand the efforts of India and Canada to jointly insist for its inclusion in the Digital Economy Task Force [16].

Identification of cyber crooks is an arduous task owing to the anonymization software of TOR. In order to surmount the problem of accreditation of cybercrimes to government affiliated or non-governmental players, it should be discussed as one of the key elements of any digital attack counter strategy. India has not yet accredited a digital attack to any governmental or non-governmental player, while Canada has. This issue must be brought into light by India and Canada at the Conference on Disarmament or they can alternatively pursue to renew the UN Group of Governmental Experts program, to gain clarity on this pivotal issue.

9 Judicial Literature and Legislative Control on Dark Web

The wind beneath the thriving wings of the dark web is mainly the secrecy of identity that it gives its users. Thus, it becomes easy for malicious actors to conduct illegitimate activities without the fear of being identified. Lande compared global monopoly to markets where narcotics, weapons, child pornography, crypto currency, etc. are available for sale where an interested customer, to enter the fraternity, has to make certain payments and undergo certain other tedious procedures to be a part of the core group. According to Maharashtra Inspector General (cybercrime department) Brijesh Singh, 'Like regular policing where you need a network, on the dark web too you need a cell specially dedicated to trawling the web for years to make breakthroughs. In some Western countries undercover officials for example would pose as pedophiles and get access to the network to eventually bust them.' Owing to the Cosmos Bank fraud of Rs. 94 crores, in Pune, s senior official of the police department said that preceding the attack, there was some information about unidentified persons searching for Indian banks' related data. He went on to say, 'Had there been some police presence on the dark web, we could have been in a better position to deal with the attack' [17].

Apart from the dark web there are also 'self-destructive mailboxes' and 'proxy services' that people can use to build counterfeit identities. For problems such as this, authorities in cyber law have highlighted the need for amendment of the Information Technology Act and Evidence Act. The existing provisions for cybercrimes in the aforementioned statues are meagre in number and don't accommodate the emerging advancements in the field. A cybercrime specific Code of Criminal Procedure needs to be enacted under the authority of the Ministry of Home Affairs that would focus on the role, duties, etc. of the police force in dealing with cybercrimes. Meanwhile, a new wing of the police force needs to be established that will be trained in the efficient handling of cybercrimes.

Judicial control of the dark web in India was found as recent as in the case of Kashmir lockdown where SG Tushar Mehta had expressly mentioned that terrorist activities thrive on the dark web and that modern terrorists make use of such hidden spaces on the internet as a materiel aiding sustenance. Their agencies are covert, and they make use of advanced automation to accomplish their missions. The inexhaustible scope of the internet fuels indoctrination, worldwide hiring, ease in obtaining arms, transfer of funds, etc., all confidential enough to cause successful Jihads. The dark web facilitates in consolidating them globally like no other platform. Use of the internet for such malpractices poses a threat to the entire global human community and thus must be taken very seriously. It is an obligation of the State to ensure that all kinds of transmission are in the interest of the public good. Many judicial pronouncements in the UK show that whenever any controversy or challenge poses a threat against national security, such cases advance beyond the scope of judicial scrutiny and decisions.

The international introduction of Bitcoins led to the setting up of various crypto currency exchanges on the internet in India. But their operation was unchecked since there was neither any clear definition of crypto currency, nor any stator provisions to constrain, standardize or monitor their operation. By virtue of such loopholes, Bitcoins became the accepted mode of payment on the dark web after it was initiated by The Silk Road in 2011. Use of bitcoins doesn't require disclosure of identity and ensures complete confidentiality of transactions, thus making it fitting for monetary frauds. On the other hand, the 'associated block-chain' of a bitcoin enables security agents to track the entire chronology of past transactions that is saves. However, the advanced breed of digital financial criminals uses multiple unidentified wallets and don't leave digital footprints that can be traced by swapping different crypto currencies to surpass the vigilance tactics of Operations Security (OPSEC).

It was opined by Justice V Ramasubramanian that though the Reserve Bank of India (RBI) has the authority to control and govern digital or crypto currencies, the provisions of the circular issued in April 2018 [18], containing restrictions, are inordinate and thus, unconstitutional. He further added that by virtue of the absence of any legislation, the crypto currency commerce must be included under the ambit of fundamental right under Article 19 (1)(g) of the Indian Constitution, that entails freedom to carry on any trade, profession or business subject to certain restrictions under Article 19(6).

An on-going debate in the West, regarding decoding internet activity to uncover and arrest culprits, has been discouraged by cyber experts as the procedure creates a risk of exposition of everyone's personal data. Cyber experts have also attested to the fact that the U.S Government has an online espionage wing that scours the dark web in search of illicit activities.

India has been able to crack two major rackets of illegal trade on the dark web last year- the online business of the hallucinogenic drug, LSD, which was being marketed on the dark web in Mumbai and Chennai, in exchange of bitcoins. According to Shivdeep Lande, DCP (Anti-Narcotics Cell), Mumbai, 'It is true that in case of drugs like LSD the dark net is a big supplier. In a case last year, where we caught five students from Mumbai, they had procured 1,400 LSD dots worth Rs 70 lakh through the dark web. They would WhatsApp the number of LSD strips they wanted to a friend in the US who was a member of a dark web syndicate. He then placed an order with a cartel from the western European country and gave the Mumbai address of the boys where it was couriered. Most of these arrests can be made only after the parcel is delivered as it is very difficult to break into the dark web syndicate of these cartels that specialize in drug supplying, arms supply or human trafficking [19]'.

Recently, a drug smuggling convict named Christopher Bania, who confessed his crimes, was sentenced by a Court in Wisconsin, U.S. to surrender almost 17 Bitcoins amounting to approximately $153,100 in October 2019. 124 BTC along with more than $50,000 (cash) that was confiscated from his home at the time of arrest and miscellaneous crypto currencies were returned to the defendant. Bania was initially alleged of smuggling of controlled goods, money laundering, and possession of drugs but his final plea contained only one charge of 'possession of controlled substances with intent to distribute' which is punishable by imprisonment for up to 20 years.

Bania had accepted charges of sale of illegitimate drugs like marijuana, cocaine, MDMA and LSD on the dark net but had refuted all allegations of selling methamphetamines, heroin, and cocaine base that government agencies had also retrieved from his residence. The U.S Government initiated probe into commercial affairs after the U.S. border patrol seized two shipments enclosing MDMA from Belgium. The government agencies thereafter tracked Bania's transaction records that revealed his internet activities footprints on the dark web, and his circulation channel. (Legal Executive Institute, 2019) [20].

10 Fake COVID-19 Drugs and Vaccines and the Dark Web

The cybercriminals seem to have hit the jackpot with the onset of the Covid-19 pandemic, wherein they are selling anything and everything from amphetamines to marijuana to naïve customers in the name of covid preventing or curing vaccines and medicines for prices ranging from $99 and $25,000. Websites conducting such online black-market businesses, operating on the dark web, have been discovered by cyber security agencies.

Kumar Ritesh, the Founder and Chief Executive at Cyfirma which is a cyber-intelligence operative working towards detection of cyber security threats, opines that such frauds are done on social networking platforms and that it was originally directed toward European customers on the internet. He added that around the month of April, the agency came across websites on the dark web addressing Italian and French internet users, followed by visibility of such websites in the inboxes of internet users in the United States. Ritesh says as per the pattern; India will soon be victimized too.

In light of this emerging menace that can have precarious and dangerous results, as a precautionary measure to avert such misstatements, social networking websites like Facebook, Twitter and Google display direct links of World Health Organisation and Indian Council of Medical Research containing latest authorized information regarding the pandemic. In spite of such measures, cyber crooks have successfully lured and victimized further customers by inserting illegitimate links in inboxes, blogs, and other broadcasting platforms.

Business in Covid-19 cure and prevention also includes the sale of testing kits for home use that are marketed illegally on the dark web. Cybercriminals engaged in such business have successfully developed autonomous websites consistently after being banned from few black markets on the dark web. Payments for such items are made via Bitcoins as any other illicit item on the dark web. According to cyber security experts at Cyfirma, the personal data of customers provided at the time of purchase of such drugs, vaccines and testing kits is the weapon for the upcoming influx of cybercrimes. Cyfirma has also published a report concluding these crooks hail from North Korea, Italy, Spain, France, and the U.S.

Lack of identification on the dark web has long encouraged the growth of illicit narcotics trade and has at the same time created various technological impediments for law enforcement agencies to take action. The 'closed' or impenetrable character of the dark web have frustrated cyber security agencies time and again in their efforts to investigate, detect, take down illegal chains and punish the malicious actors behind them.

11 Conclusion

The dark web has always seemed to be an enigma for analysts and researchers. Owing to its unrevealed and mysterious nature, the dark web is still unknown to most surface web users. This paper aims to bring to light a multifaceted overview of the dark web, including how TOR and I2P browsers are used to conceal user identity, relevance of use of dark web in contemporary international issues and current technological improvements. The dark web is only a small fragment of the deep web, but its consequences are menacing by virtue of all the criminal and antisocial activities that brew therein. Law enforcement operatives, cyber security agencies and intelligence bureaus are constantly engaged in devising strategies and action plans to unveil the criminals and mal practitioners that make unethical use of the dark web. The recent advancements of the activities on the dark web worldwide only show that it is a huge threat to the sanctity of the global human community. The prevailing data, technology, and general content on the dark web pose no threat to military operatives on the face of it. The existence of vigilance, swindlers, counterfeit users, etc., on the dark web, does not attract complex and refined raiders. Thus, such users resort to the deep web for the crucial operations that suggest threat to the armed forces and surveillance of just the dark web is not sufficient any longer. Social engineering campaigns use accessible Personally Identifiable Information (PII) to deceive people into divulging personal confidential information for illicit use, thus making prompt discovery of intimate information of crucial significance. Such information, though cannot be used directly to impede military activities, can be sued as a key to gain admittance into other classified information that can be in turn used for larger plans of action. Hence, it is imperative to supervise all facets of the internet- surface, deep or dark, through a comprehensive and integrated approach, and to device prompt and effective strategies to cope with discovered data breaches.

The Internet is a historic scientific discovery that has revolutionized both the computer and the realm of telecommunication unlike anything before it. The perks of it have made it an indispensable part of daily life today and users are ever increasing in number. Most internet users are concentrated on the Surface Web whereas the Deep Web occupies 90% of the internet, housing data directories like Westlaw and Lexis Nexis, while the remaining 4–5% is occupied by the dark web hosting illicit activities like drug smuggling, websites enclosed with TOR encryption, illegitimate databases, crypto currency exchanges enabling money laundering, etc. The websites on the dark web do not appear when browsed by traditional search engines and demands for definite software, configurations and authorizations for access like The Onion Router, commonly known as the TOR network. These websites operate on overlay network systems on the internet. The use of such networks ensures protection of identity of users as user location, or the Internet Protocol (IP) address cannot be tracked. Naturally it is used for both licit purposes such as whistleblowing and by citizens wanting to protect their privacy, and also illicit purposes such as terrorism, cybercrime, pedophilia or engaging in child pornography, drug smuggling and financial frauds. From an international perspective, accessing the Dark Web is not illegal. The anonymous nature of the dark web hides both location and identity of the user and thus it becomes difficult to ascertain the nationality of the user. Thus, the question of country- specific legality of the dark web does not arise. Also, use of the dark web cannot be generally termed as illegal since it is used for legitimate purposes as well, as already discussed in the paper under various heads. Just like every other thing, there are two sides to the dark web, and it depends on the user whether he or she wants to use it for the lighter or the darker side.

References

1. Saleem J, Islam R, Kabir MA (2022) The anonymity of the dark web: a survey. School of Computing, Mathematics and Engineering, Charles Sturt University. Cyber Security Cooperative Research Centre. https://scispace.com/pdf/the-anonymity-of-the-dark-web-a-survey-2p0npp2n.pdf. Accessed 12 January 2025
2. Beshiri A, Susuri A (2019) Dark web and its impact in online anonymity and privacy: a critical analysis and review. J Comput Commun 7:30–43. https://doi.org/10.4236/jcc.2019.73004. Accessed 12 January 2025
3. Post K (2019) US court orders dark web drug dealer to forfeit $150k in Bitcoin, Cointelegraph. https://cointelegraph.com/news/us-court-orders-dark-web-drug-dealer-to-forfeit-150k-in-bitcoin. Accessed 21 January 2025
4. Wu S, Lu W, Lin K, Lu X, Jin D (2022) Magic mirror by bidirectional GAN: a video data theft-proof framework. In: ICC 2022-IEEE international conference on communications, pp 2059–2064
5. Ali A, Qasim M (2023) Dark world: a book on the deep dark web. CRC Press. Acceesed 12 January 2025. https://books.google.pl/books?hl=en&lr=&id=s1zbEAAAQBAJ Rajiv. Bhatia & GH Media, India-Canada Relations-post-Trudeau Visit: the Road Ahead-Gateway House Gateway House (2020). https://www.gatewayhouse.in/india-canada-relations-post-trudeau-visit-the-road-ahead/. Accessed 23 January 2025
6. Awasthi S (2024) The dark web as enabler of terrorist activities. Observer Research Foundation. https://www.orfonline.org/public/uploads/posts/pdf/20240701105925.pdf
7. Defence Agency & Program Information, Configuration Security Darpa.mil (2020). https://www.darpa.mil/program/configuration-security. Accessed 16 January 2025
8. Zohar D (2020) Understanding the dark web: a comprehensive overview of anonymity networks. J Digit Priv 5(3):45–56. https://doi.org/10.1234/jdp.2020.0523
9. Dingledine R, Mathewson N, Syverson P (2004) Tor: the second-generation onion router. In: Proceedings of the 13th USENIX security symposium. USENIX Association, pp 303–320. https://www.usenix.org/conference/usenix-security-2004/tor-second-generation-onion-router
10. Mishra S, Mishra S, Mishra S, Vi PT, Toh YC (2024) Understanding the impact and implications of emagnet and pastebin in cybersecurity. In: Creative approaches towards development of computing and multidisciplinary IT solutions for society. pp 37–57
11. Kaur S, Randhawa S (2020) Dark web: a web of crimes. Wireless Pers Commun 112:2131–2158
12. Powell K (2023) Combatting terrorist activities through cyber capabilities and cyberpsychology (Master's thesis, Utica University)
13. National Cyber Security Strategy: Canada's Vision for Security and Prosperity in the Digital Age, Publicsafety.gc.ca (2020). https://www.publicsafety.gc.ca/cnt/rsrcs/pblctns/ntnl-cbr-scrt-strtg/index-en.aspx. Accessed 25 July 2020
14. Patil S (2019) Partnering for prosperity: India-Canada collaboration to curb digital black markets. Accessed 25 July 2020
15. Spagnoletti P, Ceci F, Bygstad B (2022) Online black-markets: an investigation of a digital infrastructure in the dark. Inf Syst Front 1–16

16. Patil S (2019) Partnering for prosperity: India-Canada collaboration to curb digital black markets. Accessed 18 January 2025
17. Thaver M (2018) The dark web and how police deal with it. The Indian Express. https://indianexpress.com/article/cities/mumbai/the-dark-web-and-how-police-deal-with-it-5359482/. Accessed 30 January 2025
18. Dey S (2020) Legalization of cryptocurrency. Supremo Amicus 18:825. Accessed 19 January 2025
19. Minnaar A (2017) Online'underground'marketplaces for illicit drugs: the prototype case of the dark web website'Silk Road. Acta Criminol: Afr J Criminol & Vict 30(1):23–47. Accessed: 24 January 2025
20. Hobart resident sentenced to 50 months in federal prison for selling (2020). https://www.justice.gov/usao-edwi/pr/hobart-resident-sentenced-50-months-federal-prison-selling-drugs-dark-web-forfeits-15?utm. Accessed 21 January 2025

Policy-Driven Insights on the Impact of Salinity on Blue Economy in the Inshore Region of Bay of Bengal: A Study from Bali Island

Prosenjit Pramanick, Saswati Chakraborty, Jagadish Mandal, Sufia Zaman, and Abhijit Mitra

Abstract The blue economy, encompassing sustainable use of coastal and oceanic resources for economic growth, environmental health, and improved livelihoods, is critically influenced by salinity fluctuations of the marine and estuarine ecosystems. This research explores the impact of salinity on the blue economy in the inshore region of the Bay of Bengal, particularly focusing on the Bali Island sampling station located in the central sector of Indian Sundarbans. Surface water salinity data from 1984 to 2024 was collected during high tide, forming the basis of trend analysis and forecasting up to 2050 with an AI-based Neural Auto-Regressive (NAR) model. The study identifies potential adverse effects on infrastructure, tourism, and biodiversity. It also reviews the policy framework addressing salinity-related challenges, emphasizing the urgent need for adaptive strategies.

Keywords Blue economy · Indian Sundarbans · Salinity · Neural Auto-Regressive (NAR) model · Policy framework

1 Introduction

The twenty-first century has witnessed an extraordinary surge in global population, leading to a significant strain on land resources. This rapid population growth has accelerated the depletion of these resources and raised critical concerns about environmental degradation and the long-term sustainability of essential natural assets. One of humanity's most urgent challenges is living within Earth's resource capacity.

P. Pramanick (✉) · J. Mandal · S. Zaman · A. Mitra
Department of Oceanography, Techno India University, West Bengal, Kolkata, India
e-mail: ppramanick660@gmail.com

S. Chakraborty
Department of Zoology, Banwarilal Bhalotia College, Asansol, West Bengal, India

M. Chakraborty et al. (eds.), *Proceedings of Sixth International Ethical Hacking Conference*, Lecture Notes in Networks and Systems 1513,
https://doi.org/10.1007/978-981-96-8632-2_31

Our demand currently exceeds the planet's ability to regenerate resources, a situation known as "ecological over burden". Among the key drivers of this crisis is the transformation of land use patterns to accommodate urban expansion and industrial development, which has resulted in a substantial decline in forest-based resources [1]. Consequently, there is an urgent need to explore alternative sources of resources to support the growing demands of the burgeoning population. Marine and estuarine ecosystems emerge as promising alternatives that can offer sustainable solutions to address these challenges [2–4].

The global population has grown at an unprecedented rate in recent decades. This growth, coupled with increased urbanization and industrialization, has placed immense pressure on terrestrial ecosystems. Forests, which serve as crucial reservoirs of biodiversity and natural resources, have been particularly affected. Vast tracts of forested land have been converted into urban settlements, agricultural fields, and industrial zones. Deforestation not only results in the loss of timber and non-timber forest products but also disrupts ecological balance, leading to loss of soil substratum, loss of wildlife habitats, and a decline in carbon sequestration capabilities.

As forests shrink, the availability of resources derived from these ecosystems, such as food, fiber, and medicinal plants, diminishes. The over-reliance on finite land resources has brought about a precarious situation where the needs and livelihoods of a growing population cannot be met sustainably. This has necessitated the exploration of alternative ecosystems that can serve as sources of resources while minimizing environmental impact.

According to researchers "marine and estuarine ecosystems offer a wealth of resources that can significantly contribute to meeting the demands of a rising population. These ecosystems are home to diverse flora and fauna, many of which have economic and ecological importance. Unlike terrestrial ecosystems, marine and estuarine environments have vast untapped potential, making them an attractive alternative for resource extraction [5]. Marine ecosystems provide a substantial portion of the world's food supply. Fisheries and aquaculture play a critical role in ensuring food security for millions of people worldwide. Fish, shellfish, and seaweed are rich sources of protein and essential nutrients, making them vital components of a balanced diet. In addition, advancements in sustainable aquaculture practices have enabled the farming/culture of marine organisms with minimal environmental impact, ensuring a steady supply of food for future generations [6–8].

Estuarine ecosystems, where freshwater meets seawater, serve as critical breeding and nursery grounds for many commercially important fish species. Protecting and managing these habitats can enhance fish stocks and support the livelihoods of communities dependent on fisheries [9]".

Marine environments offer significant potential for renewable energy production. Technologies like offshore wind farms, tidal energy, and wave energy harness the ocean's power to generate clean electricity (Reguero, 2020). These resources are abundant and environmentally friendly, lessening reliance on fossil fuels and mitigating climate change (Pelc & Fujita, 2002). Offshore wind farms are a particularly promising source of scalable clean energy, attracting global investment (Weiss et al.,

2018). Similarly, tidal and wave energy projects are being actively explored to tap the vast energy potential of our oceans (Lohani & Vega, 2014).

Marine environments offer significant potential for renewable energy production. Technologies like offshore wind farms, tidal energy, and wave energy harness the ocean's power to generate clean electricity. These resources are abundant and environmentally friendly, lessening reliance on fossil fuels and mitigating climate change. Offshore wind farms are a particularly promising source of scalable clean energy, attracting global investment. Similarly, tidal and wave energy projects are being actively explored to tap the vast energy potential of our oceans [10].

Marine ecosystems are a rich source of bioactive compounds with promising medical applications. Many marine organisms, such as sponges, corals, and seaweeds, produce unique chemicals with antimicrobial, anticancer, and anti-inflammatory properties. These compounds are being actively researched for their potential to create new drugs and therapies. Pharmaceutical research in marine biotechnology has already led to the discovery of several life-saving drugs. For instance, compounds derived from marine organisms have been used to develop treatments for cancer, pain management, and infectious diseases. The exploration of marine biodiversity holds the promise of unlocking new medical breakthroughs to address emerging health challenges.

Marine resources can also support industrial and agricultural sectors. Seaweed, for example, is a versatile resource with applications in food processing, cosmetics, and biofuel production. It is also used as a natural fertilizer and soil conditioner in agriculture, promoting sustainable farming practices [11–16]. The extraction of minerals from seawater, such as magnesium and lithium, is another avenue with significant industrial potential.

Estuarine ecosystems contribute to agriculture by providing nutrient-rich sediments that enhance soil fertility. The sustainable use of these ecosystems can support agricultural productivity while maintaining ecological balance.

The blue economy holds immense potential for fostering economic growth while ensuring ecological sustainability. However, this potential is increasingly threatened by environmental stressors, with salinity emerging as a critical factor. In the extended estuarine zone of the Bay of Bengal, salinity variations are driven by a confluence of natural and anthropogenic activities, including sea-level rise, reduced freshwater inflow, siltation, and industrial discharges. Bali Island, a key location in the Sundarbans delta, serves as a microcosm for studying these impacts. Situated adjacent to the Matla estuary in the Bay of Bengal's inshore zone, the region is highly vulnerable to salinity changes, which in turn affect marine biodiversity, economic activities, and infrastructure. This research investigates the dynamics of salinity over four decades, projects future trends, and evaluates their implications for the blue economy. Salinity is the key player behind the success of blue economy which is exhibiting an increasing trend in the present sampling station due to obstruction of freshwater because of siltation. In this context policy and strict laws have important roles to play. Laws and policies regulate activities like water management, agriculture, and industrial discharge to control salinity fluctuations in coastal and marine environments. Ineffective regulations can worsen salinity, harming ecosystems, fisheries, and biodiversity that underpin the blue economy. Salinity fluctuations disrupt habitats, reduce fish stocks, and impact aquaculture, tourism, and livelihoods dependent on healthy marine systems. Robust policies are crucial to balance economic activities with sustainable resource management. Poor enforcement or gaps in legislation can amplify the adverse effects, threatening the long-term viability of the blue economy.

2 Materials and Methods

Surface water salinity data was collected at Bali Island (22°07′59.98″N; 88°44′52.93″E) using refractometer during high tide. The sampling covered a 40-year period from 1984 to 2024, with measurements taken in all the three seasons (premonsoon, monsoon and postmonsoon). Data analysis involved statistical tools to identify trends and anomalies. For forecasting, a Neural Auto-Regressive (NAR) model was employed (Fig. 1). The model was trained using real-time data of four decades (1984–2024), leveraging its ability to handle nonlinear and seasonal patterns. Forecasts were validated with cross-validation techniques to ensure accuracy. This methodological approach offers a comprehensive understanding of salinity dynamics in the region.

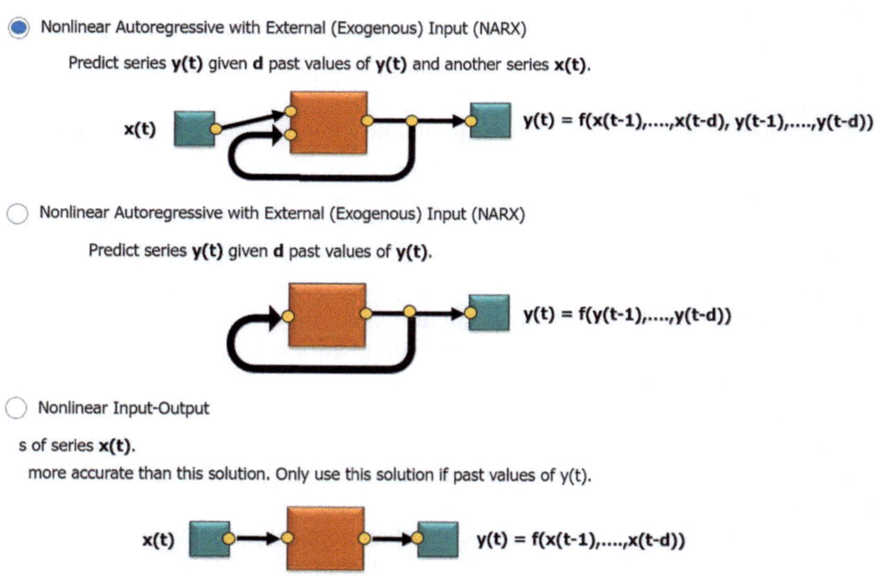

Fig. 1 Schematic representation of a nonlinear autoregressive model to analyze surface water salinity trends in the sampling station (Bali Island)

3 Results and Discussion

The present sampling station, situated in the central sector of Indian Sundarbans, exhibits an increasing trend in salinity profile, irrespective of seasons (Figs. 2, 3 and 4).

Surface water salinity levels at Bali Island displayed a gradual increase over the four decades. The salinity during high tide increased from 17.64 PSU (during monsoon) in 1984 to 29.53 PSU (during premonsoon) in 2024. Seasonal fluctuations were observed, with monsoonal freshwater inflows temporarily reducing salinity. However, the overall trend indicates an upward trajectory, attributed to reduced freshwater discharge, increased industrial effluents, and climatic factors such as rising sea levels and changing precipitation patterns.

The sudden rise of salinity in 2009 during premonsoon is the outcome of AILA, a super cyclone that hit the LGD with a speed of ~120 km/hr. This resulted in the intrusion of saline water from the Bay of Bengal in the sampling station [17, 18].

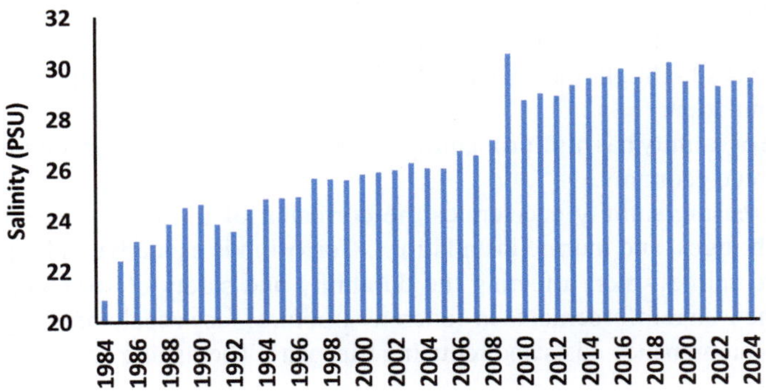

Fig. 2 Aquatic salinity of estuarine water around Bali Island during premonsoon

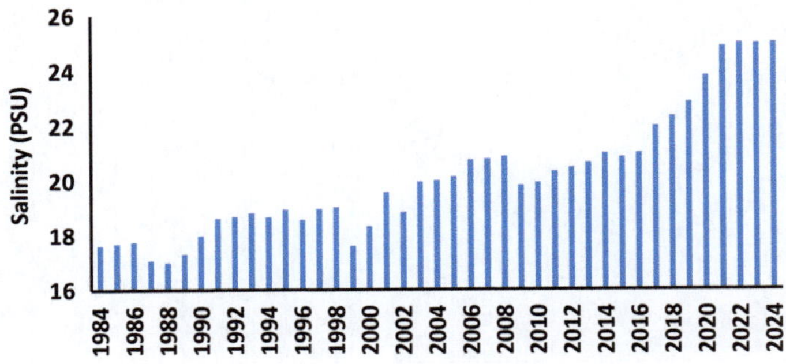

Fig. 3 Aquatic salinity of estuarine water around Bali Island during monsoon

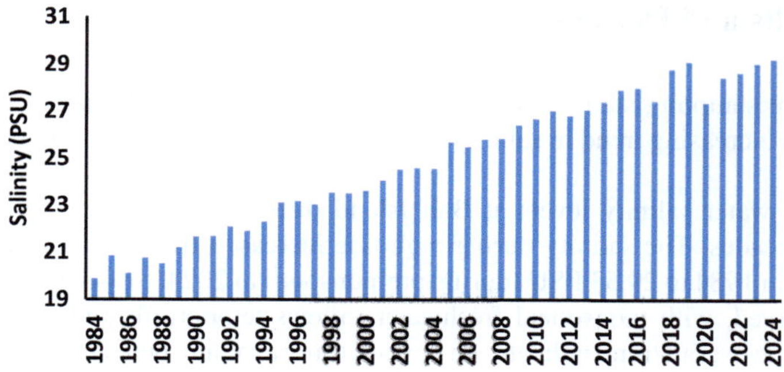

Fig. 4 Aquatic salinity of estuarine water around Bali Island during postmonsoon

Based on real time data of four decades, the NAR model predicts a continued rise in salinity, with an estimated salinity value of 35.02 PSU (during premonsoon) by 2050 (Figs. 5, 6 and 7). The projections suggest heightened variability, with occasional spikes during dry seasons. This increase poses significant risks to the blue economy, particularly in the domains of infrastructure, tourism, and biodiversity. The outputs of the forecasts highlight the utility of NAR models in environmental planning and policy formulation.

The rise of salinity has high probability to pose several adverse impacts on different sectors of blue economy.

The increasing salinity accelerates the corrosion of infrastructure, particularly buildings, bridges, and marine installations. Corrosion is exacerbated by prolonged exposure to saline water, affecting the durability and safety of structures [2]. For instance, the iron reinforcements in concrete structures face accelerated degradation, increasing maintenance costs and reducing lifespans. The fishing harbors and jetties

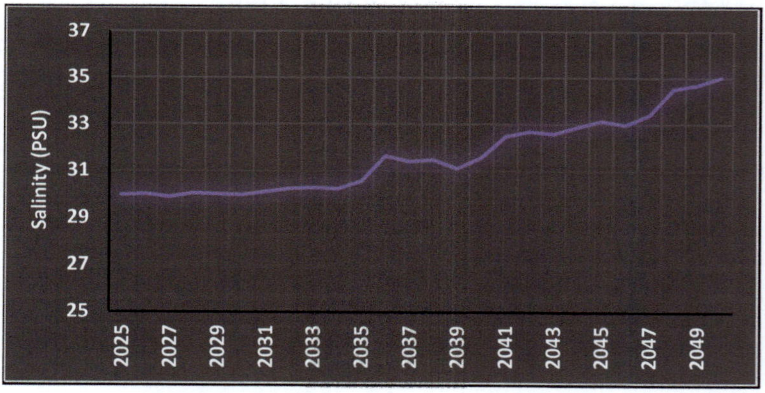

Fig. 5 Representation of a nonlinear autoregressive neural network model to predict aquatic salinity in Bali Island's estuarine waters during the premonsoon season

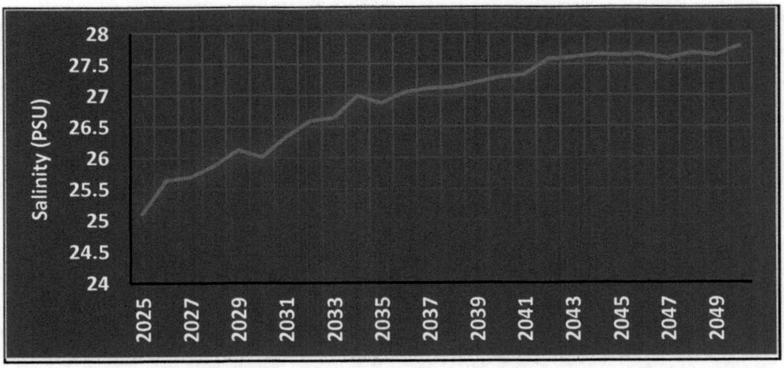

Fig. 6 Representation of a nonlinear autoregressive neural network model to predict aquatic salinity in Bali Island's estuarine waters during the monsoon season

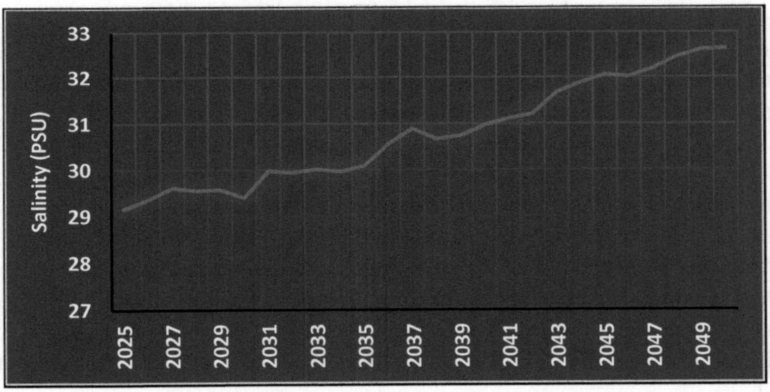

Fig. 7 Representation of a nonlinear autoregressive neural network model to predict aquatic salinity in Bali Island's estuarine waters during the postmonsoon season

at Bali Island, vital for the local economy, are particularly vulnerable, necessitating frequent repairs and replacements.

Tourism, a vital component of the blue economy, is adversely affected by rising salinity. The aesthetic and recreational value of coastal regions diminishes due to the deterioration of natural habitats, such as mangroves and faunal biodiversity. High salinity also impacts the availability of potable water, reducing the feasibility of tourism ventures. Moreover, the increased frequency of algal blooms, a byproduct of nutrient-salinity interactions, further detracts from the region's appeal to visitors.

The salinity changes disrupt the delicate balance of marine ecosystems. Species such as shrimp and certain fish, crucial for the local fishing industry, struggle to adapt to these shifts. It has been observed that in increased salinity scenario, the commercially important fishes are often replaced by trash fishes, thereby lowering the income of the local population from the fishery sector (Fig. 8).

Fig. 8 High salinity promotes the proliferation of low-priced trash fishes in the estuary which are used for poultry feed

The reduced biodiversity weakens ecosystem services, including carbon sequestration and nutrient cycling, further endangering economic and ecological stability. Reports of stunted growth of mangroves in the Lower Gangetic Delta (LGD) are available in plenty, which reduces carbon sequestration by these halophytes [19–26].

According to researchers "carbon sequestration by coastal vegetation (blue carbon) has become a key topic in the domain of climate change. In this context the Ministry of Environment, Forest and Climate Change, Government of India issued a notification on 12th October, 2023 in which a Green Credit programme has been launched at national level to leverage a competitive market-based approach to achieve Green Credit for incentivizing environmental actions of various stakeholders." Under Green Credit programme, eight activities have been referred [27] as highlighted in Table 1.

Addressing salinity impacts requires robust policies and laws. In India, several frameworks address these challenges. The Coastal Regulation Zone (CRZ) notification regulates activities in sensitive coastal areas, aiming to preserve ecological balance. The National Action Plan on Climate Change (NAPCC) and its submissions, such as the National Water Mission, emphasize sustainable water resource management to combat salinity intrusion.

Additionally, international conventions such as the Ramsar Convention on Wetlands and the United Nations Framework Convention on Climate Change (UNFCCC) provide guidelines for protecting coastal ecosystems. However, gaps in enforcement and local adaptation persist. In several pockets of the Lower Gangetic Delta, shrimp ponds have been constructed at the cost of mangroves and agricultural lands. Strengthening community involvement and integrating traditional knowledge with scientific approaches can enhance policy effectiveness. Investment in corrosion-resistant materials, desalination technologies, and eco-tourism initiatives can mitigate some of the adverse impacts identified in this study.

Table 1 Major Activities under Green Credit of Government of India

Sl. no	Activity	Details
1	Tree plantation	Promotion of increasing green cover across the country
2	Water management	Promotion of water conservation, water harvesting and water use efficiency or water savings, including treatment and reuse of waste water
3	Sustainable agriculture	Promotion of natural and regenerative agricultural practices and land restoration of improve productivity, soil health and nutritional value of food produced
4	Waste management	Promotion of circularity, sustainable and improved practices for waste management, including collection, segregation, and environmentally sound management
5	Air pollution reduction	Promotion of reducing air pollution and other pollution abatement activities
6	Mangrove conservation and restoration	Promotion of conservation and restoration of mangroves
7	Eco-mark label development	Encouragement of manufacturers to obtain eco mark label for their goods and services
8	Sustainable building and infrastructure	Encouragement of the construction of sustainable buildings and other infrastructure using environment friendly technologies and materials"

4 Conclusion

Salinity dynamics in the Bay of Bengal's inshore zone, exemplified by Bali Island and its adjacent estuary, pose significant challenges to the blue economy. Real-time and forecasted data highlight an urgent need for adaptive measures to safeguard economic, ecological, and social interests. The findings underscore the importance of integrating advanced forecasting tools, such as AI-driven NAR models, with policy frameworks to address emerging challenges. A holistic approach, involving stakeholders across sectors, is essential to ensure the sustainable growth of the blue economy in the face of rising salinity.

References

1. Mitra A et al (2023) Climate resilient innovative livelihoods in Indian Sundarban delta. Published by Springer, Hardcover ISBN: 978-3-031-42632-2, eBook ISBN 978-3-031-42633-9, pp 297, Vol. XII. https://doi.org/10.1007/978-3-031-42633-9
2. Dhar I et al (2021) Salinity: a Major environmental factor in sustainability of the Blue Carbon. J Mech Continua Math Sci 16(11):34–42. https://doi.org/10.26782/jmcms.2021.11.00004
3. Mitra A et al (2021a) Non-living resources of the sea as the pillars of blue economy. Par J Sci Edu 7(3):28–34

4. Mitra A et al (2021b) Major sectors of blue economy to trigger the growth of nation. Par J Sci Edu 7(3):16–27
5. Mitra A et al (2022) Blue Economy in Indian Sundarbans: exploring Livelihood opportunities. Published by Springer, ISBN: 978-3-031-07907-8, eBook ISBN: 978-3-031-07908-5, pp. 403, Vol. XIV. https://doi.org/10.1007/978-3-031-07908-5
6. Biswas P et al (2018) Development of fish feed incorporating with antioxidant rich mangrove floral species (*E.intestinalis*) for sustainable Tilapia culture. Techno Int J Health Engi Manag Sci 2(4):150–153
7. Pramanick P et al (2019) Probiotics: a pathway for upgrading Aquaculture sector. Int J Res Rev 6(1):94–99
8. Biswas P et al (2021) Fish quality upgradation through cutting edge feed technology research on monosex tilapia. Chapter 15, In: Mitra A, Calma MM, Chakrabarty SP (eds) Natural resources and their ecosystem services, Vol. II, published by HSRA Publication, ISBN: 978-93-5506-064-8, pp 200-208
9. Mitra A, Zaman S (2016) Basics of marine and estuarine ecology. Published by Springer New Delhi, eBook ISBN: 978-81-322-2707-6, Hardcover ISBN: 978-81-322-2705-2, pp 483, Vol. XII. https://doi.org/10.1007/978-81-322-2707-6
10. Mitra A et al (2021c) Blue economy: a new dimension to explore marine resources, published by New Millennium Graphics Publications, ISBN: 978-81-951712-6-2
11. Zodape ST (2001) Seaweeds as a biofertilizer. J Sci Indus Res 60:378–382
12. Couteau C, Coiffard L (2016) Seaweed application in cosmetics. Chapter 14, In: Fleurence J, Levine I (eds) Seaweed in Health and Disease Prevention. Published by Elsevier, pp 423–441. https://doi.org/10.1016/B978-0-12-802772-1.00014-2
13. Raghunandan BL et al (2019) Perspectives of seaweed as organic fertilizer in agriculture. In: Panpatte D, Jhala Y (eds) Soil fertility management for sustainable development. Springer, Singapore. https://doi.org/10.1007/978-981-13-5904-0_13
14. Singh S et al (2019) Seaweed: an alternative liquid fertilizer for plant growth. Int J Curr Microbiol App Sci 8(12):772–781. https://doi.org/10.20546/ijcmas.2019.812.101
15. López-Hortas L et al (2021) Applying Seaweed compounds in cosmetics, cosmeceuticals and nutricosmetics. Marine Drugs 19(10):552. https://doi.org/10.3390/md19100552
16. Arioli T et al (2024) Applications of seaweed extracts in agriculture: an Australian perspective. J Appl Phycol 36:713–726. https://doi.org/10.1007/s10811-023-03120-x
17. Mitra A et al (2009) Abiotic indicators of climate change in Indian Sundarbans. Indian Sci Crui 23(2):53–56
18. Mitra A et al (2011) Spatial and tidal variations of physico-chemical parameters in the lower Gangetic delta region, West Bengal, India. J Spatial Hydro 11(1):62–69
19. Mitra A et al (2004) In: The other face of mangroves. Published by Department of Environment, Govt. of West Bengal, India
20. Bhattacharjee AK et al (2013a) Impact of salinity on above ground biomass and stored carbon in a common mangrove *Excoecaria agallocha* of Indian Sundarbans. Ame J Biopharmacol Biochem Life Sci 2(2):1–11
21. Bhattacharjee AK et al (2013b) How mangroves respond to hypersaline condition? Preparedness for predicted sea level rise. Int J Sci Res 2(6):360–364
22. Sarwar MGM, Woodroffe CD (2013) Rates of shoreline change along the coast of Bangladesh. J Coast Conserv 17:515–526
23. Zaman S et al (2013) Rising water salinity: a threat to mangroves of Indian Sundarbans. In: Anwarul Abedin Md, Habiba U, Shaw R (eds) Water insecurity: a social dilemma, published by Emerald Group Publishing Limited (ISBN: 9781781908822)
24. Banerjee K, Mitra A (2015) Mangrove biomass and salinity: is the match made in heaven? Int J Biol Sci Engi, 45–54
25. Mitra A (2020) Mangrove forests in India: exploring ecosystem services. Published by Springer, e-Book ISBN: 978-3-030-20595-9, pp 361, Vol. XV. https://doi.org/10.1007/978-3-030-205 95-9

26. Rahman MS et al (2020) Salinity drives growth dynamics of the mangrove tree Sonneratia apetala Buch.-Ham. in the Sundarbans, Bangladesh. Dendrochronologia 62:125711
27. Mitra A et al (2024) Mangrove carbon trading in the lower gangetic delta: drawbacks and the potential. Published by Springer, Hardcover ISBN 978-3-031-76315-1, eBook ISBN: 978-3-031-76316-8, pp 282, Vol XIII. https://doi.org/10.1007/978-3-031-76316-8

Addressing CO_2 Emissions in Kolkata: Policies for Sustainable Solutions

Sana Ahmed, Debatri Banerjee, Sufia Zaman, and Abhijit Mitra

Abstract This study explores the alarming rise in atmospheric CO_2 levels in Kolkata, driven by urbanization, industrial activities, vehicular emissions, and inadequate waste management, with a focus on two distinct sites, namely Duttabagan and Barasat. Using more than three decades of real-time data (1984–2022) and the Neural Auto-Regressive (NAR) model, the research predicts a steeper rise in CO_2 levels in Duttabagan, attributed to its dense urbanization and commercial activities like presence of the biggest fish selling units in the megacity of Kolkata, compared to Barasat. The findings underscore the need for localized interventions, such as improved waste management, afforestation, and sustainable urban planning. Policies like India's National Action Plan on Climate Change (NAPCC), the Air (Prevention and Control of Pollution) Act, 1981, Water (Prevention and Control of Pollution) Act, 1974, and the emission trading schemes are critical to addressing these challenges. Advanced technologies related to waste-to-energy systems are recommended to enhance emission reduction strategies. The study emphasizes the importance of public participation, effective governance, and targeted policies to curb emissions while ensuring economic and environmental sustainability, with the aim to position the megacity of Kolkata as a model for urban climate resilience.

Keywords Kolkata city · Atmospheric CO_2 level · Neural Auto-Regressive (NAR) model · National Action Plan on Climate Change (NAPCC)

1 Introduction

The megacity of Kolkata, the capital of the maritime state of West Bengal, has become increasingly vulnerable in recent times due to several factors like (i) an exceptionally high population density of 24,000 people per square kilometer, (ii)

S. Ahmed · D. Banerjee (✉) · S. Zaman · A. Mitra
Department of Oceanography, Techno India University, Kolkata, West Bengal, India
e-mail: bdebatri91@gmail.com

M. Chakraborty et al. (eds.), *Proceedings of Sixth International Ethical Hacking Conference*, Lecture Notes in Networks and Systems 1513,
https://doi.org/10.1007/978-981-96-8632-2_32

horizontal urban expansion at the expense of urban forests, (iii) unplanned industrial development, (iv) a rapid increase in vehicles reliant on petrol and diesel, (v) significant energy consumption by overcrowded offices in specific parts of the city, and (vi) decomposing waste in various areas, which releases carbon dioxide into the atmosphere.

Research studies highlight that while several investigations have examined the carbon sequestration potential of urban trees in Kolkata [1–8], little attention has been given to the spatio-temporal variations in near-surface atmospheric carbon dioxide levels in and around the city [9]. Against this backdrop, the present study provides a preliminary analysis of atmospheric carbon dioxide levels across two selected sites in Kolkata, namely Duttabagan (Station 1), and Barasat (Station 2). Spanning over three decades of real-time data (1990–2022), this research aims to forecast the carbon dioxide levels across two selected sites during 2050.

According to the International reports "atmospheric carbon dioxide is the most significant anthropogenic greenhouse gas, with its global concentration rising from 280 ppm during the pre-industrial era to over 400 ppm today. This increase is a key driver of global warming and climate change. Approximately 70% of anthropogenic carbon dioxide emissions originate from the combustion of fossil fuels and industrial activities, further exacerbating the problem. Currently, the global atmospheric temperature is 1.5 °C higher than pre-industrial levels, with projections suggesting that climate extremes may intensify between 2030 and 2050 due to the continuous rise in carbon dioxide concentrations [10]".

To mitigate the escalation of atmospheric carbon dioxide levels, numerous nations are adopting both technological and nature-based solutions. Developing a robust, long-term database of carbon dioxide concentrations is essential to formulating cost-effective policies for controlling the rise of this greenhouse gas.

The surge in atmospheric carbon dioxide is predominantly caused by human activities. Burning fossil fuels like coal, oil, and natural gas for industry, heating, power generation, and cement manufacturing are major drivers. Deforestation and altered land use patterns also play a significant role, diminishing the effectiveness of natural carbon sinks like forests to absorb emissions. Furthermore, warmer ocean temperatures reduce their capacity to absorb carbon dioxide. The result is a substantial increase—around 45%—in atmospheric carbon dioxide levels compared to pre-industrial times.

This increased carbon dioxide concentration acts like a blanket, trapping heat radiating from the Earth's surface and preventing it from escaping into space. This phenomenon is causing global warming, which in turn leads to melting polar ice and glaciers, rising sea levels, and warmer rivers and oceans. These changes are linked to a rise in extreme weather events, such as heatwaves, intense droughts, powerful hurricanes, severe blizzards, and torrential rainfall.

The effects of climate change, driven by these elevated carbon dioxide levels, are being felt worldwide, and Kolkata is experiencing its share of these concerning trends.

India has taken steps to address the challenge of rising carbon dioxide and mitigate climate change through various policies and legislation. Acts like the Air (Prevention

Table 1 Geographical location of the selected sites

Stations	Coordinates
Duttabagan	22°36′28.2″N; 88°23′28.1″E
Barasat	22°43′23.3″N; 88°29′13.8″E

and Control of Pollution) Act of 1981 and the Environment Protection Act of 1986 provide a legal foundation for regulating emissions from industrial and transportation sectors. The National Action Plan on Climate Change (NAPCC) prioritizes the development and use of renewable energy sources, energy efficiency, and sustainable urban development, including missions like the National Solar Mission. Policies such as the Energy Conservation Act, 2001, and fuel efficiency standards aim to reduce energy consumption and associated carbon dioxide emissions. Afforestation initiatives under the National Afforestation Programme enhance natural carbon sinks, while the Municipal Solid Waste Rules, 2016, promote waste segregation and treatment to curb emissions from decomposition. India has also piloted Emission Trading Schemes (ETS) to encourage industries to cap-and-trade emissions. On a global scale, India's commitment to the Paris Agreement reflects its pledge to reduce carbon intensity and transition to renewable energy sources, thus aligning domestic policies with international goals.

2 Materials and Methods

2.1 Site Selection

Kolkata was the site of the present study that selected two locations, chosen to represent varying levels of human activity (Table 1).

2.2 Sampling and Statistical Analysis

Near-surface atmospheric carbon dioxide concentrations were measured at two sites using a portable CO_2 analyzer (Lutron CO_2 meter) throughout the study period (1984–2022). Data collection was conducted in April of each selected year, aligning with the premonsoon season in West Bengal, a time marked by elevated air temperatures and considerable humidity. To ensure data quality, 10 readings were taken at each site during afternoon hours (12.30 pm–5.30 pm), spaced 10–15 m apart, with the average value calculated for further statistical analysis in terms of forecasting (Fig. 1).

Data analysis included statistical tool to forecast the level of carbon dioxide in the near surface atmosphere of the selected stations. For forecasting, the Neural

Fig. 1 Monitoring near surface atmospheric carbon dioxide during the afternoon hours (12.30 pm to 5.30 pm)

Auto-Regressive (NAR) model was employed, leveraging more than three decades of real-time data (1984–2022). The model's capability to capture nonlinear patterns provided robust projections, which were validated using cross-validation techniques to enhance accuracy.

This methodological framework is pivotal for informing policy formulation aimed at mitigating CO_2 emissions. By understanding spatial and temporal dynamics of CO_2 levels, policymakers can design targeted interventions, such as urban planning, vehicular emission controls, and green infrastructure development, to effectively address localized and regional carbon footprints. The integration of advanced forecasting models like NAR supports data-driven decision-making and ensures proactive responses to emerging environmental challenges.

3 Results

The comparison of CO_2 concentration trends between Barasat and Duttabagan reveals significant differences in their atmospheric patterns over the years. Duttabagan exhibits a higher average CO_2 level of approximately 390.41 ppm compared to Barasat's 385.61 ppm, reflecting its more urbanized and industrialized nature. While Barasat shows a modest total increase of about 10 ppm from 1984 to 2050, Duttabagan experiences a sharper rise of 35 ppm over the same period, indicating more intensive anthropogenic activities. Considering the entire time span from 1984 to 2050, the annual increase in CO_2 levels is also notably higher in Duttabagan

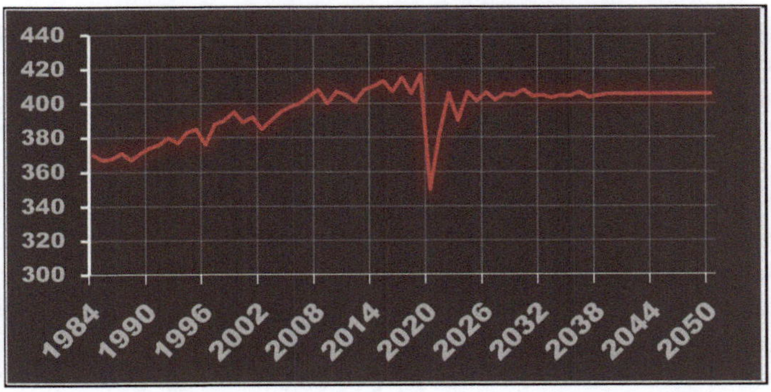

Fig. 2 Near-surface atmospheric carbon dioxide levels at Duttabagan in April (premonsoon) were predicted using a Nonlinear Autoregressive Neural Network model. The model was trained using real-time data spanning from 1984 to 2022

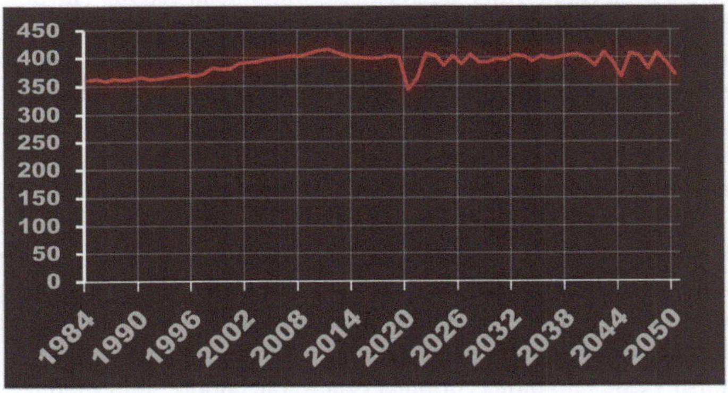

Fig. 3 Near-surface atmospheric carbon dioxide levels at Barasat in April (premonsoon) were predicted using a Nonlinear Autoregressive Neural Network model. The model was trained using real-time data spanning from 1984 to 2022

(Fig. 2) at 0.53 ppm/year, compared to Barasat's (Fig. 3) at 0.15 ppm/year. These findings underscore the need for tailored mitigation strategies, with Duttabagan requiring more robust interventions to curb its steep CO_2 escalation.

4 Discussion

Kolkata, the capital of West Bengal, is one of the most densely populated cities in the world, with a population density of 24,000 persons per square kilometer. The rapid pace of urbanization, industrial growth, and vehicle increase in Kolkata has led to a significant rise in atmospheric carbon dioxide (CO_2) levels [9]. This discussion explores the major contributors to this phenomenon, including population growth, vehicular emissions, industrial activities, deforestation, and urban development [11–15]. Furthermore, it emphasizes the policies and strategies required to mitigate CO_2 emissions while ensuring sustainable urban development. Duttabagan, a vibrant and densely populated locality in Kolkata, is a microcosm of the city's urban challenges, particularly concerning the rising levels of atmospheric carbon dioxide (CO_2). One notable feature of this area is the Patipukur fish market, a significant hub for fish trading and associated activities. The decomposition of fish carcasses, combined with the accumulation of fish-related waste, generates substantial amounts of CO_2 and other greenhouse gases, making it a localized source of emission that exacerbates the environmental issues in Duttabagan [16]. The organic decomposition of fish waste in such markets is a natural process that releases methane (CH_4), carbon dioxide (CO_2), and other volatile organic compounds (VOCs) into the atmosphere [17–21]. The high temperatures and humidity characteristic of Kolkata further accelerate this decomposition, worsening the release of harmful gases. When compared to Barasat, a suburban area with relatively less commercial activity and better natural ventilation, the elevated levels of CO_2 in Duttabagan become even more pronounced, highlighting the pressing need for intervention both at technological and policy levels. These emissions, coupled with other urban stressors, contribute to localized air quality degradation, creating a cascade of health, environmental, and socio-economic impacts. The high CO_2 levels in Duttabagan underscore the need for comprehensive waste management strategies that address emissions from fish markets and other organic waste hotspots. Implementing efficient systems for the immediate collection and disposal of fish waste is essential to minimize the duration of exposure to decomposition processes. Regular composting is a viable solution, turning organic waste into valuable fertilizer while significantly reducing emissions during decomposition. Advanced waste-to-energy technologies can further enhance this effort by converting fish waste into biogas or other renewable energy sources, thus offering a dual benefit of emission reduction and resource utilization. Alongside waste management, increasing vegetation cover in Duttabagan can be instrumental in minimizing CO_2 emissions. Trees and other green infrastructure not only absorb CO_2 but also provide shade and reduce urban heat [7], which slows down the rate of organic waste decomposition. Developing green belts around high-emission areas like the Patipukur fish market and promoting vertical gardens on residential and commercial buildings can further enhance the area's carbon sequestration potential.

Urban planning in Duttabagan must also incorporate sustainable practices that address the root causes of emissions. Policies that mandate sustainable waste disposal practices at markets at frequent intervals, enforce strict penalties for improper waste management, and incentivize eco-friendly technologies can create a more accountable framework for emission control. Encouraging fish market vendors to segregate waste at the source can simplify the process of recycling and composting, further reducing emissions. Local authorities need to ensure the availability of appropriate waste management infrastructure, including collection points, composting units, and biogas plants, to streamline this process. Integrating these measures with city-wide urban development plans can ensure that emission hotspots like Duttabagan are addressed systematically rather than in isolation. The differing CO_2 levels observed between Duttabagan and Barasat suggest that urban emission management strategies should be tailored to the specific characteristics of each area. While Barasat benefits from a more dispersed population and greater natural green cover, Duttabagan's high density, commercial activity, and limited vegetation exacerbate its environmental challenges. Policies that focus on decentralizing commercial activities and reducing the concentration of high-emission zones can help redistribute environmental stress more evenly across the city. For instance, relocating parts of the fish market to suburban areas with better waste management facilities can reduce the burden on Duttabagan while promoting balanced urban development.

Another critical aspect of mitigating CO_2 emissions in Duttabagan is leveraging technology for real-time monitoring and analysis. Geographic Information Systems (GIS) and satellite imaging can be used to map emission hotspots, track changes in green cover, and monitor the effectiveness of mitigation measures. Artificial Intelligence (AI) can optimize waste collection routes, reducing the fuel consumption and emissions associated with waste transport. Smart sensors placed in emission hotspots like the Patipukur fish market can provide real-time data on CO_2 levels, enabling authorities to respond quickly to critical situations. These technologies can also enhance public awareness by providing residents and stakeholders with accessible information on air quality and the environmental impacts of their activities.

Public participation is crucial for the success of any policy aimed at reducing CO_2 emissions in Duttabagan. Educating residents about the health and environmental consequences of improper waste disposal and high CO_2 levels can foster a culture of responsibility and proactive engagement. Community-based initiatives, such as neighborhood composting programs and tree-planting drives, can empower residents to take an active role in emission reduction. Collaborations between local authorities, businesses, and non-governmental organizations (NGOs) can further enhance these efforts by pooling resources and expertise. For example, fish market associations can partner with environmental NGOs to implement sustainable waste management practices and raise awareness among vendors and customers.

In addition to addressing localized emissions, policies must also tackle the broader systemic issues contributing to Kolkata's CO_2 levels. Kolkata's dependence on fossil fuels for energy and transportation necessitates urgent action. A shift towards renewable energy sources, like solar and wind power, offers a substantial opportunity to decrease the city's carbon footprint. Providing incentives for the adoption of energy-efficient appliances and solar panels in Duttabagan can encourage residents and businesses to make more sustainable choices. In the transportation sector, expanding public transit options like electric buses and metro lines can reduce the dependency on private vehicles, which are a major source of CO_2 emissions. Promoting non-motorized transport options like cycling and walking, along with creating dedicated lanes for such modes of transport, can further enhance the city's sustainability.

The role of policy and governance cannot be masked in achieving these goals. Effective implementation of existing policies, such as the National Action Plan on Climate Change (NAPCC), must be complemented by local-level initiatives tailored to Duttabagan's unique challenges. Strict enforcement of pollution control laws like

Air (Prevention and Control of Pollution) Act, 1981, Water (Prevention and Control of Pollution) Act, 1974, including penalties for non-compliance with waste management and emission norms, can ensure accountability among businesses and residents. Urban planning regulations should mandate the inclusion of green infrastructure and sustainable building practices in all new developments. Public–private partnerships can play a pivotal role in mobilizing the financial and technical resources required for large-scale projects like waste-to-energy plants and green belt development.

The socio-economic benefits of reducing CO_2 emissions in Duttabagan extend beyond environmental improvement. Lower emissions can lead to better air quality, reducing the prevalence of respiratory and cardiovascular diseases among residents. This, in turn, can decrease healthcare costs and improve productivity, contributing to the overall economic well-being of the community. Enhancing green cover and reducing emissions can also boost the area's aesthetic appeal, making it more attractive for tourism and real estate development. These positive outcomes can create a virtuous cycle where environmental and economic goals reinforce each other, paving the way for sustainable urban development.

Duttabagan's challenges are emblematic of the broader environmental issues faced by Kolkata and other rapidly urbanizing cities. The lessons learned from addressing Duttabagan's CO_2 emissions can inform strategies for other areas with similar challenges. For instance, the integration of waste management, green infrastructure, and sustainable urban planning can serve as a model for tackling emissions in other urban hotspots. By adopting a holistic approach that combines local actions with city-wide policies, Kolkata can position itself as a pioneer in sustainable urban development and climate change mitigation.

Thus, the elevated CO_2 levels in Duttabagan, driven by emissions from the Patipukur fish market and other urban stressors, highlight the urgent need for targeted interventions. Comprehensive waste management strategies, increased green cover, sustainable urban planning, and the adoption of advanced technologies coupled with strict enforcement of existing laws can significantly reduce emissions that can be the green lanes for improving living conditions for residents. Public participation and effective governance are crucial for the success of these measures, ensuring that environmental and economic goals are achieved in tandem. By addressing the challenges in Duttabagan, Kolkata can set a precedent for sustainable urban development that balances growth with environmental stewardship. This holistic approach not only mitigates CO_2 emissions but also enhances the resilience of urban communities in the face of climate change, creating a more sustainable future for all.

5 Conclusion

The significant rise in atmospheric CO_2 levels in Kolkata, driven by factors such as urbanization, industrial expansion, vehicular emissions, and poor waste management, highlights the urgent need for integrated and sustainable mitigation strategies. The study underscores the critical role of green infrastructure, effective urban planning, and advanced technologies like NAR models in addressing localized CO_2 emissions, as evidenced by disparities between Duttabagan and Barasat. Tailored interventions such as efficient waste management, enhanced vegetation, and the adoption of renewable energy systems can effectively curb emissions while fostering economic growth. Furthermore, the findings emphasize the importance of public participation, policy enforcement, and leveraging AI-based tools for real-time monitoring and data-driven decision-making. By addressing the challenges at micro and macro levels, Kolkata can set a benchmark for sustainable urban development and climate resilience.

References

1. Banerjee R et al (2015) Impact of urban vegetation on offsetting carbon emission: a case study from the city of Kolkata. J Environ Sci Comp Sci Engi Tech 4(3):814–818
2. Mitra A et al (2017) Oxygen generation by dominant urban trees: a case study from Konnagar municipality, West Bengal, India. Bio J Scienti Tech Res 9(1):1–9
3. Chakrabarti SS et al (2019) A comparative study of reducing carbon footprint by urban and coastal vegetation. J Emerg Tech Innov Res 6(4):410–415
4. Agarwal S et al (2021) Stored carbon in urban trees: ground zero observation from the Konnagar Area of West Bengal, India. In: Filho WL, Luetz JM, Ayal DY (eds) Handbook of climate change management (Research, Leadership, Transformation), published by Springer Nature eBook ISBN: 978-3-030-22759-3, pp 1–23, https://doi.org/10.1007/978-3-030-22759-3_284-1
5. Dutta J et al (2021) Lead- and cadmium-induced toxicity on environment and human health with special reference to a Peri-Urban Ramsar site of India. Chapter 18. In: Prasad R, Jhariya MK, Banerjee A (eds) Advances in sustainable development and management of environmental and natural resources. ISBN: 9781774630679 2, pp 163–192
6. Ahmed S et al (2023) Study on floral carbon storage potential in the megacity of Kolkata: a roadmap towards net-zero carbon emission. J Environ Sociobiol 19(2):113–123
7. Dutta P et al (2024) Role of urban trees in minimizing the atmospheric carbon dioxide level of the megacity of Kolkata, India. Chapter 33, In: Urban forests, climate change and environmental pollution. Published by Springer. https://doi.org/10.1007/978-3-031-67837-0_33
8. Ahmed S et al (2024) Regulation of ambient air temperature through trees: a comparison between industrial and rural belt in the State of West Bengal, India. Chapter 36, In: Urban forests, climate change and environmental pollution. Published by Springer. https://doi.org/10.1007/978-3-031-67837-0_36
9. Bhattacharyya S et al (2024a) Estimation of carbon dioxide –equivalent of common urban trees from two adjoining sites of the city of Kolkata, India. J Environ Sociobiol (Accepted)
10. Bhattacharyya S et al (2024b) Time series analysis of atmospheric carbon dioxide in the megacity of Kolkata, India. World Wide J Multidis Res Dev 10(2):15–20
11. Solaymani S (2019) CO_2 emissions patterns in 7 top carbon emitter economies: the case of transport sector. Energy 168:989–1001
12. Hockstad L, Hanel L (2018) Inventory of US greenhouse gas emissions and sinks. In.: Environmental system science data infrastructure for a virtual ecosystem
13. Jin C et al (2022) Ecological consequences of China's regional development strategy: evidence from water ecological footprint in Yangtze River Economic Belt. Environ Dev Sustain, pp 1–16
14. Konc T et al (2021) The social multiplier of environmental policy: application to carbon taxation. J Environ Econ Manag 105:102396
15. Ritchie H et al (2020) CO_2 and greenhouse gas emissions, Our world in data
16. Mitra A (2023). In: Impact of COVID-19 lockdown on Environmental health: Exploring the situation of the Lower Gangetic Delta. Springer, pp 1–310

17. Staley BF et al (2006) Release of trace organic compounds during the decomposition of municipal solid waste components. Environ Sci Technol 40:5984–5991
18. Wu T et al (2010) Emission of volatile organic sulfur compounds (VOSCs) during aerobic decomposition of food wastes. Atmos Environ 44(39):5065–5071
19. Colón J et al (2017) Characterization of odorous compounds and odor load in indoor air of modern complex MBT facilities. Chem Engi J 313:1311–1319
20. Liu Y et al (2020) VOCs released from municipal solid waste at the initial decomposition stage: Emission characteristics and an odor impact assessment. J Environ Sci 98:143–150
21. González D et al (2022) Odors emitted from biological waste and wastewater treatment plants: a mini-review. Atmosphere 13(5):798

Policy Interventions to Mitigate Acidification in Coastal Waters of Digha

Nabonita Pal, Nagaraj Narayan Sannabhadti, Shambhu Prasad Chakrabarty, Sufia Zaman, and Abhijit Mitra

Abstract The Digha coast, a prominent tourism hub in the maritime state of West Bengal, faces a growing challenge from acidification of its coastal waters. This phenomenon, primarily driven by anthropogenic factors like untreated wastewater from domestic sources, tourism units and shrimp farms, and industrial effluents, accelerates the corrosion of infrastructure critical to the local tourism economy. Seasonal variations in pH levels reveal significant fluctuations, with highest value in premonsoon (mean value = 8.28), followed by postmonsoon (mean value = 8.27), and monsoon (mean value = 8.24). Using a comprehensive dataset of more than three decades (1984–2022) spanning across three seasons (premonsoon, monsoon, and postmonsoon), this paper examines the impact of acidification on Digha's tourism units along with forecasting the scenario of aquatic pH in 2050. The paper also emphasizes policy interventions to combat the adverse impact of acidification. This study highlights the urgent need for strict enforcement of the Coastal Regulation Zone (CRZ) notifications to achieve ecological stability and sustain the economic viability of Digha's tourism sector. Recommendations include strengthening mangrove conservation, promoting sustainable waste management practices, and enhancing CRZ compliance.

Keywords Acidification · Digha coast · Tourism sector · Coastal Regulation Zone (CRZ)

N. Pal · N. N. Sannabhadti · S. Zaman · A. Mitra
Department of Oceanography, Techno India University, West Bengal, Kolkata, India

S. P. Chakrabarty (✉)
Department of Law, (IIIJS), University of Engineering and Management, Kolkata, India
e-mail: shambhuprasadc@gmail.com

M. Chakraborty et al. (eds.), *Proceedings of Sixth International Ethical Hacking Conference*, Lecture Notes in Networks and Systems 1513,
https://doi.org/10.1007/978-981-96-8632-2_33

1 Introduction

Digha, a coastal town in West Bengal, is renowned for its picturesque beaches and serves as a major tourism destination. The influx of tourists, coupled with unregulated developmental activities, has intensified environmental stress on this region [1]. Coastal acidification, caused by the dissolution of excess carbon dioxide and untreated wastewater into seawater, poses a critical challenge by altering aquatic chemistry and degrading infrastructure [2, 3].

The phenomenon of acidification lowers the seawater pH, leading to long-term ecological and economic consequences. In regions like Digha, where tourism infrastructure heavily relies on durable coastal constructions, acidification accelerates the corrosion of materials, jeopardizing safety, and aesthetics [4–6]. Seasonal patterns in pH variation, as observed in this study, underscore the urgent need for mitigation strategies. While premonsoon seasons occasionally show respite due to reduced pollution levels, monsoon and postmonsoon seasons exhibit alarming trends of acidification [7].

Mangrove ecosystems play a pivotal role in buffering acidification through their carbon sequestration abilities [8, 9]. However, the degradation of these natural assets, coupled with inadequate implementation of CRZ notifications, exacerbates the challenges posed by acidification [10, 11]. The problem will be intensified as revealed from our AI-based nonlinear autoregressive (NAR) models to predict the trend of aquatic pH (proxy used to evaluate the magnitude of acidification). This paper explores the multifaceted impacts of acidification on Digha's tourism infrastructure and advocates for stringent policy measures to ensure ecological and economic sustainability.

2 Materials and Methods

Data for this study were compiled from extensive field surveys and secondary sources, including reports from the Oceanography Dept., at Techno India University, Salt Lake, WB. Seasonal pH levels were monitored during 1984 to 2022 (real-time data used for training the NAR model) across premonsoon, monsoon, and postmonsoon periods using a portable pH meter with an accuracy of ± 0.02. Five surface water samples were collected during high tide within 500 m of each other, ensuring uniformity in sampling techniques since 1984.

CRZ compliance was evaluated based on existing documentation and observational data, focusing on waste management practices and land-use patterns. The study also integrated long-term pH data with AI-based nonlinear autoregressive models to predict future trends in coastal acidification under varying scenarios of CRZ enforcement.

3 Results

Analysis of pH levels revealed significant variations as stated here and highlighted in Figs. 1, 2, and 3.

- Premonsoon: The pH levels averaged 8.28, indicating a relatively stable aquatic environment. This stability was attributed to the temporary halt in anthropogenic activities during the COVID-19 lockdown, reducing pollutant discharge.
- Monsoon: The pH dropped to an average of 8.24, highlighting increased acidification due to runoff from land-based sources, including untreated wastewater and agricultural residues.

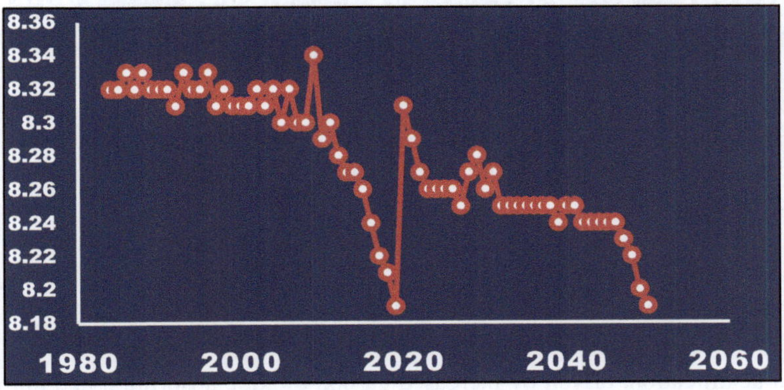

Fig. 1 pH level of coastal water in Digha during premonsoon

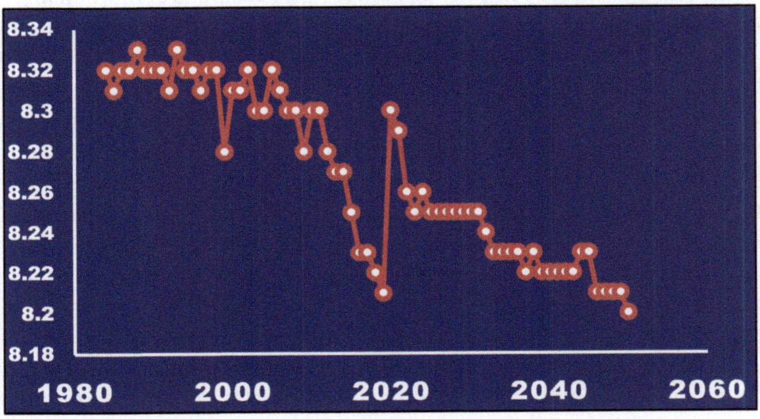

Fig. 2 pH level of coastal water in Digha during monsoon

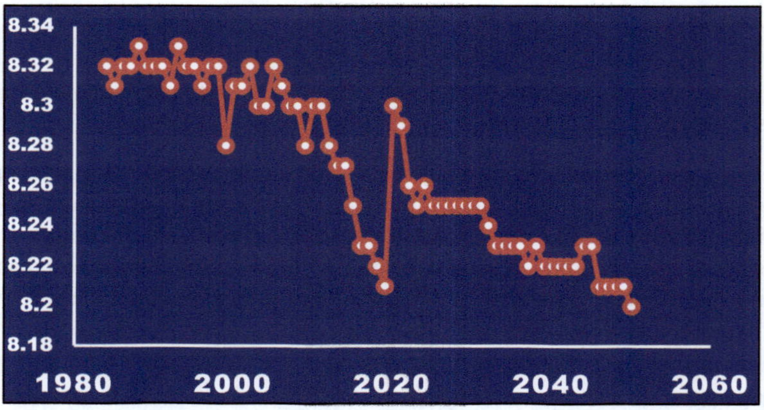

Fig. 3 pH level of coastal water in Digha during postmonsoon

- Postmonsoon: The mean pH value was 8.27, reflecting cumulative effects of monsoonal inputs and prolonged exposure to industrial and tourism-related discharges.

4 Discussion

Coastal acidification, driven by increased carbon dioxide absorption and untreated wastewater, significantly impacts coastal infrastructure and biodiversity. Acidification lowers seawater pH, accelerating the corrosion of materials used in seawalls, jetties, and buildings, thereby increasing maintenance costs and compromising safety.

Biodiversity, especially calcifying organisms like mollusks, faces severe threats. In the estuarine water, where the pH becomes low, almost less than 8.00–820, the mollusk communities face difficulties to keep their calcareous shell intact. This disrupts their ecological role as filter feeders, reducing water quality and altering food webs. The decline of molluscan populations also affects fisheries and aquaculture industries, causing economic losses for coastal communities. Acidification impacts the reproductive cycles and larval development of mollusks, leading to long-term population declines. Effective policy interventions, such as mangrove conservation and waste management, are crucial to mitigate these impacts and protect both infrastructure and biodiversity.

The findings of the present study emphasize the critical role of natural (primarily through mass scale mangrove afforestation and subsequent conservation) and policy-driven interventions in mitigating acidification impacts. Mangroves, through their organic carbon uptake and nutrient cycling, serve as natural stabilizers of aquatic pH [12–17]. However, their degradation due to unregulated tourism, expansion of shrimp farming activities, and urbanization diminishes their buffering capacity [18–21]. Understanding the salinity preferences of mangroves for optimal growth is crucial to enhancing their buffering capacity, enabling them to effectively mitigate the impacts of acidification.

Species like *Avicennia marina*, *A. officinalis*, *Avicennia alba*, and *Phoenix paludosa* thrive best in hypersaline environments (Figs. 4, 5, 6 and 7).

Excoecaria agallocha (Fig. 8) demonstrates robust distribution across varying salinity regimes, highlighting its broad salinity tolerance. However, its reduced biomass limits its effectiveness in addressing the challenge of acidification.

Fig. 4 *Avicennia marina* proliferates within the optimum salinity between 15–32 parts per thousand

Fig. 5 *Avicennia officinalis* proliferates within the optimum salinity between 15–32 parts per thousand

Fig. 6 *Avicennia alba* proliferates within the optimum salinity between 10–28 parts per thousand

Fig. 7 *Phoenix paludosa* proliferates within the optimum salinity between 5–28 parts per thousand

Thus, salinity should be given priority while planting mangroves as Nature-Based solutions.

The comparative stability of aquatic pH observed during the premonsoon season demonstrates the potential benefits of reducing anthropogenic activities. Temporary policy measures like lockdowns can significantly slow down acidification rates, offering a glimpse into long-term strategies for ecological balance.

The CRZ Notification of the Government of India can check the unbalanced human activities along with their negative impacts in the coastal zone of the country.

Fig. 8 *Excoecaria agallocha* proliferates within the optimum salinity between 5 and 30 parts per thousand

The Notification first issued in 1991 states that "no industrial operations and development can be initiated in specified coastal areas. This is applicable to coastal stretches of seas, bays, estuaries, creeks, rivers, and backwaters influenced by tidal action up to 500 m from the High Tide Line (HTL) and the land between the Low Tide Line (LTL); the notification mandated State Governments and Union Territories to develop Coastal Zone Management Plans (CZMPs) for identifying and classifying CRZs. Developmental activities were regulated under four zones: CRZ-I included ecologically sensitive areas such as national parks, sanctuaries, and heritage sites, prohibiting construction within 500 m of the HTL except for essential facilities like treated effluent pipelines and cooling systems. CRZ-II covered developed areas near the shoreline, allowing construction and reconstruction of buildings consistent with local laws and architectural styles. CRZ-III encompassed rural and less developed areas, introducing a 'No Development Zone' up to 200 m from the HTL while permitting limited activities like agriculture and forestry, and construction of hotels and resorts between 200 and 500 m with specific guidelines. CRZ-IV pertained to islands like Andaman and Nicobar, Lakshadweep, and smaller islands, imposing restrictions on construction within 200 m of the HTL and prohibiting activities like coral and sand extraction, dredging, and underwater blasting around coral formations. Certain projects, such as those related to defense, ports, thermal power plants, and investments exceeding five crores, need prior permission from the Ministry of Environment and Forests". A schematic representation of the CRZ policy is highlighted in Fig. 9.

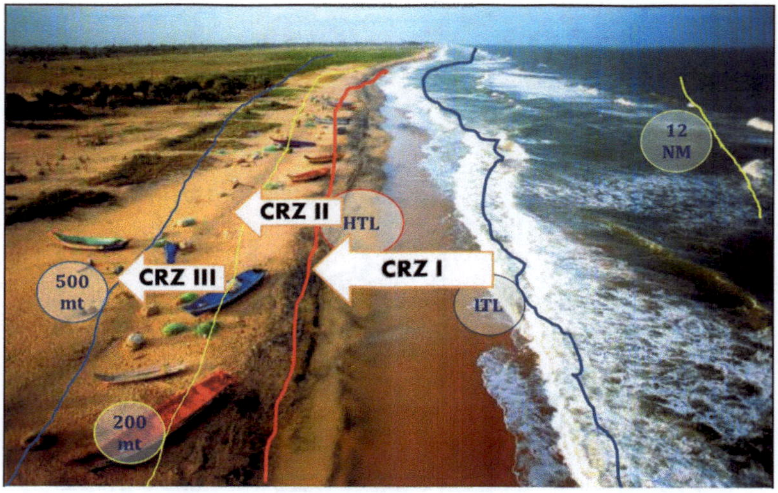

Fig. 9 Schematic representation of CRZ

Despite the existence of CRZ notifications, gaps in implementation and enforcement have allowed unchecked pollution and unregulated construction. Strengthening CRZ compliance, particularly concerning waste management and sustainable tourism practices, is imperative. Policies must also focus on community-based mangrove restoration programs to enhance natural resilience against acidification.

Integrating technological solutions like AI-based monitoring can complement policy measures. Predictive models can guide resource allocation and enforcement efforts by identifying the temporal pattern of acidification.

5 Conclusion

The coastal acidification observed along the Digha coast poses a significant threat to its tourism infrastructure and endemic biodiversity preferably the molluscan community. Seasonal variations underscore the interplay between natural buffering mechanisms and anthropogenic stressors. While premonsoon conditions offer a reprieve due to reduced activities, monsoon and postmonsoon seasons exhibit heightened acidification, emphasizing the need for sustained interventions.

Mangroves, as natural buffers, play an invaluable role in mitigating acidification but require targeted conservation efforts to maximize their efficacy. Strict enforcement of CRZ notifications, combined with sustainable waste management and responsible tourism practices, can pave the way for long-term ecological and economic sustainability. The integration of AI-based monitoring and predictive modeling offers a promising tool for adaptive management.

That can lead to environment sustainability simultaneously with the growth of the local economy. By prioritizing mangrove conservation and policy enforcement, the Digha coast can serve as salinity-based model.

References

1. Mitra A (2013) Sensitivity of mangrove ecosystem to changing climate, vol XIX. Springer, New Delhi, Heidelberg, New York, Dordrecht, London, p 323. ISBN: 10: 8132215087; ISBN-13: 978-8132215080. ISBN: 978-81-322-1509-7. https://doi.org/10.1007/978-81-322-1509-7
2. Cai WJ, Wang Y (1998) The chemistry, fluxes, and sources of carbon dioxide in the estuarine waters of the Satilla and Altamaha Rivers, Georgia. Limnol Oceanogr, 4:657–668. https://doi.org/10.4319/lo.1998.43.4.0657
3. Agarwal S et al (2019). Seasonal variability of acidification in major estuaries of Indian Sundarbans. Global J Engi Sci Res 6(4):493–498
4. Dutta P et al (2020) Reversing the phenomenon of acidification in the River Ganges: a ground—zero observation. NUJS J Regu Studies (Special Edition), pp 89–92
5. Saha A et al (2021) Inter-relationship between dissolved lead and sediment lead in the backdrop of climate change induced acidification. Chapter 14, In: Natural resources and their ecosystem services, Vol. II, Edited by Abhijit Mitra, Monruskin M. Calma and Shambhu Prasad Chakrabarty, published by HSRA Publication, ISBN: 978-93-5506-064-8, pp 195–199
6. Roy S et al (2021) Sewage disposal situation at Digha coast: a time series analysis. J Mech Continua Math Sci 16(11):62–68. https://doi.org/10.26782/jmcms.2021.11.00006
7. Mitra A, Zaman S (2016). Basics of marine and estuarine ecology. Published by Springer New Delhi, eBook ISBN: 978-81-322-2707-6, Hardcover ISBN: 978-81-322-2705-2, XII, pp 483. https://doi.org/10.1007/978-81-322-2707-6
8. Ghosh V et al (2023) Decadal variation of acidification of the estuarine water in the Indian Sundarban. J Environ Sociobio 20(2):101–111
9. Ghosh V et al (2024) Mangroves of Indian Sunderban: nature based buffer against estuarine acidification. Int J Advance Life Sci Res 7(3):130–139
10. Mitra A et al (2023) Climate resilient innovative livelihoods in Indian Sundarban delta. Published by Springer, Hardcover ISBN: 978-3-031-42632-2, eBook ISBN: 978-3-031-42633-9, pp 297, Vol XII. https://doi.org/10.1007/978-3-031-42633-9
11. Mitra A et al (2024) Mangrove carbon trading in the lower gangetic delta: drawbacks and the potential. Published by Springer, Hardcover ISBN 978-3-031-76315-1, eBook ISBN: 978-3-031-76316-8, pp 282, Vol XIII. https://doi.org/10.1007/978-3-031-76316-8
12. Singh G et al (2005) Nutrient cycling in mangrove ecosystem: a brief overview. Int J Ecol Environ Sci 31(3):231–244
13. Alongi DM (2014) Carbon cycling and storage in mangrove forests. Annu Rev Mar Sci 6:195–219
14. Kida M, Fujitake N (2020) Organic carbon stabilization mechanisms in mangrove soils: a review. Forests 11(9):981. https://doi.org/10.3390/f11090981
15. Mitra A, Zaman S (2021) Estuarine acidification: exploring the situation of mangrove dominated Indian Sundarban estuaries. Springer eBook ISBN: 978-3-030-84792-0, pp 402, Vol XII. https://doi.org/10.1007/978-3-030-84792-0
16. Arsad S et al (2022) Estimation of organic content and carbon uptake in Sonneratia alba (Smith) mangroves in aquaculture areas: a case study in the marine and brackish water cultivation area of Probolinggo City, Indonesia. Pol J Environ Stud 31(2):1019–1024. https://doi.org/10.15244/pjoes/140566
17. Padmakumar V, Shanthakumar M (2024) Mangroves: an organic carbon stabilizer. Chapter 2 In: Mangroves in a changing world: adaptation and resilience, Wetlands: Ecology, Conservation and Management. Springer, 13. https://doi.org/10.1007/978-3-031-67691-8_2

18. Ashton EC (2008) The impact of shrimp farming on mangrove ecosystems. CAB Reviews: Perspect Agri Vet Sci Nutri Nat Res 3(3). https://doi.org/10.1079/PAVSNNR20083003
19. Jusoff K (2013) Malaysian mangrove forests and their significance to the coastal marine environment. Pol J Environ Stud 22(4):979–1005
20. Prudente TA et al (2017) Degradation of mangrove ecosystems due to shrimp farming activities and its conflicts of interest in the State of Bahia, Brazil. Modern Environ Sci Engi 3(4):224–230
21. Mishra S (2023) Impact of urbanization on declining mangrove health of Goa. Int J Res Publi Rev 4(1):1566–1574. https://doi.org/10.55248/gengpi.2023.4144

"Saffron and White Terror on the Internet and Use of Artificial Intelligence". Trends of Right-Wing Extremism in India and America

Surodeep Sanyal

Abstract The study centers around the rising trends of hatred toward particular communities bolstered by conniving governments giving a fillip to extreme right-wing terrorism and propaganda. The divide between communities is deep and pervasive and can be traced to racist attitudes where historically there has been segregation, based on origin, culture and advantage. It had been the manifesto of the Bharatiya Janata Party to digitalize India, but it was a camouflage for promoting Hindutva through use of the internet and smart phones. With the Government looking the other way, extreme right '*saffronists*' began to spew venom on the minorities by use of artificial intelligence to ridicule and discredit the Muslim populace keeping well within the limits of freedom of speech though violence is encouraged. The West, especially USA, became the target of extremist Muslim groups like Al Quada. These provided the casus belli for the creation of Extreme Right Wing, Alt Right. In the USA, artificial intelligence has been used by the Alt Right for organizing programmes of hatred and extremism without abatement. The right to freedom of speech allows uncurbed rights of expression even if it encourages violence and extremism. Monitoring and censoring artificially created content on the internet are a mammoth task which suffers owing to various factors including bias and imprecise classification of data.

Keywords Right-wing extremism · Artificial Intelligence · Hate speech · Alt-right · Social media

1 Introduction

This paper aims to explore the ways in which artificial intelligence (AI) and internet technologies—particularly social media—are being leveraged by extremist groups to disseminate propaganda and attract recruits. It highlights the serious risks these

S. Sanyal (✉)
Lecturer, Jindal Global Law School, O.P. Jindal Global University, Sonipat, Haryana, India
e-mail: surodeep.sanyal@jgu.edu.in; surodeep80@gmail.com

M. Chakraborty et al. (eds.), *Proceedings of Sixth International Ethical Hacking Conference*, Lecture Notes in Networks and Systems 1513,
https://doi.org/10.1007/978-981-96-8632-2_34

technologies pose when misused, alongside the current gaps in regulation, ethical oversight and polarizing agenda of conniving governments. By shedding light on this critical issue, the paper encourages thoughtful discussion among policymakers, technologists and scholars about how to respond to these emerging threats.

It is tough to imagine life disconnected from the digital infosphere and social media. The ways in which our interactions, information access and approach to daily life are shaped [17]. That is the use of artificial intelligence. Human existence and the digital environment are so merged that it seems unlikely to conceive them separately [17]. Social media can be defined as 'a conglomeration of internet—based applications that build on the ideological and technological foundations of Web 2.0 and allow the exchange after creation of User Generated Content' [11].

Artificial intelligence is the product of the human brain—to simulate the designs of the creator as closely as possible. The design laid is through programming made by individuals which is its heart and soul. It can generate business and engage in a host of salutary applications including replicating real-time cinema personalities, but it has potential to engender monsters when put to radical use to foment hatred. The key issue in this essay is the use of artificial intelligence and internet for radicalization and indoctrination to extremist beliefs by targeting groups that have been identified for chastisement and effacement. By use of artificial intelligence violent fallout postings, signs, memes and humor are created that are calculated to harm the targeted races or religious groups. The uncensored and uncontrolled use of the internet by the extremist group is to influence like—minded individuals and groups to plan and programme assaults, whether mental, psychological or physical. The juxtaposition of the right to free speech and expression against the calculatedly indiscriminate use of 'hate' as a tool is to oppress an outgroup by an in group. In India, the outgroup would be Muslims, Dalits, Christians and Hindu liberals while in USA it would comprise of Muslims, Jews, dark-skinned and immigrants. The in-group in India are those that have been indoctrinated into the Hindutva or Saffron ideology and in the United States, it is the Whites. Interestingly, there is a middle group common to both countries which tends to be more rational in their approach and eschews violence. The moderated group 'alt lite' and the 'lite Hindutva' or "Raitas".

2 Causes for the Extremist Genre in India and the USA

Imbalances in races exist in both countries though Constitutional provisions strictly vouch equality in its expanded sense. Dissensions are voiced against immigrants, Muslims and people of other races who have settled in the USA largely through artificial intelligence (AI). Historically, there is a difference between the two. In the USA, the dark skinned and the aborigines were subdued by white skinned settlers while India was invaded by Muslims from Persia, conquered and subdued. There had been latent discontent, but the equality clause was brandished to keep right wing saffron extremism at bay till the Bharatiya Janata Party gained conclusive majority in Parliament. This unleashed the Hindu backlash against perceived Muslim threats of a merger with Pakistan, fueling separation of Kashmir; the attack on the Taj Hotel in Mumbai; the ineptitude of Governments to ensure a uniform civil code; the proclamation by General Pervez Musharaf of the future Islamization of India and the protection of property through the Muslim Auquf. Hatred against the Christian minority is less perceived and is attributable to conversion schemes fostered by English missionaries in colonial times, bestowing advantages and to the present missionary efforts of foreign funding to alleviate the sufferings of the poor, offering free quality education and placements. In the USA, radical right-wing sentiments are a natural corollary to hate messages of Al Qaeda and its regional affiliates, AQIM and AQAP which constantly craft messages highlighting battles of civilization against the USA [11]. The militant antagonism to coloureds is the perceived equality in status granted to those that the Whites enslaved before the union of the North and the South. This in turn yielded to the coloured, equal voting rights and a say in the Government and its laws. The immigrants have come under the wrath for have settled in a country which belongs to whites, having equal protection, a contender in the job market and a creating a decisive stake in the economy. In the United States, it is the ascendency of the white race as identifiable with Caucasian blood [18] while Savarkar preached it in his book. Both, White Right and Saffron Extremists have drawn on Aryan supremacy and origin to militate against those identified as belonging to subordinate races. Saffron extremists have sympathizers in the West.

"Racists exist in societies that are racially demarcated with history of racist oppression and in hierarchies where race is associated with privilege and liability. Racially conflicted societies clearly exist where the remnants of colonial regimes have left unresolved inequalities. Racially associated resentments can continue for many generations where it has proved as yet impossible to ensure equity between racial groups" [10]. This is absolutely the case in India which had been colonized by the British for two hundred years, and a wedge was driven between Hindus and Muslims. India was partitioned in 1947 and Pakistan was created as dedicated Muslim State while India adopted secularism. Right-Wing extremism existed in India as Hindutva in 1920 and beyond with Savarkar's exposition of "Who is a Hindu" in 1923 in a book named "Hindutva" [19]. Berwick [2] A European declaration of independence, pointed out that the only positive thing about the Hindu right wing is that they dominate the streets. Tolerance for current injustice is absent and often when things get out of control there are riots and attacks on Muslims, usually after the Muslims disrespect and degrade Hinduism repeatedly … India will continue to perish unless the Indian nationalists consolidate properly and plan a strike to win. It is of utmost importance that the European and Indian resistance movements learn from each other and cooperate as much as possible. Their vision & goals are mostly identical.

2.1 Digitalization of India a Façade for Promoting Hindutva

Digitalization of India was a manifesto of the Bharatiya Janata Party and internet services received a fillip after the present Government came to power in India. The manifesto camouflaged the basic goal of establishing Hindutva through the internet and artificial intelligence. With the boost of digitalization, the agenda to propagate Hindutva by Saffron extremists was facile, far from the archaic struggle to spread transnational fascist ideals of the Third Reich and Mussolini's "Black Shirts" in India concept through covert gatherings and secretly printed manifestos [13]. "Right-wing extremism in Europe emerges from philosophical exchanges in the nineteenth century that gave rise to the development and spread of fascist" trends in India [13]. Alt right and the Saffron Reich use common techniques utilizing the supersonic, cost-effective internet to further ideologies and create virtual sympathizers and groups "avoiding" "gatekeeping" and filtration by "journalists" [11]. The journalist is created through AI and the content is unbound, without scrutiny and verification.

2.2 The Moderates Are Pilloried

The rise of an extremely militant organization, organic in nature, was never conceived of by the BJP when it came to power. They are the 'traditionalists' self-styled warriors with narrow interpretation of Hindutva—the actual saffron terror. They won't stop at anything and justify all hate actions as a retaliation. This group assimilates young upper caste Hindus to hate Muslims and Dalits. It breaks the structure of the BJP and often Modi is lampooned. The Government makes concessions to minorities and downtrodden Dalits by flashing the reasonable classification card under Article 14 of the Indian Constitution. The difference between the Trads and the Alt-White is that while Trump was popularized and praised by the Neo Alt-Right, Modi is disliked as being "soft". The BJP is accused of downplaying the roles of Hindu nationalists like Nathuram Godse (who shot Mahatma Gandhi), Dara Singh (wrestler and actor) and many others. This is the Neo-Nazi Indian extremist group which, in its intensity, shames the moderates. Youths join, not for profit but only because they hate Muslims and Dalits.

There is a divide between this group and the alt-lite who are helpless to stop the surge of hatred. The Extreme group has its parallel in the USA and work on similar lines. The extremists call themselves 'Trad' as opposed to the 'Raitas', the moderates. The Raitas are considered as botching up the Hindutva movement and most Raitas are loath to associate with them. In fact, the apprehension and censure of the 'Trad' like in the case of 'Bulli Bai' and 'Sulli' cases are applauded by the Raitas. Trad, Raitas, Alt Right and White nationalists are of the same genre with varying inflexions.

3 The Government Connives in the Creation and Use of Hateful AI

It was the agenda of the Bharatiya Janata Party to promote the use of smart phones by every citizen which included laborers and villagers. It was an essential tool to promote Hindutva and rightist saffron propaganda. In a country with a high rate of illiteracy, the propagation of Islamophobia through platforms which depends on the internet was facetious and speedy. Smart phones supported AI created memes, visuals and images. The Indian corporate sectors were brought in to provide mobile and internet connections with the purpose of lowering internet rates which is now the cheapest in the world. With a conniving Government in power, saffron extremists engineered AI and flooded the internet, WhatsApp, Facebook, twitter and YouTube with extreme right-wing propaganda to distort and project Hindu supremacy. Artificial intelligence is of a relatively later origin and has been pressed into service for electioneering as well as by the extreme rightists to warp, twist and disseminate information. The villager and the common Hindu on the street easily digested the AI circulated visuals and memes easily available on WhatsApp and YouTube in their smart phones.

All Muslim rulers, including the Moghuls, are discredited as plunderers. The Saffron Rightists used conventional methods of Al Queda and other violent extremist organizations that use Information and Communication Technologies (ICT) to transmit ideological purity and vision, a way to radicalize, based on vision-based leadership where emotions are systematically exploited to project past greatness and a past ideal state [9]. The Ideal State of mythical Lord Rama has been projected and purity is espoused by worship of the 'Holy Cow'! AI images are created exalting Rama's reign and deifying the cow to woo extreme right-wing followers, the Trads.

3.1 *Misinformation and Morphed Images*

Violent Extremist Groups utilize the internet to establish virtual communities and sympathizers [9]. On 4th January 2021, two young people, 18 years old, Shweta Singh and her male friend Vishal Jha, 21 posted artificially morphed twitter pictures of Muslim women who were to be sold at the auction. The posting termed it as the 'deal of the day'. Another similar instance was the Sulli deals as 'an open—source project' which posted on GitHub personal details of around one hundred Muslim women who were to be auctioned.

3.2 They Use Conventional Methods to Distort

The Alt Right and Saffron extremists use conventional methods of recruitment and exhortation to create mass support and loyalists by use of YouTube, Facebook, WhatsApp, etc. Artificial intelligence is developed to keep track of the interest of the person. Once the net is open and something is viewed, there is constant reference to similar articles, visuals or memes. People wanting to absorb information of a particular genre are invited to go deeper. Inclination to suggestions of the glory of Hindutva or Alt Right by repeatedly foisting visuals, memes and propaganda does radicalize, even though the messages have been concocted.

3.3 Replication Fosters Duplication

Persistent replication of content based on past predilections of the user by repeating messages several times [11] so that "we expose ourselves to the possible radicalization of our positions and undermining our deliberative capacity" [17]. Persistence replicability fosters duplication, altering content which invites different interpretations [11]. By use of artificial intelligence 'Pepe the Frog' is distorted and has been put to twisted uses as the Right-Wing Terrorists have done in the USA [12]. Artificial intelligence creates and depicts the green frog with a skull cap to mimic Muslims and the blue frog signifies "Bhimtas" (in derogation of the Dalits).

"On YouTube, the far right form a shared identity and get denser and more centralised over time by centring around problematic topics such as conspiracy theories…" [11]. The conspiracy, as perceived by the Trads is the proclivity of the Muslims towards procreation and multiplication so that majority is soon achieved in a democratic state and Islamization, a reality. The 'Trads' encourage Hindu upper caste women to have more children to maintain the Aryan ethnicity and are against the 'softness' of the Raitas. Trad humor is hateful and provocative. It is designed to send shock waves through its targets. Artificial intelligence is used to create memes of Muslims being beheaded, caricatures of Muslims being mowed by cars and Dalits cockroaches being gassed by aerosol. Dalits and Muslims are projected subjects of elimination in concentration camps, being urinated upon by saffronized Pepe the Frog. Artificial intelligence mimics poor Pepe and reduces it into an object of hatred and ridicule: it could not have suffered a worse mortification!

3.4 Scalability—The Wooing for a Loyalist Base

It is the endeavor of the far right to add loyalists and increase its base. "Scalability plays an important role, especially in the far right's ability to reach new audiences on social media. With posts for the broader public, they aim to recruit new members to support their world view” [11]. The Hindu extreme has educated right-wing extremists who are technologically adept and can manufacture and post everything they want by using artificial intelligence. There have been emojis on WhatsApp showing Muslims as demons being pitted against orange flags depicting the might of Hindutva. Altered pictures, created through AI, blame Muslims for attacking Hindus and sacking their homes without reference to the time and place. Fake live visuals are created through AI to distort and gain attention.

3.5 Emulation

In line with the Alt-Right Extremists, the Trads have begun to openly resort to trolls and doxing, but the alarming issue is the adoption of the Neo Nazi methods of violence and its praise thereafter. A group of Muslims gathered to protest against intolerance at the Jamia Masjid in Delhi. The protests at Jama Masjid and Shaheen Bagh were prolonged and during its continuance, Trad believers sneaked in and shot at the Muslims. Before the shooting, one of them streamed live suicidal messages on social media and one of his posts read “Shaheen Bagh, game is over.” They were later applauded for the deed which is in line with the Christchurch incident or the Pittsburgh Synagogue attack. The parallels are inescapable and the influence of the Alternative Right-Wing Terrorist and the Neo Nazi in India cannot be ignored, though the narrative reveals that it is the West which has fomented the extreme right hatred in India. A parallel is drawn between MAGA (Making America Great Again)—Trump’s electoral slogan and the trad’s agenda of overhauling the Indian Constitution.

When the Covid 19 was raging, Muslims inappropriately congregated for three days in New Delhi, infamously known as the “Tablighi Jamaat”. It was a religious congregation of Muslims from all over the world. Permission must have been granted for such congregation as Muslims from foreign countries gathered for the three-day event. The immigration formalities said so. The opportunity was seized by the saffron extremists to vilify Muslims as the ‘super spreaders’ and ‘Corona Jihadists’ on twitter and WhatsApp [14]. False news regarding Muslims’ attempt of infecting otherwise healthy people was fanned: it spread like wildfire [16]. Memes were the offspring of artificial intelligence to portray Muslim mischief.

3.6 AI is Useful for Electioneering

News Website “Wire’ revealed that BJP operatives used artificial intelligence ‘Tek Fog’ to hijack inactive WhatsApp accounts to disseminate propaganda at the political level. The extremist propaganda machine seizes all opportunities on the Internet and its associated platforms because of the speed and easy diffusion amongst the mass. In the run up to the 2019 elections, 4 researchers gathered 2 million WhatsApp posts from 5000 political groups which was then narrowed down to 2700 useful non span posts. Politically, the saffron right has tacit Government support. In the run up to the 2019 elections, negative postings were made to derail the Congress challenge. AI was used to mimic Congress and other leaders so that they were presented as fools, suggesting that the leader did not have enough intelligence to run the country. It was again the conventional tool adopted by the BJP which used "covert discriminatory or neutral messages that align with community guidelines or link to more overt discriminatory content elsewhere. By doing so, they encourage radical thoughts and hate speech in user comments while avoiding liability for it" [11]. Instead of saying that Muslims are traitors, the posts said that by 2050, the Muslims would be in the majority in India. The BJP was ahead of all others in posting WhatsApp messages and the fear messages suggested that Muslims incite disharmony and marry Hindu girls to undermine Hinduism [3]. Most of the messages are AI curated to spread fear amongst Hindus. Fear is the key and polarizes voters in a democracy that ensures majority rule and laws that reflect majority views. Negative propaganda is sublime—it hurts but does not offend. The Constitution of India upholds free speech and expression in Article 19(1)(a) but such right is subject to reasonable restrictions and free speech cannot be construed to violate other Constitutional rights but if AI is put to service, it depicts and creates visual images and memes which distort the actual picture and instead of hurting the other community, it visualizes the actions of the opposing community in such a way so that resorting to saffron extremism and violence is vindicated.

4 Helplessness Owing to the Right of Free Speech

The Aryan Nation Liberty Net, originated in 1984 under the auspices of Louis Bean of the leaderless resistance who created a Bulletin Board System (BBS). It was accessible via telephone in Idaho, Texas and North Carolina. Anyone having a computer and a modem could dial up the BBS and gain access to the hate propaganda and leave his own message on it. The names and addresses of those who were enemies could be found on it. It was perhaps the world's first doxing attempt (collection of private data of groups or individuals with requests to harass them online or for real world purposes). Around the same time, two others founded Right Wing Extremist propaganda machinery through BBS. They were George Dietz, owner of Liberty Bell Publications and Tom Metzger, founder of White Aryan Resistance. The White Aryan Resistance set up WAR newspaper and television. It was with the introduction of the World Wide Web in 1990 that hate propaganda really began [5]. Contemporary racism was built through the net over the years because of certain advantages:

i. Freedom over control facilitates technologies that magnify the anonymity of those who espouse the cause.
ii. Freedom of speech and expression in the USA has been zealously defended and therefore censorship of hate speech could not be done.

iii. Internet allows billions to interact individually or in groups directly without exposure, maintaining anonymity [10].
iv. Our experience in social media can be so repetitive and self-validating that it promotes the development of epistemic bubbles: we do not have access to all the relevant data we need, but only the data that adhere to our interests. Anything that does not collimate with these interests is cut off [17].

4.1 The Right is Abused

Stormfront's Don Black, a website created in 1990 for RWE has a membership of over 343,000 members and 13 million posts and is the oldest to survive till date. It has diversified into 'Stormfront Downunder', 'Stormfront en Francais' and 'Stormfront Italia' to pepper White supremacy [8]. From 1990 onwards, common enemies have been targeted by other websites like American Renaissance and VDARE. Andrew Anglin of the 'Daily Stormer' gained notoriety when he wrote an article abusing and mocking Heather Heyer on the website. Heyer was killed by an RWE fanatic. This was followed by a call to harass the mourners of Heather. Major web hosting companies refused to host the site and it disappeared into the Dark Web only to reappear in 2018 as 'The most censored Publication in History.' It diversified further in Greek, Italian and Spanish language sections. The site is antisemitic, anti-Islam, anti-immigrant and incites whites against dark skinned. There are similar websites like Breitbart News, Alex Jones and InfoWars. RWE's heavily outlink from stable and traditional websites like RT, Fox News and Daily Mail [5]. With a pan European circulation on twitter, the RWE's group known as Generation Identity has about 70,000 official accounts and Facebook, which is more private, is accessed by the RWE through the public pages as Britain First, Patriotic Europeans, Occident (Pegida) and Soldiers of Odin (SoO). [5]. The RWE has a major share in YouTube postings through White Music [6]. AI is used again to launch attacks by the Alt- right.

4.2 The Channels Compete to Outbid

Censorship hurts the hosting website owing to withdrawal of viewers. This happened with 4chan which in which the meme culture of the RWE found fertile postings. 4chan suspended an extremist posting and banned the account which led to criticism from other RWE accounts. Seizing this opportunity, 8chan was conceived as an extreme version of 4chan without limits. Immediately, RWE accounts surged in 8chan.

8chan has been implicated directly in three major hate cases which are high profile RWE attacks. On March 15, 2019, Brenton Tarrant killed 50 Muslims in mosques in Christchurch and before the attack he posted a manifesto on 8chan announcing his intentions. John Earnest, who killed a woman in a synagogue in Poway and August Patrick Crusius, responsible for the murder of 22 shoppers in a Walmart in

El Paso had all posted their manifestos on 8chan. Along with 8chan and practically for promoting itself, Gab was established in 2016 when other major hosts cracked down and suspended RWE accounts [15]. Robert Bowers who created a mayhem at a Pittsburgh synagogue and killed 11 worshippers had made 998 postings on Gab expressing his intentions beforehand. His profile picture featured the neo-Nazi code '1488'. An analysis of the Gab posts showed Bower's extreme anti—semitic sentiments, referring to them as 'kike' and 'jew' [5]. Gab was unavailable after the synagogue shooting but resurfaced with anti-semitic and anti-Islam conspiracy theories.

4.3 A Dreadful Community is Born

More gruesome are the sentiments of the RWE towards their perceived enemies and even death does not end the recital. The killers were hailed by the extremists, as in India, with shit posting rating their 'achievements". Tarrant was given the highest points and hailed like a saint with halo for killing 50 Muslims while the Poway killer was ridiculed for his low score and less time being spent in preparations on 8chan. Owing to there being little or no control over the internet, postings are absolutely uncontrolled.

The Alt-Right heavily depends on AI to use sarcasm, humor, memes and jokes, all intended to hurt their perceived enemies. Pepe the Frog, an innocent simple meme has been twisted out of proportions to balloon anti—semitic, anti-Islam and anti-immigrant sentiments. They make use of 'shitposting' which include posting of hate speech and support for racist sentiments [7]. Online image boards like 4chan were full of humorous texts. Before ploughing his car through the crowd of protesters at Charlottesville, James Alex Fields Jr. posted images of a car ploughing through protesters with the caption "You have the right to protest, but I'm late for work" [5]. The images were artificial. The same posting was made by Saffron extremists before the shooting in Shanbagh. Notoriety percolates and is transnational. A dreadful community is born.

4.4 Moderates Lampooned as Alt-Lite

In 2010, Richard Spencer floated a website which he named Alternative Right and though the center was the USA it spread to the Netherlands and Germany quickly. Alternative Right philosophy is heavily influenced by Nouvelle Droite movement in France. The ideology is based on cultural awareness as with the Hindutva in India [1]. A distinction can be drawn, if possible, between RWE and White Nationalism, the same distinction almost between 'Trads' and 'Raitas' in India. White Nationalists denounce Alt-Right as fundamental racists though the leaders still maintain anti-immigration and pro-western beliefs in less radical tones. In India, the moderate

Hindutvas are referred to as people who botch up and in USA, the New White Right are derisively referred to as 'Alt—lite'. The BJP used 'Tek fog' to usurp inactive WhatsApp accounts before elections in 2019. The New Alt-Right used the twitter handle with four overlapping themes:

i. support for the President, Donald Trump
ii. support for White Nationalism
iii. opposition to immigrants, Muslims and minorities
iv. primary devotion to transgressive trolling and harassment.

MAGA (Make America Great Again) Trump's publicized policy was aggressively used as in India, the "India First". There were other features that the New Alt-Right circulated on twitter. They were:

(a) Pro Trump content with perceived attacks on the critics of Trump
(b) white nationalist content
(c) general far right content
(d) anti-immigration and anti-Muslim sentiments
(e) trolling and shit posting which were deliberately aggressive and provocative set in a humorous tone
(f) conspiracy and fake news.

Through artificial intelligence, the appetite of the voter is gauged and he is fed on that which the Alt Right or the Saffron extremists believe is ideologically in line. Most of these were common to the election scenario in India in 2019 and shit posting of fearful Muslim ascendency was common on WhatsApp as was fake news [1].

5 Warping Pluralistic Thinking

Tim Brenners-Lee, the originator of the World Website admits that his dream of bringing together families and building bonds across nations has produced a monster of hate [10]. Burke Jason said "Technology is Terrorism's Most Effective Ally. It Delivers a Global Audience." The goal is to increase radical mass base through recruitment. "Messages that are overtly radical can lead to rejection from the public sphere, along with evolvement of echo chambers that no longer reach the general public but instead—aided by individual societal and technological level influences—lead to an increased radicalisation due to lack of opposing opinions" [11].

5.1 Little Control Over AI

There is little control over AI generated terrorist propaganda on the net because the USA religiously maintains its policy of free speech though, as indicated earlier, India is not so bound. There might be lack of control over the internet and AI but at the same time there is the possibility of supervision and in most cases, action was taken after the incident occurred. All fundamentalist racial killers from Pittsburgh, Charlotteville, El Paso, Poway to Shanbagh had announced their manifesto, propagated AI created memes and visuals, to publicize their intention of carrying out violent attacks. But they were not apprehended because suspicion cannot be used to detain a person. Curbing and censoring the flow of extreme racist messages calculated to harass and harm can be intercepted. This has been done after the Christchurch attacks where about 12 employees of YouTube spent the whole night censoring the video of the massacre which were being posted and replicated every minute under different tags using AI. After the Christchurch incident the New Zealand Government ensured 'a commitment by governments and tech companies to eliminate terrorist and violent extremist content online' [4]. The USA lagged behind because of the First Amendment which ensures free speech [20]. In India, there has been a willingness lately to curb extremist content from the net as was witnessed in the Bulli Bai and Sulli cases. "As VEO content appears online it swims in an ocean of other unrelated content, leaving the noise to signal ratio very high. To solve the detection problem, manual approaches like content coding, or labelling have been standard. On one hand manual approaches have proven to be inappropriate to the task of identifying and treating content at a scale. Human understanding is the "gold -standard" of classification, but with respect to VEO content it is a poor strategy due to aspects like coder burnout and potential errors and bias due to inexact classification definitions" [9].

Hate speech and AI generated content on the internet can be controlled but it appears that major web companies are in competition to liberalize their use and gain more accounts as has been noticed with 8Chan and Gab. "What is becoming increasingly clear is that the machine learning algorithms currently used to select content in social media tend to discourage the development of critical pluralistic thinking due to the arbitrary selection of data to which we have access" [17].

6 Conclusion

International legislation promulgated through agreements respecting the comity of nations may be a way forward to fathom the use of AI as a weapon to target communities through downright rank hate speeches that instigate terrorist attacks and polarize communities. Its success depends upon a consenting international community with the caveat that there will always be the rogue to sabotage the process. Artificial intelligence is based on programming and preset data. The Polarization agenda, sponsored by democratic governments, based on the will of the majority is invariably the will of those yielding power by stifling the common voice. Artificial intelligence is doctored to achieve a goal, an agenda or a purpose and it needs taming like a shrew. Often hacking is resorted to for decoding dangerous material on the internet and if that be so, extremist material on AI can be hacked and nullified. If it is possible to preset extremist data on AI, there is also the counterpossibility of the destruction of such data through intelligence fed in the programming. The choice and panacea lie in Government policies alive to the declaration that Governments are "of the people."

To conclude, this paper examines the growing concerns surrounding the use of AI and social media platforms in promoting extremist ideologies. While identifying the challenges is vital, it is equally important to consider practical steps forward. One such step involves implementing more robust global regulations that require technology companies to be transparent about the workings of their algorithms and more accountable for the content they allow to circulate.

Additionally, fostering research into ethical AI development must become a priority. Developers should be incentivized—and where necessary, mandated—to integrate safeguards against bias and misuse into their systems. Education also plays a pivotal role: enhancing digital literacy across communities can help individuals better navigate online spaces and critically evaluate the content they encounter.

Lastly, collaboration is essential. Governments, academic institutions, tech companies and civil society organizations must work together to design policies and tools that are resilient against digital extremism. Only through such cooperative and cross-sectoral efforts can we begin to address the risks posed by AI-driven radicalization and online misinformation.

References

1. Berger J (2018) The alt-right twitter census, defining and describing the audience for alt-right content on twitter. VOX-Pol Network of Excellenc.
2. Berwick A 1475 (2083) A European declaration of independence, viewed 5 May 2022. https://info.publicintelligence.net/AndersBehringBreivikManifesto.pdf
3. Bhatia R (2017) The Year of Love Jihad in India. The New Yorker, viewed 3 May 2022. https://www.newyorker.com/culture/2017-in-review/the-year-of-love-jihad-in-india
4. Christchurch Call 2019, Christchurch call to eliminate terrorist & violent extremist content online, viewed 6 May 2022. https://www.christchurchcall.com/
5. Conway M, Scrivens R, Macnair L (2019) 'Right-wing extremists' persistent online presence: history and contemporary trends. ICCT Policy Brief
6. Ekman M (2014) The dark side of online activism: Swedish right-wing extremist video activism on youtube. J Media Commun Res 56(1):79–99
7. Evans R (2019) Shitposting, inspirational terrorism, and the Christchurch mosque massacre. Bellingcat, viewed 29 April 2022. https://www.bellingcat.com/news/rest-of-world/2019/03/15/shitposting-inspirational-terrorism-and-the-christchurch-mosque-massacre
8. Galloway B, Scrivens R (2018) The hidden face of hate groups online: an insider's perspective. VOX-Pol network of excellence blog, viewed 30 April 2022. https://www.voxpol.eu/hidden-face-hate-groups-online-formers-perspective
9. Hall M, Logan M, Ligon GS, Derrick DC (2019) Do machines replicate humans? Toward a unified understanding of radicalizing content on the open social web. Policy Internet 12(1):109–140. https://doi.org/10.1002/poi3.223
10. Jakubowicz A (n.d.) Alt right lite: trolling, hate speech and cyber racism on social media. Viewed 5 May 2022. https://doi.org/10.51/ccs.v913.5655
11. Kakavand AE (2024) Far-right social media communication in the light of technology affordances: a systematic literature review. Annals of the international communication association

12. Koebler J (2016) Hillary Clinton is right: pepe is a white supremacist. Motherboard, viewed 3 May 2022. https://motherboard.vice.com/en_us/article/hillary-clinton-is-right-pepe-is-a-white-supremacist
13. Leidig E (2020) Hindutva as a variant of right-wing extremism. Patterns of Prejudice 54(3):215–217, viewed 4 May 2022. https://doi.org/10.1080/003132XX.2020.1759861
14. Nabeel F, Raashed M (2021) Online and offline manifestation of Hindutva ideology in India During the COVID-19 spread. Margalla Papers, issue 1
15. Phillips M, Bagavathi A, Reid S, Valasik M, Krishnan S (2018) The daily use of gab is climbing. Which talker might become as violent as the Pittsburgh synagogue gunman?. The Washington Post, viewed 27 April 2022, https://www.washingtonpost.com/news/monkey-cage/wp/2018/11/29/the-daily-use-of-gab-is-climbing-which-talker-might-become-as-violent-as-the-pittsburgh-synagogue-gunman
16. Prasad A (2020) The organization of ideological discourse in times of unexpected crisis: explaining How COVID-19 is exploited by populist leaders. Leadership 16(3):294–302
17. Rodilosso E (2024) Philosophy and technology. Philos Technol 37(71). https://doi.org/10.1007/s13347-024-00758-4
18. Roose K (2017) This was the alt-right's favorite chat app. then came Charlottesville. The New York Times, viewed 3 May 2022. https://www.nytimes.com/2017/08/15/technology/discord-chat-app-alt-right.html
19. Savarkar V (1923) Hindutva: who is a Hindu?, Veer Savarkar Prakashan, Bombay
20. Taylor H (2019) Domestic terrorism and hate crimes: legal definitions and media framing of mass shootings in the United States. J Policing Intell Counter Terrorism 14

Symbiosis of Forest and Indigenous People: A Micro Level Analysis with AI Integration in West Bengal

Santanu Panda and Barnali Chakrabarty

Abstract Forests, as invaluable natural resources, are a gift from nature that should be utilized equitably while safeguarding biodiversity for future generations. Indigenous communities are deeply connected to forests, relying on their resources for livelihood and sustenance. Their close relationship with the environment positions them as exemplary stewards, having conserved and managed forest resources through sustainable practices for generations. Agriculture, particularly shifting cultivation, and the collection of forest products such as timber, honey, wax, and fruits form the core of their subsistence and culture. However, with advancements in technology, artificial intelligence (AI) has emerged as a powerful tool for managing natural resources, including forests. This paper explores the profound relationship between forests and indigenous communities and highlights the potential of AI in enhancing forest management. The study reveals a strong interdependence between indigenous people and forests while emphasizing the need to incorporate AI-driven solutions for effective conservation and sustainable resource management.

Keywords Indigenous people · Forest · Natural resources · Utilization · Artificial Intelligence

1 Introduction

India is a land of diverse geographical features that attracts people from all over the world. The country boasts of six major relief features: the mountains, norther and coastal plains, desert, Plateau, and the Islands. Each of these regions offers unique landscapes and contributes to India's rich biodiversity and cultural heritage. When

S. Panda (✉)
SACT 1 Sukumar Sengupta Mahavidyalay, Midnapore, India
e-mail: santanupanda@nujs.edu

B. Chakrabarty
Bikash Bharati Law College, University of Calcutta, Kolkata, India

M. Chakraborty et al. (eds.), *Proceedings of Sixth International Ethical Hacking Conference*, Lecture Notes in Networks and Systems 1513,
https://doi.org/10.1007/978-981-96-8632-2_35

we talk about the bio diversity it is worthy to mention that according to the ISFR 2021 report, the total forest coverage of India is 7,13,789 km^2 of India [1].

After the agricultural sector, the forest sector is the second-largest in terms of land use on which approximately 300 million tribal and rural communities depend. For nearly 70% of the country's non-urban population, fuelwood is the primary source of domestic energy, and forests provide crucial resources for 100 million people, offering cash income from fuelwood, non-timber forest products (NTFP), and construction materials.

Forests not only serve as a source of energy but also provide food, including raw products like fruits and edible plants. Many indigenous communities also cook the food they gather. Their affinity for forests is deeply rooted in their cultural and spiritual beliefs, often tied to traditions and practices passed down for generations. Many tribes believe that abandoning their traditional ways could invoke divine punishment, reflecting their inseparable bond with the forest ecosystem.

India is home to 45,000 wild plant species, of which 9,500 have ethno-botanical importance. Among these, 7,500 species are used in indigenous medicinal practices, highlighting the forests' role in traditional healthcare. Furthermore, 3,900 plant species are used as food, including 145 species of roots and tubers, 521 leafy vegetables, 101 bulbs and flowers, and 647 fruit varieties. Additionally, forests provide 525 species for fiber, 400 for fodder, 300 for chemical preparation (natural insecticides and pesticides), and 300 species for gum, resin, dyes, and perfume extraction.

This diversity underscores the critical role forests play in sustaining the livelihoods, culture, and traditions of indigenous communities while emphasizing the need for sustainable forest management to preserve these invaluable resources.

- **Present Scenario of forest in India**

India has a total of328.7 million hectares of land, where 142.5 million hectares are used for agriculture, while forests occupy 76.5 million hectares (23.27%). According to the State of Forest Report (FSI, 1997), the actual forest cover stands at 63.34 million hectares (19.27%), with 26.13 million hectares classified as degraded (NFAP, 1999). However, forest areas are depleting under the immense pressure exerted by a growing population on land resources.

A World Bank report highlighted that during the 1970s and 1980s, India experienced deforestation and degradation at an alarming rate, with an estimated loss of up to one million hectares annually. This depletion of forest resources has sparked concern and activism, particularly among the rural poor, and has led to spontaneous popular movements for forest conservation.

Today, the conservation and protection of forestland have become critical priorities for India's sustainable development, addressing the pressing need to balance ecological preservation with human and livestock demands.

- **Livelihood opportunity through Joint Forest management (JFM)**

Forest management (FM) offers an opportunity to enhance the productivity and availability of forest resources. Forests provide a wide range of direct benefits, including physical products like fuel, food, medicines, including various other essential materials. Additionally, they offer significant benefits that contribute to environmental enrichment.

Furthermore, small-scale, forest-based enterprises can contribute significantly to rural employment opportunities, improving the economy of rural populations, including forest-dependent communities and tribal communities. The rehabilitation and sustainable management of forests are vital to enhancing the quality of life in rural areas. This goal aligns with the primary objective of JFM, which focuses on conserving forests while addressing the socio-economic needs of local communities [2].

- **Forest and tribal**

The 1971 Census Report states that the majority of tribal communities resided in rural areas and relied predominantly on agriculture. Economically, these tribes could be categorized as semi-nomadic groups, jhum and settled cultivators, who were heavily dependent on forest resources for their livelihood. Forests served as their primary source of sustenance, providing food, timber, and bamboo for constructing homes, firewood for cooking and warmth during winter, grass for fodder, and materials like leaves for making plates. Additionally, they utilized forest products such as harre and behra for dyeing and tanning purposes. Non-tribal populations also inhabited forested regions and similarly relied on these resources for fuel, fodder, and other needs.

2 Economic Problem

India's national forest policy has struggled to effectively safeguard the ecosystem. As per a UN report, approximately 50% of India's total land area is severely affected by water and wind erosion, resulting in the annual displacement of about 6 billion tons of fertile soil. This loss significantly depletes the country's overall supply of plant nutrients. Historical evidence suggests that the priority should be protecting forest wealth from forest dwellers rather than from exploitative contractors. Furthermore, the adverse impacts of ecosystem disturbances—such as floods and landslides—on the lives and livelihoods of forest dwellers are often overlooked. Deforestation programs tend to prioritize the cultivation of fast-growing tree species, primarily to serve as raw materials for forest-based industries, often at the expense of ecological considerations [3].

3 Literature Review

According to Das N. et al. (2014) The tribal communities of Karkatia village in Purulia district heavily rely on non-timber forest products (NTFPs) due to the scarcity of agricultural land, infertile soil, inadequate irrigation, and poor connectivity. These challenges compel the indigenous population to depend almost entirely on the forest ecosystem to meet their daily needs. They gather and sell various types of NTFPs both for domestic use and commercial purposes. Additionally, they collect seasonal forest products to fulfill their requirements for food, fodder, and medicinal purposes, demonstrating their deep reliance on the forest for sustenance and livelihood [4].

According to Rai R. et al. (2013) This study highlights the significance of forest products for the tribal communities in the Bundu block of Ranchi district, Jharkhand. The research is based on personal interviews conducted with a selected group of individuals to gather data. The findings reveal the average extent of forest-based, directly paid employment among the sampled population, as well as the mean income generated from these activities [5].

According to Sadashivappa P. et al. The study highlights that the availability of women and child labor significantly impacts decisions related to participation in forest gathering and the income derived from these activities. The interdependence between agriculture and forestry necessitates a holistic approach to forest management programs, with a focus on supporting the development of the local farming sector [6].

According to Basavaraja H. et al. (2010) This paper seeks to identify the demographic and economic factors influencing participation in non-timber forest product (NTFP) gathering in the Western Ghat forests of Karnataka, South India. The study observes that higher agricultural income and wage earnings reduce the likelihood of households engaging in NTFP collection [7].

Article by Kumar V. (2015) This study examines the livelihoods of indigenous communities residing in protected areas and explores the various factors influencing their dependence on forests. Non-wood forest products (NWFPs) hold significant potential for generating economic benefits, thereby contributing to poverty alleviation of indigenous populations in underdeveloped regions of the country. The research focuses on a backward region in Gujarat, highlighting the role of NWFPs in shaping the economic well-being of these communities [8].

The Study by Hegde R. et al. (2000) This study was conducted in the Mudumalai Wildlife Sanctuary and the adjacent Sigur Plateau in Tamil Nadu, India. It focuses on the role of forests in the economic aspects of local-communities.

The study by Meher R. (2010)... This paper examines the challenges posed by the model of economic liberalization, privatization, and globalization in India, which has led to the displacement of tribal communities and agriculture-dependent poor populations from their traditional, sustainable livelihoods. The unregulated expansion of mineral-based industries in tribal regions has disrupted local ecosystems and resource access. Despite exploiting these resources, mining and other industries have largely failed to offer improved or sustainable livelihood opportunities for the affected communities [9].

A recent study has been done by Pati C. et al. (2017) This paper focuses on the Medinipur District of West Bengal, where a geo-botanical study was conducted to gather data on non-timber forest products (NTFPs). The study provides insights into what NTFPs are collected, who collects them, their quantities, and their uses among indigenous communities. The findings conclude that NTFPs are integral to this sector and are acknowledged as valuable resources for promoting sustainable livelihoods [10].

4 Role of Various Acts to Manage the Forest

4.1 Forest Rights Act, 2006

Local communities are often considered a marginalized and deprived section of society, facing significant social and economic challenges. Their economy is heavily dependent on forest products, which are essential for their survival. However, the unsustainable use of these forest resources has led to the extinction of numerous species of flora and fauna. The objective of the relevant "Act is to strengthen local self-governance and address issues related to the conservation and management of natural resources in India." Additionally, the Act emphasizes addressing poverty and ensuring livelihood security for tribal communities [11].

The Forest Conservation Act of 1980 imposes restrictions on the Adivasis and local communities against encroachment in forest areas across India. The Act treats tribal communities as adversaries of the forest, which many view as an unjust approach toward these people. The Act also aims to tackle the issue of poverty alleviation among tribal communities in India, while addressing livelihood security concerns. Applicable throughout the country, the Act was further amended in 2012 (Table 1).

4.2 Major Findings

A detailed study was carried out to understand how tribal communities in South Bengal live and earn their livelihood. Most of these tribes live in forested areas and rely on the forest for their daily needs.

After India gained independence, the **Zamindari Abolition Act** was passed, and in 1955, took control of forest lands under private control to the control of forest departments of the respective states. The Indian government followed the European industrial development model, aiming to build a socialist democratic state. The belief was that this new system would reduce traditional social and economic inequalities.

Table 1 Study area

South Bengal Districts	Name of the Block	Studied villages	No, of household
Purulia	Bagmundi	Bereria	17
		Bhupatipally	45
	Jhalda 1	Mahultanrd	14
	Bandwan	Ghatihuli	13
		Popo	22
		Pargora	28
Bankura	Ranibandh	Ghagra Haludkanali	31
		Routora	11
		Sarengsukra	25
Jhargram	Binpur II	Sarisabasa	35
		Dadra	27
		Sankhabhanga	21
		Amlasole	16
		Chakadoba	46
	Nayagram	Parasia	144
		Bansiasole	56
		Bnaksole	61
	Binpur I	Bhumidhansole	62
		Purnapani	24
Paschim Medinipur	Salboni	Jugdiha	24
		Burisole	31
	Medinipur Sadar	Lohatikri	90
		Muchiberiya	44
		Fulpahari	42
Total			930

It was also assumed that rapid industrialization would help improve the lives of underprivileged communities. As part of this plan, forests were mainly used for industrial purposes to generate revenue.

This study focuses on the districts of **Purulia, Bankura, Jhargram, and Medinipur** in South Bengal.

The data collection was conducted close to the forest with the focus to protect the traditional aspects of local communities.

In South Bengal, we found that most tribal houses are located on forest land. However, some families have legal ownership documents (forest patta), while others do not.

The **Lodha/Sabar** and **Birhor** tribes are indigenous communities that depend on the forest for their survival. According to members of the Lodha/Sabar tribe in South

Bengal. "Forest is our economic resources and generation wise we are depending forest produce collection, the children of our families went to forest from the age of five to collect minor forest produce". They are also further added "medicinal plant, fire wood, furniture wood, various leaf, root, and fruit, mushroom, potato, Bowla, skin of tree, creeper and hunting animal and bird are our forest product. We are surviving them with full utilization of forest product day by day". Around 21% were found to sell honey.

5 Discussion

Forests and tribal communities in South Bengal have a deep cultural and traditional connection. Tribals have lived in forest environments for generations, shaping their way of life and society. Most of their traditional knowledge comes from the forest.

Living in hills, dense forests, and coastal areas, they have developed valuable skills to survive environmental challenges and sustain their livelihoods. Their wisdom is seen in water harvesting, traditional irrigation systems, cane bridge construction in hilly areas, desert adaptation, and the use of forest plants for medicine. They also have a deep understanding of weather patterns. This traditional knowledge is precious and should be properly recorded and preserved before it disappears due to modernization.

Under the **Joint Forest Management (JFM)** program, local people are supposed to be part of forest committees. However, we did not find a single village where members of the **Lodha/Sabar or Birhor** tribes were included in these committees.

We are gathering opinions from eight villages about their experience selling **Minor Forest Produce (MFP)**. So far, most people from each village have expressed dissatisfaction with the process.

6 Pictures Depicting the Present Scenario

After collection of Piyal fruit

The Lodhas using the forest product for furniture

After collecting *Sal* leaf, they are dried up

Dried Babui Grass

After collecting fire wood women returning home

Broom prepared by June stick

7 Need of Artificial Intelligence in Forest Management

Artificial Intelligence (AI) has become a crucial tool in modern forest management because of the capability to predict and provide valuable insights after processing a large volumes of data. Forests are facing numerous threats, such as deforestation, illegal logging, climate change, and loss of biodiversity. AI helps monitor forest health through technologies like satellite imagery, drones, and sensor data, allowing for real-time tracking of deforestation and forest fires. Machine learning algorithms can identify patterns to predict risks and improve conservation strategies. Additionally, AI aids in mapping forest resources and indigenous territories, ensuring sustainable resource distribution and the protection of community rights. By automating tasks like data collection and analysis, AI improves efficiency and allows policymakers and conservationists to focus on strategic decision-making. The application of AI in forest management supports not only ecological preservation but also sustainable development along with livelihoods opportunities of those dependent on forest-produce.

8 Use of Ai in Forest Management

AI offers a range of innovative methods to enhance forest management, ensuring sustainability, conservation, and resilience. Key AI-related approaches include:

- **Forest Monitoring**: AI-powered systems process satellite imagery, drone data, and ground sensors to monitor forest cover, tree health, and biodiversity in real-time. This allows for early identification of variations, such as deforestation or habitat degradation.
- **Predictive Analytics**: The algorithm models analyses relevant historical and ecological data to forecast risks like wildfires, pest outbreaks, and climate impacts, enabling proactive management strategies.
- **Reforestation Efforts**: AI-guided drones can map degraded areas, plant seeds, and monitor their growth, making reforestation and afforestation initiatives more efficient and scalable.
- **Sustainable Resource Use**: Predictive models optimize timber harvesting schedules and locations to minimize ecological disruption while maximizing economic returns.
- **Wildlife Protection**: AI analyzes data from camera traps and acoustic sensors to monitor wildlife movements, detect poaching activities, and ensure biodiversity conservation.
- **Carbon Sequestration Analysis**: AI tools calculate carbon storage and emissions, helping to integrate forests into carbon offset and climate action plans.
- **Policy Support and Planning**: AI models simulate the impacts of various policies and land-use changes, aiding decision-makers in developing effective forest management strategies.

- **Community Collaboration**: AI platforms can integrate indigenous knowledge and local data to create management plans that respect traditional practices while leveraging modern technology.

9 Conclusion

Aboriginal communities, particularly in rural areas, have historically lived in isolation, relying entirely on forest products for their daily needs and income. These communities remain dependent on forests year-round, and according to the FR Act, 2006, they have the right to access forests as needed. Forests serve as crucial natural resources for these populations, with their livelihood intricately linked to forest products. Medicinal plants and non-timber forest resources are essential for survival, with tribes like the Lodha selling medicinal plants to higher caste communities. However, the knowledge of traditional medicine among the Lodhas is rapidly diminishing, with concerns that it may be lost entirely. Environmental studies on deforestation, soil erosion, pollution, disaster management, biodiversity, and human-animal conflicts further underscore the importance of preserving forest ecosystems.

Forest and forest produce plays a pivotal role in sustaining village development and livelihoods. Natural resources within these ecosystems maintain the environmental balance. The current study highlights the ecological consequences of the commercialization of forest products and the challenges of sustainable extraction due to shifts in tribal culture. Findings suggest that many tropical species are at risk of extinction, and the transition from subsistence agriculture to commercial exploitation poses significant threats to forest sustainability. A biodiverse forest is better able to withstand commercial exploitation if tribal culture remains intact. However, as younger generations of tribal communities seek opportunities outside the forest, this shift in culture, coupled with rising population densities and clustered settlements, facilitates easier commercial exploitation of forest resources. The complex relationship between commercialization, changes in tribal culture, and ecosystem sustainability makes it challenging to quantify the exact impact of commercialization on both tribal ways of life and the environment.

Moreover, the use of the AI in forest management had played a vital role in safeguarding the natural resource like forest which serves both for the tribal and non-tribal people.AI-based technologies, such as Artificial Neural Networks for Forest Management, Automated Forest Fire Prediction, Bird's Eye View Monitoring, and AI-Based Carbon Cycle Mapping are recently in use to protect the forest.

AI-based approaches to forest management hold immense potential for sustainable development and efficient management, but they require careful consideration of privacy and security concerns to realize their full benefits.

10 Recomendations

Some important recommendations which can assist to bridge the challenges are as follows:

1. Inclusion of tribal and local communities as permanent members in JFM committee.
2. Adhering and implementing the directives of JMF.
3. Immediate measures to prevent deforestation.
4. Organizing awareness programs by establishment of permanent awareness camps in forest adjacent villages.
5. Use of AI in forest management must be done but one must keep in mind regarding the national security and privacy.

References

1. Ministry of Environment, Forest and Climate Change, "Total Forest and tree cover increased by 2261 square kilometer in India as per the India State of Forest Report (ISFR) 2021", 13 MAR 2023, PIB Delhi available at https://pib.gov.in/PressReleasePage.aspx?PRID=1906388. Accessed 24 May 2024
2. Biswas PK (1988) Sociological issues in forestry. Indian J Soc Sci 1(2)
3. Chandramohan BP, Villalan TKS, Munirathanam J (2010) Impact of commercialization on tribal culture and forest ecosystem sustainability. J Human Ecol 5(2)
4. Dolui G, Chatterjee S, Chatterjee ND (2014) The importance of non-timber forest products in tribal livelihood: a case study of Santal community in Purulia District, West Bengal. Indian J Geogr Environ 13. ISSN: 0972-7388
5. Ajaz-ul-Islam M, Sulaiman Quli SM, Rai R, Sofi PA, livelihood contributions of forest resources to the tribal communities of Jharkhand. Indian J Fundamental Appl Life Sci. ISSN: 2231-6345 (Online)
6. Sadashivappa P (2006) Participation behavior of indigenous people in non-timber forest products extraction and marketing in the dry deciduous forests of South India. Tropentag 2006 University of Bonn, October 11–13, 2006 conference on international agricultural research for development
7. Bharath Kumar LB, Patil BL, Basavaraja H, Mundinamani SM, Mahajanashetty SB, Megeri SN (2011) Participation behavior of indigenous people in non-timber forest products extraction in western ghats forests. Karnataka J Agric Sci 24(2):170–172
8. Kumar V (2015) Role of non wood forest products (NWFPS) on tribal economy of Gujarat, India. Int J Usuf Mngt 16(1):67–75. ISSN: 0972-3927
9. Meher R, Globalization, displacement and the livelihood issues of tribal and agriculture dependent poor people: the case of mineral-based industries in India 25(4):457–480
10. Shit PK, Pati CK (2017) Non-timber forest products for livelihood security of tribal communities: a case study in Paschim Medinipur District, West Bengal, Published online: 24 Oct 2017 pp 149–156
11. Draft national policy on tribals, Ministry of Tribal Affairs, published by: Indigenous Law Centre, Law School, University of New South Wales visited on

Artificial Intelligence and Tribal Knowledge Systems: Exploring Pathways for Protection Within Indigenous Communities of India and Beyond

Vijoy Kumar Sinha, Jayanta Ghosh, and Satyendra Kumar Srivastav

Abstract This study explores the convergence of artificial intelligence (AI) with tribal knowledge systems, with an emphasis on protecting and preserving indigenous languages within Indian communities and beyond. As AI technologies advance rapidly, they offer both possibilities and hurdles for safeguarding tribal knowledge. The paper investigates AI's potential in supporting indigenous languages, specifically through AI-driven language learning platforms, automated translation systems, and speech recognition tools. Moreover, it delves into the ethical aspects of AI interventions, such as ensuring cultural sensitivity and working with indigenous communities to make certain AI applications uphold and preserve the linguistic and cultural integrity of these languages. By reviewing case studies and present AI initiatives, the paper highlights both the opportunities and the challenges associated with integrating AI in the support and revitalization of indigenous languages. Ultimately, it emphasizes the necessity of a collaborative effort between AI developers and indigenous communities to create a sustainable, culturally aware, and effective framework for language preservation.

Keywords Artificial Intelligence · Tribal knowledge system · Indigenous communities · India · Indigenous language

---

V. K. Sinha (✉)
Centre for Regulatory Studies, Governance and Public Policy, The West Bengal National University of Juridical Sciences, Kolkata, India
e-mail: vks161991@gmail.com

J. Ghosh
The West Bengal National University of Juridical Sciences, Kolkata, India

S. K. Srivastav
Department of Computer Science and Engineering, Asansol Engineering College, Asansol, India

M. Chakraborty et al. (eds.), *Proceedings of Sixth International Ethical Hacking Conference*, Lecture Notes in Networks and Systems 1513,
https://doi.org/10.1007/978-981-96-8632-2_36

1 Introduction

As technological advancement occurs at breakneck speed, we are in an age of a neo-industrial revolution where Artificial Intelligence (AI) is overlaying and shaping industries, societies and cultures. It adds value to the category that is healthcare, removes all human tasks box, and revolutionizes education category that scratches this but not the end [1]. Yet the urgency of technical progress is shadowed by important ethical questions about marginalised groups in particular, long-knowledge traditions among indigenous people and other groups to whom this knowledge has been transmitted for centuries. However, one of the major issues that these communities face today, and of this law is primarily the fact that their cultural and intellectual property needs to be protected, since in the current digital era, traditional knowledge can be easily exploited and misappropriated.

At this intersection of worlds one will find one of the world's most diverse indigenous communities and tribal groups, India, where the collision between its diverse traditional tribal knowledge and modern AI technology, takes the form of one of the world's most fruitful and productive usually marginalized and overlooked collaborative cross-pollination. Such communities specifically adapted to concrete ecosystems and social constructions for centuries have acquired broad information about sustainable agriculture, medicinal plants, rituals, arts and languages. Native knowledge can be displaced and commercialised without acknowledgement and benefit sharing due to this widespread adaptation of digitalisation and AI [2]. We will need to have urgent discussions about how AI can protect and respect indigenous knowledge, create avenues for its sustainable use and instill respect for its rightful place and against appropriation and exploitation.

The tribal knowledge, is something that is vastly different and embedded in the ecological, spiritual and culture of indigenous nations [3]. No written record exists of this type of knowledge, which is transmitted orally through practices, rituals, or experiences [4]. This makes formal recognition and protection under traditional intellectual property laws, which are meant for material, commercially viable inventions, problematic. The failure to acknowledge indigenous peoples has been exploited in many fields including biotechnology, agriculture and arts where commercial companies have taken the knowledge of tribes without giving credit or compensation, leading to rampant "theft" of indigenous knowledge [5]. All of which means that tribal communities could lose control over their intellectual property, with no venues for compensation or reparations.

Even though AI technologies offer unique opportunities and threats for protecting these kinds of knowledge systems. AI can analyze massive data sets, recognize patterns and generate insights capabilities that can be harnessed to document, archive and safeguard indigenous knowledge in ways previously difficult to fathom. AI, for example, can support the systematization of oral traditions, databases of medicinal plants and traditional agriculture or the preservation of cultural assets in a digital format. By converting tribal knowledge to a digital format, we can make sure that it does not get lost to a transient generation or overcome from outside the world.

Additionally, AI tools may help in identifying instances of knowledge appropriation, enabling indigenous rights holders to monitor and enforce their intellectual property rights.

This opportunities come with challenges as well. There's a risk of even more commodification and exploitation. Digitalization and AI can safeguard, but also lure to misuse. Indigenous Peoples are at risk of the dilution of their authenticity because the exploitation of indigenous knowledge through AI will enable commodification or extraction of traditional knowledge without proper remuneration, proper value, and community consent [6]. Moreover, there is absolutely no want for AI systems to mirror Indigenous understandings and those understandings are as often about collective ownership and the idea that some knowledge is sacred whether that refers to knowledge about caring for the environment, for rituals, for spirituality.

An important question then is how can AI engage in a constructive way with the development of ethical frameworks and solutions that respect the rights and cultural integrity of Indigenous Peoples and provide tools to protect their knowledge systems? This involves rethinking IP laws and policies to reflect the nature of tribal knowledge and ensuring that AI works for us and not against us. Equally as important, though, is that indigenous peoples become integral to the design and deployment of AI technologies and able to govern their knowledge, and dictate how it is used in ways that are aligned with their social, cultural, and holistic priorities.

This study investigates how AI can be utilized in the preservation of tribal languages, particularly as applied to indigenous communities in India and other parts of the world. This time, researchers examine the potential of AI to preserve, document, and revive native dialects and the ethics that surround the practice. By exploring methods for bridging this gap following up on the work done in the present, this aims to present pathways towards collaborations between the world of AI developers and indigenous communities alike, creating spaces for tribal languages to thrive in their digital landscapes.

2 Indigenous Language

Across the globe, Indigenous languages are disappearing at an alarming rate. Linguistic experts say nearly 40% of the estimated 7,000 languages spoken today will die out, and most of them are indigenous [2]. India, has many tribes with well over 700 different languages along with a diverse population [7]. Most of these languages were spoken by relatively few people in isolated communities. More than just a mode of communication, these languages encompass their culture, tradition, history, and identity. But these are under threat from globalisation, urbanisation, and the dominance of languages such as Hindi and English, making tribal dialects marginalised and endangered.

For Indigenous peoples, losing their language is more than a matter of syntax; it is a matter of culture and spirit. Language underlies much of indigenous identity which is still the main vessel of storytelling, oral traditions and ecological intelligence.

When these languages disappear, their community's fundamental worldview and cultural legacy do as well. Therefore, preserving these languages is a very important concern, which needs to come up with innovative plan combining traditional cultures experience with modern technologies. AI can help as well as hinder transmission of indigenous languages.

3 Artificial Intelligence and Indigenous Language

Recent decades have seen much advancement in artificial intelligence, which is now being applied to sectors ranging from healthcare to education to finance to agriculture. In the case of native languages, AI has the potential to document, preserve, and revive the native languages in ways never thought possible. Tools designed by AI make it possible for speakers of endangered languages, linguists, and members of indigenous groups to work together to record, transcribe and digitize a language that had been handed down through oral tradition over many generations to help ensure it does not vanish for good [8]. But employing AI to preserve tribal languages comes with its share of challenges, including finding ways to deploy the technology ethically, consistent with the values and consent of indigenous communities, and in a way that respects their cultural integrity [8].

The tribal languages in India such as Santhali, Gondi and Bhili are primarily used as linguistic channels of communication, but they are also the way these people engage with their natural environment, preserve their knowledge, and maintain their social networks. These deaths arising from national and global language dominance, lacking educational resources and scarce technological application for preservation are happening faster with each passing year [9]. In many cases, we do not teach tribal languages in formal stages, so it alienates their speakers with the mainstream society, leading to the loss of language.

In this context, AI has a transformative potential. The main application of AI is about building automated systems for documenting languages. Transcribing, translating and digitally preserving indigenous languages is also facilitated by AI-powered tools and they include speech recognition, machine translation, text-to-speech, etc. [10]. Specifically, AI can provide assistance in transcribing oral stories, rituals, songs and other spoken forms of cultural knowledge critical to the linguistic heritage of tribal communities [10]. AI helps identify and document less-understood dialects and linguistic structures that can be built into databases and analyzed using natural language processing algorithms, which are essential in preserving less-documented languages [10].

AI-enabled applications can also create language learning resources for young indigenous individuals. AI, therefore, presents a unique alternative as language revitalization becomes an important support system in the protection of these languages, and platforms for building education resources such as language apps for learning, tribal digital dictionaries, and online courses would contribute to the accessibility of tribal languages [11]. As an example, AI might create tailored and personalized language learning programs to match the learner's pace and manner of learning, benefiting those younger generations who need to have the language passed down to them.

While these applications have a great potential, AI also focuses on the question of ownership and control in the context of indigenous languages. A key challenge to be navigated in integrating AI into the work of language preservation is ensuring the technology aligns with how indigenous communities view cultural values [12] that now demand languages with architectures that are closest to gibberish logics of indigenous languages that are engaged in mutating the digitization of, or resonate the disruption of, indigenous languages, and so therefore, are the most highly designed system to prevent the distortion, or misarticulation of such language. This matters because many tribal languages are inextricably interwoven with cultural traditions, rituals and ecological knowledge that cannot be separated out from being spoken. Without a culturally aware understanding of these nuances, AI systems can easily commodify or appropriate language.

Additionally, Indigenous communities must be incorporated in all stages of AI language protection tool development. Involvement of tribal leaders, language experts, and community members in the design and implementation of AI technologies is a key to success. The community must also participate in creating these partnerships, so that the technology is aligned with the community's values and also to ensure that the rights to the content fed into the system doesn't transfer to an outside party. Without this involvement, AI could further marginalize these communities rather than empower them.

Addressing these issues will require the development of ethical frameworks for the implementation of AI that honour indigenous sovereignty. This means ensuring that AI is used with the community's informed consent, stopping the exploitation of language and knowledge for commercial purposes without remuneration, and protecting both the cultural value of the language. Furthermore, efforts to preserve and revitalize tribal languages should be seen as part of a broader cultural protection movement, one that also seeks to protect indigenous traditions, practices, and worldviews.

4 India's Initiative

The intersection of technology and language education in the context of India is inextricable from the complicated trajectory of the country's policy development regarding its minority or Indigenous languages. The basis of linguistic minority rights in India is provided by Article 19 of the Constitution [13]. India hosts 1,369 languages [14], many no longer spoken, and makers of the constitution regard its protection as vital to the survival of cultural diversity, especially in the face of laws like Article 351 [15]. On the other hand, this has made Hindi an overbearingly influential language among linguistic minorities. Digital resuscitation of minority tongues is thus a strategic counter to these forces.

To rectify this linguistic imbalance, the Supreme Court of India defined a linguistic minority as a group speaking a language with or without a written script [16]. This was followed by the formation of the National Commission for Minorities Act 1992 [17] (for the protection of linguistic minorities), the National Framework for Action on Minorities 2005 (advocating the inclusion of minority languages in education). These initiatives to recognize language rights demonstrate the political nature of language itself, both within India's tribal movements and more broadly, reflecting the loss of linguistic diversity currently experiencing worldwide Indigeneity [18].

India has embarked on monumental initiatives to protect and promote indigenous, tribal, and minority languages, with a major focus on technology, especially AI in the promotion of language learning, revitalization and education. To address issues of language endangerment and facilitate access to education, these initiatives operate at various levels, including government agencies, private corporations, and non-profit organizations. Here are some major initiatives.

4.1 Microsoft's Project ELLORA

Microsoft's Project ELLORA (Enabling Low Resource Languages with Technology) employs artificial intelligence and machine learning to develop tools for low-resource languages such as Gondi and Mundari, the dialects spoken by marginalized tribal communities in India [19]. It aims to build language datasets, language processing tools (speech recognition, machine translation), and increase access to digital platforms for these communities. By working with local communities and language experts, they give these languages a home in the digital space, helping to preserve them culturally [19].

4.2 Aripana Foundation (Maithili Language)

Basal to the roots of the Aripana Foundation, located in the rural landscape of Bihar is an effort to revive Maithili language—one of India's regional language—through tech tools and educational digital resources [20]. The foundation was launched with the dream to create the digital footprint of Maithili, especially due to the lack of proper educational content and regular use of Maithili in everyday digital conversation [20]. In collaboration with Google also, it made a digital Maithili script and several tools like Tirhuta Keyboard to type in Maithili. It is also translating children's books into Maithili to promote literacy and education in rural areas, and collaborating with organisations like Pratham Digital to develop educational tools and digital content [20].

4.3 Pratham Digital

An initiative of the Pratham Education Foundation, Pratham Digital leverages AI and machine learning to improve education in some of the poorest parts of India, including rural and tribal areas. AI-driven technology, such as voice recognition and automatic grading, is being used to make education for children in disadvantaged communities more inclusive. Pratham Digital collect children's voice data to improve AI models as very few languages are available in the digital space, and also partner with the local communities, and guide the tool development to make it culturally relevant and effective. Such a platform is vital in providing tech-based solutions to regions without a conventional educational setup [20].

4.4 Rajasthani Bhasha Academy

The Rajasthani Bhasha Academy is primarily concerned with the promotion, care and protection of the Rajasthan language, which is not recognized as a national language of India and has over 22 dialects in Rajasthan [20]. The Academy uses digital platforms to standardize and promote Rajasthani learning, making sure it is taught in schools and in other learning institutions. In response to the problems of language loss and the diversity of variation within the Rajasthani language, the Academy creates video lectures, gamified vocabulary tasks, and automated quizzes in Rajasthani so that speakers can engage with and grow their language skills [20].

4.5 BR Ambedkar University (Multiple Regional Languages)

BR Ambedkar University in New Delhi has taken up research and development for promoting low-resource regional languages through AI and other digital tools. The university's focus includes developing assistive technology for teaching and preserving regional languages, particularly those that are marginalized or are at risk of disappearing. Their work consists of creating linguistic databases for different regional languages, which are then used to create AI models for educational use. They also create tools for Indigenous language education that can be used in schools, where there is a lack of qualified language teachers [20].

4.6 Mundari Language Project of IIT Kharagpur

The Indian Institute of Technology (IIT) Kharagpur has announced an AI-powered app Mundari App to promote and preserve Mundari, a tribal language of West Bengal and nearby regions [21]. The Mundari App project aims at making Mundari available in the digital world by developing voice recognition systems and other language resources enabling the speakers of the language to access content as well as learn the language. IIT Kharagpur has developed a Mundari digital script and compiling linguistic data for the corpus to make functional the AI models. The application with a chat bot is assisting for Mundari learning and communication which is very much required for education and daily use [21].

4.7 National Translation Mission

The National Translation Mission (NTM), a government initiative to preserve and develop India's regional languages, is translating different materials in these languages [22]. The mission focuses on the translation of scientific, technical, and educational texts into Indian languages, making information and knowledge available to people in their native languages. NTM partners with technology companies to develop these tools and embed AI into the translation tools to expedite and optimize the translation process for the ITM languages. NTM contributes to the representation of these languages in educational and scientific discussions by building up a corpus of translated texts [22].

4.8 National Strategy on Artificial Intelligence (NITI Aayog)

The National Strategy for Artificial Intelligence was formulated by NITI Aayog, the Indian government's policy think tank, which proposes the use of AI in all sectors including education [23]. This strategy underscores AI's ability to address business challenges within historically low-resource languages, particularly those of tribal and minority significance. If we support AI model building for these languages, we shall be making education more accessible and inclusive: NITI Aayog Through this framework India wants to develop technologies that will ensure regional languages are not pushed aside in the digital world which will only aid their national development role [23].

4.9 Bhashini

Under the AI4Bharat project, Bhashini is a MeitY (Ministry of Electronics and Information Technology) government initiative that aims to develop an AI-based National Language Translation system [24]. Its intention is to establish a strong language ecosystem in India that supports interaction in multiple languages through AI technologies including speech recognition, text to speech and machine translation.

The core agenda of Bhashini is to remove the language barriers in digital space, empowering people to use their languages not only to access the information or communicate but also use digital services [24].

4.10 AI Kosha

To accelerate the pace of AI research, India has launched AI Kosha, an open platform with edited and structured, non-personal datasets, said to help in training AI models in Indian languages [25]. Powered by 316 datasets language translation tools, health data, satellite imagery and pollution data, to name a few AI Kosha is part of the IndiaAI Mission with an outlay of Rs 10,370 crore [25]. These datasets are an important part of the IndiaAI Datasets Platform's strategy to spur AI development in India. Union IT Minister Ashwini Vaishnaw revealed that an additional 14,000 GPUs would help train AI models and more would be coming. Increased interest and potential for India's AI ecosystem from AI startups AI Kosha builds upon the Open Governance Data Platform, which hosts more than 12,000 datasets and promotes sharing of data to spur innovation and AI research [25].

India's efforts clearly demonstrate a significant investment in the digital conservation, revitalization and promotion of Indigenous and marginalized languages. These efforts are much more than producing technical tools, they build collaboration between communities, linguists, educators and technology experts. AI in language education could bridge the gap in education and uplift communities while preserving India's versatile linguistic legacy in the digital age. Expectations such as these when combined together can not only help India create an inclusive educational ecosystem but also provide an opportunity for even the most marginalised languages to survive.

5 Other Countries

5.1 Canada [26]

The First Peoples' Map of B.C. launched on 27 February, 2020, is an interactive, engaging resource that honours the rich, intricate cultures of the Indigenous peoples of British Columbia. Developed in partnership with Indigenous communities, Elders, Knowledge Keepers, language advocates, cultural workers and artists across Canada, this tool was forged in Canada (in the province of British Columbia) by the First Peoples' Cultural Council (FPCC). This is not just a map, it is a journey through the languages, art and cultural expressions that bound Indigenous people to their land and bond each other to one another.

This map shows the 36 Indigenous languages of British Columbia, including their language regions, place names and important community sites. By interacting with this map, users can listen to pronunciations of the names, greetings and places in their native tongue. It's a deeply immersive experience that offers an opportunity to hear languages that have been spoken for thousands of years.

The mapping is the work of the First Peoples' Cultural Council done in partnership with Indigenous communities, Elders, Knowledge Keepers, language advocates and artists. This map is the culmination of these community specialists, and with additional support from FPCC staff. It is like a dynamic, mutable resource—one that grows and changes as new knowledge and stories are added.

Visitors also have several fun ways to explore the map: They can search for specific places, scroll through the sidebar or search using keywords and filters. And whether your interests are in language, art or cultural heritage, the map encourages exploration of the interconnections in the histories of Indigenous cultures throughout the region for an exciting, hands-on approach to learning about and connecting with the customs and traditions of British Columbia's First Nations people in Canada.

5.2 *Australia [27]*

The 50 Words Project is a campaign to promote the cultural diversity of Australia's Indigenous languages through an interactive online map lined with audio recordings of words, recorded by native speakers. The project wants to provide 50 words for every Indigenous language in the country, from those that are spoken today to those being revived, and promote linguistic diversity along the way. The words include ubiquitous phrases such as "hello," "goodbye" and "what is your name?" or, words for aspects of the family and animals, nature.

From the University of Melbourne's Research Unit for Indigenous Language (RUIL), this project partners closely with Indigenous communities to create a sustainable resource to help Australians connect with their cultural heritage. It showcases the unique and varied sonic landscapes of these languages, most of which share common features despite spanning vast territories.

In addition to being an educational tool, this site offers teachers resources to use in the classroom. Over 60 languages are currently accessible on the internet and the team keeps actively looking for contributions from communities to expand the map and guarantee that every Indigenous Australian language is present.

5.3 *New Zealand [28]*

Indigenous languages face rolling decline, with about 3,000 of them at risk of extinction by the end of the century, according to estimates by UNESCO. In response to this threat, New Zealand-based Māori language broadcaster Te Hiku Media is harnessing ethical AI to help preserve and revive te reo Māori, the native Polynesian language of the Māori people.

Automatic Speech Recognition (ASR) models are being developed by Te Hiku Media leveraging both the open-source NVIDIA NeMo toolkit and NVIDIA A100 Tensor Core GPUs. They also had an accuracy of 92% for transcribing te reo, and 82% for bilingual speech (English and te reo). These tools have been built, by and for the Māori community, and are important for the preservation and amplification of Māori stories and language.

The organization started as a radio station more than three decades ago to ensure people could hear te reo on air. So as technology developed, they felt they needed to embrace the digital age and still control their content. Rather than accept global platforms with restrictive terms, they built their own content platform, Whare Kōrero ("house of speech"), which now hosts more than 30 years' worth of audiovisual content in te reo. Now served via an app and used by 20 Māori radio stations, this archive collects recordings from native speakers, second-language learners and bilingual Māori.

Te Hiku Media's adoption of trustworthy AI started in 2016 to solve the resource-heavy job of transcribing massive amounts of content. With the support of community elders and a crowdsourcing campaign by the name of Kōrero Māori, they collected 300 + hours of speech data, which was labelled and used to train their ASR models. It worked so well, in fact, that the project now powers the newly launched Kaituhi transcription service, which allows Māori speakers to transcribe te reo online.

The work being done by Te Hiku Media goes far beyond revitalising te reo Māori; it's also helping other Indigenous nations to navigate the ethical application of AI for language preservation. This work inspired other efforts by Native Hawaiians and the Mohawk Nation, showing how Indigenous-led efforts in trusted AI can help empower marginalized communities.

5.4 *Iceland [29]*

By ensuring that indigenous and small languages are represented among that data set, Iceland hopes to counter a tendency among many large AI models to favour big languages, such as English, due to the vast amount of data available. Iceland, a country with a population of around 350,000, has done a great job of preserving and revitalizing the Icelandic language as a successful collaboration with OpenAI brought Icelandic to their models, the GPT models.

Under the guidance of the Ministry of Culture and Business Affairs, in conjunction with Miðeind, a guideline was produced on how to support smaller languages in the AI context. This is a guide to Iceland's attempts, including building a healthy language tech infrastructure, that made this small language with a small speech community part of LLMs. Iceland's initiatives stand as a model for other small language communities the world over, and raise important lessons on the role of AI in linguistic and cultural preservation.

Iceland has fared better in preserving language in large part because of government investment and commitment. In 2014, the Icelandic Parliament had passed a resolution to emphasize the importance of Icelandic. This marked the beginning of a concerted, national industry drive to ensure the language can survive the forces of globalization and the maw of English. This included funding the Icelandic Language Technology Programme co-monitized by state and private sectors. It was made possible by an effort made collectively by universities, public institutions and private companies to generate the necessary language resources.

Data gathering was key to Iceland's success. Speech and text data of Icelandic speakers were collected and published as open-source to enable developers and organisations to add Icelandic into their products. Moreover, crowdsourcing campaigns improved the quality and quantity of this data, making it more available for use in language technology.

Working alongside OpenAI, Iceland saw a step forward towards its digital language preservation efforts with a GPT integration of their language. In fact over 4 billion words of Icelandic text was available to train the models that run OpenAI, which proves that AI can be extremely successful with small languages, as long as there is an effort and a commitment. Augmenting the previous, this partnership demonstrates the magnitude of technological collaborations in advancing language reclaiming.

The successes in Iceland provide a roadmap for other small and indigenous language groups. Miðeind has this guide calling for international project in support of Multilingual AI Development This initiative is centred on creating open data sets, establishing standards for data collection and licensing, and designing benchmarks to measure the performance of AI in minor languages. It further assists research and funding efforts that benefit small language communities.

Unlocking the Potential of Multilingual AI through the Icelandic Model By investing in language technologies and working with AI companies, small language minorities can preserve a digital presence for their languages. Benefits driven by real-time translation, speech-to-text and other AI technologies can make content and services available in the local languages, such handcrafted translation services could be one drive on inclusion and access to be opened, for example, in both native and second-language learners.

The Iceland example underscores the importance of continued investment in language technology, open data practices and partnerships with A.I. companies. These tours are meant to demonstrate that AI can respect and promote linguistic diversity, and serve as an example for other countries whose languages are at risk of disappearing in favour of a more dominant one.

The guide calls for joint international action by governments, AI companies and the language community to help keep AI inclusive for all. AI has played and continues to have a huge role to play in helping ensure that its developments are leveraged for everyone in all communities, through support for approaches that promote minority languages and preserve cultural heritage. This joint effort will also usher in a future in which a few human languages cannot cost humanity the beautiful variety of human languages and cultures.

5.5 *Africa [30]*

Masakhane—which means "we build together" in native Nguni languages—is an audacious open-source initiative to create machine translation systems for African languages, a key prerequisite to improved communication and access to technology across the continent. Based on the fact that Africa accounts for more than 2,140 languages, some of which have not been written down, and most of which are hardly ever used for this purpose, millions of people remain in the dark and fail to see the potential of gaining access to critical information and technology. With this spirit of cooperation, the project—known as Masakhane, isiZulu for "We Build Together"—aims to bridge the gap between African languages and the global digital space.

At the core of Masakhane's mission lies neural machine translation systems for African languages, a crucial piece for the development of other natural language processing (NLP) use cases such as speech recognition, sentiment analysis or text prediction. Low availability of language data or corpora for African languages is one of the major barriers. While the major languages have large, high-quality datasets available to them, the African languages, and especially the low-resource languages, are often neglected. To address this problem, Masakhane's team leverages public resources like government documents, literature, religious texts, and news articles, which they use to design open-source models and datasets usable with machine learning. This project allows smaller African startups, NGOs and researchers to access the opportunity of creating digital resources for their communities.

The reach of the project transcends preserving the linguistically obscure and serves as a means for inclusivity. If it can unlock how to automatically translate African languages, then it can change the game for sectors like healthcare, education and crisis management. More concretely, in low-resource language contexts, machine translation supports the provision of emergency services and critical assistance. Moreover, education in a native language leads to better learning, and the removal of language barriers allows Africa's citizens to participate more fully in the global economy. As of now, language barriers prevent many Sub-Saharan Africans from accessing global markets; in fact, 63% of them cannot participate because of this issue. Masakhane's efforts directly address these barriers, embedding African languages into digital services and tools.

The project is also in line with a larger push for AI development in the African continent. For much of AI's history, a majority of its research and development resided outside of Africa, but initiatives such as Masakhane are allowing Africans to help determine their technological destiny. This project allows African solutions for African problems, reducing external software biases and culturally irrelevant solutions by helping build AI models that can operate in African languages. Such efforts are part of a wider wave of innovation, including companies like Google and Mozilla also investing in A.I. solutions for African needs. In particular efforts like Google's AI research lab in Accra, Ghana and Mozilla's work building voice enabled products for African languages are putting a spotlight on Africa's rising role as a leader in factoring innovation into AI development.

The impact of the Masakhane project has already been felt across the continent, despite being just in its early stages. It promotes collaboration across researchers in different countries in Africa and invites more contributors. Then build a pan-African strategy in order to grow AI and provide a space for African languages in the world stage. With continued investment and collaboration, Africa's young and rapidly growing population could allow it to dominate the AI landscape. The Masakhane project is this important step, enabling African communities with the necessary tools to construct their own digital futures and proactively engage with the international tech ecosystem.

6 Challenges in Using AI for Language Preservation

6.1 Data Scarcity

However, there are numerous issues regarding its implementation, although AI has great potential for language preservation. One of the main challenges is the absence of data—many of the threatened languages don't have extensive written or audio information in them for AI models to learn from, if they exist at all. There are no regularized grammatical rules or spellings and even that will make it even harder for AI to train. Even different dialects of the same language need own approaches to collect and manage your data.

6.2 Diverse Grammatical Structures

Any use of AI should respect the conventions and desires of native speakers as well as the fact that non-native speakers are not necessarily designated figures—they can engage in only so much use before they are accused of being parasitic or abusive. Involving communities in the development process is critical to ensuring AI systems embody their aspirations and cultural values. This highlights a need for communities to retain control of their linguistic resources to avoid exploitation or commercialisation, echoing similar ethical concerns around data ownership and accessibility.

6.3 Dialect Variations

Generative AI is an interesting approach, but you also have to work with linguists, native speakers, software engineers, etc. Without this dialogue, the risk is that the tools simply will not be useful for the communities they are meant to help. Which leads to the next important factor, which is linguistic justice; in other words, making sure that AI-powered tools are made widely available across demographics and location.

7 Conclusion and Suggestions

Particularly for the preservation of our languages, AI provides significant opportunities for the protection and revitalization of tribal knowledge systems. Indigenous languages have deep ties to the cultural and spiritual practices of tribes and generations of knowledge. Unfortunately, many of these languages are endangered or dead, losing important cultural, historic and worldviews. Emerging AI technologies, including NLP and machine learning, offer fresh opportunities to document, analyze, and preserve these languages. AI can capture, translate, and revive tribal languages, so they are preserved and available for future generations.

Yet the future of AI in cultural sustainability is equally concerned with being responsible in applying these technologies and having the highest regard for the indigenous communities whose knowledge we seek to protect. These projects need involvement of tribal communities at all levels. Ensuring AI tools align with community values, goals, and needs must begin with community consent, participation, and leadership. Otherwise, the risk is of misrepresentation, appropriation or the development of resources which do not accurately represent a language or culture.

In this context, ethical AI use includes addressing issues associated with data sovereignty, ensuring indigenous communities have the right to determine how their languages, stories, and traditions are used in the creation and deployment of computational and AI systems. The communities should own what they collect and have the agency to share their knowledge and distribute data. AI algorithms must recognize that tribal knowledge is often written, spoken, and traditional, but so much of it is based in orality and spirituality and is connected to cultural practices and generations of people.

The intersection of these two domains holds transformational potential for cultural preservation. By using AI as a protector of languages and traditions, indigenous people will help defend their heritage from vanishing in modern-day society. Language models infused with artificial intelligence can yield interactive instruments for language learning for the younger generations to engage their heritage language in ways that matter to them. Besides, AI may help create a digital archive of oral stories, rituals, songs, or other cultural expressions that can both ensure their preservation and make them more easily accessible to a wider public.

The end goal should be to wield AI as a tool of empowerment, where indigenous peoples determine and have control over how their cultures are represented and sustained through the digital, transnational era. This is because AI can potentially give a voice to indigenous communities, allowing them to participate in the global dialogue while still upholding their cultural heritage. As such, AI can connect with the past, the present, and the future so that tribal knowledge systems are preserved for the benefit of people in an ever increasingly digital world.

7.1 *Suggestions*

7.1.1 Prioritize Indigenous Communities in AI Research and Development

Indigenous communities should lead AI research efforts to help ensure that the technologies that impact them are designed by them. Native peoples must be included in the creation, testing and deployment of AI systems. The concerns of the people being affected should be prioritized so that AI technologies respect and reflect the users needs, values and epistemological systems. To put things right, financially support indigenous-led AI projects, so these communities maintain ownership over technology, instead of it co-opting their cultural preservation and/or future. This not only ensures that AI is beneficial, but also culturally appropriate for indigenous populations.

7.1.2 Collaborate with Indigenous Leaders to Ensure Cultural Sensitivity

Indigenous leaders importantly can help to ensure culturally sensitive AI technologies are created. These leaders know the old ways, their traditions and languages, their lifestyles and ways of being, the ones who can spin AI in a respectful, accurate way. Including them from the start helps avoid misrepresentation, stereotypes and the appropriation of indigenous knowledge. Their wisdom in turn then informs ethical guidelines that can help shape how through practice AI systems operate so they comport to community values and beliefs. Indigenous leaders work with developers of AI systems to strengthen, not endanger, their communities, putting in place sustainable partnerships grounded in trust and collaboration.

7.1.3 Develop AI Models Aligned with Indigenous Values and Ethics

AI models should be pointed to a secure platform of ethics and values that rests upon indigenous traditional cultures, which are usually based on respect and care for nature, and where collective welfare and interdependence are common features, and AI systems need to cover these fundamentals. These values will enable the AI solutions to be environmentally sustainable, culturally respectful and protect indigenous knowledge. Mending the frameworks with conservation, protection of sacred sites, biodiversity maintenance in an indigenous way—the model will all be with native vision. This alignment leads to culturally appropriate solutions that support, rather than detract, from indigenous ways of living.

7.1.4 Create Platforms for Indigenous People to Share and Protect Their Knowledge

Indigenous knowledge sharing platforms as a cultural heritage preservation approach in the age of Big Data. Through these platforms, communities can safely document and share their oral traditions and practices, languages, stories, many of which we share—or say quietly in our own lives. AI-enabled tools translate data collection and organization into ways that matter to the communities. These platforms have to integrate widely based protections against the exploitation or abuse of knowledge, upholding data sovereignty. Enabling cross generational knowledge transfer and create an opportunity for younger generations to learn from their ancestors in a world where tradition is blended with technology.

7.1.5 Focus on Language Documentation, Revitalization, and Transmission Using AI Tools

Indigenous cultures can place a heavy emphasis on language, and AI tools are crucial to recording, revitalizing and otherwise transmitting these languages. One area where AI can be helpful for an indigenous language is in creating dictionaries and pronunciation guides or education materials based on machine learning and natural language processing, ultimately making indigenous languages more accessible. Similarly, useful applications include language models that contain translation and transcription tools, overcoming solving the communication gap. Use of AI for language revitalization efforts—AI used in efforts for language revitalization (e.g., Ensuring these languages are not lost in order to preserve them for future generations and empower the communities to teach the younger generations; allowing a sense of pride in cultural identity, cultural continuity).

References

1. Panesar A (2019) Machine learning and AI for healthcare, vol 10. Apress, Coventry, UK
2. Malapane OL, Chanza N, Musakwa W (2024) Transmission of indigenous knowledge systems under changing landscapes within the vhavenda community. South Africa. Environ Sci Policy 161:103861
3. Dudgeon RC, Berkes F (2003) Local understandings of the land: traditional ecological knowledge and indigenous knowledge. Nature across cultures: Views of nature and the environment in non-Western cultures. Springer, Netherlands, Dordrecht, pp 75–96
4. Hare J (2012) 'They tell a story and there's meaning behind that story': Indigenous knowledge and young indigenous children's literacy learning. J Early Child Lit 12(4):389–414
5. Burger J (1987) Report from the frontier: the state of the world's indigenous peoples (pp viii+-310pp)
6. Stefano ML (2024) Threats and approaches to the safeguarding of intangible cultural. Handbook on intangible cultural practices as global strategies for the future, p143
7. Victoriamma K (2018) Cultural diversity of Indian tribes. IOSR J Humanities Soc Sci 99–23(6):101

8. AI-Powered Preservation of Endangered Languages, HISTORICA (2025). https://www.historica.org/blog/ai-powered-preservation-of-endangered-languages
9. Benedikter T (2013) Minority languages in India. EurAsia Net Partners
10. Swargiary K (2024) How AI revolutionizes regional language education. Scholar Press
11. Pinhanez C, Cavalin P, Storto L, Finbow T, Cobbinah A, Nogima J, Vasconcelos M, Domingues P, Mizukami PDS, Grell N, Gongora M (2024) Harnessing the power of artificial intelligence to vitalize endangered indigenous languages: technologies and experiences. arXiv preprint arXiv: 2407.12620
12. Nanduri DK (2024) Exploring the role of generative artificial intelligence in culturally relevant storytelling for native language learning among children (Master's thesis, University of Maryland, College Park)
13. Article 19, Constitution of India, 1950
14. Census of India 2011
15. Article 351, Constitution of India, 1950
16. Tyagi Y (2003) Some legal aspects of minority languages in India. Social scientist, pp 5–28
17. The National Commission for Minorities Act, 1992
18. Shulist S (2018) Transforming indigeneity: urbanization and language revitalization in the Brazilian Amazon. University of Toronto Press
19. Bali K, Choudhury M, Sitaram S, Seshadri V (2019) Ellora: enabling low resource languages with technology. In: Proceedings of the 1st international conference on language technologies for all, pp 160–163
20. Pradhan U, Dey J (2023) Language, artificial education, and future-making in indigenous language education. Learning, media and technology, pp 1–14
21. IIT Set to launch app in Mundari to keep indigenous language relevant, TIMES OF INDIA (2019). https://timesofindia.indiatimes.com/city/kolkata/iit-set-to-launch-app-in-mundari-to-keep-indigenous-language-relevant/articleshow/68071254.cms.
22. National Translation Mission. https://www.ntm.org.in (last visited Mar. 19, 2025)
23. Aayog NITI (2018) National strategy for artificial intelligence. Paper. June, pp.2019–01
24. Bhashini—national language translation mission. https://bhashini.gov.in (last visited Mar. 19, 2025)
25. Government Launches AI Kosha to Strengthen AI Development, ENTREPRENEUR (2025). https://www.entrepreneur.com/en-in/news-and-trends/government-launches-ai-kosha-to-strengthen-ai-development/488155
26. First Peoples' Cultural Council—Language Maps. https://maps.fpcc.ca (last visited Mar. 19, 2025)
27. Words Project, FIRST LANGUAGES AUSTRALIA, https://www.firstlanguages.org.au/50words (last visited Mar. 19, 2025).
28. Angie Lee, Māori Speech AI Model helps preserve and promote new Zealand indigenous language, NVIDIA Blog, https://blogs.nvidia.com/blog/te-hiku-media-maori-speech-ai/ (last visited Mar. 19, 2025)
29. Linda Heimisdóttir, The Icelandic approach: preserving and revitalizing linguistic and cultural diversity in AI, Miðeind. https://mideind.is/en/greinar/islenska-adferdin-hvernig-studla-ma-ad-fjolbreytni-tungumala-og-menningar-a-svidi, (last visited Mar. 19, 2025)
30. Masakhane. https://www.masakhane.io (last visited Mar. 19, 2025)

Author Index

M. Chakraborty et al. (eds.), *Proceedings of Sixth International Ethical Hacking Conference*, Lecture Notes in Networks and Systems 1513,
https://doi.org/10.1007/978-981-96-8632-2

MIX
Papier aus verantwortungsvollen Quellen
Paper from responsible sources
FSC® C105338

If you have any concerns about our products,
you can contact us on
ProductSafety@springernature.com

In case Publisher is established outside the EU,
the EU authorized representative is:
Springer Nature Customer Service Center GmbH
Europaplatz 3, 69115 Heidelberg, Germany

Printed by Libri Plureos GmbH
in Hamburg, Germany